T0135440

Self-organizing control of networked systems

Dissertation zur Erlangung des Grades
eines Doktor-Ingenieurs
der Fakultät für Elektrotechnik und Informationstechnik
an der Ruhr-Universität Bochum

von
René Schuh
geboren in Teterow

Bochum, 2016

Gutachter: Prof. Dr.-Ing. J. Lunze
Prof. Dr.-Ing. O. Stursberg

Dissertation eingereicht am: 12. Oktober 2016
Tag der mündlichen Prüfung: 02. Juni 2017

Bibliografische Information der Deutschen Nationalbibliothek

Die Deutsche Nationalbibliothek verzeichnet diese Publikation in der Deutschen Nationalbibliografie; detaillierte bibliografische Daten sind im Internet über http://dnb.d-nb.de abrufbar.

ISBN 978-3-8325-4540-6

Logos Verlag Berlin GmbH
Comeniushof, Gubener Str. 47,
10243 Berlin
Tel.: +49 (0)30 42 85 10 90
Fax: +49 (0)30 42 85 10 92
INTERNET: http://www.logos-verlag.de

Acknowledgements

This thesis is the result of almost five years of research at the Institute of Automation and Computer Control of Prof. Jan Lunze at the Ruhr-University Bochum which was an everlasting experience.

First of all, I would like to thank Prof. Jan Lunze for giving me the chance to work at his institute to broaden my knowledge in control theory. Moreover, he proposed me a very interesting and challenging research topic and was an excellent supervisor with a lot of advices and support at all times. I also thank Prof. Olaf Stursberg for accepting to review this thesis and his constructive comments regarding the manuscript.

I owe a lot of gratitude to my scientific co-workers at the Institute of Automation and Computer Control: Sven Bodenburg, Ozan Demir, Sebastian Drüppel, Fabian Just, Daniel Lehmann, Andrej Mosebach, Yannick Nke, Tobias Noeßelt, Jörg Pfahler, Sebastian Pröll, Kai Schenk, Prof. Christian Schmid, Michael Schwung, Christian Stöcker, Michael Ungermann, Daniel Vey, Philipp Welz, Christian Wölfel, and Markus Zgorzelski. Thank you not only for discussions about research and proofreading this thesis but most of all for the pleasant working atmosphere at the institute. For supporting me with the daily technical, administrative and other tasks I have to thank Dr. Johannes Dastych, Kerstin Funke, Susanne Malow, Andrea Marschall, Rudolph Pura, and Udo Wieser.

This thesis would have never been possible without the permanent encouragement of my parents Elke and Werner over all the years.

Finally, I would like to thank the two most important persons in my life, my wife Melanie and my daughter Marie. Melanie, thank you not only for proofreading my thesis and the inspiring discussions about work but most of all for your enduring support in all situations. Marie, for being such a dear child that I could write this thesis.

Lengerich, Juni 2017 René Schuh

Contents

Abstract

This thesis presents a novel distributed control paradigm for networked control systems in which the local control units of the subsystems exchange information, whenever this is necessary to fulfill an overall control aim. The local control units act in a self-organized way, which means that they adapt their communication structure depending on the current situation of the subsystems based on locally available information only.

A new controller structure is proposed. The local control units are divided into three components fulfilling universal tasks to generate a situation-dependent communication structure: The *feedback unit* performs a local feedback by using local measurements to fulfill basic performance requirements. The *observation unit* detects the current situation of the subsystem by evaluating locally available information. The *decision unit* decides about the transmission of information from the corresponding subsystem to other local control units. This structure is consistently used in all five presented concepts developed for controlling physically interconnected systems and multi-agent systems.

Two self-organizing controllers for physically interconnected systems in which the local control units adapt the communication among each other depending on the current disturbances are introduced. The first concept bounds the disturbance propagation within the overall system by exchanging locally determined coupling signals in case of large coupling impacts. The second kind of controller mimics the behavior of a centralized controller by exchanging locally determined control signals whenever their values have a significant amount on the overall control signal. The applicability of the concepts is demonstrated by means of a water supply system.

Three novel self-organizing controllers for synchronizing multi-agent systems within leader-follower structures by adapting the communication structure to situations like set-point changes, disturbances and communication faults are proposed. A concept for improving the synchronization performance compared to a fixed communication structure is introduced, where the local control units request additional information in case of set-point changes at the leader. To bound the disturbance propagation within the overall systems, a concept is presented in which the local controllers interrupt the communication whenever there are large disturbance impacts on the corresponding subsystem. Furthermore, a concept is derived, where the local control units perform a local reconfiguration of the communication structure in case of vanishing communication links in order to preserve a desired basic synchronization performance. The concepts are applied to control a robot formation.

A main result of the thesis is the development of a general structure for local control units divided into a feedback unit, an observation unit and a decision unit to perform a situation-dependent information exchange. This structure forms the basis for the construction of five proposed self-organizing control concept fulfilling different control aims.

1 Introduction to self-organizing control of networked systems

This chapter gives an introduction to self-organizing control, its structure and its desired properties. It presents the main task, gives a literature survey on related results and summarizes the main contributions of the thesis.

Chapter contents

1.1 Self-organizing control of distributed networked systems

This thesis concerns networked control of distributed systems (Fig. 1.1).

The plant consist of locally distributed subsystems Σ_i, ($i = 1, \ldots, N$), which may or may not be physically coupled. Every subsystem Σ_i is controlled by a *local control unit* C_i which generates a local control input to its corresponding subsystem Σ_i based on the locally measured output. If necessary, the local control units C_i exchange information over a common communication network.

Due to the development of modern means of communication, this kind of control structure has become of major interest in the last decade. The communication network has become flexible and, therefore, the local control units C_i can easily exchange information whenever the information link may improve the overall system performance. However, for complexity reasons an information exchange among all local control units C_i is not desirable. Furthermore, in many cases the communication networks are not only used for control purposes, but simultaneously act as a communication medium for other tasks. Therefore, the local control units C_i should only exchange information whenever this is necessary for the given control task.

This thesis elaborates methods for the situation-dependent selection of admissible communication structures. The local control units C_i decide about their information exchange based on locally available information only. Since there exists no central coordinator, the communication structure results from *self-organization*. This kind of control strategy is reasonable since in many applications it is not appropriate or possible to have a central unit which receives all information to decide about the communication structure. In contrast, previous results in literature mainly deliver methods for designing the local control units C_i such that the closed-loop system is tolerant with respect to communication imperfection (delays, dropouts or packet loss).

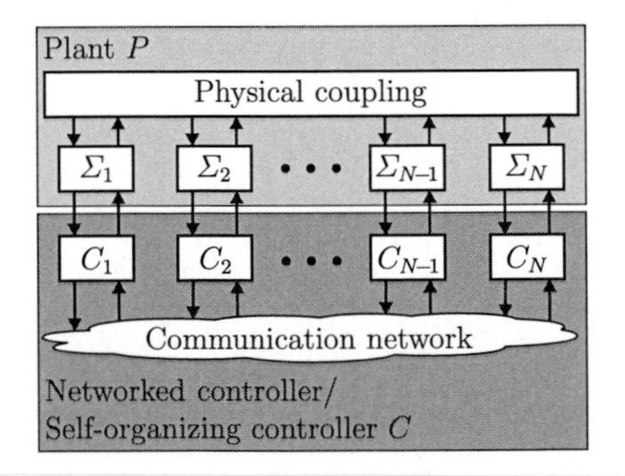

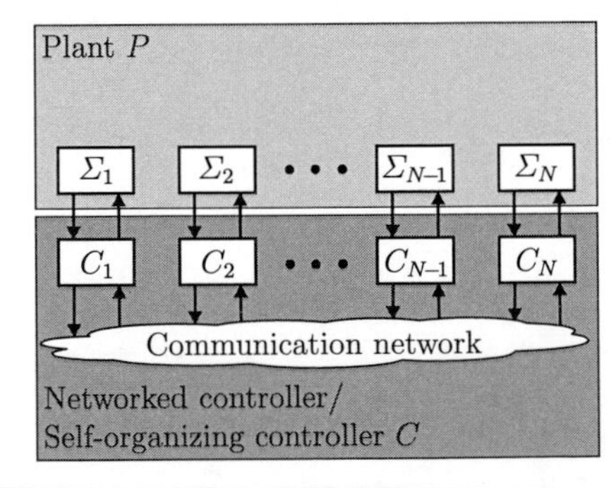

Figure 1.1: Networked control of physically interconnected systems (left) and multi-agent systems (right).

The following statement summarizes the main task of this thesis:

> Develop local control units C_i which adapt their communication structure depending on the current situation of the subsystems based on locally available information to guarantee a desired performance of the overall closed-loop system.

For example, the following control strategies for different situations are presented:

- In case of a **disturbance** affecting a subsystem Σ_i, the corresponding local control unit C_i transmits information to other local control units to bound the propagation of the disturbance (Chapter 4).
- For a **set-point change** the local control units request additional information to improve the control performance compared to a fixed basic communication structure (Chapter 8).
- If there are **faulty links** within the communication structure, the local control units perform a local reconfiguration of the communication to preserve the desired control performance (Chapter 10).

1.2 Properties of self-organizing controlled systems

The plant P together with the *self-organizing controller* C consisting of the communicating local control units C_i builds the overall closed-loop system which is denoted as *self-organizing control system* Σ, i.e.,

$$\text{Self-organizing control system } \Sigma : \begin{cases} \text{Plant } P \\ \text{Self-organizing controller } C. \end{cases}$$

The expression "self-organizing" results from the fact that Σ is desired to have the following characteristic properties of self-organizing systems, where the first one is the most important property:

1. **Flexibility**: The communication structure is adapted in response to the current situation of the subsystems to guarantee a desired overall system performance. The decision of transmitting information is made individually by each local control unit C_i only based on information that is available at C_i. There is no central coordinator.
2. **Scalability**: The local control units C_i use only model information of their corresponding subsystem Σ_i and, therefore, allow an easy integration of new subsystems into a present overall system.

3. **Fault-tolerance**: The local control units C_i bound the fault propagation to fault-free subsystems or recover the overall system by diagnosing the fault and reconfiguring the self-organizing controller C.

All three properties are related to biological, chemical and physical self-organizing systems which reorganize their structure in a changed situation, make autonomous decisions and can easily integrate new subsystems [21].

1.3 Physically interconnected systems and multi-agent systems

This thesis considers two different types of plants P:

- **Physically interconnected systems** consist of *coupled* subsystems Σ_i which generally have individual control aims (Fig. 1.1 left). In many cases there is no need to exchange information among the local control units C_i (decentralized control) to stabilize the overall closed-loop system. However, in case of disturbances, set-point changes or subsystem faults the overall system performance can be considerably improved by such an information exchange. Application examples for physically interconnected systems are, e.g., electrical power grids, chemical plants, power plants, water supply systems or mechanical multibody systems in vehicles as well as robots.

- **Multi-agent systems** are composed of physically *uncoupled* subsystems Σ_i which are generally denoted as agents (Fig. 1.1 right). The agents usually have a cooperative control goal like synchronization or consensus. These goals necessitate communication among the local control units C_i of the agents Σ_i. Generally a sparse temporal communication is sufficient to achieve synchronization. However, in case of disturbances, set-point changes or communication faults it is necessary to adapt the communication structure for achieving a desired overall performance. Application examples for multi-agent systems are, e.g., traffic control, the synchronization of shafts in a paper machine, or formation problems for vehicles, aircrafts, multi-copters and robots.

Both types of plants require different kinds of controllers with different strategies of adapting the communication structure. Note that there exist several control methods in the literature with a fixed communication structure to achieve the different control goals. Despite the differences between both types of plants, this thesis proposes a **consistent structure for the local control units** C_i which adapt their communication to the current situation of the subsystems (cf. Section 1.5).

1.4 Literature survey on self-organizing aspects in networked control systems

This section provides a brief summary on topics related to the concept of self-organizing control proposed in this thesis. The main focus on this literature survey lies on control concepts which have one of the mentioned properties of self-organization (flexibility, scalability, fault-tolerance) or which build the foundation to obtain these properties.

Networked control systems. The starting point for the idea of self-organizing control are networked systems which enable a flexible communication among the local control units, but also bring the challenge to save energy resources at the distributed sensors, actuators and controllers.

An overview of recent results on networked control systems give [24, 65, 93, 132] with different focus, e.g., analysis and controller synthesis [24, 65, 93, 132], decentralized control [24], distributed control [24, 93], distributed estimation [24, 93], distributed optimization [24], predictive control [93, 132], event-based control [24, 93], multi-agent systems [93, 132] or control over wireless networks [24, 93, 132].

Control of large-scale systems. Methods for the control of so-called *large-scale systems* within a decentralized architecture have been developed already in the 1980's [70, 88, 120, 121]. The main topics were, e.g., structural analysis of controllability and observability, stabilization by a decentralized state feedback or output feedback, distributed implementation of a centralized controller or observer, hierarchical control structures or decomposition of the plant.

These topics play an important role for developing self-organizing control methods since decentralized control and distributed control are desired operation modes of the self-organizing controller C.

Controller design for a given communication structure. Consider a distributed controller architecture with a given communication structure, where the aim is to design the local controllers such that the overall closed-loop system has a desired performance. There exists different approaches for the two classes of systems to be considered:

- **Physically interconnected systems**: Methods for designing an overall feedback gain matrix subject to structural constraints to obtain a sparse communication structure are given in [54, 86]. The design problems are formulated as an optimization problem, where [54] uses Newton iterations and [86] an augmented Lagrangian method to find a solution. The design of a distributed controller with a communication structure that has the same interconnection pattern as the plant is concerned in [46, 96]. The procedure in [96] is a multiobjective optimization under linear matrix inequality constraints with system norms as performance indices. [46] presents a method for symmetrically interconnected subsystems with identical dynamics, where the resulting distributed controller guarantees a better performance than the corresponding decentralized controller.

- **Multi-agent systems**: Designing local controllers for agents within a given communication structure to guarantee asymptotic synchronization is a basic aim for controlling multi-agent systems. The use of graph theory to describe the information coupling of vehicle formations was introduced in [55]. In [83, 89, 90, 134] it is shown that agents with individual dynamics can only be synchronized if they satisfy the internal-reference principle, which says that all extended agents have to include an identical reference system. The design concepts can be divided into different groups. There are approaches for agents with identical dynamics [30, 31, 55, 99, 102, 142] and individual dynamics [59, 83, 89, 90, 101, 134]. Synchronization of the states [30, 31, 101, 142] or the outputs [55, 59, 83, 89, 90, 99, 102, 134] of the agents is concerned. The underling communication graph may be undirected [30, 31, 102] or directed [55, 59, 83, 89, 90, 99, 101, 134, 142]. Furthermore, leaderless structures [30, 31, 55, 59, 83, 90, 101, 102, 134] and leader-follower structures [89, 142] are investigated.

All these concepts show that the design of a distributed controller for a given structured communication is a non-trivial task. In this thesis the communication among the local control units is not only structured, but the communication links are switched on and off depending on the situation of the subsystems (flexibility).

Design of the communication structure. To design the communication structure of a distributed controller for a given plant and a desired control aim is a challenging topic. The structure and the dynamics of the plant have to be merged with the desired dynamics of the closed-loop system to decide which communication link is appropriate to obtain a desired behavior. In the following a brief overview on concepts for physically interconnected systems and multi-agent systems is given:

- **Physically interconnected systems**: The design of an appropriate communication structure for the distributed control of physically interconnected systems is investigated in several publications. A method for designing an overall state feedback which guarantees an optimal relation between the system performance and the communication effort is presented in [61, 63]. The method in [61] is extended for considering link failures in [60]. [64] extends the results of [63] to time-delayed systems. An hierarchical distributed control with a two-layer state feedback controller is considered in [71], were the controller gains and the communication structure are designed simultaneously by solving an optimization problem. In [72] a decomposition approach for solving a similar optimization problem is presented for a special structure of the physical coupling among the subsystems, where the aggregation of the subsystems into clusters reduces the complexity of the optimization problem. The approach in [71] is extended to a special class of jump Semi-Markov systems in [73].

The design of a distributed controller with a sparse communication is concerned in [53, 75, 87], where the emphasis lies on identifying favorable communication structures without any prior assumptions on the sparsity patterns of the overall feedback gain matrix as in [54, 86]. A survey on recent results in that area is given in [75].

- **Multi-agent systems**: Several publications focus on the design of the communication structure for agents with integrator dynamics. In [103] it is shown that the algebraic connectivity and, hence, the transient response can be considerably increased by random rewiring of the agents. A genetic-algorithm methodology to obtain the long-range link configuration of a small-world communication network with faster consensus speed is proposed in [135]. The design of the communication structure for strongly balanced digraphs using a mixed integer semidefinite program is concerned for minimizing the communication cost [111] or the convergence speed [35, 111], respectively. In [136] the weights of the communication links for a given communication structure are chosen such that the average convergence time of the agents is minimized. The design of the topology in sensor networks is investigated in [78, 79].

  A method for designing the communication structure for agents with individual dynamics controlled within a leader-follower structure is proposed in [91]. A measure on the delay between the transient behavior of the leader and the followers is used to find a trade-off between the number of communication links and the quality of the synchronization performance.

These concept are a good starting point for the concepts presented in this thesis since they provide methods deciding which communication links are necessary to achieve a desired behavior of the closed-loop system. Nevertheless, this thesis proposes methods which do not only decide which communication links are necessary but also in which situations these links are necessary to obtain a desired behavior (flexibility). Furthermore, the decision is made by local information only, whereas most of the optimal control approaches require a central coordinator.

Time-variant communication structure due to network properties. The robustness of the distributed controller due to a time-varying or switching communication structure is of major interest in the field of networked control systems. Due to network properties (delays, dropouts, packet losses) or faults, the communication among the local controllers might change over time. The design of a distributed controller is studied for both types of plants considered in this thesis:

- **Physically interconnected systems**: The authors in [60, 68, 74] study the design of an optimal controller for unreliable communication links. The aim in [68] is to stabilize the system while minimizing a quadratic performance criterion when the information flow between sensors, controllers and actuators is disrupted due to link failures or packet losses. [60] provides distributed control laws for the case that some communication links are prone to failure and that a controller may reconfigure itself if it is directly affected by a link failure. This approach is extended to time-varying communication structures modeled by Markov chains in [74].
- **Multi-agent systems**: The problem of consensus among multiple agents in the presence of a dynamically changing communication structure due to nearest neighbor rules is investigated in [69], where the simulated convergence results in [131] are explained by a theoretical analysis. The results for undirected graphs in [69] are extended to directed graphs in [112]. The authors in [105] introduce a common Lyapunov function that guarantees asymptotic convergence of single integrator systems with a linear consensus protocol in networks with switching topology. The synchronization of identical agents described by a linear state-space model within time-varying communication structures is investigated in [109, 117], whereas [100] considers agents with individual nonlinear dynamics.

All these results have in common that the change of the communication structure is assumed to be given. In contrast to that, in this thesis the communication structure is changed on purpose by local decision rules to guarantee a desired performance of the overall closed-loop system (flexibility).

Autonomy and cooperation in networked systems. There are a few concepts which concern the switching between two different control modes [45, 47, 92]. Within the *autonomous mode* there is either no communication or only basic communication among the local controllers to guarantee basic control aims (e.g.: stability, synchronization). The *cooperative mode*, in which the local controllers exchange information, is activated at certain situations to guarantee a desired performance of the overall system. There exists different approaches for the two concerned types of plants:

- **Physically interconnected systems**: In [47] the control switches between a decentralized controller (autonomous mode) and a distributed controller (cooperative mode), where the structure of the communication within the distributed controller is identical to the physical interconnection structure among the subsystems. The results in [47] for identical symmetrically interconnected subsystems are extended to subsystems with similar dynamics in [2].

- **Multi-agent systems**: The synchronization of agents with a minimum number of communication events is concerned in [92]. It is shows that only $2n$ communication events are necessary to synchronize the agents, where n denotes the dynamical order of the agents. If the synchronization is completed, the cooperative communication is deactivated and there is no communication exchange among the local controllers (autonomous mode). [45] presents a concept where the communication switches between a serial structure (autonomous mode) to a structure with additional information from the leading agent (cooperative mode).

These concepts guarantee the desired self-organizing property of flexibility since the communication is adapted to the current situation of the subsystems. However, generally the decision of activating the cooperative control mode is made by using global information rather than local information of the subsystems. Furthermore, the mentioned concepts switch only between two communication structures. In contrast, this thesis presents concepts where the local control units only decide about their own incoming and outgoing transmissions such that various different communication structures may emerge.

Adaption of the network topology depending on the system state. The adaption of the edge weights for a given communication graph to improve the system performance is investigated in [38–44, 141] for oscillators and systems with integrator dynamics. The edge weights vary over time depending on the difference between the state or the output of the agents. A local and a global adaption law are proposed in [43], whereas [39] presents a proof of the asymptotic synchronization for the local adaption law while using Lyapunov functions. In [38] the edge weights are set to be either 0 or 1 depending on a given threshold. Compared to this approach a bistable potential function with the minima 0 and 1 is introduced in [41] which is called "edge snapping". [44] extends the concept in [41] to a potential function with three minima. The adaption of only a subset of the coupling weights is concerned [141], where an estimate for the minimum number of weights to be controlled in order to synchronize the network is provided.

All of these concepts have the desired self-organizing property of flexibility since the communication structure is adapted to the current situation of the agents. Compared to the control concept presented in this thesis the potential functions of the aforementioned concepts cut off the edges of the communication graph only if time goes to infinity.

Event-based control. The aspect of communication reduction has been of major interest in the field of distributed event-based control, where the sampling of the output of the plant and the updating of the control signals is non-periodical. There exist concepts for the stabilization of physically interconnected systems and for the synchronization of multi-agent systems:

- **Physically interconnected systems**: There are approaches which propose decentralized triggering mechanisms for the communication among sensors, controllers and actuators within a centralized control structure [51, 97, 110]. The distributed implementation of centralized control laws is investigated in [57, 80, 140]. Distributed event-based control concepts are proposed in [62, 107, 124, 133], where the control input of the subsystems is a function of the local state and the transmitted state of some other subsystems.
- **Multi-agent systems**: Many concepts on synchronization and consensus in multi-agent systems consider single integrator systems [50, 52, 58, 98] or double-integrator systems [119, 137]. Furthermore, there exist event-based control methods for identical agents [48, 49, 143] and individual agents [139] with more general dynamics.

The approaches in this thesis change the communication structure, thus, the structure of the control law changes among time intervals and not at specific time instances. In most approaches for event-based control the triggering condition depends on the deviation of the last subsystem state and the current subsystem state which is broadcasted to all local control units of neighboring subsystems. With the approaches in this thesis information is exchanged among certain local control units as long as its transmission is necessary (flexibility).

Flocking and swarming. In the area of flocking/swarming, the agents are desired to fulfill three aims [113]:

- Cohesion: The agents attempt to stay close to nearby flockmates.
- Alignment: The agents attempt to match velocity with nearby flockmates.
- Separation: The agents avoid collisions with nearby flockmates.

The agents (flocks) are locally distributed and are generally considered to have double integrator dynamics. Every agent communicates with neighbors within its communication range. Hence, there is no fixed communication among the agents. In fact there is a state-dependent switching of the communication depending on the communication range and position of the agents. The pioneering work [113] studies the swarming behavior of animals and defines the three “swarming rules”: separation, alignment and cohesion. Simulation studies of the "swarming rules" are presented in [131], where special group leaders are considered. First stability analysis methods for fixed and switching network topologies are derived in [127, 128]. These results are extended in [104] to the usage of a reference agent and to obstacle avoidance. An approach for flocking of agents with a virtual leader is proposed in [126].

In contrast to the proposed concept, in this thesis generally a global reference agent is necessary to guarantee flocking and the communication topology is not chosen, but depends on the communication range of the agents (flexibility).

Integration of new subsystems. If a new subsystem shall be integrated into a present closed-loop system, the change of the distributed controller should be as small as possible. In particular, only those local control units should be redesigned which are affected by the new subsystem to preserve the desired overall system performance. There exist results for physically interconnected systems as well as for multi-agent systems:

- **Physically interconnected systems**: An approach for adding new sensors and actuators to an existing system is proposed in [125]. The focus is on the identification of the new hardware and the adjustment of the controller based on the identified model to make use of the new device. A design method using a model predictive controller with limited model information has been proposed in [114] in order to guarantee global stability and state constraint satisfaction. The authors in [27, 28] introduce local design agents for each subsystem for reconfiguring the local controllers. Each design agent has only local model information of its corresponding subsystem. In [28] the design agent of the new subsystems and the design agents of the present subsystems exchange information for redesigning their local controller, whereas in [27] only the local controller of the new subsystem has to be designed.

- **Multi-agent systems**: A method for integrating new agents to an existing network topology by individual selection of two leaders to provide a minimal persistent graph and a maximal local convergence rate is proposed in [67]. The agents are considered to have integrator dynamics and the integration of the new agent into the present communication graph should lead to a minimal persistent graph and a maximal local convergence rate.

All these concepts concern the desired self-organizing property of scalability which requires a control structure that allows an easy integration of new subsystems into the present overall closed-loop system. Note that in this thesis scalability is an additional property of the proposed control concepts which is desired to be fulfilled besides the main requirement of a situation-dependent communication (flexibility).

Fault-tolerant control in distributed systems. Fault-tolerant control in locally distributed systems is a non-trivial task since generally there is no central control unit which can diagnose faults within the subsystems and which performs an appropriate reconfiguration of the local controllers. In general, fault-tolerant control approaches can be classified into two classes: passive approaches and active approaches [26]. Typical passive approaches use unchangeable robust controllers throughout the nominal/faultless cases and failure cases (see, e.g., the topic "Time-variant communication structure due to network properties"). The following brief survey for physically interconnected systems and multi-agent systems focuses on active fault-tolerant control:

- **Physically interconnected systems**: [77, 106] provide an overview on fault-tolerant control in distributed networked systems. The reconfiguration of a network of actuators after actuator failures has been studied in [129], where the actuators are reconfigured by solving a distributed optimization problem. The decentralization of the reconfiguration task is considered in [130]. The authors propose a decentralized fault-tolerant control strategy in which the local controller of the faulty subsystem transmits information to the other local controllers to compensate the effect of the fault on the fault-free subsystems. The authors in [29] introduce local design agents for each subsystem for reconfiguring the local controllers, where each design agent has only local model information of its corresponding subsystem. The design agent of the faulty subsystem collects model information from the design agents of the fault-free subsystems to reconfigure only the local controller of the faulty subsystem. A method which re-distributes the common control aim for all subsystems if a controlled subsystem can not achieve its local control aim due to a fault is presented in [118].

- **Multi-agent systems**: [23, 33, 116] concern the problem of actuator faults within agents. A hierarchical cooperative recovery control design method is proposed in [23] which consists of three levels, where even the lowest level receives information from all agents. A decentralized fault-tolerant control protocol is introduced in [33] which only adapts the control input of the faulty agent. [116] concerns multiple faulty agents and proposes a reconfiguration strategy for which the feedback gains of the agents are adapted in such a manner that healthier agents contribute more control efforts to compensate for the controller degradations of faulty agents. More general faults which change the dynamics of the agents are investigated in [138]. The weights of the communication edges are adjusted to preserve the synchronization of the agents, whereas the communication structure maintains.

All presented methods fulfill the property of fault-tolerance (cf. Section 1.2). In this thesis the local control units C_i decide on their own how to react in case of faults by adapting the communication structure which is only concerned in a few of these concepts.

1.5 Structure of the local control units

To obtain the self-organizing properties of flexibility, scalability and fault-tolerance, the local control units C_i are divided into three components (Fig. 1.2):

- The **feedback unit** F_i performs a local feedback by using local measurements to fulfill basic performance requirements for the subsystems Σ_i which are, e.g, asymptotic stability for physically interconnected systems or asymptotic synchronization for multi-agent systems. In addition to this local feedback, the local control input includes an input which is generated by using the received information transmitted by the other local control units. Hence, the structure of the overall control law changes with the change of the communication structure.

- The **observation unit** O_i detects the current situation of the subsystem Σ_i and provides it to the decision unit D_i. For the situation detection the behavior of Σ_i is evaluated based on locally available information.

- The **decision unit** D_i decides about the transmission of information about subsystem Σ_i to other local control units C_i. Which information is transmitted and which other local control units C_i receive information depends on the current situation of the subsystems and is decided based on local rules. Furthermore, the decision unit D_i preprocesses the received information from the other local control units for the feedback unit F_i.

The actual realization of the local control units C_i depends on the properties of the plant P to be controlled (cf. Section 1.3) and the control aims, e.g., disturbance attenuation, set-point following or fault tolerance.

1.6 Fundamental questions

This section states fundamental questions that are answered in this thesis in order to obtain a closed-loop system Σ which has the desired self-organizing properties of flexibility, scalability and fault-tolerance stated in Section 1.2. The questions are assigned to the possibly occurring situation, the actual realization of the components of the local control units C_i and the desired self-organizing properties for the closed-loop system.

- Occurring situations
  - In which situations is an adaption of the communication structure appropriate to improve the overall system performance or to guarantee a desired control aim?

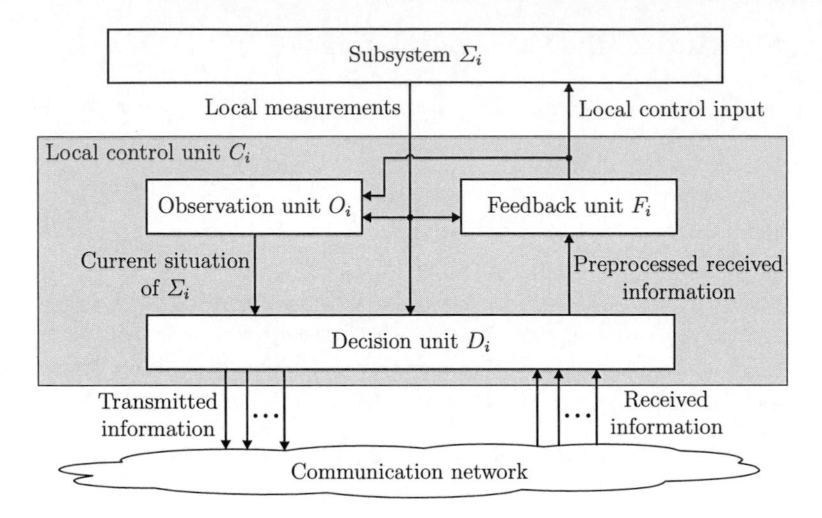

Figure 1.2: General structure of the local control units C_i.

  - How do the local control units have to behave when disturbances or faults affect the subsystems or when set-point changes occur?
  - Are the approaches for a specific situation identical for both considered classes of plants (physically interconnected systems and multi-agent systems)?

- Feedback unit F_i
  - How should the feedback unit F_i be designed to guarantee basic control aims like stability of the physically interconnected subsystems or synchronization of the agents?
  - Which basic information is necessary to achieve the basic control aims and how can the situation-dependently transmitted information be used to adapt the local control input?

- Observation unit O_i
  - Which states or signals characterize the current situation of the subsystems Σ_i?
  - How can the characteristic states or signals be determined by using locally available information?

- Decision unit D_i
  - Which information has to be exchanged among the local control units to achieve the desired performance of the overall closed-loop system?

- When and to which other local control units should information be transmitted?

- Self-organizing properties
  - Does the communication structure adapt to the current situation of the subsystems (Flexibility)?
  - Do the local control units C_i only use model information of their corresponding subsystem Σ_i (Scalability)?
  - Does the overall system recover from a fault within a subsystem or is at least the propagation of the fault bounded (fault tolerance)?

1.7 Main contributions of the thesis

The main contribution of the thesis can be summarized as follows:

> This thesis provides a general structure for the local control units C_i consisting of a feedback unit F_i, an observation unit O_i and a decision unit D_i which is appropriate to develop self-organizing controllers with a situation-dependent communication exchange. This structure is used to derive concepts for controlling physically interconnected systems and multi-agent systems which bound the disturbance propagation as well as the fault propagation, improve the system performance, mimic a centralized controller or recover the system from a fault.

Table 1.1 gives an overview on the contributions of the thesis by listing the developed self-organizing control concepts designed for different situations in physically interconnected systems and multi-agent systems.

There are **two concepts** for the self-organizing control of physically interconnected systems considering disturbances (Chapter 4 and Chapter 5) and faults (Chapter 4) affecting the subsystems. The situation of set-point changes is not analyzed for the concepts in Chapter 4 and Chapter 5 since the effects are similar as for disturbances. A subsystem can be forced to leave its equilibrium by a disturbance or a set-point change, the result is the same. The local control units exchange information to either bound the propagation of the disturbance or the effect of a set-point change on the neighboring subsystems (Chapter 4) or try to mimic the behavior of a centralized controller in case of a disturbance or a set-point change (Chapter 5).

There are **three concepts** for multi-agent systems: one for bounding the disturbance propagation within the overall system (Chapter 9), one for improving the system performance by requesting additional information in case of set-point changes (Chapter 8) and one for pre-

Table 1.1: Overview of the structure and the contribution of the thesis

<table>
<tr><th>Plant / Situation</th><th>Physically interconnected systems (Part I)</th><th>Multi-agent systems (Part II)</th></tr>
<tr><td rowspan="2">Disturbances</td><td>Mimicry of a centralized controller (Chapter 5)</td><td rowspan="2">Disturbance attenuation by communication interruption (Chapter 9, from [95])</td></tr>
<tr><td rowspan="2">Disturbance attenuation by compensating physical couplings (Chapter 4)</td></tr>
<tr><td>Faults</td><td>Preservation of the performance by local reconfiguration of the faulty communication structure (Chapter 10)</td></tr>
<tr><td rowspan="2">Set-point changes</td><td>Mimicry of a centralized controller (Chapter 5)</td><td rowspan="2">Performance improvement by requesting additional information (Chapter 8)</td></tr>
<tr><td>Disturbance attenuation by compensating physical coupling impacts (Chapter 4)</td></tr>
</table>

serving the system performance in case of faulty communication links by performing a local reconfiguration of the communication structure (Chapter 10).

Note that the basic idea of structuring the local control units in three components was first presented in [95] for disturbance attenuation in multi-agent systems. This concept is described in Chapter 9 using the notation of this thesis to give a complete overview on current self-organizing control concepts.

Structure of the local control units. This thesis proposes a new structure for distributed controllers which divides the local control units C_i into components fulfilling universal task to generate a communication among the local control units C_i which is adapting to the current situation of the agents (Section 1.5). This structure is consistently used in all developed control concepts. Furthermore, the modular structure of the local control unit C_i enables to use the same components for different concepts as well as an easy combination of the components to simultaneously fulfill different control aims by using one controller.

The **feedback unit** F_i generally includes classical control devices from literature, e.g., a local state feedback. The basic behavior obtained by the feedback units F_i is influenced in a targeted way by the novel developed **observation unit** O_i and **decision unit** D_i such that the closed-loop system has the desired self-organizing properties. Hence, the components in F_i are not new, but through their combination with O_i and D_i a **new type of local controller** arises.

Concepts for physically interconnected systems (Part I). Part I presents two novel distributed controllers for physically interconnected systems in which the local control units C_i adapt the communication among each other depending on the current disturbance situation. The feedback units F_i perform a permanent decentralized feedback using local measurement. The observation unit O_i and the decision unit D_i differ for both concepts. Conditions for the plant with respect to practical stabilization (Theorem 4.2 and Theorem 5.3) and asymptotic stabilization (Theorem 4.3 and Theorem 5.4) are derived for both concepts:

- **Disturbance attenuation by compensating physical couplings** (Chapter 4): Every local control unit C_i determines the coupling input to subsystems on which its corresponding subsystem Σ_i has a physical coupling impact. C_i transmits the coupling input to C_j in order to bound the disturbance propagation within the overall system, whenever the coupling input on Σ_j exceeds a switching threshold ϵ_{ji}. The concept can be used if the coupling inputs on the subsystems can be completely compensated by an applicable control input and the isolated subsystems are stabilizable (Theorem 4.1).

  The main result is that the control error of the undisturbed subsystems can be made arbitrarily small by the choice of the switching thresholds (Theorem 4.5 and Theorem 4.6). The analysis of the behavior of the controlled subsystem and the design of the local control units can be performed by using only local model information. Due to this property, a decentralized procedure for integrating new subsystems is presented (Table 4.8). Furthermore, the propagation of a fault affecting a subsystem is bounded within the overall system (Proposition 4.3).

- **Mimicry of a centralized controller** (Chapter 5): A centralized controller C_{C} that stabilizes the physically interconnected subsystems is implemented in a distributed way. The local control units exchange the locally determined control signals whenever their values exceed a given switching threshold ϵ_{ji} to obtain a similar performance as with C_{C}. The concept can be used whenever the overall plant consisting of the physically interconnected subsystems is stabilizable (Theorem 5.2).

  The main result is that the difference between the behavior of the given centralized control system Σ_{C} and the behavior of the resulting self-organizing control system Σ can be arbitrarily adjusted by the choice of the switching thresholds (Theorem 5.1 and Theorem 5.5)

Algorithms for designing the local control units C_i to guarantee a maximal desired disturbance propagation (Algorithm 4.1) and desired difference behavior (Algorithm 5.1) are given.

Concepts for multi-agent systems (Part II). Three novel distributed controllers for synchronizing agents within a leader-follower structure in which the communication among the local control units C_i is adapted to situations like set-point changes, disturbances and communication faults are presented. The feedback units F_i are identical for all three concepts, where F_i tracks its corresponding local reference signal based on a given basic communication structure $\mathcal{G}_{\mathrm{B}}$. The observation unit O_i and the decision unit D_i differ for the particular concepts. The main difference between concepts is how the decision units D_i adapt the communication structure depending on the current situation of the agents detected by the observation units O_i.

- **Performance improvement by requesting additional information** (Chapter 8): The incoming communication links to local control units C_i are changed from links within a basic communication graph $\mathcal{G}_{\mathrm{B}}$ with a low communication effort but a bad synchronization performance to links within a adjusted communication graph $\mathcal{G}_{\mathrm{A}}$ with a high communication effort but a good synchronization behavior, whenever the local control error $e_i(t)$ exceeds the switching threshold ϵ_i to guarantee a desired synchronization performance. Hence, the communication structure adapts to the current difference between the situation of the leader and the agents represented by the local control error $e_i(t)$.

  The main result is that the ability to adjust the maximal control error of the agents by the choice of the adjusted communication graph $\mathcal{G}_{\mathrm{A}}$ and the switching thresholds ϵ_i (Theorem 8.2) can be used to derive algorithms for designing the adjusted communication graph $\mathcal{G}_{\mathrm{A}}$ in a centralized way (search algorithm in Fig. 8.8) as well as in a decentralized way (Algorithm 8.1). The self-organizing controlled agents always asymptotically synchronize with the leader (Theorem 8.1).

- **Disturbance attenuation by communication interruption** (Chapter 9, from [95]): Whenever the disturbance impact on an agent Σ_i exceeds the switching threshold ϵ, the corresponding local control unit C_i interrupts the transmission of the output to its basic followers. Hence, the communication structure is adapting to the currently acting disturbances.

  The main result is that the maximal control error of the undisturbed subsystems can be arbitrary adjusted by the choice of the switching threshold ϵ (Theorem 9.2 and Theorem 9.3). Despite the interruption of the communication, the agents asymptotically synchronize with the leader after the disturbances have vanished (Theorem 9.1).

  Note that this concept was developed by the supervisor of the author.

- **Preservation of the performance by local reconfiguration of the faulty communication structure** (Chapter 10): Whenever a faulty communication link leads to a communication structure which does not guarantee a desired basic synchronization performance, the local control units C_i compensate the missing predecessors by choosing admissible new predecessors to preserve the desired performance. Hence, the communication structure is adapting to the currently faulty communication links.

The main result is that the maximal basic control error remains for the fault-free agents (Theorem 10.2 and Corollary 10.2). Asymptotic synchronization to the leader can be achieved by agents even if there is faulty communication (Theorem 10.1 and Corollary 10.1).

Each chapter presents an algorithm for designing the parameters of the local control units C_i with respect to their desired control aims (Algorithm 8.2, Algorithm 9.1 and Algorithm 10.1).

1.8 Structure of the thesis

Chapter 2 introduces the general notation of the thesis, explains the properties of the situation-dependent communication among the local control units C_i and presents general definitions on stability, controllability and observability used throughout the thesis. Finally, possible faults affecting the subsystems are discussed.

The rest of the thesis is divided into three parts (Table 1.1):

- **Part I** introduces two different self-organizing control concepts for physically interconnected systems:
  - **Chapter 3** presents the modeling framework for the interconnected subsystems, introduces a basic structure of the self-organizing controller for the two control concepts, derives the desired control aims and presents two running examples for physically interconnected systems.
  - **Chapter 4** introduces a self-organizing controller C which bounds the disturbance propagation within the overall system by a situation-dependent exchange of coupling signals among the local control units C_i. State feedback as well as output feedback is concerned. Afterwards, the fault-tolerant behavior of the resulting self-organizing control system is analyzed.
  - **Chapter 5** presents a self-organizing controller C that mimics the behavior of a centralized controller designed to stabilize the overall plant.
  - **Chapter 6** compares the control concepts in Chapter 4 and Chapter 5 for physically interconnected systems with respect to their similarities and differences.
- **Part II** introduces three controllers with a situation-dependent information exchange among the local control unit C_i among agents within a leader-followers structure:

- **Chapter 7** presents the model of the agents with individual dynamics, derives a basic structure of the self-organizing controller C which is identical for all three concepts, introduces the desired control aims, performs an analysis of a general closed-loop multi-agent system and presents a robot formation as a running example.
- **Chapter 8** gives the structure of a self-organizing controller with a situation-dependent request of information to improve the overall synchronization performance of the agents. Furthermore, two methods for designing the communication structure for this controller are derived.
- In **Chapter 9** a self-organizing controller C is introduced which bounds the disturbance propagation within the overall closed-loop system by deactivating the outgoing communication in case of large disturbance effects.
- **Chapter 10** presents a control strategy where faulty communication links are replaced by admissible new links to preserve a desired basic synchronization performance.
- In **Chapter 11** the structure and the properties of the three control concepts for multi-agent systems presented in Chapter 8, Chapter 9 and Chapter 10 are compared.

- **Part III** concludes the thesis:
  - **Chapter 12** finally compares the structure and the main properties of the presented control concepts for physically interconnected systems and multi-agent systems.
  - **Chapter 13** briefly discussed possible future research directions.

2 Preliminaries

This chapter introduces the general notation of the thesis. Definitions on stability, controllability and observability used throughout the thesis are presented. Furthermore, properties of the situation-dependent communication among the local control units are introduced, where the resulting communication structure is modeled by a graph. Finally, possible faults affecting the subsystems or the local control units are briefly discussed.

Chapter contents

2.1 Notation

Throughout this thesis scalars are denoted by italic letters (s), vectors by bold italic letters ($\boldsymbol{x}$) and matrices by upper-case bold italic letters ($\boldsymbol{A}$). Note that vectors and matrices with special meaning retain their writing style even if they become a scalar or a vector in a certain application example.

$\mathbb{R}$ denotes the set of real numbers and $\mathbb{R}_+ := \{s \in \mathbb{R} \mid s \geq 0\}$. $\mathbb{C}$ is the set of complex numbers, $\mathbb{C}_+ := \{s \in \mathbb{C} \mid \mathrm{Re}(s) \geq 0\}$ and $\mathbb{C}_- := \{s \in \mathbb{C} \mid \mathrm{Re}(s) < 0\}$. $\mathbb{N}$ is the set of natural numbers and $\mathbb{N}_0 := \mathbb{N} \cup \{0\}$.

The transpose of a vector $\boldsymbol{x}$ or a matrix $\boldsymbol{A}$ is denoted by $\boldsymbol{x}^\mathrm{T}$ or $\boldsymbol{A}^\mathrm{T}$, respectively. $\boldsymbol{I}_n$ denotes the identity matrix of size n. $\boldsymbol{O}_{n\times m}$ is the zero matrix with n rows and m columns and $\boldsymbol{0}_n$ represents the zero vector of dimension n. The dimensions of $\boldsymbol{O}$ and $\boldsymbol{0}$ are omitted if they are clear from the context.

Consider the matrices $\boldsymbol{A}_1, \ldots, \boldsymbol{A}_N$. The notation $\boldsymbol{A} = \mathrm{diag}(\boldsymbol{A}_1, \ldots, \boldsymbol{A}_N)$ is used to denote a block diagonal matrix

$$\boldsymbol{A} = \mathrm{diag}(\boldsymbol{A}_1, \ldots, \boldsymbol{A}_N) = \begin{pmatrix} \boldsymbol{A}_1 & & \\ & \ddots & \\ & & \boldsymbol{A}_N \end{pmatrix}.$$

The i-th eigenvalue of a square matrix $\boldsymbol{A} \in \mathbb{R}^{n\times n}$ is denoted by $\lambda_i(\boldsymbol{A})$.

Sets are denoted by calligraphic letters ($\mathcal{P}$). $|\mathcal{P}|$ denotes the cardinality of the set $\mathcal{P}$.

Consider a time-dependent matrix $\boldsymbol{G}(t)$ and vector $\boldsymbol{u}(t)$. The asterisk $*$ is used to denote the convolution-operator, e.g.,

$$\boldsymbol{G} * \boldsymbol{u} = \int_0^t \boldsymbol{G}(t-\tau)\boldsymbol{u}(\tau)\mathrm{d}\tau.$$

The inverse of a square matrix $\boldsymbol{H} \in \mathbb{R}^{n\times n}$ is symbolized by $\boldsymbol{H}^{-1}$. Matrix $\boldsymbol{H}^+$ is the pseudoinverse of $\boldsymbol{H}$, where $\boldsymbol{H}^+$ fulfills the four Moore-Penrose conditions (cf. [84]). If $\boldsymbol{H}^{n\times m}$ has full rank, then $\boldsymbol{H}^+$ is defined by

$$\boldsymbol{H}^+ = \begin{cases} \left(\boldsymbol{H}^\mathrm{T}\boldsymbol{H}\right)^{-1}\boldsymbol{H}^\mathrm{T} & \text{if } \mathrm{rank}(\boldsymbol{H}) = m \\ \boldsymbol{H}^\mathrm{T}\left(\boldsymbol{H}^\mathrm{T}\boldsymbol{H}\right)^{-1} & \text{if } \mathrm{rank}(\boldsymbol{H}) = n. \end{cases} \tag{2.1}$$

If $\boldsymbol{H}$ has no full rank, then $\boldsymbol{H}^+$ can be constructed from a singular value decomposition (cf. [84]).

For two vectors $\boldsymbol{v}, \boldsymbol{w} \in \mathbb{R}^n$ the relation $\boldsymbol{v} > \boldsymbol{w}$ ($\boldsymbol{v} \geq \boldsymbol{w}$) holds element-wise, i.e., $v_i > w_i$

$(v_i \geq w_i)$ is true for all $i = 1, \ldots, n$, where v_i and w_i refer to the i-th element of the vectors $\boldsymbol{v}$ and $\boldsymbol{w}$, respectively. Furthermore, the relation $\boldsymbol{v} \overset{\exists}{\geq} \boldsymbol{w}$ with the newly defined symbol $\overset{\exists}{\geq}$ holds true if there exists some element v_i for which the relation $v_i \geq w_i$ holds true, $(i \in \{1, \ldots n\})$. Accordingly, for two matrices $\boldsymbol{V}, \boldsymbol{W} \in \mathbb{R}^{n \times m}$ where $\boldsymbol{V} = (v_{ij})$ and $\boldsymbol{W} = (w_{ij})$ are composed of the elements v_{ij} and w_{ij} for $i = 1, \ldots, n$ and $j = 1, \ldots m$ the relation $\boldsymbol{V} > \boldsymbol{W}$ $(\boldsymbol{V} \geq \boldsymbol{W})$ refers to $v_{ij} > w_{ij}$ $(v_{ij} \geq w_{ij})$. For a scalar s, $|s|$ denotes the absolute value. For a vector $\boldsymbol{x} \in \mathbb{R}^n$ or a matrix $\boldsymbol{A} = (a_{ij}) \in \mathbb{R}^{n \times m}$ the $|\cdot|$-operator holds element-wise, i.e.,

$$|\boldsymbol{x}| = \begin{pmatrix} |x_1| \\ \vdots \\ |x_n| \end{pmatrix}, \qquad |\boldsymbol{A}| = \begin{pmatrix} |a_{11}| & \ldots & |a_{1m}| \\ \vdots & \ddots & \vdots \\ |a_{n1}| & \ldots & |a_{nm}| \end{pmatrix}.$$

If the real part of all eigenvalues $\lambda_i(\boldsymbol{A}) \in \mathbb{C}$, $(i = 1 \ldots n)$, of the matrix $\boldsymbol{A} \in \mathbb{R}^{n \times n}$ satisfies

$$\mathrm{Re}(\lambda_i) < 0, \quad i = 1 \ldots n,$$

then $\boldsymbol{A}$ is called *Hurwitz matrix* or *stability matrix* [81].

A spectrum $\sigma(A) := \{\lambda_i(\boldsymbol{A}), i = 1 \ldots n\}$ is the set of all eigenvalues $\lambda_i(\boldsymbol{A}) \in \mathbb{C}$, $(i = 1 \ldots n)$, of the matrix $\boldsymbol{A} \in \mathbb{R}^{n \times n}$ (cf. [84]).

Norms. The following definitions and results on norms for the vector $\boldsymbol{x} = (x_i) \in \mathbb{R}^n$ and the matrix $\boldsymbol{A} = (a_{ij}) \in \mathbb{R}^{m \times n}$ are taken from [66]. $\|\cdot\|$ denotes the Euclidean vector norm and $\|\cdot\|_\infty$ refers to the uniform norm which are defined by

$$\|\boldsymbol{x}\| := \sqrt{\sum_{i=1}^{n} |x_i|^2}$$
$$\|\boldsymbol{x}\|_\infty := \max_{i \in \{1,\ldots,n\}} |x_i|.$$

For matrices the symbol $\|\cdot\|$ denotes the spectral norm and $\|\cdot\|_{\mathrm{F}}$ refers to the Frobenius norm which are defined by

$$\|\boldsymbol{A}\| := \sqrt{\lambda_{\max}(\boldsymbol{A}^{\mathrm{H}}\boldsymbol{A})}$$
$$\|\boldsymbol{A}\|_{\mathrm{F}} := \sqrt{\sum_{j=1}^{m}\sum_{i=1}^{n} |a_{ji}|^2}, \tag{2.2}$$

where $\lambda_{\max}(\boldsymbol{A}^{\mathrm{H}}\boldsymbol{A})$ is the maximal eigenvalue of the matrix $\boldsymbol{A}^{\mathrm{H}}\boldsymbol{A}$ and $\boldsymbol{A}^{\mathrm{H}}$ is the conjugate transpose of $\boldsymbol{A}$. For the spectral norm, the relation

$$\|\boldsymbol{A}\| = \max_{\boldsymbol{x}\neq\boldsymbol{0}} \frac{\|\boldsymbol{A}\boldsymbol{x}\|}{\|\boldsymbol{x}\|}$$

holds true which implies

$$\|\boldsymbol{A}\boldsymbol{x}\| \leq \|\boldsymbol{A}\| \, \|\boldsymbol{x}\| \, .$$

Furthermore, the relation

$$\|\boldsymbol{A}\| \leq \|\boldsymbol{A}\|_{\mathrm{F}} \leq \sqrt{n}\,\|\boldsymbol{A}\| \tag{2.3}$$

holds true.

For a compact set $\mathcal{A} \subset \mathrm{I\!R}^n$,

$$\mathrm{dist}(\boldsymbol{x}, \mathcal{A}) := \inf\left\{\|\boldsymbol{x}-\boldsymbol{z}\| \;\middle|\; \boldsymbol{z} \in \mathcal{A}\right\}$$

denotes the point-to-set distance from $\boldsymbol{x} \in \mathrm{I\!R}^n$ to $\mathcal{A}$ [123].

$\|\boldsymbol{G}(s)\|_{H_\infty}$ denotes the standard $\mathcal{H}_\infty$ norm of the transfer matrix $\boldsymbol{G}(s)$, defined as

$$\|\boldsymbol{G}(s)\|_{H_\infty} := \max_{s\in\mathbb{C}+} \sigma_{\max}(\boldsymbol{G}(s)),$$

where $\sigma_{\max}(\cdot)$ is the largest singular value of its argument.

Graph theory. A directed time-varying graph

$$\mathcal{G}(t) = (\mathcal{V}, \mathcal{E}(t)) \tag{2.4}$$

is defined by the set $\mathcal{V} \subset \mathrm{I\!N}_0$ of vertices and the time-varying set $\mathcal{E}(t) \subset \mathcal{V}\times\mathcal{V}$ of directed edges $(i \rightarrow j)$, $(i, j \in \mathcal{V})$. The *set* $\mathcal{P}_i(t) \subset \mathcal{V}$ *of predecessors* of the vertex i includes all vertices j from which there is a directed edge $(j \rightarrow i)$ towards the vertex i. Hence, the set $\mathcal{E}(t)$ of edges and the *sets* $\mathcal{F}_i(t) \subset \mathcal{V}$ *of followers* can be expressed by means of the sets $\mathcal{P}_i(t)$ of predecessors

$$\mathcal{E}(t) = \{(j \rightarrow i) \mid j \in \mathcal{P}_i(t), i \in \mathcal{V}\} \tag{2.5}$$

$$\mathcal{F}_i(t) = \{j \in \mathcal{V} \mid i \in \mathcal{P}_j(t)\}, \quad i \in \mathcal{V}. \tag{2.6}$$

$|\mathcal{P}_i(t)|$ represents the in-degree and $|\mathcal{F}_i(t)|$ the out-degree of vertex i. It is assumed that no vertex has a self-loop $(i \rightarrow i) \notin \mathcal{E}$, $(\forall i \in \mathcal{V})$. The set $\mathcal{P}_i(t)$ of predecessors and the set $\mathcal{F}_i(t)$ of

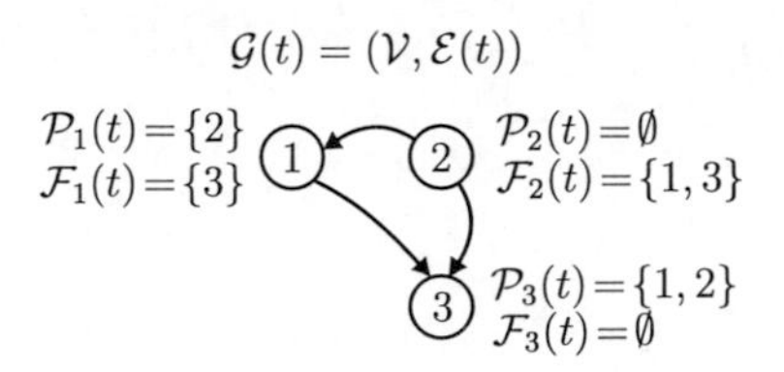

Figure 2.1: Local description of the graph $\mathcal{G}(t) = (\mathcal{V}, \mathcal{E}(t))$ by the the set $\mathcal{P}_i(t)$ of predecessors and the set $\mathcal{F}_i(t)$ of followers.

followers are a local description of the graph $\mathcal{G}(t) = (\mathcal{V}, \mathcal{E}(t))$ which is necessary to describe the local perspective of the vertices representing the subsystems and the local control units (Fig. 2.1).

A path in $\mathcal{G}(t)$ starts from vertex i and ends in vertex j traversing several vertices and is denoted by

$$Path(i \rightarrow j) = ((k_0 \rightarrow k_1), \ldots, (k_{l-1} \rightarrow k_l))$$

with

$$\begin{aligned} &(k_p \rightarrow k_{p+1}) \in \mathcal{E}(t), \quad p = 1, \ldots, l-1, \quad k_p \in \mathcal{V} \\ &k_0 = i, \quad k_l = j, \quad k_m \neq k_n \text{ for } m, n = 1, \ldots, l \text{ and } m \neq n. \end{aligned}$$

M-Matrix. A matrix $\boldsymbol{A} \in \mathbb{R}^{n \times n}$ that can be expressed in the form $\boldsymbol{A} = s\boldsymbol{I} - \boldsymbol{B}$, where $\boldsymbol{B} = (b_{ii})$ with $b_{ii} > 0$, $i \geq 1$, $j \leq n$, and $s > \rho(B)$ is a nonsingular *M-matrix*. $\rho(B)$ is the maximal eigenvalue of $\boldsymbol{B}$, (cf. [25, 108]). The following two conditions are equivalent to the statement: $\boldsymbol{A}$ is an M-Matrix. First, the inverse of a nonsingular M-Matrix always exists and is non-negative, that is

$$\boldsymbol{A}^{-1} \geq \boldsymbol{O}.$$

Second, a nonsingular M-Matrix is monotone, that is

$$\boldsymbol{A}\boldsymbol{x} \geq \boldsymbol{0} \Rightarrow \boldsymbol{x} \geq \boldsymbol{0}, \quad \forall \boldsymbol{x} \in \mathbb{R}^n. \tag{2.7}$$

2.2 Definitions on stability, controllability and observability

The following definitions and results concern the stability, controllability and observability of a linear time-invariant system

$$\Sigma : \begin{cases} \dot{\boldsymbol{x}}(t) = \boldsymbol{A}\boldsymbol{x}(t) + \boldsymbol{B}\boldsymbol{u}(t), \quad \boldsymbol{x}(0) = \boldsymbol{x}_0 \\ \boldsymbol{y}(t) = \boldsymbol{C}\boldsymbol{x}(t) \end{cases} \tag{2.8}$$

with the state $\boldsymbol{x}(t) \in \mathrm{IR}^n$, the initial state $\boldsymbol{x}_0 \in \mathrm{IR}^n$, the input $\boldsymbol{u}(t) \in \mathrm{IR}^m$, the output $\boldsymbol{y}(t) \in \mathrm{IR}^r$ and the matrices $\boldsymbol{A} \in \mathrm{IR}^{n \times n}$, $\boldsymbol{B} \in \mathrm{IR}^{n \times m}$ as well as $\boldsymbol{C} \in \mathrm{IR}^{r \times n}$. Since for linear systems there is no difference between local asymptotic stability and global asymptotic stability (cf. [122]), a system is called asymptotically stable if it is global asymptotically stable. Furthermore, due to the linearity there will be no distinction between local asymptotic controllability and global asymptotic controllability as well as between local asymptotic observability and global asymptotic observability (cf. [122]). The system Σ is called completely controllable or completely observable if it is global asymptotically controllable or global asymptotically observable, respectively.

In the following, definitions on the stability, controllability and observability of the linear system Σ defined in (2.8) are given, the difference between asymptotic stability and practical stability is described and a condition on the stability of Σ with an uncertain output feedback is presented.

Controllability and observability of linear systems. The following definition concerns the controllability and observability of the linear system Σ defined in (2.8):

Definition 2.1 ([88] **Controllability and observability**) *The system Σ defined in* (2.8) *is said to be completely controllable if for every initial state $\boldsymbol{x}_0$ there exists a finite time T and a control input $\boldsymbol{u}(t)$, $(t \in [0, T])$, such that $\boldsymbol{x}(T) = \boldsymbol{0}$ holds. The system Σ defined in* (2.8) *is said to be completely observable if for $\boldsymbol{u}(t) = \boldsymbol{0}$, $(t \geq 0)$, every initial state $\boldsymbol{x}_0$ can be reconstructed from the measurements of $\boldsymbol{y}(t)$, $(t \in [0, T])$, for some T.*

Conditions for testing the controllability of Σ on the pair $(\boldsymbol{A}, \boldsymbol{B})$ and observability of Σ on the pair $(\boldsymbol{A}, \boldsymbol{C})$ are given, e.g., in [88, 94, 122]. In the following it is said that the pair $(\boldsymbol{A}, \boldsymbol{B})$ is controllable if Σ is completely controllable and the pair $(\boldsymbol{A}, \boldsymbol{C})$ is observable if Σ is completely observable.

Comparison between asymptotic stability and practical stability. The main difference between asymptotic stability and practical stability is that an asymptotically stable system Σ converges to the equilibrium point and a practically stable system Σ converges into an area around

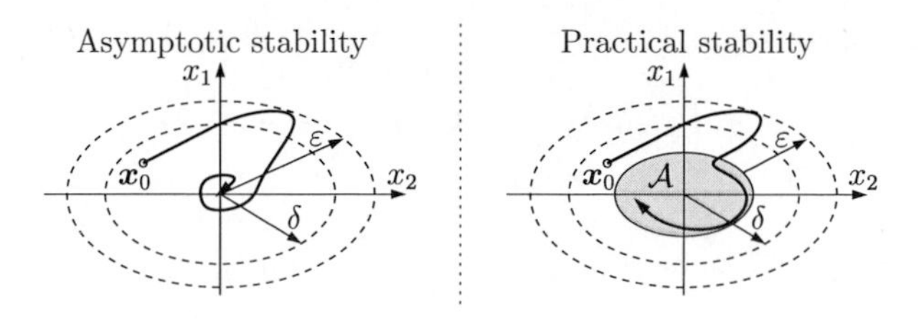

Figure 2.2: Comparison between asymptotic stability and practical stability.

the equilibrium point (Fig. 2.2).

In the following a general definition for asymptotic stability of Σ defined in (2.8) is given, where Σ is considered as an autonomous system, i.e., $\boldsymbol{u}(t) = \boldsymbol{0}$ for all $t \geq 0$.

Definition 2.2 ([81] **Asymptotic stability**) *The autonomous system Σ defined in* (2.8) *with the input $\boldsymbol{u}(t) = \boldsymbol{0}$ for all $t \geq 0$ is said to be asymptotically stable with respect to the equilibrium point $\boldsymbol{x} = \boldsymbol{0}$ if for each $\varepsilon > 0$ there exists $\delta > 0$ such that $\|\boldsymbol{x}_0\| \leq \delta$ implies $\|\boldsymbol{x}(t)\| \leq \varepsilon$, $(\forall t \geq 0)$, and the condition $\lim_{t\to\infty} \|\boldsymbol{x}(t)\| = 0$ holds.*

The following definition of practical stability is closely related to input-to-state practical stability:

Definition 2.3 (**Practical stability**) *The system Σ defined in* (2.8) *is said to be practically stable with respect to the compact set $\mathcal{A} \in \mathrm{I\!R}^n$ if for each $\varepsilon > 0$ there exists $\delta > 0$ such that $||\boldsymbol{x}_0|| \leq \delta$ implies $\mathrm{dist}(\boldsymbol{x}(t), \mathcal{A}) \leq \varepsilon$, $(\forall t \geq 0)$, and $\lim_{t\to\infty} \mathrm{dist}(\boldsymbol{x}(t), \mathcal{A}) = 0$ holds.*

Compared to the definition of asymptotic stability in Definition 2.2, for practical stability an input $\boldsymbol{u}(t) \neq \boldsymbol{0}$ is considered (cf. [123]).

The comparison of Definition 2.2 and Definition 2.3 shows that for asymptotic stability the state $\boldsymbol{x}(t)$ of the linear system Σ has to converge to the equilibrium point $\boldsymbol{x} = \boldsymbol{0}$ for $t \to \infty$ and for practical stability $\boldsymbol{x}(t)$ has to converge to the compact set $\mathcal{A}$ for $t \to \infty$, where $\mathcal{A}$ defines an area around the equilibrium point $\boldsymbol{x} = \boldsymbol{0}$.

Stabilization and observation of linear systems. A well known condition for the asymptotic stability of linear systems is presented in the following theorem.

Theorem 2.1 ([122] **Condition on asymptotic stability**) *The equilibrium point $\boldsymbol{x} = \boldsymbol{0}$ of the autonomous system Σ defined in* (2.8) *with the $\boldsymbol{u}(t) = \boldsymbol{0}$ for all $t \geq 0$ is asymptotically stable if and only if $\boldsymbol{A} \in \mathrm{I\!R}^{n\times n}$ is a Hurwitz matrix.*

Consider the linear system Σ defined in (2.8) controlled by a linear state feedback

$$\boldsymbol{u}(t) = -\boldsymbol{K}\boldsymbol{x}(t) \tag{2.9}$$

with the feedback gain matrix $\boldsymbol{K} \in \mathbb{R}^{m\times n}$ which leads to the closed-loop system

$$\bar{\Sigma}: \begin{cases} \dot{\boldsymbol{x}}(t) = (\boldsymbol{A} - \boldsymbol{B}\boldsymbol{K})\boldsymbol{x}(t), \quad \boldsymbol{x}(0) = \boldsymbol{x}_0 \\ \boldsymbol{y}(t) = \boldsymbol{C}\boldsymbol{x}(t). \end{cases} \tag{2.10}$$

Considering Theorem 2.1 the following condition on the stabilization of Σ by the state feedback (2.9) results:

Theorem 2.2 ([36] **Stabilization by state feedback**) *If the pair* $(\boldsymbol{A}, \boldsymbol{B})$ *is controllable, a gain matrix* $\boldsymbol{K} \in \mathbb{R}^{m\times n}$ *can be chosen to arbitrarily place the eigenvalues of the closed-loop system matrix* $\boldsymbol{A} - \boldsymbol{B}\boldsymbol{K}$*. Consequently, every completely controllable linear system* Σ *defined in* (2.8) *can be stabilized by the linear state feedback* (2.9) *which means that the closed-loop system* $\bar{\Sigma}$ *defined in* (2.10) *can be made asymptotically stable.*

In the following, the main ideas of two approaches for designing the feedback gain matrix $\boldsymbol{K}$ in (2.9) which are used in this thesis are briefly explained. First, with methods for the so-called "Pole-Assignment" presented in [36, 94, 122] the feedback gain matrix $\boldsymbol{K}$ in (2.9) is determined such that the system matrix $\boldsymbol{A} - \boldsymbol{B}\boldsymbol{K}$ of the closed-loop system $\bar{\Sigma}$ defined in (2.10) has the desired eigenvalues λ_i^*

$$\lambda_i(\boldsymbol{A} - \boldsymbol{B}\boldsymbol{K}) = \lambda_i^*, \quad i = 1, \ldots n.$$

Second, for designing the state feedback matrix $\boldsymbol{K}$ in (2.9) consider the optimization problem

$$\min_{\boldsymbol{u}(t),\, \boldsymbol{x}_0} \; J := \int_0^\infty \left(\boldsymbol{x}^{\mathrm{T}}(t)\boldsymbol{Q}\boldsymbol{x}(t) + \boldsymbol{u}^{\mathrm{T}}(t)\boldsymbol{R}\boldsymbol{u}(t)\right) \mathrm{d}t \tag{2.11}$$
$$\text{s.t. } \Sigma \text{ defined in (2.8)},$$

where $\boldsymbol{Q}$ and $\boldsymbol{R}$ are symmetric positive definite matrices. The solution of the optimization problem in (2.11) yields a state feedback of the form (2.9) which is called linear quadratic regulator (LQR) leading to an asymptotically stable closed-loop system $\bar{\Sigma}$ defined in (2.10) (cf. [94, 122]).

Linear Observers. Consider that only the output $\boldsymbol{y}(t)$ of the system Σ can be measured. The state $\boldsymbol{x}(t)$ of the linear system Σ defined in (2.8) that is controlled by the state feedback (2.9) can be reconstructed by a linear observer of the form

$$\tfrac{\mathrm{d}}{\mathrm{d}t}\hat{\boldsymbol{x}}(t) = \boldsymbol{A}\hat{\boldsymbol{x}}(t) + \boldsymbol{B}\boldsymbol{u}(t) - \boldsymbol{K}_{\mathrm{O}}(\boldsymbol{C}\hat{\boldsymbol{x}} - \boldsymbol{y}(t)), \quad \hat{\boldsymbol{x}}(0) = \hat{\boldsymbol{x}}_0 \tag{2.12}$$

with the estimated state $\hat{\boldsymbol{x}}(t) \in \mathrm{I\!R}^n$, the initial state $\hat{\boldsymbol{x}}_0 \in \mathrm{I\!R}^n$ and the observer gain matrix $\boldsymbol{K}_\mathrm{O} \in \mathrm{I\!R}^{n\times r}$. This leads to the observed closed-loop system

$$\tilde{\Sigma}: \begin{cases} \begin{pmatrix} \dot{\boldsymbol{x}}(t) \\ \dot{\boldsymbol{e}}(t) \end{pmatrix} = \underbrace{\begin{pmatrix} \boldsymbol{A}-\boldsymbol{B}\boldsymbol{K} & \boldsymbol{B}\boldsymbol{K} \\ \boldsymbol{O} & \boldsymbol{A}-\boldsymbol{K}_\mathrm{O}\boldsymbol{C} \end{pmatrix}}_{=\bar{\boldsymbol{A}}} \begin{pmatrix} \boldsymbol{x}(t) \\ \boldsymbol{e}(t) \end{pmatrix}, \quad \begin{pmatrix} \boldsymbol{x}(0) \\ \boldsymbol{e}(0) \end{pmatrix} = \begin{pmatrix} \boldsymbol{x}_0 \\ \boldsymbol{x}_0 - \hat{\boldsymbol{x}}_0 \end{pmatrix} \\ \boldsymbol{y}(t) = \begin{pmatrix} \boldsymbol{C} & \boldsymbol{O} \end{pmatrix} \begin{pmatrix} \boldsymbol{x}(t) \\ \boldsymbol{e}(t) \end{pmatrix} \end{cases} \tag{2.13}$$

with the observation error $\boldsymbol{e}(t) = \boldsymbol{x}(t) - \hat{\boldsymbol{x}}(t)$. With this, the following theorem gives a condition for designing an observer (2.12) for reconstructing $\boldsymbol{x}(t)$ that leads to an asymptotically stable closed-loop system $\tilde{\Sigma}$:

Theorem 2.3 ([36] **Observation by linear observer**) *If the pair $(\boldsymbol{A}, \boldsymbol{C})$ is observable, an observer gain matrix $\boldsymbol{K}_\mathrm{O} \in \mathrm{I\!R}^{n\times r}$ can be chosen to arbitrarily place the eigenvalues of the closed-loop matrix $\boldsymbol{A}-\boldsymbol{K}_\mathrm{O}\boldsymbol{C}$. Consequently, a linear asymptotic observer can be elaborated for every completely observable system Σ defined in* (2.8) *which means that the observed closed-loop system $\tilde{\Sigma}$ defined in* (2.13) *can be made asymptotically stable.*

Furthermore, for the observed closed-loop system $\tilde{\Sigma}$ defined in (2.13) the following separation theorem is relevant:

Theorem 2.4 ([88] **Separation theorem**) *The spectrum $\sigma(\bar{\boldsymbol{A}})$ of the system matrix $\bar{\boldsymbol{A}}$ of the observed closed-loop system $\tilde{\Sigma}$ defined in* (2.13) *consists of the spectrum $\sigma(\boldsymbol{A}-\boldsymbol{B}\boldsymbol{K})$ of the system matrix $\boldsymbol{A}-\boldsymbol{B}\boldsymbol{K}$ of the closed-loop system $\bar{\Sigma}$ defined in* (2.10) *and the spectrum $\sigma(\boldsymbol{A}-\boldsymbol{K}_\mathrm{O}\boldsymbol{C})$ of the system matrix $\boldsymbol{A}-\boldsymbol{K}_\mathrm{O}\boldsymbol{C}$ of the linear observer defined in* (2.10)

$$\sigma(\bar{\boldsymbol{A}}) = \sigma(\boldsymbol{A}-\boldsymbol{B}\boldsymbol{K}) \cup \sigma(\boldsymbol{A}-\boldsymbol{K}_\mathrm{O}\boldsymbol{C}).$$

Theorem 2.4 shows that the feedback gain matrix $\boldsymbol{K}$ in (2.9) can be designed independently of observer gain matrix $\boldsymbol{K}_\mathrm{O}$ in (2.12) to guarantee desired eigenvalues for the observed closed-loop system $\tilde{\Sigma}$ defined in (2.13).

Due to the duality of the design problem, the presented methods for designing the feedback gain matrix $\boldsymbol{K}$ in (2.9) can also be used to design the observer gain matrix $\boldsymbol{K}_\mathrm{O}$ in (2.12) (cf. [36, 94, 122]).

Stability of uncertain systems. Consider the linear system Σ defined in (2.8) with the output-feedback

$$\boldsymbol{u}(t) = \boldsymbol{K}_{\mathrm{u}}(t)\boldsymbol{y}(t), \quad \text{with } \|\boldsymbol{K}_{\mathrm{u}}(t)\| \leq 1, \tag{2.14}$$

where the feedback gain matrix $\boldsymbol{K}_{\mathrm{u}}(t)$ is time-variant. Then, the closed-loop uncertain system $\bar{\Sigma}_{\mathrm{u}}$ results from (2.8) and (2.14) to

$$\bar{\Sigma}_{\mathrm{u}}: \ \dot{\boldsymbol{x}}(t) = (\boldsymbol{A} + \boldsymbol{B}\boldsymbol{K}_{\mathrm{u}}(t)\boldsymbol{C})\boldsymbol{x}(t), \quad \boldsymbol{x}(0) = \boldsymbol{x}_0, \quad \text{with } \|\boldsymbol{K}_{\mathrm{u}}(t)\| \leq 1. \tag{2.15}$$

The following theorem gives a small gain condition for the asymptotic stability of $\bar{\Sigma}_{\mathrm{u}}$ defined in (2.15).

Theorem 2.5 ([82, 85] **Stability of uncertain linear systems**) *The uncertain closed-loop system defined in* (2.15) *is asymptotically stable if it satisfies the following conditions:*

1. *The matrix $\boldsymbol{A}$ is a Hurwitz matrix.*

2. $\|\boldsymbol{C}(s\boldsymbol{I} - \boldsymbol{A})^{-1}\boldsymbol{B}\|_{H_\infty} < 1$.

Input-output representation of a linear system. The linear system Σ defined in (2.8) with $\boldsymbol{x}_0 = \boldsymbol{0}$ can be written in the input-output representation

$$\Sigma: \ \boldsymbol{y}(t) = \boldsymbol{G} * \boldsymbol{u}$$

with the impulse response matrix $\boldsymbol{G}(t) = \boldsymbol{C}\mathrm{e}^{\boldsymbol{A}t}\boldsymbol{B}$.

Theorem 2.6 ([88] **Stability within input-output representation**) *Consider that the linear system Σ defined in* (2.8) *with $\boldsymbol{u}(t) = \boldsymbol{0}$ for all $t \geq 0$ is completely controllable and completely observable, then Σ is asymptotically stable if and only if the condition*

$$\int_0^\infty \|\boldsymbol{G}(t)\| \, \mathrm{d}t < \infty$$

holds true.

The condition in Theorem 2.6 is also a test for the input-output stability of a system (cf. [88, 122]).

2.3 Communication among the local control units

The local control units exchange information via a communication network (Fig. 1.1). This section states the general modeling of the situation-dependent communication structure by a time-varying graph and defines the properties of the communication network.

Modeling of the situation-dependent communication structure. The situation-dependent communication structure of the self-organizing controller is represented by the *communication graph*

$$\mathcal{G}_{\mathrm{C}}(t) = (\mathcal{V}_{\mathrm{C}}, \mathcal{E}_{\mathrm{C}}(t)) \tag{2.16}$$

according to the general definition of a graph in (2.4). The *set* $\mathcal{V}_{\mathrm{C}} \subset \mathbb{N}_0$ *of communicational vertices* represents the local control units C_i, $(\forall i \in \mathcal{V}_{\mathrm{C}})$, and the *set* $\mathcal{E}_{\mathrm{C}}(t) \subset \mathcal{V}_{\mathrm{C}} \times \mathcal{V}_{\mathrm{C}}$ *of communicational edges* represents the communication links among the local control units C_i. A directed edge $(i \rightarrow j) \in \mathcal{E}_{\mathrm{C}}(t)$ shows that information is communicated from the local control unit C_i to the local control unit C_j. The *set* $\mathcal{P}_{\mathrm{C}i}(t) \subset \mathcal{V}_{\mathrm{C}}$ *of communicational predecessors* of C_i includes the index j of all local control units C_j from which there is a directed communication link $(j \rightarrow i)$ towards C_i. According to (2.5) and (2.6), the set $\mathcal{E}_{\mathrm{C}}(t)$ of communicational edges and the *sets* $\mathcal{F}_{\mathrm{C}i}(t) \subset \mathcal{V}_{\mathrm{C}}$ *of communicational followers* can be expressed by means of the sets $\mathcal{P}_{\mathrm{C}i}(t)$ of communicational predecessors

$$\mathcal{E}_{\mathrm{C}}(t) := \{(j \rightarrow i) \mid j \in \mathcal{P}_{\mathrm{C}i}(t), i \in \mathcal{V}_{\mathrm{C}}\} \tag{2.17}$$

$$\mathcal{F}_{\mathrm{C}i}(t) := \{j \in \mathcal{V}_{\mathrm{C}} \mid i \in \mathcal{P}_{\mathrm{C}j}(t)\}, \quad i \in \mathcal{V}_{\mathrm{C}}. \tag{2.18}$$

For simplification of the notation, the terms set $\mathcal{P}_{\mathrm{C}i}(t)$ of communicational predecessors and the set $\mathcal{F}_{\mathrm{C}i}(t)$ of communicational followers are often called set of predecessors and set of followers, respectively.

Limitiation of the communication links. Consider a time-invariant *maximal communication graph* $\bar{\mathcal{G}}_{\mathrm{C}}$ and a *minimal communication graph* $\underline{\mathcal{G}}_{\mathrm{C}}$ defined by

$$\bar{\mathcal{G}}_{\mathrm{C}} = (\mathcal{V}_{\mathrm{C}}, \bar{\mathcal{E}}_{\mathrm{C}}) \tag{2.19}$$

$$\underline{\mathcal{G}}_{\mathrm{C}} = (\mathcal{V}_{\mathrm{C}}, \underline{\mathcal{E}}_{\mathrm{C}}), \tag{2.20}$$

where $\bar{\mathcal{E}}_{\mathrm{C}} \subset \mathcal{V}_{\mathrm{C}} \times \mathcal{V}_{\mathrm{C}}$ is the *set of maximal communicational edges* and $\underline{\mathcal{E}}_{\mathrm{C}} \subset \mathcal{V}_{\mathrm{C}} \times \mathcal{V}_{\mathrm{C}}$ is the *set of minimal communicational edges*. According to (2.17) and (2.18), the set $\bar{\mathcal{E}}_{\mathrm{C}}$ and the *sets* $\bar{\mathcal{F}}_{\mathrm{C}i} \subset \mathcal{V}_{\mathrm{C}}$ *of maximal communicational followers* can be expressed by means of the *sets*

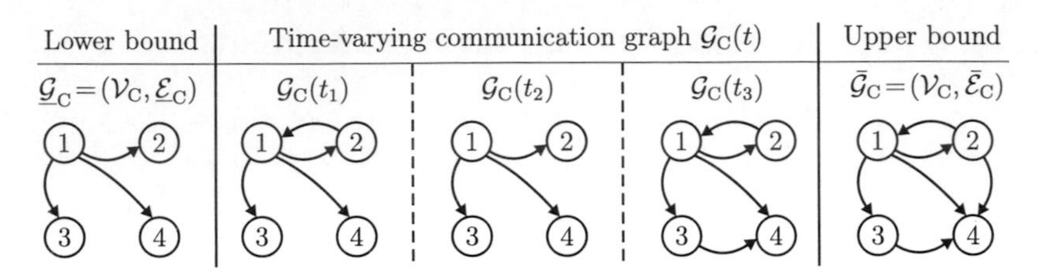

Figure 2.3: Limitation of the set of communicational edges $\mathcal{E}_C(t)$ by a set of maximal communicational edges $\bar{\mathcal{G}}_C$ and a set of minimal communicational edges $\underline{\mathcal{G}}_C$.

$\bar{\mathcal{P}}_{Ci} \subset \mathcal{V}_C$ *of maximal communicational predecessors* as follows:

$$\bar{\mathcal{E}}_C = \{(j \to i) \mid j \in \bar{\mathcal{P}}_{Ci}, i \in \mathcal{V}_C\} \tag{2.21}$$

$$\bar{\mathcal{F}}_{Ci} = \{j \in \mathcal{V}_C \mid i \in \bar{\mathcal{P}}_{Cj}\}, \quad i \in \mathcal{V}_C. \tag{2.22}$$

Similarly, the set $\underline{\mathcal{E}}_C$ and the *sets* $\underline{\mathcal{F}}_{Ci} \subset \mathcal{V}_C$ *of minimal communicational followers* can be expressed by means of the *sets* $\underline{\mathcal{P}}_{Ci} \subset \mathcal{V}_C$ *of minimal communicational predecessors*:

$$\underline{\mathcal{E}}_C = \{(j \to i) \mid j \in \underline{\mathcal{P}}_{Ci}, i \in \mathcal{V}_C\} \tag{2.23}$$

$$\underline{\mathcal{F}}_{Ci} = \{j \in \mathcal{V}_C \mid i \in \underline{\mathcal{P}}_{Cj}\}, \quad i \in \mathcal{V}_C. \tag{2.24}$$

Figure 2.3 illustrates that the set $\mathcal{E}_C(t)$ of communication links within the communication graph $\mathcal{G}_C(t) = (\mathcal{V}_C, \mathcal{E}_C(t))$ is assumed to be bounded according to

$$\underline{\mathcal{E}}_C \subseteq \mathcal{E}_C(t) \subseteq \bar{\mathcal{E}}_C, \quad \forall t \geq 0. \tag{2.25}$$

The following chapters will show that these upper and lower bound vary for the developed self-organizing control concepts.

Properties of the communication network. The communication network is assumed to be ideal in the following sense.

Assumption 2.1 *The transmission of information by the local control units C_i over the communication network happens instantaneously, without delays and without packet losses. There are no collisions while transmitting information simultaneously.*

Obviously, in reality this assumption is hard to be satisfied. However, in this thesis the communication network is considered to transmit information much faster compared to the dynamical behavior of the control system which justifies Assumption 2.1.

2.4 Faults affecting the self-organizing control system

A self-organizing control system should have the property of fault-tolerance as mentioned in Section 1.2. This section defines possible faults that can affect the subsystems Σ_i and the local control units C_i.

In [34] a fault is defined as "*An unpermitted deviation of at least one characteristic property or parameter of the system from the acceptable/usual/standard condition*". A fault can change the system structure or the system parameters in a dynamical system (cf. [26]). Therefore, a fault could case a failure of the overall system.

The classification of faults in [26] is used to define the faults affecting a subsystem or an agent. Furthermore, also faults in the local control units are considered. The classification of the faults is as follows:

- **Subprocess fault:** Such faults change the input-output behavior of a subprocess (Fig. 2.4).
- **Actuator fault:** The subprocess properties are not affected, but the influence of the local control unit C_i on the subprocess is interrupted or modified.
- **Sensor fault:** The subprocess properties are not affected, but the sensor readings have substantial errors.
- **Communication fault:** The subsystem is not affected, but the communication among the local control units is interrupted.

Obviously also the physical coupling among the subsystems could change (be faulty) which would be a process fault of the overall plant P. Such kind of fault is not considered in this thesis. Rather, the physical coupling is considered to be fixed.

Compared to the other faults, a communication fault is a fault within the self-organizing controller, where usually in the literature the controllers are considered as fault-free (cf. [26]). Note that a communication fault does not violate Assumption 2.1 on the ideality of the communication network. A communication fault may occur due to a faulty transmission unit or receiver unit in the local control units, whereas Assumption 2.1 classifies the properties/behavior of the communication among the local control units in the fault-free case. In the following no precise model of the transmission unit and receiver unit will be given. Faults within these components are represented, for example, by the fact, that information is not transmitted (faulty transmission unit) or can not be received (faulty receiver unit) by the local control unit. Note that the control structure within the local control units is assumed to be fault-free.

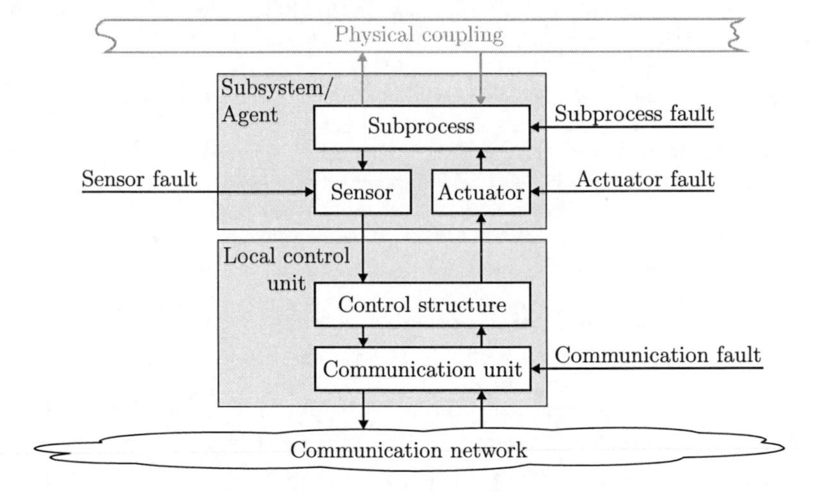

Figure 2.4: Distinction between subprocess faults, actuator faults, sensor faults and communication faults.

Part I

Self-organizing control of physically interconnected systems

3 Modeling and control of physically interconnected systems

This chapter gives an introduction to the self-organizing control structure for physically interconnected systems considered in this part of the thesis. The model of the coupled subsystems is presented. A general model for the self-organizing control system is introduced which has specific constraints on the communication structure. A formal definition of the control aims concerning stability, disturbance attenuation and a desired difference behavior compared to a centralized control system are presented. Finally, two running examples are introduced to demonstrate the properties of the developed control concepts.

Chapter contents

3.1 Introduction to self-organizing control of physically interconnected systems

This part of the thesis concerns the control of physically interconnected systems, whose plant P consists of subsystems Σ_i that are interconnected by a physical coupling (Fig. 3.1). It is assumed that each subsystem Σ_i is controlled by a local control unit C_i which can exchange information over the communication network. The local control units C_i with their situation-dependent communication build the self-organizing controller C. In summary, the physically coupled subsystems Σ_i (plant P) and the communicating local control units C_i (self-organizing controller C) build the overall self-organizing control system Σ.

Due to the physical coupling, the behavior of a subsystem affects the behavior of other subsystems. For example, a disturbance affecting only one subsystem can propagate over the physical coupling to other subsystems. Hence, for the analysis and the controller design it is not enough to take the dynamics of the individual subsystems into account. The impact on the other subsystems and its feedback on the own subsystem always has to be considered.

The local control units C_i generate the input $\boldsymbol{u}_i(t)$ to the subsystems Σ_i based on the local measured output $\boldsymbol{y}_i(t)$ of Σ_i and, if available, information about other subsystems transmitted by their corresponding local control units. It seems reasonable to exchange information among the local control units, whenever there are great physical coupling impacts among the subsystems, e.g., in case of a disturbance affecting a subsystem. Hence, the main task for the self-organizing control of physically interconnected subsystems can be summarized as follows:

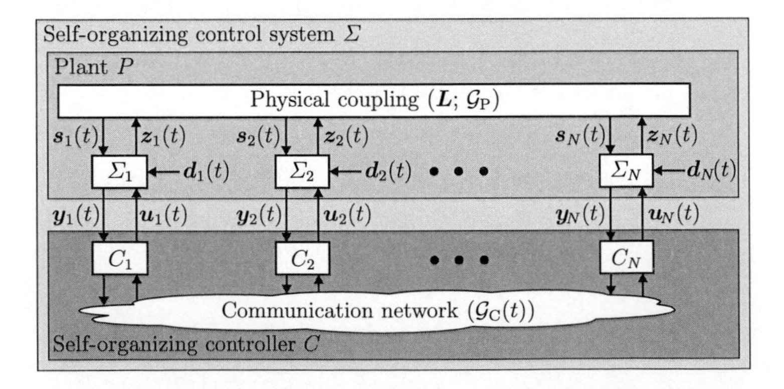

Figure 3.1: Self-organizing control of physically interconnected systems.

> Develop local control units C_i that can recognize great coupling impacts among the subsystems to counteract these impacts by exchanging information to guarantee a desired performance of the overall closed-loop system Σ.

Due to the physical coupling among the subsystems, to guarantee a desired performance of the overall closed-loop system Σ the local control units C_i can generally not be designed with local model information of the corresponding subsystem Σ_i only. At least the physical coupling with other subsystems or even the model of the overall plant P has to be known.

Before explaining the actual control concepts in Chapters 4-5, at first the following general aspects are introduced:

- The model of the physically interconnected subsystem Σ_i building the overall plant P (Section 3.2).
- A general model of the self-organizing control system Σ which includes the restriction of the communication among the local control units C_i (Section 3.3).
- The control aims claimed for the resulting self-organizing control system Σ (Section 3.4).
- Two different running examples of physically interconnected systems, a small one with two subsystem and a large one with 23 subsystems (Section 3.5).

3.2 Modeling of the physically interconnected subsystems

This section derives a detailed model of the overall plant P (Section 3.2.3), where at first the model of the subsystems (Section 3.2.2) and the model of their physical coupling (Section 3.2.2) are introduced. Possible faults affecting the subsystems are modeled in Section 3.2.4. Section 3.2.5 summarizes the basic assumptions and properties of the plant P.

3.2.1 Model of the subsystems

The subsystems Σ_i, $(i \in \mathcal{N} = \{1, \ldots, N\})$, are described by the linear state-space model

$$\Sigma_i : \begin{cases} \dot{\boldsymbol{x}}_i(t) = \boldsymbol{A}_i \boldsymbol{x}_i(t) + \boldsymbol{B}_i \boldsymbol{u}_i(t) + \boldsymbol{E}_i \boldsymbol{s}_i(t) + \boldsymbol{G}_i \boldsymbol{d}_i(t), \quad \boldsymbol{x}_i(0) = \boldsymbol{x}_{0i} \\ \boldsymbol{y}_i(t) = \boldsymbol{C}_i \boldsymbol{x}_i(t) \\ \boldsymbol{v}_i(t) = \boldsymbol{C}_{\mathrm{v}i} \boldsymbol{x}_i(t) \\ \boldsymbol{z}_i(t) = \boldsymbol{C}_{\mathrm{z}i} \boldsymbol{x}_i(t) \end{cases} \tag{3.1}$$

with the state $\boldsymbol{x}_i(t) \in \mathbb{R}^{n_i}$, the initial state $\boldsymbol{x}_{0i} \in \mathbb{R}^{n_i}$, the input $\boldsymbol{u}_i(t) \in \mathbb{R}^{m_i}$, the measured output $\boldsymbol{y}_i(t) \in \mathbb{R}^{r_i}$, the performance output $\boldsymbol{v}_i(t) \in \mathbb{R}^{r_{\mathrm{v}i}}$, the coupling input $\boldsymbol{s}_i(t) \in \mathbb{R}^{m_{\mathrm{s}i}}$, the coupling output $\boldsymbol{z}_i(t) \in \mathbb{R}^{r_{\mathrm{z}i}}$, the disturbance $\boldsymbol{d}_i(t) \in \mathbb{R}^{n_{\mathrm{d}i}}$ and the matrices $\boldsymbol{A}_i \in \mathbb{R}^{n_i \times n_i}$, $\boldsymbol{B}_i \in \mathbb{R}^{n_i \times m_i}$, $\boldsymbol{E}_i \in \mathbb{R}^{n_i \times m_{\mathrm{s}i}}$, $\boldsymbol{G}_i \in \mathbb{R}^{n_i \times n_{\mathrm{d}i}}$, $\boldsymbol{C}_i \in \mathbb{R}^{r_i \times n_i}$, $\boldsymbol{C}_{\mathrm{v}i} \in \mathbb{R}^{r_{\mathrm{v}i} \times n_i}$ as well as $\boldsymbol{C}_{\mathrm{z}i} \in \mathbb{R}^{r_{\mathrm{z}i} \times n_i}$. In the model (3.1) of Σ_i it is assumed that there is no direct throughput of $\boldsymbol{u}_i(t)$, $\boldsymbol{s}_i(t)$ and $\boldsymbol{d}_i(t)$ towards $\boldsymbol{y}_i(t)$, $\boldsymbol{v}_i(t)$ and $\boldsymbol{z}_i(t)$.

A so-called *isolated subsystem* denotes a subsystem Σ_i with no coupling input and vanishing output, i.e., $\boldsymbol{z}_i(t) = \boldsymbol{0}$ and $\boldsymbol{s}_i(t) = \boldsymbol{0}$ for all $t \geq 0$.

The disturbances $\boldsymbol{d}_i(t)$, $(\forall i \in \mathcal{N})$, affecting the subsystems Σ_i are assumed to be bounded, as stated in the following assumption.

Assumption 3.1 *The disturbances $\boldsymbol{d}_i(t)$ are upper bounded by the vector $\bar{\boldsymbol{d}}_i \in \mathbb{R}_+^{n_{\mathrm{d}i}}$*

$$|\boldsymbol{d}_i(t)| \leq \bar{\boldsymbol{d}}_i, \quad \forall i \in \mathcal{N}, \quad t \geq 0. \tag{3.2}$$

With Assumption 3.1 and the definition of the maximal disturbance $\bar{d}_i := \left\|\bar{\boldsymbol{d}}_i\right\|$ it follows that

$$\|\boldsymbol{d}_i(t)\| \leq \bar{d}_i, \quad \forall i \in \mathcal{N}, \quad t \geq 0,$$

since $\|\boldsymbol{d}_i(t)\| = \| \, |\boldsymbol{d}_i(t)| \, \| \leq \left\|\bar{\boldsymbol{d}}_i\right\| = \bar{d}_i$ holds true.

Furthermore, the subsystems Σ_i defined in (3.1) are assumed to be completely controllable and completely observable (cf. Section 2.2).

Assumption 3.2 *The pairs $(\boldsymbol{A}_i, \boldsymbol{B}_i)$ are controllable and the pairs $(\boldsymbol{A}_i, \boldsymbol{C}_i)$ are observable for all $i \in \mathcal{N}$.*

If not otherwise stated, it is assumed that the complete state $\boldsymbol{x}_i(t)$ can be measured, i.e., $\boldsymbol{y}_i(t) = \boldsymbol{x}_i(t)$ for all $i \in \mathcal{N}$ and all $t \geq 0$.

Assumption 3.3 *The relation $\boldsymbol{C}_i = \boldsymbol{I}_{n_i}$ holds true for all $i \in \mathcal{N}$.*

3.2.2 Physical coupling structure

The relations

$$\boldsymbol{s}_i(t) = \sum_{j=1}^{N} \boldsymbol{L}_{ij} \boldsymbol{z}_j(t), \quad \forall i \in \mathcal{N} \tag{3.3}$$

between the coupling outputs $\boldsymbol{z}_j(t)$, $(\forall j \in \mathcal{N})$, and the coupling inputs $\boldsymbol{s}_i(t)$, $(\forall i \in \mathcal{N})$, describe the physical interconnection among the subsystems Σ_i, where the matrices $\boldsymbol{L}_{ij} \in \mathbb{R}^{m_{\mathrm{s}i} \times r_{\mathrm{z}j}}$ build the *physical interconnection matrix* $\boldsymbol{L} = (\boldsymbol{L}_{ij})$, $(i, j \in \mathcal{N})$. The coupling input $\boldsymbol{s}_i(t)$ is the sum of all coupling inputs $\boldsymbol{s}_{ij}(t) := \boldsymbol{L}_{ij}\boldsymbol{z}_j(t) = \boldsymbol{L}_{ij}\boldsymbol{C}_{\mathrm{z}j}\boldsymbol{x}_j(t)$ from subsystem Σ_j to subsystem Σ_i

$$\boldsymbol{s}_i(t) = \sum_{j=1}^{N} \boldsymbol{s}_{ij}(t), \quad \forall i \in \mathcal{N}. \tag{3.4}$$

Assumption 3.4 *The relation $\boldsymbol{L}_{ii} = \boldsymbol{O}_{n_i \times n_i}$ holds for all $i \in \mathcal{N}$, i.e., the coupling input $\boldsymbol{s}_i(t)$ does not directly depend upon the coupling output $\boldsymbol{z}_i(t)$.*

This assumption is weak and can always be fulfilled by modeling all internal couplings $\boldsymbol{s}_{ii}(t) = \boldsymbol{L}_{ii}\boldsymbol{C}_{\mathrm{z}i}\boldsymbol{x}_i(t)$ in the matrices $\boldsymbol{A}_i$, $(\forall i \in \mathcal{N})$. With Assumption 3.4 the structure of the interconnection matrix used in this thesis is given as follows

$$\boldsymbol{L} = \begin{pmatrix} \boldsymbol{O} & \boldsymbol{L}_{12} & \cdots & \boldsymbol{L}_{1N} \\ \boldsymbol{L}_{21} & \boldsymbol{O} & \ddots & \vdots \\ \vdots & \ddots & \ddots & \boldsymbol{L}_{N-1,N} \\ \boldsymbol{L}_{N1} & \cdots & \boldsymbol{L}_{N,N-1} & \boldsymbol{O} \end{pmatrix}$$

with the zero matrices $\boldsymbol{O}$ of appropriate dimensions.

According to (3.3), the overall physical coupling relation results to

$$\boldsymbol{s}(t) = \boldsymbol{L}\boldsymbol{z}(t) \tag{3.5}$$

with the overall coupling input $\boldsymbol{s}(t) \in \mathbb{R}^{m_{\mathrm{s}}}$ and the overall coupling output $\boldsymbol{z}(t) \in \mathbb{R}^{r_{\mathrm{z}}}$ defined by

$$\boldsymbol{s}(t) = \begin{pmatrix} \boldsymbol{s}_1(t) \\ \vdots \\ \boldsymbol{s}_N(t) \end{pmatrix}, \quad \boldsymbol{z}(t) = \begin{pmatrix} \boldsymbol{z}_1(t) \\ \vdots \\ \boldsymbol{z}_N(t) \end{pmatrix}.$$

Physical coupling graph. Based on the general definition of a graph in (2.4), the physical interconnection among the subsystems is represented by the time-invariant *physical coupling graph*

$$\mathcal{G}_{\mathrm{P}} = (\mathcal{V}_{\mathrm{P}}, \mathcal{E}_{\mathrm{P}}) \tag{3.6}$$

with the *set* $\mathcal{V}_{\mathrm{P}} = \mathcal{N}$ *of physical vertices* representing the subsystems Σ_i, $(\forall i \in \mathcal{N})$ and the *set* $\mathcal{E}_{\mathrm{P}} \subset \mathcal{N} \times \mathcal{N}$ *of physical edges* representing the physical coupling among the subsystems Σ_i.

$$\boldsymbol{L} = \begin{pmatrix} \boldsymbol{O} & \boldsymbol{L}_{12} & \boldsymbol{O} & \boldsymbol{O} \\ \boldsymbol{L}_{21} & \boldsymbol{O} & \boldsymbol{O} & \boldsymbol{O} \\ \boldsymbol{L}_{31} & \boldsymbol{L}_{32} & \boldsymbol{O} & \boldsymbol{O} \\ \boldsymbol{L}_{41} & \boldsymbol{L}_{42} & \boldsymbol{L}_{43} & \boldsymbol{O} \end{pmatrix} \Rightarrow \mathcal{G}_{\mathrm{P}} :$$

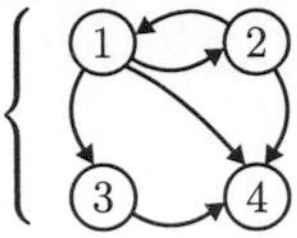

Figure 3.2: Construction of the physical coupling graph $\mathcal{G}_{\mathrm{P}}$ from the physical interconnection matrix $\boldsymbol{L}$, where $\boldsymbol{L}_{12}$, $\boldsymbol{L}_{21}$, $\boldsymbol{L}_{31}$, $\boldsymbol{L}_{32}$, $\boldsymbol{L}_{41}$, $\boldsymbol{L}_{42}$ and $\boldsymbol{L}_{43}$ are non-zero matrices.

The graph $\mathcal{G}_{\mathrm{P}}$ in (3.6) is time-invariant due to the fact that the physical couplings are assumed to be persistent. A directed edge $(i \rightarrow j) \in \mathcal{E}_{\mathrm{P}}$ means that Σ_i has a physical impact on Σ_j, i.e., $\|\boldsymbol{L}_{ji}\| \neq 0$, and corresponds to the coupling input $\boldsymbol{s}_{ji}(t)$. The *set* $\mathcal{P}_{\mathrm{P}i}$ *of physical predecessors* of Σ_i includes the indices j of all subsystems Σ_j from which there is a physical impact towards Σ_i

$$\mathcal{P}_{\mathrm{P}i} := \{j \in \mathcal{N} \mid \|\boldsymbol{L}_{ij}\| \neq 0\}, \quad \forall i \in \mathcal{N}. \tag{3.7}$$

According to (2.5) and (2.6), the set $\mathcal{E}_{\mathrm{P}}$ of physical edges and the *sets* $\mathcal{F}_{\mathrm{P}i}$, $(\forall i \in \mathcal{N})$, *of physical followers* can be expressed subject to the sets $\mathcal{P}_{\mathrm{P}i}$ of physical predecessors

$$\mathcal{E}_{\mathrm{P}} = \{(j \rightarrow i) \mid \|\boldsymbol{L}_{ij}\| \neq 0,\ i, j \in \mathcal{V}_{\mathrm{C}},\ i \neq j\} = \{(j \rightarrow i) \mid j \in \mathcal{P}_{\mathrm{P}i}, i \in \mathcal{N}\} \tag{3.8}$$

$$\mathcal{F}_{\mathrm{P}i} = \{j \in \mathcal{N} \mid \|\boldsymbol{L}_{ji}\| \neq 0\} = \{j \in \mathcal{N} \mid i \in \mathcal{P}_{\mathrm{P}j}\}, \quad i \in \mathcal{N}. \tag{3.9}$$

An example on how to construct $\mathcal{G}_{\mathrm{P}}$ defined in (3.6) based on the physical interconnection matrix $\boldsymbol{L}$ is given in Fig. 3.2.

3.2.3 Model of the overall plant

The overall plant P consists of the subsystems Σ_i, $(\forall i \in \mathcal{N})$, defined in (3.1) and their physical interconnection defined in (3.5) and is given by

$$P : \begin{cases} \dot{\boldsymbol{x}}(t) = \boldsymbol{A}_{\mathrm{P}} \boldsymbol{x}(t) + \boldsymbol{B} \boldsymbol{u}(t) + \boldsymbol{G} \boldsymbol{d}(t), & \boldsymbol{x}(0) = \boldsymbol{x}_0 \\ \boldsymbol{y}(t) = \boldsymbol{C} \boldsymbol{x}(t) \\ \boldsymbol{v}(t) = \boldsymbol{C}_{\mathrm{v}} \boldsymbol{x}(t) \end{cases} \tag{3.10}$$

with

$$\boldsymbol{A}_{\mathrm{P}} = \boldsymbol{A} + \boldsymbol{E} \boldsymbol{L} \boldsymbol{C}_{\mathrm{z}} \in \mathbb{R}^{n \times n}, \qquad \boldsymbol{A} = \mathrm{diag}(\boldsymbol{A}_1 \ldots \boldsymbol{A}_N) \in \mathbb{R}^{n \times n}$$

$$\boldsymbol{B} = \mathrm{diag}(\boldsymbol{B}_1 \ldots \boldsymbol{B}_N) \in \mathbb{R}^{n \times m}, \qquad \boldsymbol{E} = \mathrm{diag}(\boldsymbol{E}_1 \ldots \boldsymbol{E}_N) \in \mathbb{R}^{n \times m_{\mathrm{s}}}$$

$$G = \mathrm{diag}(G_1 \dots G_N) \in \mathbb{R}^{n \times m_\mathrm{d}}, \qquad C = \mathrm{diag}(C_1 \dots C_N) \in \mathbb{R}^{r \times n}$$
$$C_\mathrm{v} = \mathrm{diag}(C_{\mathrm{v}1} \dots C_{\mathrm{v}N}) \in \mathbb{R}^{r_\mathrm{v} \times n}, \qquad C_\mathrm{z} = \mathrm{diag}(C_{\mathrm{z}1} \dots C_{\mathrm{z}N}) \in \mathbb{R}^{r_\mathrm{z} \times n}$$
$$x(t) = \begin{pmatrix} x_1(t) \\ \vdots \\ x_N(t) \end{pmatrix}, \quad x_0 = \begin{pmatrix} x_{01} \\ \vdots \\ x_{0N} \end{pmatrix}, \quad y(t) = \begin{pmatrix} y_1(t) \\ \vdots \\ y_N(t) \end{pmatrix}$$
$$u(t) = \begin{pmatrix} u_1(t) \\ \vdots \\ u_N(t) \end{pmatrix}, \quad d(t) = \begin{pmatrix} d_1(t) \\ \vdots \\ d_N(t) \end{pmatrix}, \quad v(t) = \begin{pmatrix} v_1(t) \\ \vdots \\ v_N(t) \end{pmatrix},$$

where $x(t) \in \mathbb{R}^n$, $x_0 \in \mathbb{R}^n$, $y(t) \in \mathbb{R}^r$, $u(t) \in \mathbb{R}^m$, $d(t) \in \mathbb{R}^{n_\mathrm{d}}$ and $v(t) \in \mathbb{R}^{r_\mathrm{v}}$ are denoted as the overall state, the overall initial state, the overall measured output, the overall input, the overall disturbance and the overall performance output, respectively.

The plant P defined in (3.10) is assumed to be completely controllable and completely observable.

Assumption 3.5 *The pair (A_P, B) is controllable and the pair (A_P, C) is observable.*

Note that Assumption 3.2 is not implied by Assumption 3.5, since subsystems that are completely controllable and completely observable do not necessary yield to an overall plant which is completely completely controllable and completely observable (cf. [56]).

According to Assumption 3.1, the overall disturbance $d(t)$ is bounded by

$$|d(t)| < \bar{d}, \quad \forall t > 0 \tag{3.11}$$

with

$$\bar{d} := \begin{pmatrix} \bar{d}_1 \\ \vdots \\ \bar{d}_N \end{pmatrix} \in \mathbb{R}^{n_\mathrm{d}}.$$

Hence, the following holds true

$$\max_{t \geq 0} \|d(t)\|_\infty \leq d_{\max}. \tag{3.12}$$

with $d_{\max} := ||\bar{d}||_\infty$.

3.2.4 Modeling of faults within the subsystems

The model of the subsystems $\Sigma_{\mathrm{F}i}$, $(\forall i \in \mathcal{N})$, that might be faulty is similar to the model of the nominal subsystems Σ_i in (3.1)

$$\Sigma_{\mathrm{F}i}: \begin{cases} \dot{\boldsymbol{x}}_{\mathrm{F}i}(t) = \boldsymbol{A}_{\mathrm{F}i}\boldsymbol{x}_{\mathrm{F}i}(t) + \boldsymbol{B}_{\mathrm{F}i}\boldsymbol{u}_{\mathrm{F}i}(t) + \boldsymbol{E}_{\mathrm{F}i}\boldsymbol{s}_{\mathrm{F}i}(t) + \boldsymbol{G}_{\mathrm{F}i}\boldsymbol{d}_i(t), \quad \boldsymbol{x}_{\mathrm{F}i}(0) = \boldsymbol{x}_{\mathrm{F}0i} \\ \boldsymbol{y}_{\mathrm{F}i}(t) = \boldsymbol{C}_{\mathrm{F}i}\boldsymbol{x}_{\mathrm{F}i}(t) \\ \boldsymbol{v}_{\mathrm{F}i}(t) = \boldsymbol{C}_{\mathrm{v}i}\boldsymbol{x}_{\mathrm{F}i}(t) \\ \boldsymbol{z}_{\mathrm{F}i}(t) = \boldsymbol{C}_{\mathrm{Fz}i}\boldsymbol{x}_{\mathrm{F}i}(t) \end{cases} \tag{3.13}$$

with the state $\boldsymbol{x}_{\mathrm{F}i}(t) \in \mathbb{R}^{n_i}$, the initial state $\boldsymbol{x}_{\mathrm{F}0i} \in \mathbb{R}^{n_i}$, the input $\boldsymbol{u}_{\mathrm{F}i}(t) \in \mathbb{R}^{m_i}$, the measured output $\boldsymbol{y}_{\mathrm{F}i}(t) \in \mathbb{R}^{r_i}$, the performance output $\boldsymbol{v}_{\mathrm{F}i}(t) \in \mathbb{R}^{r_{\mathrm{v}i}}$, the coupling input $\boldsymbol{s}_{\mathrm{F}i}(t) \in \mathbb{R}^{m_{\mathrm{s}i}}$, the coupling output $\boldsymbol{z}_{\mathrm{F}i}(t) \in \mathbb{R}^{r_{\mathrm{z}i}}$ and the matrices $\boldsymbol{A}_{\mathrm{F}i} \in \mathbb{R}^{n_i \times n_i}$, $\boldsymbol{B}_{\mathrm{F}i} \in \mathbb{R}^{n_i \times m_i}$, $\boldsymbol{E}_{\mathrm{F}i} \in \mathbb{R}^{n_i \times m_{\mathrm{s}i}}$, $\boldsymbol{G}_{\mathrm{F}i} \in \mathbb{R}^{n_i \times n_{\mathrm{d}i}}$, $\boldsymbol{C}_{\mathrm{F}i} \in \mathbb{R}^{r_i \times n_i}$ as well as $\boldsymbol{C}_{\mathrm{Fz}i} \in \mathbb{R}^{r_{\mathrm{z}i} \times n_i}$ belonging to $\Sigma_{\mathrm{F}i}$.

Consider the classification of faults in Section 2.4. Comparing the model of a subsystem $\Sigma_{\mathrm{F}i}$ defined in (3.13) that might be faulty with the nominal subsystem model Σ_i defined in (3.1) leads to the following definition of the faults within a subsystem (Fig. 3.3):

- A subsystem $\Sigma_{\mathrm{F}i}$ is said to have a **subprocess fault**, if the following holds true

$$\boldsymbol{A}_{\mathrm{F}i} \neq \boldsymbol{A}_i \;\vee\; \boldsymbol{E}_{\mathrm{F}i} \neq \boldsymbol{E}_i \;\vee\; \boldsymbol{G}_{\mathrm{F}i} \neq \boldsymbol{G}_i \;\vee\; \boldsymbol{C}_{\mathrm{Fz}i} \neq \boldsymbol{C}_{\mathrm{z}i}, \quad i \in \mathcal{N}. \tag{3.14}$$

- A subsystem $\Sigma_{\mathrm{F}i}$ is said to have an **actuator fault**, if the following holds true

$$\boldsymbol{B}_{\mathrm{F}i} \neq \boldsymbol{B}_i, \quad i \in \mathcal{N}. \tag{3.15}$$

- A subsystem $\Sigma_{\mathrm{F}i}$ is said to have a **sensor fault**, if the following holds true

$$\boldsymbol{C}_{\mathrm{F}i} \neq \boldsymbol{C}_i, \quad i \in \mathcal{N}. \tag{3.16}$$

These faults have different effects on the behavior of the closed-loop system. Where a change of the matrix $\boldsymbol{C}_{\mathrm{z}i}$ leads to a different physical impact on the other subsystems, the change of the system matrix $\boldsymbol{A}_i$ could cause an unstable controlled isolated subsystem, since a local feedback is synthesized for $\boldsymbol{A}_i$.

In the following the subsystems $\Sigma_{\mathrm{F}i}$, $(\forall i \in \mathcal{N})$, defined in (3.13) that might be faulty are divided into two groups:

- The **faulty subsystems** $\Sigma_{\mathrm{F}j}$, $(\forall j \in \mathcal{N}_{\mathrm{F}})$, are the subsystems for which either relation (3.14), (3.15) or (3.16) holds true, where $\mathcal{N}_{\mathrm{F}}$ is the *set of faulty subsystems* with $\mathcal{N}_{\mathrm{F}} \subseteq \mathcal{N}$.

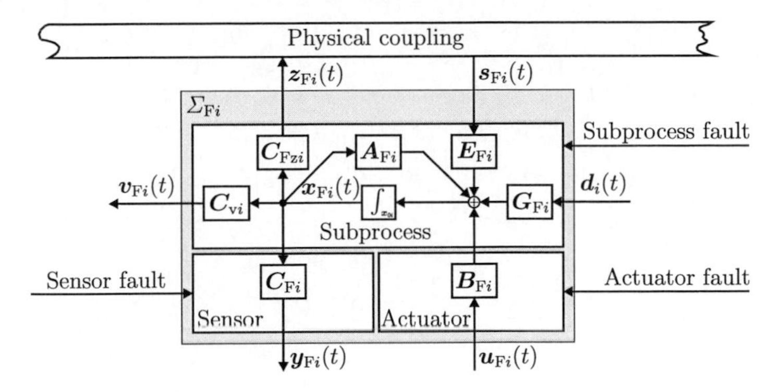

Figure 3.3: Distinction between subprocess faults, actuator faults and sensor faults for the subsystem Σ_{Fi} defined in (3.13).

- The **fault-free subsystems** Σ_{Fk}, $(\forall k \in \mathcal{N} \backslash \mathcal{N}_F)$, are the subsystems for which neither of the relations (3.14), (3.15) and (3.16) holds true. For those subsystems the relation

$$\Sigma_{Fk} \equiv \Sigma_k, \quad \forall k \in \mathcal{N} \backslash \mathcal{N}_F$$

holds true.

Since the physical interconnection is assumed to be fault-free (cf. Section 2.4) the overall physical coupling relation

$$\boldsymbol{s}_F(t) = \boldsymbol{L}\boldsymbol{z}_F(t) \tag{3.17}$$

is similar to the nominal one defined in (3.5) with the corresponding overall coupling input $\boldsymbol{s}_F(t) \in \mathbb{R}^{m_s}$ and the overall coupling output $\boldsymbol{z}_F(t) \in \mathbb{R}^{r_z}$ defined by

$$\boldsymbol{s}_F(t) = \begin{pmatrix} \boldsymbol{s}_{F1}(t) \\ \vdots \\ \boldsymbol{s}_{FN}(t) \end{pmatrix}, \quad \boldsymbol{z}_F(t) = \begin{pmatrix} \boldsymbol{z}_{F1}(t) \\ \vdots \\ \boldsymbol{z}_{FN}(t) \end{pmatrix}.$$

The model of the faulty plant P_F results from (3.17) and (3.13) and is given by

$$P_F : \begin{cases} \dot{\boldsymbol{x}}_F(t) = \boldsymbol{A}_{FP}\boldsymbol{x}_F(t) + \boldsymbol{B}_F\boldsymbol{u}_F(t) + \boldsymbol{G}_F\boldsymbol{d}(t), \quad \boldsymbol{x}_F(0) = \boldsymbol{x}_{F0} \\ \boldsymbol{y}_F(t) = \boldsymbol{C}_F\boldsymbol{x}_F(t) \\ \boldsymbol{v}_F(t) = \boldsymbol{C}_v\boldsymbol{x}_F(t) \end{cases} \tag{3.18}$$

with

$$\begin{aligned}
\boldsymbol{A}_{\mathrm{FP}} &= \boldsymbol{A}_{\mathrm{F}} + \boldsymbol{E}_{\mathrm{F}}\boldsymbol{L}\boldsymbol{C}_{\mathrm{Fz}} \in \mathbb{R}^{n\times n}, & \boldsymbol{A}_{\mathrm{F}} &= \mathrm{diag}(\boldsymbol{A}_{\mathrm{F1}}\dots\boldsymbol{A}_{\mathrm{F}N}) \in \mathbb{R}^{n\times n}\\
\boldsymbol{B}_{\mathrm{F}} &= \mathrm{diag}(\boldsymbol{B}_{\mathrm{F1}}\dots\boldsymbol{B}_{\mathrm{F}N}) \in \mathbb{R}^{n\times m}, & \boldsymbol{E}_{\mathrm{F}} &= \mathrm{diag}(\boldsymbol{E}_{\mathrm{F1}}\dots\boldsymbol{E}_{\mathrm{F}N}) \in \mathbb{R}^{n\times m_{\mathrm{s}}}\\
\boldsymbol{G}_{\mathrm{F}} &= \mathrm{diag}(\boldsymbol{G}_{\mathrm{F1}}\dots\boldsymbol{G}_{\mathrm{F}N}) \in \mathbb{R}^{n\times m_{\mathrm{d}}}, & \boldsymbol{C}_{\mathrm{F}} &= \mathrm{diag}(\boldsymbol{C}_{\mathrm{F1}}\dots\boldsymbol{C}_{\mathrm{F}N}) \in \mathbb{R}^{r\times n}\\
\boldsymbol{C}_{\mathrm{Fz}} &= \mathrm{diag}(\boldsymbol{C}_{\mathrm{Fz1}}\dots\boldsymbol{C}_{\mathrm{Fz}N}) \in \mathbb{R}^{r_{\mathrm{z}}\times n}
\end{aligned}$$

$$\boldsymbol{x}_{\mathrm{F}}(t)=\begin{pmatrix}\boldsymbol{x}_{\mathrm{F1}}(t)\\ \vdots\\ \boldsymbol{x}_{N}(t)\end{pmatrix},\quad \boldsymbol{x}_{\mathrm{F0}}=\begin{pmatrix}\boldsymbol{x}_{\mathrm{F01}}\\ \vdots\\ \boldsymbol{x}_{\mathrm{F0}N}\end{pmatrix},\quad \boldsymbol{y}_{\mathrm{F}}(t)=\begin{pmatrix}\boldsymbol{y}_{\mathrm{F1}}(t)\\ \vdots\\ \boldsymbol{y}_{\mathrm{F}N}(t)\end{pmatrix}$$

$$\boldsymbol{u}_{\mathrm{F}}(t)=\begin{pmatrix}\boldsymbol{u}_{\mathrm{F1}}(t)\\ \vdots\\ \boldsymbol{u}_{\mathrm{F}N}(t)\end{pmatrix},\quad \boldsymbol{v}_{\mathrm{F}}(t)=\begin{pmatrix}\boldsymbol{v}_{\mathrm{F1}}(t)\\ \vdots\\ \boldsymbol{v}_{\mathrm{F}N}(t)\end{pmatrix},$$

where $\boldsymbol{x}_{\mathrm{F}}(t) \in \mathbb{R}^n$, $\boldsymbol{x}_{\mathrm{F0}} \in \mathbb{R}^n$, $\boldsymbol{y}_{\mathrm{F}}(t) \in \mathbb{R}^r$, $\boldsymbol{u}_{\mathrm{F}}(t) \in \mathbb{R}^m$ and $\boldsymbol{v}_{\mathrm{F}}(t) \in \mathbb{R}^{r_{\mathrm{v}}}$ are the corresponding overall state, overall initial state, the overall measured output, overall input and overall performance output, respectively.

3.2.5 Summary of the basic properties and assumptions

The following list summarizes the main assumptions and properties of the plant P (3.10) consisting of physically interconnected subsystem Σ_i defined in (3.1):

- The disturbances $\boldsymbol{d}_i$ affecting the subsystems are assumed to be bounded (Assumption 3.1). This assumption is required in Chapter 4 as well as in Chapter 5.
- The subsystems are assumed to be completely controllable and completely observable (Assumption 3.2). This assumption is required in Chapter 4.
- Generally, it is assumed that the complete subsystem state $\boldsymbol{x}_i(t)$ can be measured (Assumption 3.3).
- The plant P defined in (3.10) is assumed to be completely controllable and completely observable (Assumptions 3.5). This assumption is required in Chapter 5.
- The structure of the physical coupling among the subsystems is represented by the time-invariant physical coupling graph $\mathcal{G}_{\mathrm{P}}$ defined in (3.6) which is illustrated in Fig. 3.5.

3.3 General model of the self-organizing control system

This section presents properties of communication within the self-organizing controller C (Section 3.3.1). This leads to a model which describes the general behavior of the self-organizing control system Σ independent of the actual self-organizing control concept (Section 3.3.2). In Section 3.3.3 a decentralized control system Σ_{D} and a centralized control system Σ_{C} are deduced from the general self-organizing control system Σ.

3.3.1 Constrained communication among the local control units

Recall that the self-organizing controller C generates the time-varying communication graph $\mathcal{G}_{\mathrm{C}}(t) = (\mathcal{V}_{\mathrm{C}}, \mathcal{E}_{\mathrm{C}}(t))$ defined in (2.16). According to (2.25), the set $\mathcal{E}_{\mathrm{C}}(t)$ of communication links is bounded by $\underline{\mathcal{E}}_{\mathrm{C}}$ and $\bar{\mathcal{E}}_{\mathrm{C}}$. For controlling the physically interconnected subsystems the set $\mathcal{E}_{\mathrm{C}}(t)$ is bounded as follows

$$\underline{\mathcal{E}}_{\mathrm{C}} = \emptyset \subseteq \mathcal{E}_{\mathrm{C}}(t) \subseteq \bar{\mathcal{E}}_{\mathrm{C}}, \quad \forall t \geq 0 \tag{3.19}$$

illustrated in Fig. 3.4. The set $\bar{\mathcal{E}}_{\mathrm{C}}$ of maximal communication links will be specified in more detail for the proposed control concepts presented in Chapter 4 and Chapter 5. The set $\underline{\mathcal{E}}_{\mathrm{C}}$ of minimal communication links is defined to be the empty set which is equivalent to a decentralized control of the subsystem. A basic model of a decentralized control system Σ_{D} is presented in Section 3.3.3. Hence, the minimal communication effort for controlling physically interconnected subsystems is no communication.

With this, the basic model of the self-organizing controller C for physically interconnected systems results to

$$C : C_i, (\forall i \in \mathcal{N}), \text{ generating } \mathcal{G}_{\mathrm{C}}(t) = (\mathcal{V}_{\mathrm{C}}, \mathcal{E}_{\mathrm{C}}(t)) \text{ with } \mathcal{E}_{\mathrm{C}}(t) \text{ bounded according to (3.19).} \tag{3.20}$$

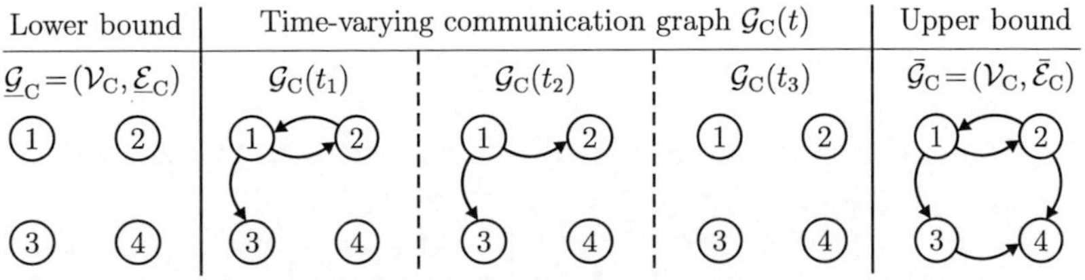

Lower bound	Time-varying communication graph $\mathcal{G}_{\mathrm{C}}(t)$			Upper bound
$\underline{\mathcal{G}}_{\mathrm{C}} = (\mathcal{V}_{\mathrm{C}}, \underline{\mathcal{E}}_{\mathrm{C}})$	$\mathcal{G}_{\mathrm{C}}(t_1)$	$\mathcal{G}_{\mathrm{C}}(t_2)$	$\mathcal{G}_{\mathrm{C}}(t_3)$	$\bar{\mathcal{G}}_{\mathrm{C}} = (\mathcal{V}_{\mathrm{C}}, \bar{\mathcal{E}}_{\mathrm{C}})$
1 2 3 4	1 2 3 4	1 2 3 4	1 2 3 4	1 2 3 4

Figure 3.4: Limitation of the communication graph $\mathcal{G}_{\mathrm{C}}(t)$ for the self-organizing control of physically interconnected subsystems.

Note that the actual local control units C_i are specified in the Chapters 4-5 for the two different control concepts.

3.3.2 General model of the overall closed-loop system

The resulting general model of self-organizing control system Σ, that consist of the plant P and the self-organizing controller C is given by

$$\Sigma : \begin{cases} P \text{ defined in (3.10)} \\ C \text{ defined in (3.20).} \end{cases} \tag{3.21}$$

In summary, the structure of the self-organizing control system Σ is represented by two graphs: The plant P is described by the time-invariant physical coupling graph $\mathcal{G}_{\text{P}}$ defined in (3.6) and the situation-dependent communication is described by the time-varying communication graph $\mathcal{G}_{\text{C}}(t)$ defined in (2.16) with bounded communication links according to (3.19). Figure 3.5 illustrates this representation of the structure of Σ by an example with four subsystems at a certain time t.

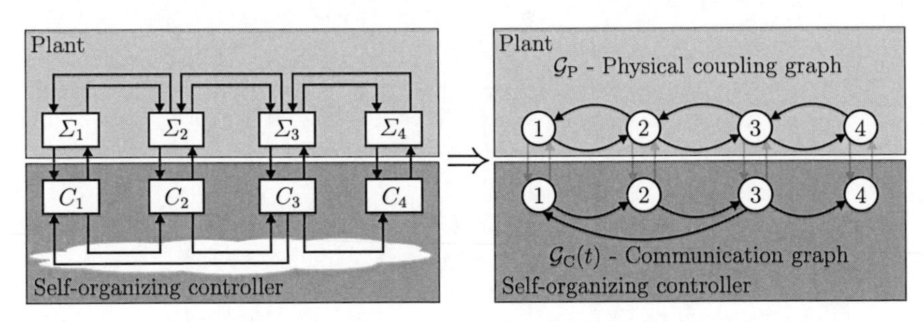

Figure 3.5: Structural representation of the plant P and the networked controller C of the controlled physically interconnected systems for an example with four subsystems at time t.

3.3.3 Comparison to a decentralized control system and a centralized control system

In the following chapters the behavior of the self-organizing control system Σ defined in (3.21) is compared to the behavior of a decentralized control system Σ_{D} and the behavior of a centralized control system Σ_{C} (Fig. 3.6). These three types of control systems only differ in the communication among the local control units C_i, $(\forall i \in \mathcal{N})$, while the plant P and the local control units C_i are identical. Therefore, three different types of controllers are distinguished:

- **Decentralized controller** C_{D}: No exchange of information among the local control units C_i, $(\forall i \in \mathcal{N})$, i.e.,

$$C_{\mathrm{D}} : C_i, (\forall i \in \mathcal{N}) \text{ with } \mathcal{G}_{\mathrm{C}}(t) = (\mathcal{V}_{\mathrm{C}}, \mathcal{E}_{\mathrm{C}}(t) = \emptyset) \text{ for all } t \geq 0 \tag{3.22}$$

- **Self-organizing controller** C: Situation-dependent exchange of information among the local control units C_i, $(\forall i \in \mathcal{N})$ with the time-varying set $\mathcal{E}_{\mathrm{C}}(t)$ of edges within the communication graph $\mathcal{G}_{\mathrm{C}}(t)$ according to (2.16) and (3.19).

- **Centralized controller** C_{C}: Permanent exchange of information among the local control units C_i, $(\forall i \in \mathcal{N})$, i.e.,

$$C_{\mathrm{C}} : C_i, (\forall i \in \mathcal{N}) \text{ with } \mathcal{G}_{\mathrm{C}}(t) = (\mathcal{V}_{\mathrm{C}}, \mathcal{E}_{\mathrm{C}}(t) = \mathcal{E}_{\mathrm{Cmax}}) \text{ for all } t \geq 0. \tag{3.23}$$

  with $\mathcal{E}_{\mathrm{Cmax}} := \{(i \to j) \mid i \neq j, \;\; i, j \in \mathcal{V}_{\mathrm{C}}\}$.

The centralized controller C_{C} is a distributed realization of a centralized controller rather than a single control unit that gets all information and generates the inputs to all subsystems. Furthermore, within the centralized controller C_{C} there is a broadcast communication among all local control units, no matter if the information is used at the local control units or not. Hence, the correct name of C_{C} would be "controller with permanent communication among all local control units". Nevertheless, the naming of C_{C} as centralized controller is used, to keep the notation simple.

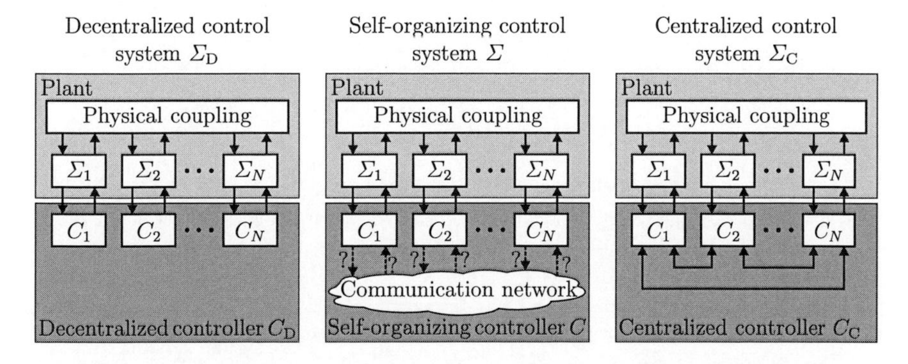

Figure 3.6: Distinction between a self-organizing control system Σ, a decentralized control system Σ_{D} and a centralized control system Σ_{C}.

Consequently, the *decentralized control system* Σ_{D} is defined by

$$\Sigma_{\mathrm{D}}: \begin{cases} P \text{ defined in (3.10)} \\ C_{\mathrm{D}} \text{ defined in (3.22)} \end{cases} \tag{3.24}$$

and the *centralized control system* Σ_{C} is defined by

$$\Sigma_{\mathrm{C}}: \begin{cases} P \text{ defined in (3.10)} \\ C_{\mathrm{C}} \text{ defined in (3.23).} \end{cases} \tag{3.25}$$

3.4 Control aims: Stability, disturbance attenuation and difference behavior

The local control units C_i shall be designed such that the self-organizing control systems Σ satisfies the following four main control aims.

The first two control aims concern the stability of Σ and are essential for designing the self-organizing controller C.

Control aim 3.1 (Asymptotic stability) *The self-organizing control system Σ defined in* (3.21) *shall be asymptotically stable according to Definition 2.2.*

Due to disturbances and the hybrid character of the self-organizing control system Σ, the first control aim might need to be relaxed to a weaker requirement as practical stability which claims that the state of Σ converges to an area around the equilibrium point $\boldsymbol{x} = \mathbf{0}$.

Control aim 3.2 (Practical stability) *The self-organizing control system Σ defined in* (3.21) *shall be practically stable w.r.t. to a compact set according to Definition 2.3.*

Chapter 4 concerns the situation that the disturbance propagation shall be bounded in such a way that the performance output $\boldsymbol{v}_i(t)$ of the undisturbed subsystems is smaller than a desired upper bound.

Control aim 3.3 (Disturbance attenuation) *Consider the situation that disturbances affect the self-organizing control system Σ defined in* (3.21) *with $\boldsymbol{x}_0 = \mathbf{0}$. The performance output $\boldsymbol{v}_i(t)$ of the undisturbed subsystems Σ_i,* $(\forall i \in \mathcal{U})$, *has to be bounded by a* desired maximal performance output $\bar{v}_i^*$, *i.e.*,

$$\|\boldsymbol{v}_i(t)\| \leq \bar{v}_i^*, \quad \forall i \in \mathcal{U}, \quad t \geq 0, \tag{3.26}$$

where

$$\mathcal{U} = \{i \in \mathcal{N} \mid \boldsymbol{d}_i(t) = 0, \forall t \geq 0\} \tag{3.27}$$

is the set of undisturbed subsystems.

The following control aim applies to the control concept in Chapter 5. It requires a desired difference behavior of the self-organizing control system Σ to the corresponding centralized control system Σ_{C}.

Control aim 3.4 (Desired difference behavior compared to centralized control system) *The deviation between the performance of the self-organizing control system Σ defined in* (3.21) *and the performance of the centralized control system Σ_{C} defined in* (3.25) *shall be bounded. That is, the difference performance output $\boldsymbol{v}_{\Delta i}(t) = \boldsymbol{v}_i(t) - \boldsymbol{v}_{\mathrm{C}i}(t)$ has to be bounded by*

$$\|\boldsymbol{v}_{\Delta i}(t)\| \leq \bar{v}_{\Delta i}^*, \quad \forall i \in \mathcal{N}, \quad t \geq 0, \tag{3.28}$$

where $\boldsymbol{v}_{\mathrm{C}i}(t)$ is a performance output of a subsystem $\Sigma_{\mathrm{C}i}$ of Σ_{C} and $\bar{v}_{\Delta i}^*$ is the desired maximal difference performance output.

The desired maximal difference performance outputs $\bar{v}_{\Delta i}^*$, $(\forall i \in \mathcal{N})$, in Control aim 3.4 represent a measure on how good the self-organizing control system Σ with situation-dependent communication mimics the behavior of the centralized control system Σ_{C} with permanent communication. It is expected that the smaller the desired maximal difference performance outputs $\bar{v}_{\Delta i}^*$, $(\forall i \in \mathcal{N})$, are, the more communication among the local control units is necessary to guarantee the desired difference behavior.

3.5 Running examples of physically interconnected systems

This section defines two running examples. A water supply system with 23 subsystems (Section 3.5.1) and a small academic example with two symmetrically interconnected subsystems (Section 3.5.2).

3.5.1 Water supply system

The water supply system consists of 23 subsystems, where the physical construction of one subsystem Σ_i is depicted in Fig. 3.7 and the physical coupling graph of the overall system is presented in Fig. 3.8. The complete water supply system does not exist and is a fictive example. The physical coupling structure is fictional, but the parameters of the subsystem belong to an existing pilot plant at the Institute of Automation and Computer Control at Ruhr-University Bochum, Germany (cf. [124]). Each subsystem Σ_i has a Tank TAi and Tank TBi each of which is fed by a water supply $q_{\mathrm{A\_SUP}i}(t)$ and $q_{\mathrm{B\_SUP}i}(t)$, respectively. The inflow can be controlled by means of the opening angles $u_{\mathrm{TA}i}(t)$ and $u_{\mathrm{TB}i}(t)$ of the respective valves. The disturbances are caused by the change of the water extraction $q_{\mathrm{EXT}i}(t)$ via the valve positions $\boldsymbol{d}_i(t) = d_{\mathrm{TB}i}(t)$. The control aim is to keep the level $l_{\mathrm{TB}i}(t)$ constant in all tanks TBi, $(\forall i \in \mathcal{N})$. The subsystems are coupled with other subsystems over the inflow $q_{\mathrm{IN}i}(t)$ and the outflow $q_{\mathrm{OUT}i}(t)$.

Model of the subsystems. The states $\boldsymbol{x}_i(t)$, $(i = 1, \ldots, 23)$, result to

$$\boldsymbol{x}_i(t) = \begin{pmatrix} l_{\mathrm{TA}i}(t) \\ l_{\mathrm{TB}i}(t) \end{pmatrix}.$$

The inputs are the valve positions restricting the water supply

$$\boldsymbol{u}_i(t) = \begin{pmatrix} u_{\mathrm{TA}i}(t) \\ u_{\mathrm{TB}i}(t) \end{pmatrix}.$$

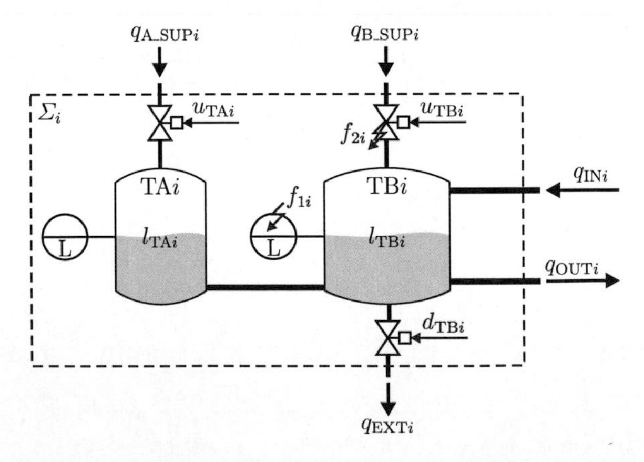

Figure 3.7: Physical construction of the subsystem Σ_i of the water supply system.

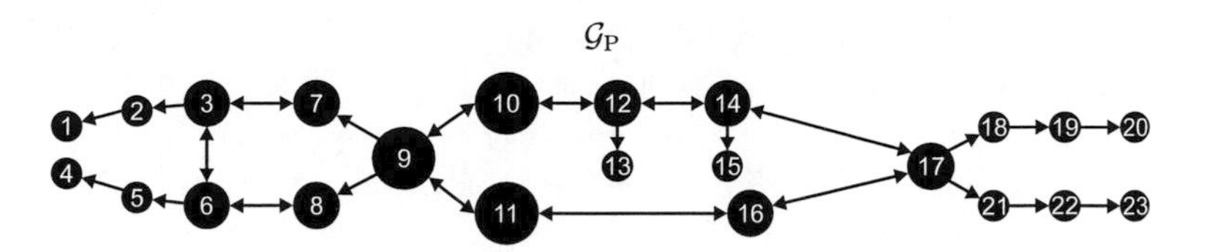

Figure 3.8: Physical coupling graph $\mathcal{G}_{\mathrm{P}}$ of the water supply system.

The coupling input and the coupling output

$$\boldsymbol{s}_i(t) = q_{\mathrm{IN}i}(t), \quad \boldsymbol{z}_i(t) = q_{\mathrm{OUT}i}(t)$$

are the inflow and the outflow to and from the other subsystems, respectively.

The model of the subsystems Σ_i, $(i = 1, \dots, 23)$, are linearized in the operating points $\bar{l}_{\mathrm{TA}i}$ and $\bar{l}_{\mathrm{TB}i}$ and are defined by

$$\boldsymbol{A}_i = \begin{pmatrix} a_{\mathrm{AA}i} & a_{\mathrm{BA}i} \\ a_{\mathrm{BA}i} & a_{\mathrm{BB}i} \end{pmatrix}, \quad \boldsymbol{B}_i = \begin{pmatrix} b_{\mathrm{A}i} & 0 \\ 0 & b_{\mathrm{B}i} \end{pmatrix}, \quad \boldsymbol{G}_i = \begin{pmatrix} 0 \\ -b_{\mathrm{B}i} \end{pmatrix}, \quad \boldsymbol{E}_i = \begin{pmatrix} 0 \\ b_{\mathrm{B}i} \end{pmatrix}$$

$$\boldsymbol{C}_i = \begin{pmatrix} 1 & 0 \\ 0 & 1 \end{pmatrix}, \quad \boldsymbol{C}_{\mathrm{v}i} = \begin{pmatrix} 0 & 1 \end{pmatrix}, \quad \boldsymbol{C}_{\mathrm{z}i} = \begin{pmatrix} 0 & c_{\mathrm{B}i} \end{pmatrix}.$$

The deviations from these operating points are considered as $l_{\mathrm{TA}i}(t)$ and $l_{\mathrm{TB}i}(t)$. The corresponding parameters are listed in Table 3.1. The subsystems have different dynamics due to the size of the tanks and size of the valves. The subsystems are classified in large ($\mathcal{N}_{\mathrm{large}}$), middle ($\mathcal{N}_{\mathrm{middle}}$) and small ($\mathcal{N}_{\mathrm{small}}$) representing the size of the corresponding tanks with

$$\begin{aligned} \mathcal{N}_{\mathrm{large}} &:= \{9, 10, 11\} \\ \mathcal{N}_{\mathrm{middle}} &:= \{3, 6, 7, 8, 12, 14, 16, 17\} \\ \mathcal{N}_{\mathrm{small}} &:= \{1, 2, 4, 5, 13, 15, 18, 19, 20, 21, 22, 23\}. \end{aligned}$$

Table 3.1: Parameters of the tanks.

$i \in$	$a_{\mathrm{AA}i}$	$a_{\mathrm{BA}i}$	$a_{\mathrm{BB}i}$	$c_{\mathrm{B}i}$	$b_{\mathrm{A}i}$	$b_{\mathrm{B}i}$	$\boldsymbol{L}_{ji}$
$\mathcal{N}_{\mathrm{large}}$	-0.006	0.003	-0.005	0.4	0.080	0.090	4.4
$\mathcal{N}_{\mathrm{middle}}$	-0.010	0.005	-0.010	0.5	0.040	0.041	4.4
$\mathcal{N}_{\mathrm{small}}$	-0.019	0.010	-0.022	0.8	0.030	0.035	8

Physical coupling structure. The physical coupling graph $\mathcal{G}_\mathrm{P}$ is depicted in Fig. 3.8. The size of the circles represents the dynamics of the subsystem. Small circles represent subsystems Σ_i, ($\forall i \in \mathcal{N}_\mathrm{small}$), with small tanks and large circles represent subsystems Σ_j, ($j \in \mathcal{N}_\mathrm{large}$), with large tanks. The corresponding sets $\mathcal{P}_{\mathrm{P}i}$ of physical predecessors and the sets $\mathcal{F}_{\mathrm{P}i}$ of physical followers are listed in Table 3.2.

Performance requirement of the centralized control system. The performance requirement on the behavior of the subsystems controlled by local control units with permanent communication (centralized control system Σ_C) is represented by the matrices

$$\boldsymbol{Q}_i = \mathrm{diag}(50, 100), \qquad \forall i \in \mathcal{N} \qquad (3.29a)$$

$$\boldsymbol{R}_i = \mathrm{diag}(3, 1), \qquad \forall i \in \mathcal{N} \qquad (3.29b)$$

Table 3.2: The sets $\mathcal{P}_{\mathrm{P}i}$ of physical predecessors and the sets $\mathcal{F}_{\mathrm{P}i}$ of physical followers of the tanks.

i	$\mathcal{P}_{\mathrm{P}i}$	$\mathcal{F}_{\mathrm{P}i}$
1	$\{2\}$	$\emptyset$
2	$\{3\}$	$\{1\}$
3	$\{6,7\}$	$\{2,6,7,\}$
4	$\{5\}$	$\emptyset$
5	$\{6\}$	$\{4\}$
6	$\{3,8\}$	$\{3,5,8\}$
7	$\{3,9\}$	$\{3\}$
8	$\{6,9\}$	$\{6\}$
9	$\{10,11\}$	$\{7,8,10,11\}$
10	$\{9,12\}$	$\{9,12\}$
11	$\{9,16\}$	$\{9,16\}$
12	$\{10,14\}$	$\{10,13,14\}$
13	$\{12\}$	$\emptyset$
14	$\{12,17\}$	$\{12,15,17\}$
15	$\{14\}$	$\emptyset$
16	$\{11,17\}$	$\{11,17\}$
17	$\{14,16\}$	$\{14,16,18,21\}$
18	$\{17\}$	$\{19\}$
19	$\{18\}$	$\{20\}$
20	$\{19\}$	$\emptyset$
21	$\{17\}$	$\{22\}$
22	$\{21\}$	$\{23\}$
23	$\{22\}$	$\emptyset$

which weight the subsystem states and the control inputs within the optimization problem in (2.11).

Disturbance situation. The maximal disturbances $\bar{d}_i$, $(\forall i \in \mathcal{N})$, of the tanks are given by

$$\bar{d}_6 = 0.6, \quad \bar{d}_9 = 0.9, \quad \bar{d}_{17} = 0.7 \quad \text{and} \quad \bar{d}_i \approx 0, \;\; \forall i \in \mathcal{N}\backslash\{6, 9, 17\}.$$

Since the disturbance affecting the subsystems Σ_i, $(\forall i \in \mathcal{N}\backslash\{6, 9, 17\})$, are small, the corresponding maximal disturbances are assumed to be zero, i.e., the *set of disturbed subsystems* is given by $\mathcal{D} = \{6, 9, 17\}$.

Possible faults. Two possible faults are considered:

- A **sensor fault** f_{1i} at subsystem Σ_i leads to

$$\boldsymbol{C}_{\mathrm{F}i} = \begin{pmatrix} 1 & 0 \\ 0 & 0 \end{pmatrix}, \quad i \in \mathcal{N}_{\mathrm{F}} \tag{3.30}$$

  which describes a failure of the level sensor at TBi (Fig. 3.7).

- An **actuator fault** f_{2i} at subsystem Σ_i leads to

$$\boldsymbol{B}_{\mathrm{F}i} = \begin{pmatrix} b_{\mathrm{A}i} & 0 \\ 0 & 0 \end{pmatrix}, \quad i \in \mathcal{N}_{\mathrm{F}}. \tag{3.31}$$

  which describes a blocking of the control valve at TBi in the operating point.

3.5.2 Symmetrically interconnected subsystems

Consider the symmetrically interconnected subsystems defined by

$$P: \begin{cases} \Sigma_i, \; (i \in \{1, 2\}), \text{ defined in (3.1) with } \boldsymbol{A}_i = \boldsymbol{B}_i = \boldsymbol{E}_i = \boldsymbol{G}_i = \boldsymbol{C}_i = \boldsymbol{C}_{\mathrm{v}i} = \boldsymbol{C}_{\mathrm{z}i} = 1 \\ \text{physical coupling } \boldsymbol{L} = \begin{pmatrix} 0 & 1 \\ 1 & 0 \end{pmatrix} \text{ according to (3.5),} \end{cases} \tag{3.32}$$

where the $\boldsymbol{x}_i(t)$, $\boldsymbol{x}_{0i}$, $\boldsymbol{u}_i(t)$, $\boldsymbol{y}_i(t)$, $\boldsymbol{v}_i(t)$, $\boldsymbol{s}_i(t)$, $\boldsymbol{z}_i(t)$ and $\boldsymbol{d}_i(t)$ are all scalar. The plant P consists of two identical unstable subsystems Σ_i interconnected by a symmetrical physical coupling matrix $\boldsymbol{L}$ with $\boldsymbol{L}_{12} = \boldsymbol{L}_{21} = 1$ which leads to the physical coupling graph $\mathcal{G}_{\mathrm{P}}$ depicted in

Fig. 3.9. With this, the sets of physical predecessors and sets of physical followers are given by

$$\mathcal{P}_{\mathrm{P}1} = \mathcal{F}_{\mathrm{P}1} = \{2\}, \quad \mathcal{P}_{\mathrm{P}2} = \mathcal{F}_{\mathrm{P}2} = \{1\}.$$

Figure 3.9: Physical coupling graph $\mathcal{G}_{\mathrm{P}}$ of the symmetrically interconnected subsystems defined in (3.32).

The performance requirement on the behavior of the symmetrically interconnected subsystems controlled by local control units with permanent communication (centralized control system Σ_{C}) is represented by the matrices

$$\boldsymbol{Q}_1 = \boldsymbol{Q}_2 = 3 \tag{3.33a}$$

$$\boldsymbol{R}_1 = \boldsymbol{R}_2 = 1 \tag{3.33b}$$

which weight the subsystem states $\boldsymbol{x}_i(t)$ and the control inputs $\boldsymbol{u}_i(t)$ within the optimization problem in (2.11).

4 Disturbance attenuation by compensating physical couplings

This chapter presents a control concept for disturbance attenuation in physically interconnected systems. The locally determined coupling inputs are exchanged among the local control units, whenever there is a significant coupling impact on other subsystems. The result is that the disturbance propagation on the other subsystems is bounded. A method for guaranteeing a desired maximal upper bound on the performance output of the undisturbed subsystems is presented. The effectiveness of the proposed control concept is demonstrated by its application of a water supply system. The realization of the local control units is adapted to handle the measurement of the output instead of the complete subsystem state. A method for integrating new subsystems into a present overall closed-loop system is derived. Furthermore, the fault-tolerance of the resulting self-organizing control system is briefly discussed.

Chapter contents

4.1 Situation-dependent compensation of physical couplings

Due to the permanent physical coupling among the subsystems, a disturbance $\boldsymbol{d}_i(t)$ affecting a single subsystem Σ_i might propagate within the overall plant P (Fig. 4.1). Such a propagation needs to be prevented or at least has to be reduced within the closed-loop system Σ. With this, the main **aim** of this chapter can be summarized as follows:

> Develop a self-organizing controller C that bounds the disturbance propagation within the closed-loop system Σ (Control aim 3.3).

Every subsystem Σ_i, $(i \in \mathcal{N})$, is affected by a coupling input $\boldsymbol{s}_i(t)$ and generates a coupling output $\boldsymbol{z}_i(t)$ (cf. (3.3)). According to (3.4), the coupling input $\boldsymbol{s}_i(t)$, $(i \in \mathcal{N})$, is assembled by the coupling inputs $\boldsymbol{s}_{ij}(t) = \boldsymbol{L}_{ij}\boldsymbol{C}_{zj}\boldsymbol{x}_j(t)$ generated by the subsystems Σ_j, $(\forall j \in \mathcal{P}_{\mathrm{P}i})$. A disturbance $\boldsymbol{d}_j(t)$ affecting subsystem Σ_j drives the state $\boldsymbol{x}_j(t)$ out of the operating point, i.e., $\boldsymbol{x}_j(t) \neq \boldsymbol{0}$, which causes a coupling input $\boldsymbol{s}_{ij}(t)$ from Σ_j to Σ_i. If the coupling input is small, i.e., $\boldsymbol{s}_{ij}(t) \approx \boldsymbol{0}$, then the physical impact of Σ_j on Σ_i can be neglected. For large coupling inputs, i.e., $\|\boldsymbol{s}_{ij}(t)\| \gg \boldsymbol{0}$, the coupling impact on Σ_i might be significant such that the local control unit C_i needs that information to adapt its input $\boldsymbol{u}_i(t)$. Therefore, consider the local control units C_i with a decomposition of the control input

$$\boldsymbol{u}_i(t) = \underbrace{-\boldsymbol{K}_{\mathrm{C}i}\boldsymbol{x}_i(t)}_{\text{permanent}} \underbrace{-\,\boldsymbol{K}_{\mathrm{D}i}\sum_{j\in\mathcal{N}\setminus\{i\}} \alpha_{ij}(t)\hat{\boldsymbol{s}}_{ij}(t)}_{\text{situation-dependent}} \tag{4.1}$$

into a permanent control input using the *feedback-gain matrix* $\boldsymbol{K}_{\mathrm{C}i}$ and a situation-dependent control input using the *decoupling matrix* $\boldsymbol{K}_{\mathrm{D}i}$. The scalars $\alpha_{ij}(t) \in \{0, 1\}$ indicate whether the

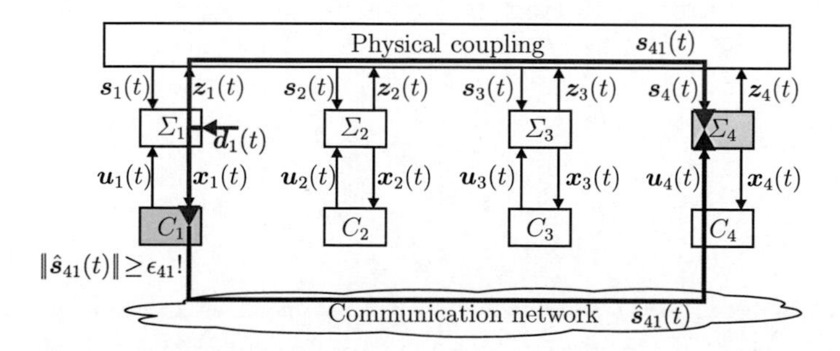

Figure 4.1: Idea of the self-organizing control concept with a compensation of physical coupling inputs.

determined coupling input

$$\hat{\boldsymbol{s}}_{ij}(t) = \boldsymbol{L}_{ij}\boldsymbol{C}_{zj}\boldsymbol{x}_j(t) = \boldsymbol{s}_{ij}(t), \quad i,j \in \mathcal{N}\backslash\{i\} \tag{4.2}$$

is transmitted from C_j to C_i ($\alpha_{ij}(t)=1$) or not ($\alpha_{ij}(t)=0$). If the decoupling matrix $\boldsymbol{K}_{\mathrm{D}i}$ fulfills the relation

$$\boldsymbol{B}_i\boldsymbol{K}_{\mathrm{D}i} = \boldsymbol{E}_i, \quad i \in \mathcal{N}, \tag{4.3}$$

the transmission of $\hat{\boldsymbol{s}}_{ij}(t)$ indicated by $\alpha_{ij}(t)=1$ compensates the corresponding coupling input $\boldsymbol{s}_{ij}(t)$ since

$$\dot{\boldsymbol{x}}_i(t) = (\boldsymbol{A}_i - \boldsymbol{B}_i\boldsymbol{K}_{\mathrm{C}i})\boldsymbol{x}_i(t) + \underbrace{\underbrace{\boldsymbol{E}_i\sum_{i=1}^{N}\boldsymbol{s}_{ij}(t) - \boldsymbol{B}_i\boldsymbol{K}_{\mathrm{D}i}\sum_{i=1}^{N}\alpha_{ij}(t)\hat{\boldsymbol{s}}_{ij}(t)}_{=\boldsymbol{E}_i\left(\sum_{i=1}^{N}\boldsymbol{s}_{ij}(t) - \alpha_{ij}(t)\hat{\boldsymbol{s}}_{ij}(t)\right)}}_{=\boldsymbol{E}_i\left(\sum_{i=1}^{N}\boldsymbol{s}_{ij}(t)(1-\alpha_{ij}(t))\right)} + \boldsymbol{G}_i\boldsymbol{d}_i(t). \tag{4.4}$$

With this implementation, there exist two extremes: full communication ($\alpha_{ij}(t) = 1, \forall i,j \in \mathcal{N}, \forall t \geq 0$), which would be identical to the centralized control system Σ_{C} according to (3.25), and no communication ($\alpha_{ij}(t) = 0, \forall i,j \in \mathcal{N}, \forall t \geq 0$), which would result in a decentralized control system Σ according to (3.24). On the one hand, full communication would induce no disturbance propagation within the closed-loop system Σ but causes a high communication effort among the local control units. On the other hand, no communication might lead to great disturbance propagation, but does not load the communication network. Hence, a rule has to be defined in which situations a transmission of $\hat{\boldsymbol{s}}_{ij}(t)$ from C_j to C_i is appropriate.

If, for example, due to a disturbance the determined coupling input $\hat{\boldsymbol{s}}_{ij}(t)$ exceeds a given switching threshold $\epsilon_{ij} \in \mathbb{R}$, the value of $\hat{\boldsymbol{s}}_{ij}(t)$ is sent from C_j to C_i ($\alpha_{ij}(t) = 1$) to compensate the coupling input $\boldsymbol{s}_{ij}(t)$. If $\hat{\boldsymbol{s}}_{ij}(t)$ falls below the threshold ϵ_{ij}, its value is not sent ($\alpha_{ij}(t)=0$), i.e.,

$$\alpha_{ij}(t) = \begin{cases} 1 & \text{if } \|\hat{\boldsymbol{s}}_{ij}(t)\| \geq \epsilon_{ij} \\ 0 & \text{else,} \end{cases} \quad \forall j,i \in \mathcal{N}, \quad j \neq i. \tag{4.5}$$

Roughly speaking, C_j decides whether $\hat{\boldsymbol{s}}_{ij}(t)$ is so large that the local control unit C_i of the neighboring subsystem Σ_i needs $\hat{\boldsymbol{s}}_{ij}(t)$ to react to the effect of the physical coupling impact and, therefore, attenuate the disturbance propagation. Hence, if no subsystem is disturbed, then no communication is necessary.

Based on these considerations, the main **idea** of the self-organizing control concept proposed in this chapter can be summarized as follows:

> Transmit the determined coupling input $\hat{\boldsymbol{s}}_{ij}(t)$ from the local control unit C_j to the local control unit C_i whenever $\hat{\boldsymbol{s}}_{ij}(t)$ exceeds the switching threshold ϵ_{ij} in order to compensate the actual coupling impact $\boldsymbol{s}_{ij}(t)$ on subsystem Σ_i by adapting the control input $\boldsymbol{u}_i(t)$.

This idea is illustrated Fig. 4.1 for the situation that subsystem Σ_1 is affected by a disturbance, i.e. $\boldsymbol{d}_1(t) \neq \boldsymbol{0}$. Due to the disturbance $\boldsymbol{d}_1(t)$, there is a coupling input $\boldsymbol{s}_{41}(t)$ from Σ_1 to Σ_4. The local control unit C_1 determines the coupling input with $\hat{\boldsymbol{s}}_{41}(t) = \boldsymbol{L}_{41}\boldsymbol{C}_{z1}\boldsymbol{x}_1(t)$. If $\hat{\boldsymbol{s}}_{41}(t)$ on Σ_4 exceeds the switching threshold ϵ_{41}, then C_1 sends the determined coupling input $\hat{\boldsymbol{s}}_{41}(t)$ to C_4 to compensate the impact of $\boldsymbol{s}_{41}(t)$ on Σ_4 by adapting $\boldsymbol{u}_4(t)$.

With this setting, the basic **tasks** of the local control units C_i can be summarized as follows:

1. Stabilize the isolated subsystems Σ_i by a permanent local feedback (cf. (4.1)).
2. Determine the coupling input $\hat{\boldsymbol{s}}_{ji}(t)$ to the neighboring subsystems Σ_j, $(\forall j \in \mathcal{F}_{\mathrm{P}i})$, according to (4.2).
3. Decide whether it is beneficial to send the determined coupling input $\hat{\boldsymbol{s}}_{ji}(t)$ to other local control units C_j, $(\forall j \in \mathcal{F}_{\mathrm{P}i})$, based on $\hat{\boldsymbol{s}}_{ji}(t)$ (cf. (4.5)).
4. Use the incoming coupling input $\hat{\boldsymbol{s}}_{ij}(t)$ in C_i to compensate the effect of the current coupling input $\boldsymbol{s}_{ij}(t)$ on Σ_i by adapting $\boldsymbol{u}_i(t)$ (cf. (4.1), (4.3) and (4.4)).

In the following, the main idea of the concept is illustrated for concerning two subsystems only. Therefore, the structure of the local control units which performs these four tasks is presented.

Structure of the self-organizing controller for two subsystems. Figure 4.2 shows the self-organizing control concept for two subsystems and its way of working in case of a disturbance affecting subsystem Σ_1. The components of the local control units C_1 and C_2 work as follows:

- The **feedback unit** F_1 uses a permanent state feedback $-\boldsymbol{K}_{\mathrm{C}1}\boldsymbol{x}_1(t)$ to stabilize the isolated subsystem Σ_1 (1. Task).
- The **observation unit** O_1 calculates the coupling input $\boldsymbol{s}_{21}(t)$ to subsystem Σ_2 with $\hat{\boldsymbol{s}}_{21}(t) = \boldsymbol{L}_{21}\boldsymbol{C}_{z1}\boldsymbol{x}_1(t)$ by using the state $\boldsymbol{x}_1(t)$ (2. Task).

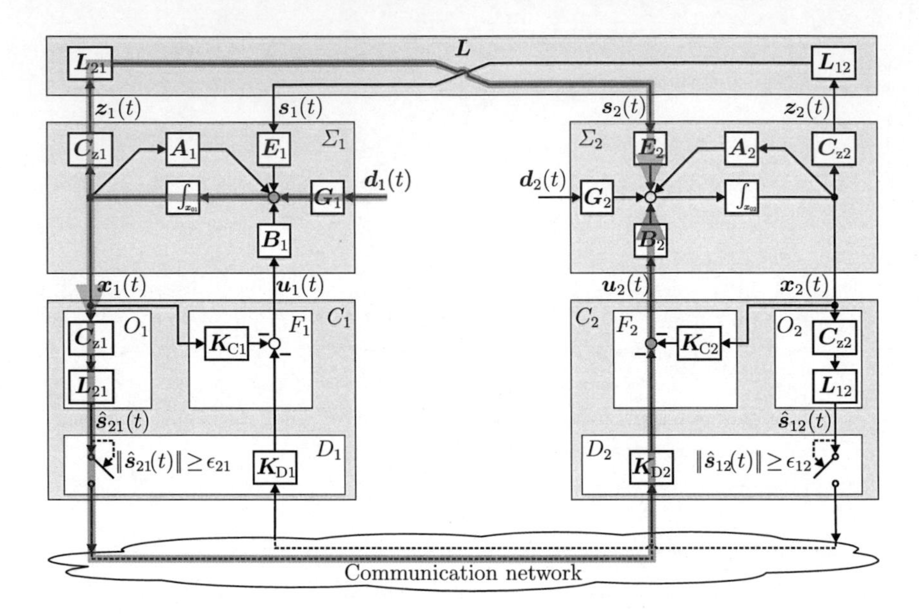

Figure 4.2: Self-organizing control concept with transmitting physical coupling signals for two subsystems.

- The **decision unit** D_1 sends the calculated coupling input $\hat{\boldsymbol{s}}_{21}(t)$ to the local control unit C_2 if $\hat{\boldsymbol{s}}_{21}(t)$ exceeds the switching threshold ϵ_{21} (3. Task), i.e.,

$$\|\hat{\boldsymbol{s}}_{21}(t)\| \geq \epsilon_{21} \quad \Leftrightarrow \quad C_1 \text{ sends } \hat{\boldsymbol{s}}_{21}(t) \text{ to } C_2 \quad \Leftrightarrow \quad \alpha_{21}(t) = 1. \tag{4.6}$$

- The **decision unit** D_2 in the local control unit C_2 uses $\hat{\boldsymbol{s}}_{21}(t)$ to compensate the effect of $\boldsymbol{s}_{21}(t)$ on Σ_2 such that

$$\boldsymbol{E}_2 \boldsymbol{s}_{21}(t) - \boldsymbol{B}_2 \boldsymbol{K}_{\text{D}2} \hat{\boldsymbol{s}}_{21}(t) = \boldsymbol{0}, \tag{4.7}$$

  if the matrix $\boldsymbol{K}_{\text{D}2}$ guarantees $\boldsymbol{E}_2 = \boldsymbol{B}_2 \boldsymbol{K}_{\text{D}2}$ (4. Task).

This structure allows to completely compensate the coupling input $\boldsymbol{s}_{21}(t)$ whenever the determined coupling input $\hat{\boldsymbol{s}}_{21}(t)$ is transmitted from C_1 to C_2. Consider the controlled subsystem

Σ_2 with

$$\dot{\boldsymbol{x}}_2(t) = (\boldsymbol{A}_2 - \boldsymbol{B}_2\boldsymbol{K}_{\mathrm{C}2})\boldsymbol{x}_2(t) + \underbrace{\underbrace{\boldsymbol{E}_2\boldsymbol{s}_{21}(t) - \boldsymbol{B}_2\boldsymbol{K}_{\mathrm{D}2}\alpha_{21}\hat{\boldsymbol{s}}_{21}(t)}_{= \boldsymbol{E}_2(\boldsymbol{s}_{21}(t) - \alpha_{21}(t)\boldsymbol{s}_{21}(t))}}_{= \boldsymbol{E}_2(\boldsymbol{s}_{21}(t)(1-\alpha_{21}(t)))} + \boldsymbol{G}_2\boldsymbol{d}_2(t)$$

and $\boldsymbol{E}_2 = \boldsymbol{B}_2\boldsymbol{K}_{\mathrm{D}2}$. According to (4.6), the difference signal $\|\boldsymbol{s}_{21}(t) - \alpha_{21}(t)\hat{\boldsymbol{s}}_{21}(t)\|$ is always bounded by ϵ_{21}, i.e.,

$$\max_{t\geq 0} \|\boldsymbol{s}_{21}(t) - \alpha_{21}(t)\hat{\boldsymbol{s}}_{21}(t)\| = \max_{t\geq 0} \|\boldsymbol{s}_{21}(t)(1 - \alpha_{21}(t))\| \leq \epsilon_{21}.$$

Therefore, the effective coupling impact from Σ_1 to Σ_2 is bounded by the switching threshold ϵ_{21}. With this, the disturbance propagation from Σ_1 to Σ_2 can be adjusted by the choice of the switching threshold ϵ_{21} since no matter how large the amount of the disturbance $\boldsymbol{d}_1(t)$ is, the effective coupling impact from Σ_1 to Σ_2 is bounded by ϵ_{21}. This property is explained in more detail in the following example.

Example 4.1 *Compensation of physical coupling inputs*

In this example the self-organizing control concept with a situation-dependent compensation of the physical coupling inputs illustrated in Fig. 4.2 is used to control the symmetrically interconnected subsystems defined in (3.32). The example is divided into two parts. First, the construction of the local control units C_1 and C_2 is presented. Second, the behavior of the resulting self-organizing control system Σ is investigated.

Parameters of the local control units C_i**.** The parameters of the components of the two identical local control units C_1 and C_2 are listed in Table 4.1. The feedback gain matrices $\boldsymbol{K}_{\mathrm{C}1} = \boldsymbol{K}_{\mathrm{C}2} = 3$ for the feedback units F_1 and F_2 result from the solutions of the optimization problem in (2.11) with the weighting matrices $\boldsymbol{Q}_1 = \boldsymbol{Q}_2 = 3$ and $\boldsymbol{R}_1 = \boldsymbol{R}_2 = 1$ according to (3.33). The parameters $\boldsymbol{L}_{ji}$ and $\boldsymbol{C}_{zi}$ of the observation units O_i for calculating $\hat{\boldsymbol{s}}_{ji}(t) = \boldsymbol{s}_j(t) = \boldsymbol{L}_{ji}\boldsymbol{C}_{zi}\boldsymbol{x}_i(t) = \boldsymbol{x}_i(t)$ result from the model of the symmetrically interconnected subsystem defined in (3.32). The decoupling matrices $\boldsymbol{K}_{\mathrm{D}1} = \boldsymbol{K}_{\mathrm{D}2} = 1$ within the decision units D_1 and D_2 guarantee the requirement in (4.3) with $\boldsymbol{E}_i = \boldsymbol{B}_i\boldsymbol{K}_{\mathrm{D}i} = 1$, $(\forall i \in \{1,2\})$, since $\boldsymbol{E}_i = \boldsymbol{B}_i = 1$ holds true. The switching thresholds are chosen to be identical, i.e., $\epsilon_{21} = \epsilon_{12} = 0.5$.

The links within the communication graph $\mathcal{G}_{\mathrm{C}}(t)$ depicted in Fig. 4.3 result from the corresponding physical coupling graph $\mathcal{G}_{\mathrm{P}}$ of the symmetrically interconnected subsystems (cf. Fig. 3.9). For example, the communication link $(1 \to 2)$ is active if the relation $\|\hat{\boldsymbol{s}}_{21}(t)\| \geq \epsilon_{21} = 0.5$ holds true.

The permanent local feedback with the feedback gain matrices $\boldsymbol{K}_{\mathrm{C}i}$, $(\forall i \in \{1,2\})$, within the feedback units F_i stabilize the isolated subsystems, i.e., $\lambda_i(\boldsymbol{A}_i - \boldsymbol{B}_i\boldsymbol{K}_{\mathrm{C}i}) = -2$. Furthermore, the decentralized

Table 4.1: Parameters of the local control units C_1 and C_2 presented in Fig. 4.2 for the control of the symmetrically interconnected subsystems.

Feedback units F_1 and F_2	Observation units O_1 and O_2		Decision units D_1 and D_2	
$\boldsymbol{K}_{\mathrm{C}1} = \boldsymbol{K}_{\mathrm{C}2} = 3$	$\boldsymbol{L}_{21} = \boldsymbol{L}_{12} = 1$	$\boldsymbol{C}_{\mathrm{z}1} = \boldsymbol{C}_{\mathrm{z}2} = 1$	$\boldsymbol{K}_{\mathrm{D}1} = \boldsymbol{K}_{\mathrm{D}2} = 1$	$\epsilon_{21} = \epsilon_{12} = 0.5$

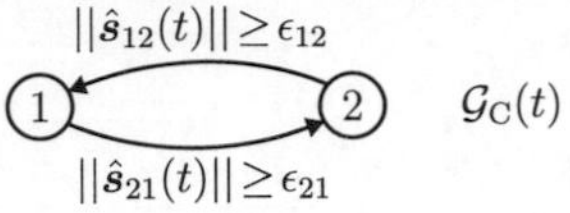

$||\hat{\boldsymbol{s}}_{12}(t)|| \geq \epsilon_{12}$

(1) (2) $\mathcal{G}_{\mathrm{C}}(t)$

$||\hat{\boldsymbol{s}}_{21}(t)|| \geq \epsilon_{21}$

Figure 4.3: Communication graph $\mathcal{G}_{\mathrm{C}}(t)$ for the situation-dependent compensation of the physical coupling inputs at the symmetrically interconnected subsystems.

controlled system Σ_{D} (c.f. (3.24)) is also stable since the overall system matrix

$$\bar{\boldsymbol{A}}_{\mathrm{P}} = \underbrace{\begin{pmatrix} 1 & 0 \\ 0 & 1 \end{pmatrix}}_{\boldsymbol{A}} + \underbrace{\begin{pmatrix} 1 & 0 \\ 0 & 1 \end{pmatrix}}_{\boldsymbol{E}} \underbrace{\begin{pmatrix} 0 & 1 \\ 1 & 0 \end{pmatrix}}_{\boldsymbol{L}} \underbrace{\begin{pmatrix} 1 & 0 \\ 0 & 1 \end{pmatrix}}_{\boldsymbol{C}_{\mathrm{z}}} - \underbrace{\begin{pmatrix} 1 & 0 \\ 0 & 1 \end{pmatrix}}_{\boldsymbol{B}} \begin{pmatrix} \overbrace{3}^{\boldsymbol{K}_{\mathrm{C1}}} & 0 \\ 0 & \underbrace{3}_{\boldsymbol{K}_{\mathrm{C2}}} \end{pmatrix} = \begin{pmatrix} -2 & 1 \\ 1 & -2 \end{pmatrix} \tag{4.8}$$

of Σ_{D} is a Hurwitz matrix.

Behavior of the self-organizing control system Σ. The behavior of Σ with $\boldsymbol{x}_{01} = \boldsymbol{x}_{02} = 0$ and a disturbance affecting subsystem Σ_1 is depicted in Fig. 4.4 (black solid line). The disturbance $\boldsymbol{d}_1(t) \neq 0$ forces the state $\boldsymbol{x}_1(t)$ out of the equilibrium point which generates a coupling input $\boldsymbol{s}_{21}(t) = \boldsymbol{x}_1(t)$ to subsystem Σ_2. Hence, $\boldsymbol{x}_2(t)$ increases. When $\boldsymbol{s}_{21}(t)$ exceeds the threshold ϵ_{21}, the local control unit C_1 sends the calculated coupling input $\hat{\boldsymbol{s}}_{21}(t)$ to the local control unit C_2 (indicated by the black bar in the lower plot) which compensates the coupling input $\boldsymbol{s}_{21}(t)$ according to (4.7) and, therefore, the state $\boldsymbol{x}_2(t)$ goes to zero. If the disturbance $\boldsymbol{d}_1(t)$ goes to zero, then $\boldsymbol{s}_{21}(t)$ falls below ϵ_{21} and, therefore, the communication from C_1 to C_2 is deactivated. In comparison to the behavior of the decentralized controlled system Σ_{D} (black dashed line in Fig. 4.4) especially the deviation of $\boldsymbol{x}_2(t)$ can be significantly

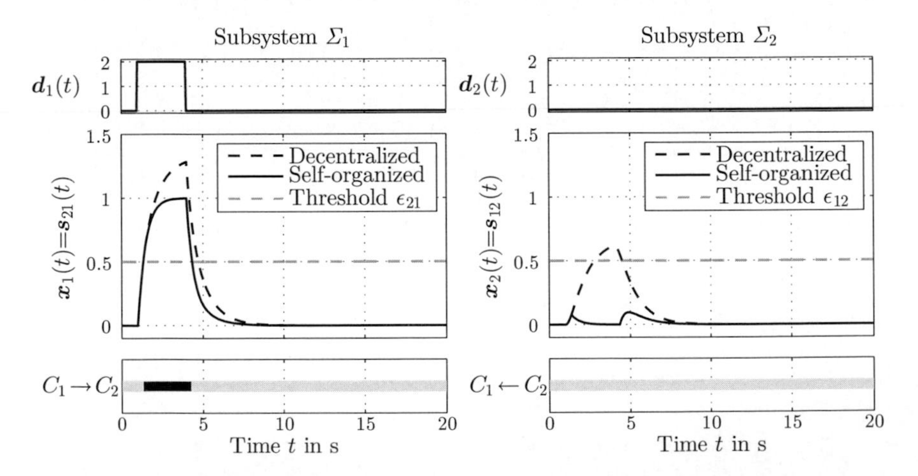

Figure 4.4: Behavior of the disturbed symmetrically interconnected subsystems defined (3.32) controlled be the local control units C_i specified in Fig. 4.2 and Table 4.1.

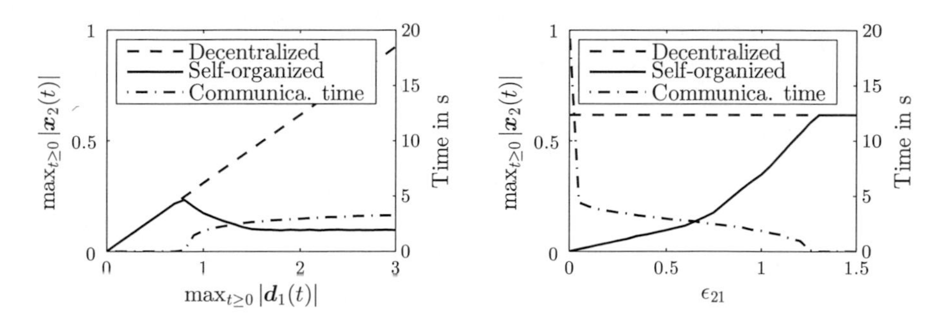

Figure 4.5: Relation between the maximal disturbance $\boldsymbol{d}_1(t)$ at Σ_1 and maximal value of $\boldsymbol{x}_2(t)$ at Σ_2 (left); relation between the switching threshold ϵ_{21} in C_1 and maximal value of $\boldsymbol{x}_2(t)$ at Σ_2 (right).

reduced with the self-organizing controller C by performing a short information transmission from C_1 to C_2.

Figure 4.5 shows interesting properties of the self-organizing control system Σ. The left plot depicts the maximal value $\max_{t\geq 0}|\boldsymbol{x}_2(t)|$ of the state of Σ_2 for different maximal values of the disturbance $\boldsymbol{d}_1(t)$ with a course as in Fig. 4.4 and the thresholds $\epsilon_{21} = \epsilon_{12} = 0.5$ as stated above. The value of $\max_{t\geq 0}|\boldsymbol{x}_2(t)|$ within the self-organizing control system Σ is bounded (solid line), whereas for the decentralized controller C_{D} (dashed line) the value of $\max_{t\geq 0}|\boldsymbol{x}_2(t)|$ increases linearly with $\max_{t\geq 0}|\boldsymbol{d}_1(t)|$. To compensate the larger disturbance, the communication time increases with $\max_{t\geq 0}|\boldsymbol{d}_1(t)|$. The *communication time*, depicted in Fig. 4.5, is the length of the time interval in which C_1 transmits information to C_2 which is the length of the black bar in the lower left plot of Fig. 4.4.

The relation between $\max_{t\geq 0}|\boldsymbol{x}_2(t)|$ and the switching threshold ϵ_{21} is depicted in the right part of Fig. 4.5. The value for $\max_{t\geq 0}|\boldsymbol{x}_2(t)|$ increases with switching threshold ϵ_{21} until $\epsilon_{21} > 1.3$ (cf. $\boldsymbol{x}_1(t)$ in Fig. 4.5), thereafter, $\max_{t\geq 0}|\boldsymbol{x}_2(t)|$ is equal to its value while using a decentralized controller C_{D}. If $\epsilon_{21} = 0$ holds, then C_1 sends $\hat{\boldsymbol{s}}_{21}(t)$ to C_2 at all time and the coupling input is compensated at all times. Even for small values of the threshold ϵ_{21} there is a huge reduction of the communication time but only a small value for $\max_{t\geq 0}|\boldsymbol{x}_2(t)|$. Hence, information has to be exchanged only for a small time interval to improve the performance compared to a decentralized control system Σ_{D}. This shows that there is a trade-off between $\max_{t\geq 0}|\boldsymbol{x}_2(t)|$ and the communication effort.

In summary, this example shows that the proposed self-organizing control concept has the following interesting properties:

- The local control units exchange information if a disturbance affects a subsystem, which is a situation-dependent communication (cf. Fig. 4.4).
- The disturbance propagation can be adjusted by the choice of the switching thresholds (cf. Fig. 4.5 right).
- There is a trade-off between the communication effort and the disturbance attenuation (cf. Fig. 4.5 right).
- The influence of one subsystem to another subsystem can be bounded by the switching threshold no matter what happens at a single subsystem (cf. Fig. 4.5 left).

□

Overview of the chapter. In the following sections it is shown that the mentioned properties of the proposed control strategy also exists if it is applied to an arbitrary number of subsystem. Therefore, the content of the next sections is as follows:

- The structure of the self-organizing controller C is extended to an arbitrary number of subsystem (Section 4.2).
- An analysis of the self-organizing control system Σ yields conditions on the stability of Σ, bounds on the difference behavior between Σ and Σ_{C} as well as estimations of the behavior of Σ in case of disturbances (Section 4.3).
- Methods for designing the switching thresholds and the communication structure for guaranteeing the control aims claimed in Section 3.4 are given (Section 4.4).
- The self-organizing control strategy is demonstrated by its application to a water supply system (Section 4.5).
- An adaption of the control concept for the case that the complete states can not be measured, i.e., $\boldsymbol{y}_i(t) \neq \boldsymbol{x}_i(t)$, is presented (Section 4.6).
- A basic concept for the integration of new subsystems to an existing self-organizing control system Σ is given (Section 4.7).
- The passive fault-tolerance of the self-organizing control system Σ is analyzed and discussed (Section 4.8).
- The self-organizing properties of the closed-loop system Σ with the situation-dependent compensation of the physical coupling inputs are evaluated (Section 4.9).

4.2 Structure of the self-organizing controller and the resulting closed-loop system

This section derives the structure of the self-organizing controller C which situation-dependently compensates the coupling inputs as illustrated for two subsystems in the previous section. The properties of the communication graph $\mathcal{G}_{\mathrm{C}}(t)$ are illustrated in Section 4.2.1 and the general structure of the local control units C_i is given in Section 4.2.2. Section 4.2.3 derives the overall model of self-organizing control system Σ.

4.2.1 Communication structure depending on the physical coupling

To compensate the coupling inputs $\boldsymbol{s}_{ij}(t)$, the local control units C_i, $(\forall i \in \mathcal{N})$, have to receive the determined coupling inputs $\hat{\boldsymbol{s}}_{ij}(t)$, $(j \in \mathcal{P}_{\mathrm{P}i})$, from the local control units C_j whose subsystems Σ_j have a physical impact on Σ_i. Hence, Σ_j, $(j \in \mathcal{P}_{\mathrm{P}i})$ is a physical predecessor of Σ_i according to the definition of $\mathcal{P}_{\mathrm{P}i}$ in (3.7). Therefore, the set $\bar{\mathcal{P}}_{\mathrm{C}i}$, $(\forall i \in \mathcal{N})$ of maximal communicational predecessors introduced in Section 2.3 is equal to the set $\mathcal{P}_{\mathrm{P}i}$ of physical predecessors, i.e.,

$$\bar{\mathcal{P}}_{\mathrm{C}i} = \mathcal{P}_{\mathrm{P}i}, \quad \forall i \in \mathcal{N}. \tag{4.9}$$

The sets $\bar{\mathcal{P}}_{\mathrm{C}i}$ define the set $\bar{\mathcal{E}}_{\mathrm{C}}$ of maximal communicational edges according to (2.21). Hence, $\bar{\mathcal{E}}_{\mathrm{C}} = \mathcal{E}_{\mathrm{P}}$ holds true. The set $\bar{\mathcal{E}}_{\mathrm{C}}$ defines the maximal communication graph $\bar{\mathcal{G}}_{\mathrm{C}} = (\mathcal{V}_{\mathrm{C}}, \bar{\mathcal{E}}_{\mathrm{C}})$ according to (2.19) and bounds the set $\mathcal{E}_{\mathrm{C}}(t)$ of communication links within the communication graph $\mathcal{G}_{\mathrm{C}}(t)$ as stated in (3.19) (cf. Fig. 3.4). Recall that the sets $\mathcal{P}_{\mathrm{P}i}$, $(\forall i \in \mathcal{N})$, also define the sets $\bar{\mathcal{F}}_{\mathrm{C}i}$ of maximal communicational followers according to (2.22), i.e., $\bar{\mathcal{F}}_{\mathrm{C}i} = \mathcal{F}_{\mathrm{P}i}$, $(\forall i \in \mathcal{N})$.

In summary, the maximal communication graph $\bar{\mathcal{G}}_{\mathrm{C}}$ results from the physical coupling graph $\mathcal{G}_{\mathrm{P}}$ which is illustrated in Fig. 4.6. Furthermore, the set $\mathcal{E}_{\mathrm{C}}(t)$ of communication links is bounded according to (2.25), (3.19) and (5.11), i.e.,

$$\underbrace{\emptyset}_{\underline{\mathcal{E}}_{\mathrm{C}}} \subseteq \mathcal{E}_{\mathrm{C}}(t) \subseteq \underbrace{\{(j \to i) \mid \|\boldsymbol{L}_{ij}\| \neq 0,\ i, j \in \mathcal{V}_{\mathrm{C}},\ i \neq j\}}_{\bar{\mathcal{E}}_{\mathrm{C}} = \mathcal{E}_{\mathrm{P}}}, \quad \forall t \geq 0. \tag{4.10}$$

$$\boldsymbol{L} = \begin{pmatrix} \boldsymbol{O} & \boldsymbol{L}_{12} & \boldsymbol{O} & \boldsymbol{O} \\ \boldsymbol{L}_{21} & \boldsymbol{O} & \boldsymbol{O} & \boldsymbol{O} \\ \boldsymbol{L}_{31} & \boldsymbol{L}_{32} & \boldsymbol{O} & \boldsymbol{O} \\ \boldsymbol{L}_{41} & \boldsymbol{L}_{42} & \boldsymbol{L}_{43} & \boldsymbol{O} \end{pmatrix} \Rightarrow$$

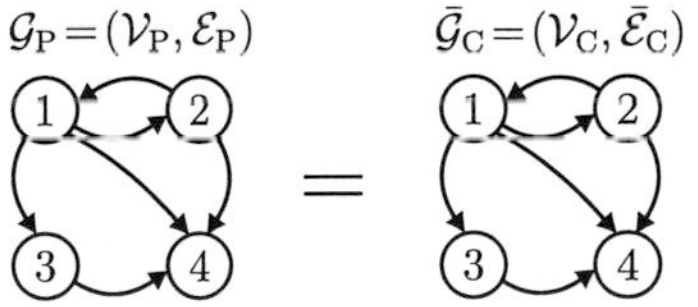

Figure 4.6: Definition of the maximal communication graph $\bar{\mathcal{G}}_{\mathrm{C}}$ for the situation-dependent compensation of physical coupling inputs.

4.2.2 Structure of the local control units

The local control units C_i consist of a feedback unit F_i, an observation unit O_i and a decision unit D_i (Fig. 4.7). The feedback unit F_i is used to stabilize the isolated subsystem Σ_i, where it is assumed that the complete state $\boldsymbol{x}_i(t)$ can be measured (Assumption 3.3). The observation unit O_i determines the current situation of the subsystem Σ_i indicated by the magnitude of the

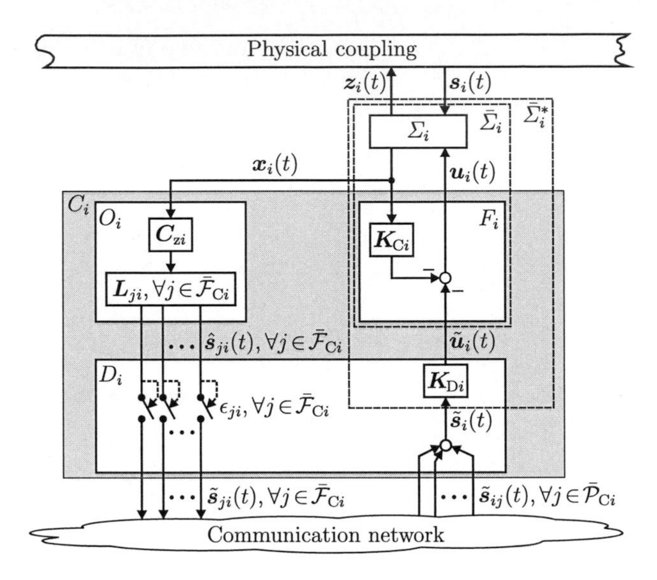

Figure 4.7: Local control unit C_i for situation-dependent compensation of the coupling inputs.

calculated coupling inputs $\hat{s}_{ji}(t)$ for the other local control units C_j, $(\forall j \in \bar{\mathcal{F}}_{Ci})$. The decision unit D_i sends the calculated coupling inputs $\hat{s}_{ji}(t)$ to C_j if $\|\hat{s}_{ji}(t)\|$ exceeds the switching threshold ϵ_{ji}. Furthermore, D_i uses the transmitted coupling signals $\tilde{s}_{ij}(t)$, $(\forall j \in \bar{\mathcal{P}}_{Ci})$, from the local control units C_j to generate the *communicated control input* $\tilde{u}_i(t)$ which compensates the coupling input $s_i(t)$ on Σ_i. The following paragraphs describe these components.

Feedback unit F_i. The task of the feedback unit F_i is to generate the input $u_i(t)$ for the subsystem Σ_i:

$$F_i : u_i(t) = -K_{Ci} x_i(t) - \tilde{u}_i(t) \quad \forall i \in \mathcal{N}, \tag{4.11}$$

using the feedback gain matrix $K_{Ci} \in \mathbb{R}^{m_i \times n_i}$ and the communicated control input $\tilde{u}_i(t) \in \mathbb{R}^{m_i}$ that is determined by the decision unit D_i in (4.13).

Observation unit O_i. The observation unit O_i provides the calculated coupling inputs $\hat{s}_{ji}(t) \in \mathbb{R}^{m_{si}}$ to D_i in (4.13):

$$O_i : \hat{s}_{ji}(t) = L_{ji} C_{zi} x_i(t), \quad \forall j \in \bar{\mathcal{F}}_{Ci}, \quad i \in \mathcal{N}. \tag{4.12}$$

Decision unit D_i. The decision unit D_i, $(i \in \mathcal{N})$, accomplishes two tasks and is defined as follows:

$$D_i : \begin{cases} \tilde{\boldsymbol{s}}_{ji}(t) = \begin{cases} \hat{\boldsymbol{s}}_{ji}(t) & \text{if } j \in \mathcal{F}_{\mathrm{C}i}(t) \\ \boldsymbol{\eta} & \text{else,} \end{cases} \quad \forall j \in \bar{\mathcal{F}}_{\mathrm{C}i} & (4.13\mathrm{a}) \\ \mathcal{F}_{\mathrm{C}i}(t) = \{ j \in \bar{\mathcal{F}}_{\mathrm{C}i} \mid \|\hat{\boldsymbol{s}}_{ji}(t)\| \geq \epsilon_{ji} \} & (4.13\mathrm{b}) \\ \tilde{\boldsymbol{u}}_i(t) = \boldsymbol{K}_{\mathrm{D}i} \tilde{\boldsymbol{s}}_i(t), \;\; \tilde{\boldsymbol{s}}_i(t) = \sum\limits_{j \in \mathcal{P}_{\mathrm{C}i}(t)} \tilde{\boldsymbol{s}}_{ij}(t) & (4.13\mathrm{c}) \\ \mathcal{P}_{\mathrm{C}i}(t) = \{ j \in \bar{\mathcal{P}}_{\mathrm{C}i} | \tilde{\boldsymbol{s}}_{ij}(t) \neq \boldsymbol{\eta} \}. & (4.13\mathrm{d}) \end{cases}$$

First, according to (4.13a)-(4.13b), D_i decides whether the calculated coupling inputs $\hat{\boldsymbol{s}}_{ji}(t)$ are transmitted from the local control unit C_i to the local control units C_j, $(j \in \bar{\mathcal{F}}_{\mathrm{C}i})$ through the transmitted coupling signal $\tilde{\boldsymbol{s}}_{ji}(t)$. The signal $\tilde{\boldsymbol{s}}_{ji}(t)$ is identical to the calculated coupling input $\hat{\boldsymbol{s}}_{ji}(t)$ if $\|\hat{\boldsymbol{s}}_{ji}\|$ exceeds the switching threshold ϵ_{ji}. The symbol $\boldsymbol{\eta}$ is a placeholder for the corresponding vector $\tilde{\boldsymbol{s}}_{ji}(t)$ and means that $\hat{\boldsymbol{s}}_{ji}(t)$ is not sent from C_i to C_j. Hence, information is sent to the local control units C_j, $(j \in \mathcal{F}_{\mathrm{C}i}(t))$, that are in the current set $\mathcal{F}_{\mathrm{C}i}(t)$ of followers of Σ_i.

Second, according to (4.13c)-(4.13d), D_i determines the communicated control input $\tilde{\boldsymbol{u}}_i(t)$ for the feedback unit F_i using the decoupling matrix $\boldsymbol{K}_{\mathrm{D}i} \in \mathbb{R}^{m_i \times m_{\mathrm{s}i}}$ and the *overall transmitted coupling input* $\tilde{\boldsymbol{s}}_i(t) \in \mathbb{R}^{m_{\mathrm{s}i}}$. Hence, C_i only receives information $(\tilde{\boldsymbol{s}}_{ij}(t) = \hat{\boldsymbol{s}}_{ij}(t))$ from local control units C_j, $(j \in \mathcal{P}_{\mathrm{C}i}(t))$ that are in the current set $\mathcal{P}_{\mathrm{C}i}(t)$ of communicational predecessors of C_i.

In summary, the local control units C_i, $(\forall i \in \mathcal{N})$, for the situation-dependent compensation of the coupling inputs are defined by

$$C_i : \begin{cases} F_i \text{ defined in (4.11)} \\ O_i \text{ defined in (4.12)} \\ D_i \text{ defined in (4.13),} \end{cases} \qquad \forall i \in \mathcal{N}. \qquad (4.14)$$

This leads to the overall definition of the self-organizing controller C which transmits physical coupling signals

$$C : \; C_i, (\forall i \in \mathcal{N}), \text{ defined in (4.14) with } \mathcal{E}_{\mathrm{C}}(t) \text{ bounded according to (4.10).} \qquad (4.15)$$

4.2.3 Overall model of the self-organizing control system

The controlled subsystems $\bar{\Sigma}_i$, $(\forall i \in \mathcal{N})$, result from (3.1) and (4.11) to

$$\bar{\Sigma}_i : \begin{cases} \dot{\boldsymbol{x}}_i(t) = \bar{\boldsymbol{A}}_i \boldsymbol{x}_i(t) - \boldsymbol{B}_i \tilde{\boldsymbol{u}}_i(t) + \boldsymbol{E}_i \boldsymbol{s}_i(t) + \boldsymbol{G}_i \boldsymbol{d}_i(t), \quad \boldsymbol{x}_i(0) = \boldsymbol{x}_{0i} \\ \boldsymbol{v}_i(t) = \boldsymbol{C}_{\mathrm{v}i} \boldsymbol{x}_i(t) \\ \boldsymbol{z}_i(t) = \boldsymbol{C}_{\mathrm{z}i} \boldsymbol{x}_i(t) \end{cases} \tag{4.16}$$

with $\bar{\boldsymbol{A}}_i = \boldsymbol{A}_i - \boldsymbol{B}_i \boldsymbol{K}_{\mathrm{C}i}$ (Fig. 4.7). The feedback matrices $\boldsymbol{K}_{\mathrm{C}i}$, $(\forall i \in \mathcal{N})$, are assumed to stabilize the isolated subsystems.

Assumption 4.1 *The feedback gain matrix* $\boldsymbol{K}_{\mathrm{C}i}$, $(i \in \mathcal{N})$, *is chosen such that* $\bar{\boldsymbol{A}}_i = \boldsymbol{A}_i - \boldsymbol{B}_i \boldsymbol{K}_{\mathrm{C}i}$ *are Hurwitz matrices.*

This assumption is weak since according to Assumption 3.2 and Theorem 2.2 there always exists a feedback gain matrix $\boldsymbol{K}_{\mathrm{C}i}$ that guarantees Assumption 4.1.

The relation (4.3) can be fulfilled if and only if $\boldsymbol{B}_i \boldsymbol{B}_i^+ \boldsymbol{E}_i = \boldsymbol{E}_i$ holds true (cf. Theorem 6.2 in [84]), where $\boldsymbol{B}_i^+$ denotes the pseudoinverse of $\boldsymbol{B}_i$ (cf. (2.1)). Then, the decoupling matrix $\boldsymbol{K}_{\mathrm{D}i}$ can be determined by

$$\boldsymbol{K}_{\mathrm{D}i} = \boldsymbol{B}_i^+ \boldsymbol{E}_i, \quad i \in \mathcal{N}. \tag{4.17}$$

Therefore, the following assumption is made.

Assumption 4.2 *The decoupling matrices* $\boldsymbol{K}_{\mathrm{D}i}$, $(\forall i \in \mathcal{N})$, *are chosen to fulfill the relation* (4.3).

If (4.3) holds true, then the *extended subsystem* $\bar{\Sigma}_i^*$ depicted in Fig. 4.7 is given by

$$\bar{\Sigma}_i^* : \begin{cases} \dot{\boldsymbol{x}}_i(t) = \bar{\boldsymbol{A}}_i \boldsymbol{x}_i(t) + \boldsymbol{E}_i \tilde{\boldsymbol{s}}_{\Delta i}(t) + \boldsymbol{G}_i \boldsymbol{d}_i(t), \quad \boldsymbol{x}_i(0) = \boldsymbol{x}_{0i} \\ \boldsymbol{v}_i(t) = \boldsymbol{C}_{\mathrm{v}i} \boldsymbol{x}_i(t) \\ \boldsymbol{z}_i(t) = \boldsymbol{C}_{\mathrm{z}i} \boldsymbol{x}_i(t) \end{cases} \tag{4.18}$$

with the difference signal

$$\tilde{\boldsymbol{s}}_{\Delta i}(t) = \boldsymbol{s}_i(t) - \tilde{\boldsymbol{s}}_i(t), \quad \forall i \in \mathcal{N} \tag{4.19}$$

as a new input. The signal $\tilde{\boldsymbol{s}}_{\Delta i}(t) \in \mathbb{R}^{m_{si}}$ is the difference between the coupling input $\boldsymbol{s}_i(t)$ and the overall transmitted coupling input $\tilde{\boldsymbol{s}}_i(t)$. Hence, $\tilde{\boldsymbol{s}}_{\Delta i}(t)$ describes the current impact of the coupling input which is still effective.

The self-organizing control system Σ with a situation-dependent compensation of the physical coupling inputs follows from (3.5), (4.18), (4.3) and (4.15) to

$$\Sigma : \begin{cases} \dot{\boldsymbol{x}}(t) = \bar{\boldsymbol{A}}\boldsymbol{x}(t) + \boldsymbol{E}\tilde{\boldsymbol{s}}_{\Delta}(t) + \boldsymbol{G}\boldsymbol{d}(t), \quad \boldsymbol{x}(0) = \boldsymbol{x}_0 \\ \boldsymbol{v}(t) = \boldsymbol{C}_{\mathrm{v}}\boldsymbol{x}(t) \\ \boldsymbol{z}(t) = \boldsymbol{C}_{\mathrm{z}}\boldsymbol{x}(t) \\ \boldsymbol{s}(t) = \boldsymbol{L}\boldsymbol{z}(t) \text{ according to (3.5)} \\ O_i, \ (\forall i \in \mathcal{N}) \text{ defined in (4.12)} \\ D_i, \ (\forall i \in \mathcal{N}) \text{ defined in (4.13)} \end{cases} \tag{4.20}$$

with $\bar{\boldsymbol{A}} = \mathrm{diag}(\bar{\boldsymbol{A}}_1 \ldots \bar{\boldsymbol{A}}_N) \in \mathbb{R}^{n \times n}$ and

$$\tilde{\boldsymbol{s}}_{\Delta}(t) := \begin{pmatrix} \tilde{\boldsymbol{s}}_{\Delta 1}(t) \\ \vdots \\ \tilde{\boldsymbol{s}}_{\Delta N}(t) \end{pmatrix} \in \mathbb{R}^{m_{\mathrm{s}}}.$$

4.3 Analysis of the self-organizing control system

In this section the self-organizing control system Σ is analyzed with respect to its stability (Section 4.3.2), its behavior compared to the corresponding centralized control system Σ_{C} (Section 4.3.2) and its behavior in case of disturbances (Section 4.3.2). Due to the situation-dependent compensation of the coupling inputs, the controlled subsystems can by analyzed individually which is presented in Section 4.3.1 beforehand.

4.3.1 Behavior of a single controlled subsystem

In this section the controlled subsystems are analyzed while using local model information only. Upper bounds for the dynamical behavior are given and local condition for the activation and deactivation of the communication are presented.

Behavior of the controlled subsystems. From the model of the extended subsystem $\bar{\Sigma}_i^*$, $(i \in \mathcal{N})$, defined in (4.18) it follows that

$$\boldsymbol{x}_i(t) = \mathrm{e}^{\bar{\boldsymbol{A}}_i t}\boldsymbol{x}_{0i} + \int_0^t \mathrm{e}^{\bar{\boldsymbol{A}}_i(t-\tau)}\boldsymbol{G}_i\boldsymbol{d}_i(\tau)\mathrm{d}\tau + \int_0^t \mathrm{e}^{\bar{\boldsymbol{A}}_i(t-\tau)}\boldsymbol{E}_i\tilde{\boldsymbol{s}}_{\Delta i}(\tau)\mathrm{d}\tau \tag{4.21}$$

holds for all $t \geq 0$ and all $i \in \mathcal{N}$. This shows that besides the disturbance $\boldsymbol{d}_i(t)$ the only input is the difference signal $\tilde{\boldsymbol{s}}_{\Delta i}(t)$. Since the disturbance $\boldsymbol{d}_i(t)$, $(i \in \mathcal{N})$, is bounded according to

Assumption 3.1, the boundedness of the difference signal $\tilde{\boldsymbol{s}}_{\Delta i}(t)$ leads to an overall bound on the inputs of the extended subsystem $\bar{\Sigma}_i^*$. An upper bound on $\tilde{\boldsymbol{s}}_{\Delta i}(t)$, $(i \in \mathcal{N})$, is presented in the following lemma.

Lemma 4.1 (Boundedness of the difference signal $\tilde{\boldsymbol{s}}_{\Delta i}(t)$) *The difference signals $\tilde{\boldsymbol{s}}_{\Delta i}$, $(\forall i \in \mathcal{N})$, within the self-organizing control system Σ defined in* (4.20) *are bounded by*

$$\max_{t \geq 0} \|\tilde{\boldsymbol{s}}_{\Delta i}(t)\| \leq \sum\nolimits_{j \in \bar{\mathcal{P}}_{\mathrm{C}i}} \epsilon_{ij} \quad \forall i \in \mathcal{N}. \tag{4.22}$$

Proof. From (3.3), (4.9), (4.12) and (4.19) it follows that

$$\tilde{\boldsymbol{s}}_{\Delta i}(t) = \sum_{j \in \bar{\mathcal{P}}_{\mathrm{C}i}} \tilde{\boldsymbol{s}}_{\Delta ij}(t), \quad \forall i \in \mathcal{N}$$

with the signals $\tilde{\boldsymbol{s}}_{\Delta ij}(t) = \boldsymbol{s}_{ij}(t) - \tilde{\boldsymbol{s}}_{ij}(t)$, $(j \in \bar{\mathcal{P}}_{\mathrm{C}i})$, and $\boldsymbol{s}_{ij}(t) = \boldsymbol{L}_{ij}\boldsymbol{C}_{\mathrm{z}j}\boldsymbol{x}_j(t) = \hat{\boldsymbol{s}}_{ij}(t)$. According to the situation-dependent communication among the local control units C_i defined in (4.14), the signal $\tilde{\boldsymbol{s}}_{\Delta ij}(t)$ can be rewritten as follows

$$\tilde{\boldsymbol{s}}_{\Delta ij}(t) = \begin{cases} \boldsymbol{0} & \text{if } \|\hat{\boldsymbol{s}}_{ij}(t)\| \geq \epsilon_{ij} \\ \boldsymbol{s}_{ij}(t) = \hat{\boldsymbol{s}}_{ij}(t) & \text{else,} \end{cases} \quad \forall j, i \in \mathcal{N}, \quad j \neq i$$

since $\boldsymbol{s}_{ij}(t) = \tilde{\boldsymbol{s}}_{ij}(t)$ holds for $\|\hat{\boldsymbol{s}}_{ij}(t)\| \geq \epsilon_{ij}$ according to (4.13). With this, the relation $\|\tilde{\boldsymbol{s}}_{\Delta ij}(t)\| \leq \epsilon_{ij}$, $(\forall t \geq 0, \forall i \in \mathcal{N}, \forall j \in \bar{\mathcal{P}}_{\mathrm{C}i})$ holds true which completes the proof. □

Since the difference signal $\tilde{s}_{\Delta i}(t)$ is bounded according to Lemma 4.1, the relation (4.21) is used to derive an upper bound on the states $\boldsymbol{x}_i$, $(\forall i \in \mathcal{N})$, of the subsystems Σ_i in the following lemma.

Lemma 4.2 (Independent boundedness of the subsystem states) *The states $\boldsymbol{x}_i(t)$, $(\forall i \in \mathcal{N})$, of subsystems Σ_i within the self-organizing control system Σ defined in* (4.20) *are bounded by*

$$\|\boldsymbol{x}_i(t)\| \leq \left\| \mathrm{e}^{\bar{\boldsymbol{A}}_i t} \right\| \|\boldsymbol{x}_{0i}\| + \bar{x}_i, \quad \forall t \geq 0, \quad \forall i \in \mathcal{N} \tag{4.23}$$

with

$$\begin{aligned} \bar{x}_i &:= m_{\mathrm{xd}i}\bar{d}_i + m_{\mathrm{xs}i} \sum\nolimits_{j \in \bar{\mathcal{P}}_{\mathrm{C}i}} \epsilon_{ij} \\ m_{\mathrm{xd}i} &:= \int_0^\infty \left\| \mathrm{e}^{\bar{\boldsymbol{A}}_i \tau} \boldsymbol{G}_i \right\| \mathrm{d}\tau \\ m_{\mathrm{xs}i} &:= \int_0^\infty \left\| \mathrm{e}^{\bar{\boldsymbol{A}}_i \tau} \boldsymbol{E}_i \right\| \mathrm{d}\tau. \end{aligned}$$

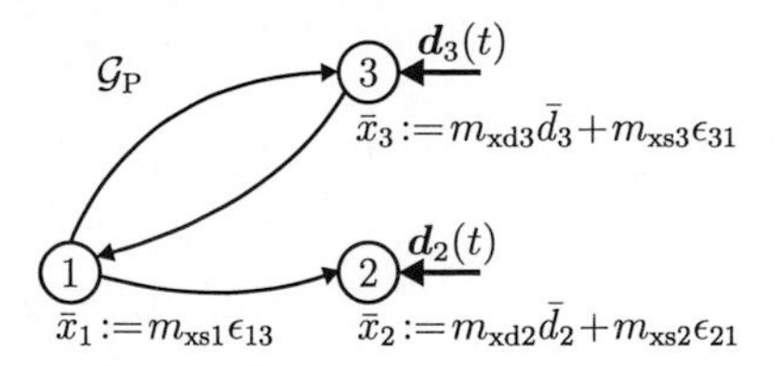

Figure 4.8: Illustration of boundedness of the subsystem states according to Lemma 4.2 for certain physically interconnected subsystems.

Proof. From (4.21) it follows that

$$\|\boldsymbol{x}_i(t)\| = \left\| \mathrm{e}^{\bar{\boldsymbol{A}}_i t}\boldsymbol{x}_{0i} + \int_0^t \mathrm{e}^{\bar{\boldsymbol{A}}_i (t-\tau)}\boldsymbol{G}_i \boldsymbol{d}_i(\tau)\mathrm{d}\tau + \int_0^t \mathrm{e}^{\bar{\boldsymbol{A}}_i (t-\tau)}\boldsymbol{E}_i \tilde{\boldsymbol{s}}_{\Delta i}(\tau)\mathrm{d}\tau \right\|, \ \forall t \geq 0, \ \forall i \in \mathcal{N}.$$

From Lemma 4.1 and Assumption 3.1 it follows that

$$\begin{aligned}\|\boldsymbol{x}_i(t)\| &\leq \left\|\mathrm{e}^{\bar{\boldsymbol{A}}_i t}\right\| \|\boldsymbol{x}_{0i}\| + \int_0^\infty \left\|\mathrm{e}^{\bar{\boldsymbol{A}}_i \tau}\boldsymbol{G}_i\right\| \mathrm{d}\tau \max_{t\geq 0} \|\boldsymbol{d}_i(t)\| \\ &\qquad + \int_0^\infty \left\|\mathrm{e}^{\bar{\boldsymbol{A}}_i \tau}\boldsymbol{E}_i\right\| \mathrm{d}\tau \max_{t\geq 0} \|\tilde{\boldsymbol{s}}_{\Delta i}(t)\|, \\ &\leq \left\|\mathrm{e}^{\bar{\boldsymbol{A}}_i t}\right\| \|\boldsymbol{x}_{0i}\| + m_{\mathrm{xd}i}\bar{d}_i + m_{\mathrm{xs}i}\sum\nolimits_{j\in\bar{\mathcal{P}}_{\mathrm{C}i}} \epsilon_{ij}.\end{aligned}$$

According to Assumption 4.1, the matrices $\bar{\boldsymbol{A}}_i$, $(\forall i \in \mathcal{N})$, are Hurwitz matrices, hence, the scalars $m_{\mathrm{xd}i}$ and $m_{\mathrm{xs}i}$ are finite which completes the proof. □

Lemma 4.2 shows that for an initial state $\boldsymbol{x}_{0i} = \boldsymbol{0}$ the state $\boldsymbol{x}_i(t)$ of the subsystem Σ_i is bounded by $\bar{x}_i$, where $\bar{x}_i$ depends on the maximal disturbance $\bar{d}_i$, $(\forall i \in \mathcal{N})$, and on the choice of the switching thresholds ϵ_{ij}, $(j \in \bar{\mathcal{P}}_{\mathrm{C}i})$. Note that the bounds $\bar{x}_i$ are independent from the dynamics of the other subsystems. Figure 4.8 illustrates that fact for certain physically interconnected subsystems. Consider that a subsystem Σ_i is not disturbed ($\boldsymbol{d}_i(t) = \boldsymbol{0}$), then the bound $\bar{x}_i$ of the subsystem state $\boldsymbol{x}_i(t)$ can be adjusted by the switching thresholds ϵ_{ij}, $(j \in \bar{\mathcal{P}}_{\mathrm{C}i})$ only. Therefore, the choice of ϵ_{ij}, $(j \in \bar{\mathcal{P}}_{\mathrm{C}i})$ defines how well disturbances on other subsystems are attenuated at Σ_i. The choice of the switching thresholds will be further investigated in Section 4.4.1.

According to (4.12) and (4.18), the behavior of the performance outputs $\boldsymbol{v}_i(t)$ and the behavior of the determined coupling inputs $\hat{\boldsymbol{s}}_{ji}(t)$ are given by

$$\boldsymbol{v}_i(t) = \boldsymbol{C}_{\mathrm{v}i}\mathrm{e}^{\bar{\boldsymbol{A}}_i t}\boldsymbol{x}_{0i} + \int_0^t \boldsymbol{C}_{\mathrm{v}i}\mathrm{e}^{\bar{\boldsymbol{A}}_i \bar{\tau}}\boldsymbol{G}_i\boldsymbol{d}_i(\tau)\mathrm{d}\tau + \int_0^t \boldsymbol{C}_{\mathrm{v}i}\mathrm{e}^{\bar{\boldsymbol{A}}_i \bar{\tau}}\boldsymbol{E}_i\tilde{\boldsymbol{s}}_{\Delta i}(\tau)\mathrm{d}\tau, \ \forall i \in \mathcal{N} \quad (4.24)$$

$$\hat{\boldsymbol{s}}_{ji}(t) = \boldsymbol{L}_{ji}\boldsymbol{C}_{\mathrm{z}i}\mathrm{e}^{\bar{\boldsymbol{A}}_i t}\boldsymbol{x}_{0i} + \int_0^t \boldsymbol{L}_{ji}\boldsymbol{C}_{\mathrm{z}i}\mathrm{e}^{\bar{\boldsymbol{A}}_i \bar{\tau}}\boldsymbol{G}_i\boldsymbol{d}_i(\tau)\mathrm{d}\tau + \int_0^t \boldsymbol{L}_{ji}\boldsymbol{C}_{\mathrm{z}i}\mathrm{e}^{\bar{\boldsymbol{A}}_i \bar{\tau}}\boldsymbol{E}_i\tilde{\boldsymbol{s}}_{\Delta i}(\tau)\mathrm{d}\tau, \quad \forall j \in \bar{\mathcal{F}}_{\mathrm{C}i}, \quad \forall i \in \mathcal{N} \quad (4.25)$$

with $\bar{\tau} = t - \tau$.

Lemma 4.3 (Boundedness of the subsystem output and determined coupling inputs) *The performance output $\boldsymbol{v}_i(t)$ of Σ_i, $(i \in \mathcal{N})$, and the determined coupling inputs $\hat{\boldsymbol{s}}_{ji}$, $(\forall j \in \bar{\mathcal{F}}_{\mathrm{C}i})$, of O_i within the self-organizing control system Σ defined in* (4.20) *are bounded by*

$$\|\boldsymbol{v}_i(t)\| \leq \left\|\boldsymbol{C}_{\mathrm{v}i}\mathrm{e}^{\bar{\boldsymbol{A}}_i t}\right\| \|\boldsymbol{x}_{0i}\| + m_{\mathrm{vd}i}\bar{d}_i + m_{\mathrm{vs}i}\sum_{l \in \bar{\mathcal{P}}_{\mathrm{C}i}} \epsilon_{il}, \quad \forall t \geq 0, \quad \forall i \in \mathcal{N} \quad (4.26)$$

$$\|\hat{\boldsymbol{s}}_{ji}(t)\| \leq \left\|\boldsymbol{L}_{ji}\boldsymbol{C}_{\mathrm{z}i}\mathrm{e}^{\bar{\boldsymbol{A}}_i t}\right\| \|\boldsymbol{x}_{0i}\| + m_{\mathrm{sd}ji}\bar{d}_i + m_{\mathrm{ss}ji}\sum_{l \in \bar{\mathcal{P}}_{\mathrm{C}i}} \epsilon_{il}, \quad \forall t \geq 0, \quad \forall j \in \bar{\mathcal{F}}_{\mathrm{C}i}, \quad \forall i \in \mathcal{N} \quad (4.27)$$

for all $t \geq 0$ with

$$m_{\mathrm{vd}i} := \int_0^\infty \left\|\boldsymbol{C}_{\mathrm{v}i}\mathrm{e}^{\bar{\boldsymbol{A}}_i \tau}\boldsymbol{G}_i\right\| \mathrm{d}\tau \quad (4.28)$$

$$m_{\mathrm{vs}i} := \int_0^\infty \left\|\boldsymbol{C}_{\mathrm{v}i}\mathrm{e}^{\bar{\boldsymbol{A}}_i \tau}\boldsymbol{E}_i\right\| \mathrm{d}\tau \quad (4.29)$$

$$m_{\mathrm{sd}ji} := \int_0^\infty \left\|\boldsymbol{L}_{ji}\boldsymbol{C}_{\mathrm{z}i}\mathrm{e}^{\bar{\boldsymbol{A}}_i \tau}\boldsymbol{G}_i\right\| \mathrm{d}\tau \quad (4.30)$$

$$m_{\mathrm{ss}ji} := \int_0^\infty \left\|\boldsymbol{L}_{ji}\boldsymbol{C}_{\mathrm{z}i}\mathrm{e}^{\bar{\boldsymbol{A}}_i \tau}\boldsymbol{E}_i\right\| \mathrm{d}\tau. \quad (4.31)$$

Proof. The proof is given in Appendix A.1. □

The bounds in Lemma 4.1, Lemma 4.2 and Lemma 4.3 are used in the following to investigate the communication activation, the stability, the disturbance propagation as well as the design of the switching thresholds.

Communication activition and deactivation. To investigate the communication effort among the local control units three questions concerning the transmission of the determined coupling input $\hat{\boldsymbol{s}}_{ji}(t)$ from the local control unit C_i to the local control unit C_j are considered:

1. Is there any communication from C_i to C_j?

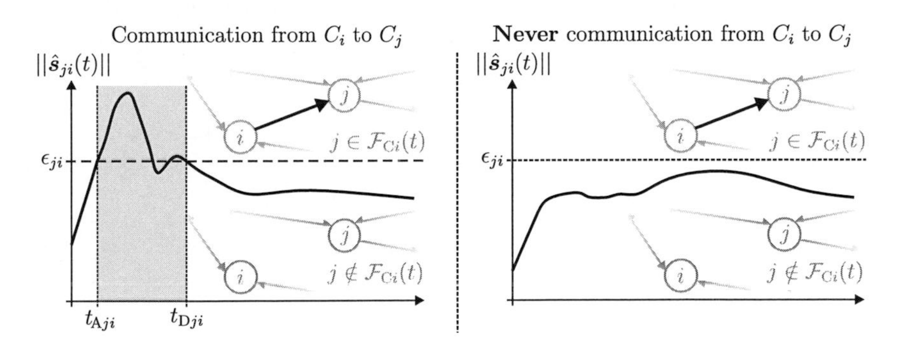

Figure 4.9: Activation and deactivation of communication within Σ defined in (4.20).

2. If there is communication from C_i to C_j, will the communication be permanently deactivated eventually?

3. At which times will the communication be activated and deactivated, respectively?

Recall that communication from C_i to C_j implies that there exists a communication link $(i \to j)$ within the set $\mathcal{E}_{\mathrm{C}}(t)$ of communicational edges, i.e., $(i \to j) \in \mathcal{E}_{\mathrm{C}}(t)$, which defines the time-varying part the communication graph $\mathcal{G}_{\mathrm{C}}(t) = (\mathcal{V}_{\mathrm{C}}, \mathcal{E}_{\mathrm{C}}(t))$.

The left part of Fig. 4.9 shows the situation in which C_i sends information to C_j, whereas in the right part no communication from C_i to C_j is present. According to the definition of the decision unit D_i in (4.13), communication from C_i to C_j is activated at a time $t_{\mathrm{A}ji}$, if the relation

$$(\exists t_{\mathrm{A}ji} \geq 0) \quad t_{\mathrm{A}ji} = \min_{t \geq 0} t \ : \ \|\hat{\boldsymbol{s}}_{ji}(t)\| \geq \epsilon_{ji}, \quad j \in \bar{\mathcal{F}}_{\mathrm{C}i}, \quad i \in \mathcal{N} \tag{4.32}$$

holds true. The communication from C_i to C_j stays deactivated after a time $t_{\mathrm{D}ji}$ if the following relation holds true

$$(\exists t_{\mathrm{D}ji} \geq 0) \quad t_{\mathrm{D}ji} = \max_{t \geq 0} t : \|\hat{\boldsymbol{s}}_{ji}(t)\| \geq \epsilon_{ji}, \quad j \in \bar{\mathcal{F}}_{\mathrm{C}i}, \quad i \in \mathcal{N}. \tag{4.33}$$

The following lemma concerns the calculation of the *activation time* $t_{\mathrm{A}ji}$ for the communication link $(i \to j) \in \bar{\mathcal{E}}_{\mathrm{C}}$.

Lemma 4.4 (Activation of the communication from C_i to C_j for Σ defined in (4.20)) *Consider the self-organizing control system Σ defined in* (4.20)*. The communication from the local control unit C_i, $(i \in \mathcal{N})$, to the local control unit C_j, $(j \in \bar{\mathcal{F}}_{\mathrm{C}i})$ is activated only if there exists a* *time*

$$\bar{t}_{\mathrm{A}ji} = \min_{t \geq 0} t: \; \bar{s}_{\mathrm{T}ji}(t) \geq \epsilon_{ji}, \quad j \in \bar{\mathcal{F}}_{\mathrm{C}i}, \quad i \in \mathcal{N} \tag{4.34}$$

which is a lower bound on time activation time $t_{\mathrm{A}ji}$, i.e., $t_{\mathrm{A}ji} \geq \bar{t}_{\mathrm{A}ji}$, with

$$\bar{s}_{\mathrm{T}ji}(t) := \left\| \boldsymbol{L}_{ji} \boldsymbol{C}_{\mathrm{z}i} \mathrm{e}^{\bar{\boldsymbol{A}}_i t} \right\| \|\boldsymbol{x}_{0i}\| + \int_0^t \left\| \boldsymbol{L}_{ji} \boldsymbol{C}_{\mathrm{z}i} \mathrm{e}^{\bar{\boldsymbol{A}}_i \tau} \boldsymbol{G}_i \right\| \mathrm{d}\tau \cdot \bar{d}_i + \int_0^t \left\| \boldsymbol{L}_{ji} \boldsymbol{C}_{\mathrm{z}i} \mathrm{e}^{\bar{\boldsymbol{A}}_i \tau} \boldsymbol{E}_i \right\| \mathrm{d}\tau \cdot \sum_{l \in \bar{\mathcal{P}}_{\mathrm{C}i}} \epsilon_{il}.$$

Proof. According to (4.25), Assumption 3.1, and Lemma 4.1, the relation

$$\|\hat{\boldsymbol{s}}_{ji}(t)\| \leq \left\| \boldsymbol{L}_{ji} \boldsymbol{C}_{\mathrm{z}i} \mathrm{e}^{\bar{\boldsymbol{A}}_i t} \right\| \|\boldsymbol{x}_{0i}\| + \int_0^t \left\| \boldsymbol{L}_{ji} \boldsymbol{C}_{\mathrm{z}i} \mathrm{e}^{\bar{\boldsymbol{A}}_i \tau} \boldsymbol{G}_i \right\| \mathrm{d}\tau \bar{d}_i + \int_0^t \left\| \boldsymbol{L}_{ji} \boldsymbol{C}_{\mathrm{z}i} \mathrm{e}^{\bar{\boldsymbol{A}}_i \tau} \boldsymbol{E}_i \right\| \mathrm{d}\tau \sum_{l \in \bar{\mathcal{P}}_{\mathrm{C}i}} \epsilon_{il} = \bar{s}_{\mathrm{T}ji}(t)$$

holds true for all $t \geq 0$, all $j \in \bar{\mathcal{F}}_{\mathrm{C}i}$ and all $i \in \mathcal{N}$. Hence, only if $\bar{s}_{\mathrm{T}ji}(t) \geq \epsilon_{ji}$ is fulfilled, the time in (4.32) exists which completes the proof. □

Lemma 4.4 gives a necessary condition for the activation of the communication from C_i to C_j, where the lower bound $\bar{t}_{\mathrm{A}ji}$ on the activation time $t_{\mathrm{A}ji}$ depends on the initial state $\boldsymbol{x}_{0i}$, $(i \in \mathcal{N})$, the bound $\bar{d}_i$ on the disturbance $\boldsymbol{d}_i(t)$ and on the choice of the switching thresholds ϵ_{il}, $(\forall l \in \bar{\mathcal{P}}_{\mathrm{C}i})$. The necessity of the condition results from the fact that the signal $\bar{s}_{\mathrm{T}ji}(t)$ is an upper bound on the norm $\|\hat{\boldsymbol{s}}_{ji}(t)\|$ of the determined coupling input. Hence, it is not guaranteed that $\|\hat{\boldsymbol{s}}_{ji}(t)\|$ exceeds ϵ_{ji} if $\bar{s}_{\mathrm{T}ji}(t)$ exceeds ϵ_{ji}. An interesting aspect of Lemma 4.4 is that the activation of each communication link can be checked individually without any model information of the other subsystems.

Lemma 4.4 and (4.27) show that for $\boldsymbol{x}_{0i} = \boldsymbol{0}$, $(i \in \mathcal{N})$, the communication from C_i to C_j, $(j \in \bar{\mathcal{F}}_{\mathrm{C}i})$ is activated only if the relation

$$m_{\mathrm{ss}ji} \sum\nolimits_{l \in \bar{\mathcal{P}}_{\mathrm{C}i}} \epsilon_{il} + m_{\mathrm{sd}ji} \bar{d}_i \geq \epsilon_{ji}, \quad j \in \bar{\mathcal{F}}_{\mathrm{C}i}, \quad i \in \mathcal{N} \tag{4.35}$$

is fulfilled. Hence, the relation

$$\bar{d}_i \geq \frac{\epsilon_{ji} - m_{\mathrm{ss}ji} \sum_{l \in \bar{\mathcal{P}}_{\mathrm{C}i}} \epsilon_{il}}{m_{\mathrm{sd}ji}}, \quad j \in \bar{\mathcal{F}}_{\mathrm{C}i}, \quad i \in \mathcal{N}$$

gives a necessary condition on the maximal disturbance $\bar{d}_i$ for which the communication from C_i to C_j is activated.

The following lemma concerns the existence of the *deactivation time* $t_{\mathrm{D}ji}$ according to (4.33).

Lemma 4.5 (Deactivation of the communication from C_i to C_j) *Consider the self-organizing control system Σ in (4.20). The communication from the local control unit C_i, $(i \in \mathcal{N})$ to the local control unit C_j, $(j \in \bar{\mathcal{F}}_{\mathrm{C}i})$ stays deactivated after some deactivation time $t_{\mathrm{D}ji}$ if the relation*

$$m_{\mathrm{ss}ji} \sum_{l \in \bar{\mathcal{P}}_{\mathrm{C}i}} \epsilon_{il} + m_{\mathrm{sd}ji} \bar{d}_i < \epsilon_{ji}, \quad j \in \bar{\mathcal{F}}_{\mathrm{C}i}, \quad i \in \mathcal{N} \tag{4.36}$$

holds true, i.e., $j \notin \mathcal{F}_{\mathrm{C}i}(t)$ for all $t > t_{\mathrm{D}ji} \geq 0$. The deactivation time $t_{\mathrm{D}ji}$ is upper bounded by the time $\bar{t}_{\mathrm{D}ji}$ which is defined by

$$t_{\mathrm{D}ji} \leq \bar{t}_{\mathrm{D}ji} = \max_{t \geq 0} t : \ \bar{s}_{\mathrm{T}ji}(t) \geq \epsilon_{ji}, \quad j \in \bar{\mathcal{F}}_{\mathrm{C}i}, \quad i \in \mathcal{N}.$$

Proof. According to Lemma 4.3, the relation

$$\lim_{t \to \infty} \|\hat{\boldsymbol{s}}_{ji}(t)\| \leq m_{\mathrm{ss}ji} \sum_{l \in \bar{\mathcal{P}}_{\mathrm{C}i}} \epsilon_{il} + m_{\mathrm{sd}ji} \bar{d}_i$$

holds true for all $t \geq 0$, all $j \in \bar{\mathcal{F}}_{\mathrm{C}i}$ and all $i \in \mathcal{N}$. Hence, if the relation in (4.36) holds true, then the condition for deactivating the communication in (4.33) is fulfilled. Since $\bar{s}_{\mathrm{T}ji}(t)$ is an upper bound on $\|\hat{\boldsymbol{s}}_{ji}(t)\|$, the time $\bar{t}_{\mathrm{D}ji}$ is an upper bound on the deactivation time $t_{\mathrm{D}ji}$. □

Lemma 4.5 gives a sufficient condition for the deactivation of the communication from C_i, $(i \in \mathcal{N})$, to C_j, where the existence of a deactivation time $t_{\mathrm{D}ji}$ only depends on the bound $\bar{d}_i$ on the disturbance $\boldsymbol{d}_i(t)$ and the choice of the switching thresholds ϵ_{il}, $(\forall l \in \bar{\mathcal{P}}_{\mathrm{C}i})$, but not on the initial state $\boldsymbol{x}_{0i}$. Since the left part of the inequality (4.36) is an upper bound on the absolute value $\lim_{t\to\infty} \|\hat{\boldsymbol{s}}_{ji}(t)\|$ of the determined coupling signal for $t \to \infty$, the condition (4.36) is sufficient but not necessary for the communication deactivation. Hence, the communication from C_i to C_j might also be deactivated even if the relation in (4.36) is not fulfilled. Furthermore, the deactivation of each communication link can be checked individually and with local model information only.

If there is no disturbance $\bar{d}_i = 0$, then the deactivation only depends on the switching thresholds and the condition

$$m_{\mathrm{ss}ji} \sum_{l \in \bar{\mathcal{P}}_{\mathrm{C}i}} \epsilon_{il} < \epsilon_{ji}, \quad j \in \bar{\mathcal{F}}_{\mathrm{C}i}, \quad i \in \mathcal{N}$$

has to be fulfilled to guarantee an eventual permanent deactivation of the communication from C_i, $(i \in \mathcal{N})$, to C_j.

The conditions for the activation and the deactivation of the communication links presented in Lemma 4.4 and Lemma 4.5 are used in the following sections to derive a condition for the asymptotic stability of the self-organizing control system Σ (Section 4.3.2), to give a condition for activating the communication w.r.t certain disturbance situations (Section 4.3.4) and to chose the communication structure for guaranteeing a desired performance (Section 4.4.2).

In summary, for $\boldsymbol{x}_0 = \mathbf{0}$, the three initial questions concerning the communication from C_i to C_j can be answered as follows:

1. The communication is activated only if the relation in (4.35) holds true.
2. The communication will be deactivated permanently if the relation in (4.36) holds true.
3. Lemma 4.4 and Lemma 4.5 derive a lower bound for the activation time $t_{\mathrm{A}ji}$ and an upper bound for the deactivation time $t_{\mathrm{D}ji}$. These bounds can be used to determine a communication interval for the link $(i \rightarrow j) \in \mathcal{E}_{\mathrm{C}}$ for the case of a specific disturbance $\boldsymbol{d}_i(t)$ affecting the subsystem Σ_i which is illustrated in more detail in the following example.

The most interesting aspect of this section is:

> The behavior of the controlled subsystems and the resulting communication among the local control units can be analyzed by using local model information only.

In the following the behavior of the self-organizing controlled symmetrically interconnected subsystems is analyzed.

Example 4.1 (cont.) *Compensation of physical coupling inputs (Analysis of subsystems)*

Figure 4.10 depicts the behavior of the symmetrically interconnected subsystems P defined in (3.32) with $\boldsymbol{x}_{01} = 0$ and $\boldsymbol{x}_{02} = 2.5$ controlled by the local control units C_i defined in (4.14) with the parameters listed in Table 4.3, where subsystem Σ_1 is affected by a disturbance. The gray area in the middle plots show the upper bounds on $\|\boldsymbol{x}_i(t)\|$ according to (4.23) in Lemma 4.2 with

$$\bar{x}_1 = m_{\mathrm{xd1}}\bar{d}_1 + m_{\mathrm{xs1}}\epsilon_{12} = 1.25$$
$$\bar{x}_2 = m_{\mathrm{xd2}}\bar{d}_2 + m_{\mathrm{xs2}}\epsilon_{21} = 0.25$$

from $m_{\mathrm{xs1}} = m_{\mathrm{s2}} = m_{\mathrm{xd1}} = m_{\mathrm{xd2}} = 0.5$, $\bar{d}_1 = 2$ and $\bar{d}_2 = 0$. The bounds also hold for $\boldsymbol{v}_i(t)$ and $\hat{\boldsymbol{s}}_{ji}(t)$ since $\boldsymbol{x}_i(t) = \boldsymbol{v}_i(t) = \hat{\boldsymbol{s}}_{ji}(t)$. Due to the change of the disturbance $\boldsymbol{d}_1(t)$ at $t = t^* = 10\,\mathrm{s}$, the analysis of the behavior of Σ_1 is divided into two time intervals. In the first time interval $T_1 = [0, 10)$ the initial state is $\boldsymbol{x}_{01} = 0$ and the maximal disturbance is $\bar{d}_1 = 2$. In the second time interval $T_2 = [10, 20]$ the initial state is the upper bound on $\|\boldsymbol{x}_1(t)\|$ from the analysis in time interval T_1, i.e., $x_{01}^* = \bar{x}_1 = 1.25 \geq \max_{t \in T_1} \|\boldsymbol{x}_1(t)\|$, and the maximal disturbance is $\bar{d}_1^* = 0$.

Figure 4.10 shows that the states $\boldsymbol{x}_1(t)$ and $\boldsymbol{x}_2(t)$ never leave the gray area, hence, the upper bound on the states according to (4.23) holds true. The bounds on the activation times and deactivation times of the communication resulting from Lemma 4.4 and Lemma 4.5 are

$$\bar{t}_{\mathrm{A21}} = 0.26\,\mathrm{s}, \quad \bar{t}_{\mathrm{A12}} = 0\,\mathrm{s}, \quad \bar{t}^*_{\mathrm{A21}} = 10\,\mathrm{s}, \quad \bar{t}_{\mathrm{D21}} = \infty, \quad \bar{t}_{\mathrm{D12}} = 1.09\,\mathrm{s}, \quad \bar{t}^*_{\mathrm{D21}} = 10.69\,\mathrm{s},$$

where the analysis for $\boldsymbol{s}_{21}(t)$ is also divided into the two time intervals T_1 and T_2. Hence, $\bar{t}^*_{\mathrm{A21}}$ is the lower bound on the time t^*_{A21} at which the communication would be activated for the analysis for T_2, where $\bar{t}^*_{\mathrm{D21}}$ is the upper bound the corresponding deactivation time t^*_{D21}. The result $\bar{t}_{\mathrm{D21}} = \infty$ occurs due to the fact that $\bar{s}_{\mathrm{T21}}(t)$ always is larger than ϵ_{21} for $t \geq t_{\mathrm{A21}}$ (cf. Lemma 4.5). The determination of the

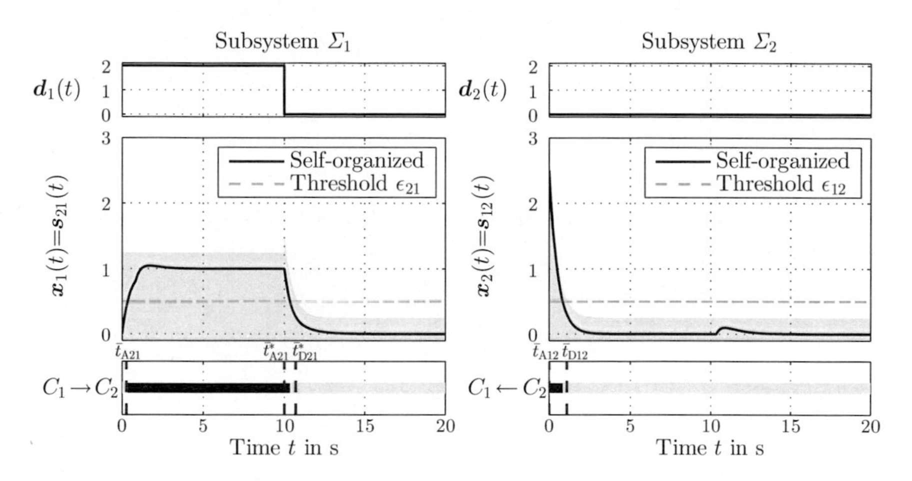

Figure 4.10: Individual analysis of the behavior of the interconnected subsystems defined in (3.32) with situation-dependent compensation of physical coupling inputs.

bounds on the activation time and the deactivation time can be used to calculate the time interval in which the communication is active while using information about the course of the maximum values of the disturbances. For the considered course of the disturbances the analysis shows that the communication from C_1 to C_2 is at most active for $\bar{t}^*_{D21} - \bar{t}_{A21} = 10.43$ s and from C_2 to C_1 for $\bar{t}_{D12} - \bar{t}_{A12} = 1.09$ s, where the actual communication intervals have a length of 9.98 s and 0.834 s, respectively.

In summary, the example shows that the results from Lemma 4.2, Lemma 4.3, Lemma 4.4 and Lemma 4.5 provide a good approximation of the actual behavior of the self organizing control system Σ. Furthermore, the analysis of the activation and the deactivation of the communication can be used to specify the communication effort for certain disturbance situations. □

4.3.2 Practical stability and asymptotic stability

This section investigates the first two control aims stated in Section 3.4. Therefore, the section is divided into two parts. The first part concerns the practical stability of the self-organizing control system Σ (Control aim 3.2). A condition on the properties of the plant P for practical stabilization using the proposed self-organizing controller C is given. Furthermore, it is shown that under the stated assumptions Σ is always practically stable. In the second part two sufficient conditions on the asymptotic stability of the undisturbed self-organizing control system Σ are derived (Control aim 3.1). The first condition results is a distributed analysis using local model information and the second condition is a central analysis using global model information.

Practical stability. In the following a condition on the properties of the plant P is derived, which guarantees that the self-organizing controller C with a situation-dependent compensation of the coupling inputs can stabilize the plant P within the overall closed-loop system Σ. For this purpose, the previous assumptions are neglected.

Theorem 4.1 (Practical stabilizability by compensation of physical coupling inputs) *The plant P defined in* (3.10) *can be practically stabilized w.r.t. to a compact set $\mathcal{A}$ by the self-organizing controller defined in* (4.15) *if all the following four conditions hold true:*

1. *The complete states $\boldsymbol{x}_i(t)$ are measurable, i.e., $\boldsymbol{C}_i = \boldsymbol{I}_{n_i}$, $(\forall i \in \mathcal{N})$.*
2. *The subsystems Σ_i can be physically decoupled, i.e., $\boldsymbol{B}_i \boldsymbol{B}_i^+ \boldsymbol{E}_i = \boldsymbol{E}_i$, $(\forall i \in \mathcal{N})$.*
3. *The subsystems Σ_i are completely controllable, i.e., the pairs $(\boldsymbol{A}_i, \boldsymbol{B}_i)$ are controllable, $(\forall i \in \mathcal{N})$.*
4. *The disturbances $\boldsymbol{d}_i(t)$ are bounded, i.e., $\|\boldsymbol{d}_i(t)\| < \infty$, $(\forall t \geq 0, \forall i \in \mathcal{N})$.*

Proof. The feedback unit F_i defined in (4.11) can only be realized with the full subsystem state $\boldsymbol{x}_i(t)$ (Condition 1). The extended subsystem $\bar{\Sigma}_i^*$ defined in (4.18) can be build if and only if relation (4.3) holds true. Therefore, according to Theorem 6.2 in [84], the relation $\boldsymbol{B}_i \boldsymbol{B}_i^+ \boldsymbol{E}_i = \boldsymbol{E}_i$ has to be fulfilled (Condition 2). Consider the bound for $\boldsymbol{x}_i(t)$ in (4.23). To fulfill the first requirement in Definition 2.3, $\lim_{t \to \infty} \left\| \mathrm{e}^{\bar{\boldsymbol{A}}_i t} \boldsymbol{x}_{0i} \right\| = 0$ has to be satisfied. Therefore, $\bar{\boldsymbol{A}}_i$ has to be a Hurwitz matrix. If $\bar{\boldsymbol{A}}_i$ is a Hurwitz matrix, then $\left\| \mathrm{e}^{\bar{\boldsymbol{A}}_i t} \right\|$ is bounded from above for all $t \geq 0$. Hence, there always exists an initial state $\|\boldsymbol{x}_{0i}\| \leq \delta$ that fulfills the second requirement in Definition 2.3. According to Theorem 2.2, if the pair $(\boldsymbol{A}_i, \boldsymbol{B}_i)$ is controllable, then the feedback gain matrix $\boldsymbol{K}_{\mathrm{C}i}$ can be chosen to make $\bar{\boldsymbol{A}}_i = \boldsymbol{A}_i - \boldsymbol{B}_i \boldsymbol{K}_{\mathrm{C}i}$ a Hurwitz matrix (Condition 3). Furthermore, according to Lemma 4.2 the value for $\max_{t \geq 0} \|\boldsymbol{d}_i(t)\|$ has to be bounded, which is fulfilled for $\|\boldsymbol{d}_i(t)\| < \infty$ (Condition 4). □

Theorem 4.1 states four sufficient conditions on the properties of the plant P defined in (3.10) such that P can be practically stabilized by the proposed self-organizing controller C with situation-dependent compensation of physical coupling signals. The first three conditions are sufficient but not necessary since some part of the undisturbed plant which could not be influenced by the controller might be stable in advance. The fourth condition is sufficient but not necessary since the effect of the disturbances $\boldsymbol{d}_i(t)$ might be compensated within Σ even if $\boldsymbol{d}_i(t)$ is not upper bounded. All conditions depend on the properties of the isolated subsystems Σ_i, $(\forall i \in \mathcal{N})$ only. This leads to the following property of the proposed self-organizing control concept:

> The practical stabilizability of the plant P to be controlled by the self-organizing controller C with a situation-dependent compensation of physical coupling inputs can be checked by using local model information of the subsystems only.

This property is very useful in complex networks of physically interconnected subsystems since the complexity of the analysis scales linearly with the number of subsystems.

Remark 4.1 *The third condition in Theorem 4.1 can be relaxed by considering that the subsystems Σ_i, $(\forall i \in \mathcal{N})$, are decomposed into a part with unstable eigenvalues and a part with stable eigenvalues. It is well known, that only the part with the unstable eigenvalues has to be controllable to guarantee that $\boldsymbol{A}_i - \boldsymbol{B}_i\boldsymbol{K}_{\mathrm{C}i}$, $(\forall i \in \mathcal{N})$, can be made a Hurwitz matrix (cf. [20, 76, 94] and, therefore, also the plant P controlled by the proposed self-organizing controller C can be practically stabilized.*

The following theorem shows that with the previous assumptions the self-organizing controller C defined in (4.15) always leads to a practically stable self-organizing control system Σ (cf. Definition 2.3).

Theorem 4.2 (Practical stability of the self-organizing control system Σ in (4.20)) *The self-organizing control system Σ defined in (4.20) is practically stable with respect to the compact set*

$$\mathcal{A} = \{\boldsymbol{x} = (\boldsymbol{x}_1^{\mathrm{T}} \dots \boldsymbol{x}_N^{\mathrm{T}})^{\mathrm{T}} \in \mathrm{I\!R}^n \mid \|\boldsymbol{x}_i\| \leq \bar{x}_i, i \in \mathcal{N}\}, \tag{4.37}$$

where the bounds $\bar{x}_i$, $(\forall i \in \mathcal{N})$, are defined in Lemma 4.2.

Proof. Consider the bound for $\boldsymbol{x}_i(t)$ in (4.23). Due to Assumption 4.1, the matrices $\bar{\boldsymbol{A}}_i$, $(\forall i \in \mathcal{N})$ are Hurwitz and, therefore, $\lim_{t\to\infty} \left\| \mathrm{e}^{\bar{\boldsymbol{A}}_i t}\boldsymbol{x}_{0i} \right\| = 0$ holds, which fulfills the first requirement in Definition 2.3. Since $\left\| \mathrm{e}^{\bar{\boldsymbol{A}}_i t} \right\|$ is bounded from above, there always exists an initial state $\|\boldsymbol{x}_{0i}\| \leq \delta$ that fulfills the second requirement in Definition 2.3, which completes the proof. □

Theorem 4.2 shows that the complexity of the stability analysis of the self-organizing control system Σ scales linearly with the number N of subsystems since each subsystem Σ_i can be analyzed individually (cf. (4.23)). For every subsystem there is an independent region defined by $\bar{x}_i$ to which the subsystem state $\boldsymbol{x}_i(t)$ converges (Fig. 4.11), where $\bar{x}_i$ depends on the dynamics of the isolated subsystem Σ_i only.

Moreover, Theorem 4.2 shows that the feedback gain matrices $\boldsymbol{K}_{\mathrm{C}i}$, $(\forall i \in \mathcal{N})$ can be designed individually to stabilize the isolated subsystems Σ_i which guarantees the practical stability of the self-organizing control system Σ. An interesting aspect is that the corresponding decentralized control system Σ_{D} defined in (3.24) does not have to be asymptotically stable since the

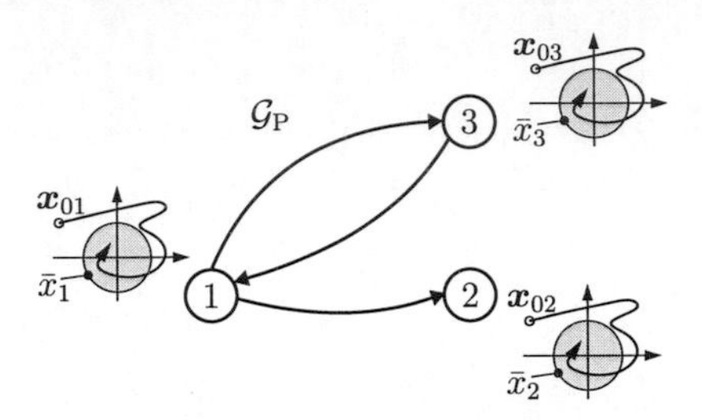

Figure 4.11: Illustration of Theorem 4.2 for certain physically interconnected subsystems.

situation-dependent communication of the control signals always keeps the state of Σ close to the equilibrium point $\boldsymbol{x} = \boldsymbol{0}$.

Asymptotic stability. In the following two conditions on the asymptotic stability of the self-organizing control system Σ defined in (4.20) are presented:

1. The first condition (Theorem 4.3) checks whether the communication among the local control units stays deactivated after a certain time t_{D}. If there is no communication among the local control units the self-organizing control system Σ behaves like the decentralized control system Σ_{D} defined in (4.38). Hence, if there exists such a deactivation time t_{D} and Σ_{D} is asymptotically stable, then Σ is also asymptotically stable.

2. For the second condition (Proposition 4.1) the model of the self-organizing control system Σ is rewritten as an uncertain system. The results on asymptotic stability of such uncertain systems in Theorem 2.5 are used to derive a condition on the asymptotic stability of the self-organizing control system Σ.

For both conditions it is assumed that the overall disturbance is zero, i.e., $\boldsymbol{d}(t) = \boldsymbol{0}$ for all $t \geq 0$.

Deactivation of all communication links (first condition). For the first condition consider the decentralized control system

$$\Sigma_{\mathrm{D}}: \ \dot{\boldsymbol{x}}_{\mathrm{D}}(t) = \bar{\boldsymbol{A}}_{\mathrm{P}} \boldsymbol{x}_{\mathrm{D}}(t), \quad \boldsymbol{x}_{\mathrm{D}}(0) = \boldsymbol{x}_{\mathrm{D0}} \tag{4.38}$$

resulting from Σ defined in (4.20) with $\mathcal{E}_{\mathrm{C}}(t) = \emptyset$ for all $t \geq 0$ according to the definition of Σ_{D} in (3.24). Hence, $\tilde{\boldsymbol{u}}_i(t) = \boldsymbol{0}$ holds true for $\forall i \in \mathcal{N}$ and $\forall t \geq 0$. The signal $\boldsymbol{x}_{\mathrm{D}}(t) \in \mathbb{R}^n$ is the overall state of Σ_{D} and the vector $\boldsymbol{x}_{\mathrm{D0}} \in \mathbb{R}^n$ is the overall initial state of Σ_{D}. For simplicity reasons the performance output $\boldsymbol{v}_{\mathrm{D}}(t) \in \mathbb{R}^{r_{\mathrm{v}}}$ is not modeled within Σ_{D} defined in (4.38).

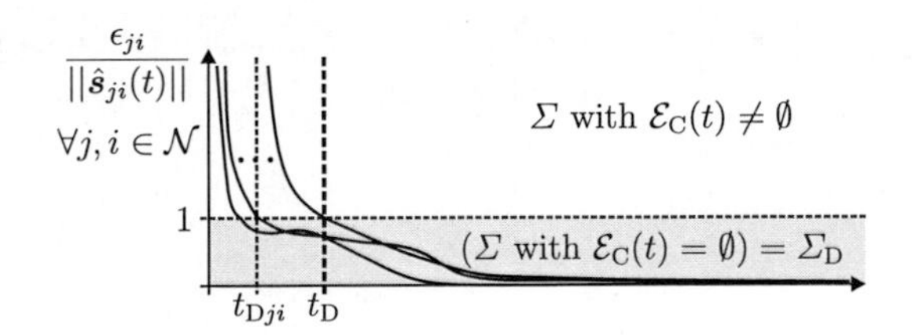

Figure 4.12: Idea for finding a condition to test the asymptotic stability of Σ defined in (4.20).

If the matrix $\bar{\boldsymbol{A}}_{\mathrm{P}} = \bar{\boldsymbol{A}} + \boldsymbol{ELC}_{\mathrm{z}}$ is a Hurwitz matrix, then the decentralized controlled system Σ_{D} defined in (4.38) is asymptotically stable (cf. Theorem 2.1).

Figure 4.12 illustrates the idea to get a condition for the asymptotic stability of Σ. A fact resulting from the definition of Σ_{D} in (4.38) is: If all communication links among the local control units C_i, $(\forall i \in \mathcal{N})$, stay deactivated after some time t_{D}, i.e., $\mathcal{E}_{\mathrm{C}}(t) = \emptyset$ for all $t \geq t_{\mathrm{D}}$, then Σ behaves like the decentralized controlled system Σ_{D} after the time t_{D}

$$(\Sigma \text{ with } \mathcal{E}_{\mathrm{C}}(t) = \emptyset, \forall t \geq t_{\mathrm{D}}) \Rightarrow (\Sigma = \Sigma_{\mathrm{D}}, \forall t \geq t_{\mathrm{D}}) \tag{4.39}$$

for $\boldsymbol{x}_{\mathrm{D}}(t_{\mathrm{D}}) = \boldsymbol{x}(t_{\mathrm{D}})$. Hence, if there exists a time $t_{\mathrm{D}} \geq 0$ after which all communication links among the local control units C_i stay deactivated, i.e., $\mathcal{E}_{\mathrm{C}}(t) = \emptyset$ for all $t \geq t_{\mathrm{D}}$, and Σ_{D} defined in (3.24) is asymptotically stable, then the self-organizing control system Σ defined in (4.20) is asymptotically stable:

$$(\Sigma_{\mathrm{D}} \text{ asympt. stable}) \wedge (\Sigma \text{ with } \mathcal{E}_{\mathrm{C}}(t) = \emptyset, \forall t \geq t_{\mathrm{D}}) \Rightarrow (\Sigma \text{ asympt. stable}). \tag{4.40}$$

According to (4.33), the overall communication among the local control units stays deactivated, i.e. $\mathcal{E}_{\mathrm{C}}(t) = \emptyset$, after the time t_{D} if the relation

$$(\exists t_{\mathrm{D}} \geq 0) \quad \|\hat{\boldsymbol{s}}_{ji}(t)\| < \epsilon_{ji}, \quad \forall t > t_{\mathrm{D}}, \quad \forall j \in \bar{\mathcal{F}}_{\mathrm{C}i}, \quad \forall i \in \mathcal{N} \tag{4.41}$$

holds true, where t_{D} is the maximum of all times $t_{\mathrm{D}ji}$, $(\forall j \in \bar{\mathcal{F}}_{\mathrm{C}i}, \forall i \in \mathcal{N})$ in (4.33).

In the following theorem the logical relation (4.40) is used to derive a condition for asymptotic stability of the self-organizing control system Σ while using the results from Lemma 4.5 on the deactivation of the communication links.

Theorem 4.3 (Testing asymptotic stability using local model information) *The self-organizing control system Σ in (4.20) with $\boldsymbol{d}(t) = \boldsymbol{0}$, ($\forall t \geq 0$) is asymptotically stable if the following condition is fulfilled:*

$$\epsilon_{ji} > m_{\mathrm{ss}ji} \sum_{l \in \bar{\mathcal{P}}_{\mathrm{C}i}} \epsilon_{il}, \quad \forall j \in \bar{\mathcal{F}}_{\mathrm{C}i}, \quad \forall i \in \mathcal{N}. \tag{4.42}$$

Proof. The proof is given in Appendix A.2. □

Theorem 4.3 gives a sufficient condition for the asymptotic stability of the self-organizing control system Σ defined in (4.20), where the bounds for the deactivation of each communication link $(i \to j)$ is checked individually by the conditions in (4.42). The proof of Theorem 4.3 shows that the fulfillment of the conditions in (4.42) implies that the decentralized control system Σ_{D} defined in (4.38) is asymptotically stable and, therefore, (4.40) holds true. The analysis scales linearly with the number N of subsystems since the conditions in (4.42) can be checked based on local model information only.

Reformulation as an uncertain system (second condition). For the second condition the self-organizing control system Σ defined in (4.20) is rewritten as an uncertain system $\bar{\Sigma}_{\mathrm{u}}$ according to (2.15)

$$\Sigma: \; \dot{\boldsymbol{x}}(t) = \left(\bar{\boldsymbol{A}} + \boldsymbol{E}\boldsymbol{L}_{\mathrm{u}}(t)\boldsymbol{C}_{\mathrm{z}}^{*}\right)\boldsymbol{x}(t), \quad \boldsymbol{x}(0) = \boldsymbol{x}_0 \tag{4.43}$$

with

$$\begin{aligned}
\boldsymbol{L}_{\mathrm{u}}(t) &:= \frac{1}{\sqrt{r_{\mathrm{z}}}\,\|\boldsymbol{L}\|}\boldsymbol{L}_{\mathrm{C}}(t) \in \mathbb{R}^{m_{\mathrm{s}} \times r_{\mathrm{z}}} \\
\boldsymbol{L}_{\mathrm{C}}(t) &:= \left(\boldsymbol{L}_{\mathrm{C}ij}(t)\right) \in \mathbb{R}^{m_{\mathrm{s}} \times r_{\mathrm{z}}} \\
\boldsymbol{L}_{\mathrm{C}ij}(t) &:= \begin{cases} \boldsymbol{O}_{m_{\mathrm{s}i} \times r_{\mathrm{z}j}} & \text{if } \|\hat{\boldsymbol{s}}_{ij}(t)\| \geq \epsilon_{ij} \\ \boldsymbol{L}_{ij} & \text{else,} \end{cases} \qquad \forall i, j \in \mathcal{N} \\
\boldsymbol{C}_{\mathrm{z}}^{*} &:= \sqrt{r_{\mathrm{z}}}\,\|\boldsymbol{L}\|\,\boldsymbol{C}_{\mathrm{z}},
\end{aligned}$$

where $\boldsymbol{L}_{ij}$ build the physical interconnection matrix $\boldsymbol{L}$ according to (3.3). The matrix $\boldsymbol{L}_{\mathrm{C}}(t)$ represents the physical coupling among the subsystems and their compensation by transmitting $\hat{\boldsymbol{s}}_{ji}(t)$ among the local control units. The definition of the local control units C_i in (4.14) shows that whenever $\|\hat{\boldsymbol{s}}_{ij}(t)\| \geq \epsilon_{ij}$ holds true, the physical coupling from Σ_j to Σ_i represented by $\boldsymbol{L}_{ij}$ has no effect on Σ_i since the coupling input $\boldsymbol{s}_{ij}(t)$ is compensated by sending $\hat{\boldsymbol{s}}_{ij}(t)$. Note that the relation

$$\boldsymbol{L}_{\mathrm{C}}(t)\boldsymbol{C}_{\mathrm{z}} = \underbrace{\frac{1}{\sqrt{r_{\mathrm{z}}}\,\|\boldsymbol{L}\|}\boldsymbol{L}_{\mathrm{C}}(t)}_{\boldsymbol{L}_{\mathrm{u}}(t)}\underbrace{\sqrt{r_{\mathrm{z}}}\,\|\boldsymbol{L}\|\,\boldsymbol{C}_{\mathrm{z}}}_{\boldsymbol{C}_{\mathrm{z}}^{*}}$$

holds true. In the following, Theorem 2.5 is reformulated for the self-organizing control system Σ.

Proposition 4.1 (Testing asymptotic stability using global model information) *Consider the self-organizing control system Σ defined in* (4.20) *which can be reformulated as the uncertain system presented in* (4.43). *Then Σ with $\boldsymbol{d}(t) = \boldsymbol{0}$, $(\forall t \geq 0)$, is asymptotically stable if the relation*

$$\left\|\boldsymbol{C}_{\mathrm{z}}^{*}(s\boldsymbol{I} - \bar{\boldsymbol{A}})^{-1}\boldsymbol{E}\right\|_{H_\infty} < 1 \tag{4.44}$$

holds true.

Proof. The relation

$$\|\boldsymbol{L}_{\mathrm{C}}(t)\| \leq \|\boldsymbol{L}\|$$

holds true since $\|\boldsymbol{L}\|_{\mathrm{F}} \leq \sqrt{r_{\mathrm{z}}}\,\|\boldsymbol{L}\|$ holds true according to (2.3), $\|\boldsymbol{L}_{\mathrm{C}}(t)\|_{\mathrm{F}} \leq \|\boldsymbol{L}\|_{\mathrm{F}}$ holds true for all $t \geq 0$ according to (2.2) and $\|\boldsymbol{L}_{\mathrm{C}}(t)\| \leq \|\boldsymbol{L}_{\mathrm{C}}(t)\|_{\mathrm{F}}$ holds true for all $t \geq 0$ according to (2.3). Hence, the relation $\|\boldsymbol{L}_{\mathrm{u}}(t)\| \leq 1$, $(\forall t \geq 0)$, claimed for $\boldsymbol{K}_{\mathrm{u}}(t)$ in (2.14) is fulfilled. The proof is completed by replacing the matrices $\boldsymbol{A}$, $\boldsymbol{B}$, $\boldsymbol{K}_{\mathrm{u}}(t)$ and $\boldsymbol{C}$ in (2.15) by $\bar{\boldsymbol{A}}$, $\boldsymbol{E}$, $\boldsymbol{L}_{\mathrm{u}}(t)$ and $\boldsymbol{C}_{\mathrm{z}}^{*}$ from (4.43), respectively, which leads to similar conditions as in Theorem 2.5. Condition 1 in Theorem 2.5 is always fulfilled since $\bar{\boldsymbol{A}}$ is a Hurwitz matrix according to Assumption 4.1. □

Proposition 4.1 presents a sufficient condition for the asymptotic stability of the self-organizing control system Σ, where the condition in (4.44) is some kind of small gain condition (cf. [82]). A drawback of Proposition 4.1 is that compared to Theorem 4.3 overall model information is necessary to check the condition in (4.44). Similar to Theorem 4.3, the condition in (4.44) implies the asymptotic stability of the decentralized control system Σ_{D} defined in (3.24) since $\bar{\boldsymbol{A}} = \mathrm{diag}(\bar{\boldsymbol{A}}_1 \ldots \bar{\boldsymbol{A}}_N)$ only includes the system matrices of the controlled subsystems $\bar{\Sigma}_i$.

Summary of the stability analysis. Figure 4.13 summarizes the results of stability analysis of the self-organizing control system Σ defined in (4.20). Theorem 4.3 and Proposition 4.1 hold also if there exists a time $t^* \geq 0$ after which the overall disturbance vanishes, i.e., $\boldsymbol{d}(t) = \boldsymbol{0}$ for all $t \geq t^*$. If the disturbances vanish, then Theorem 4.3 checks the asymptotic stability of

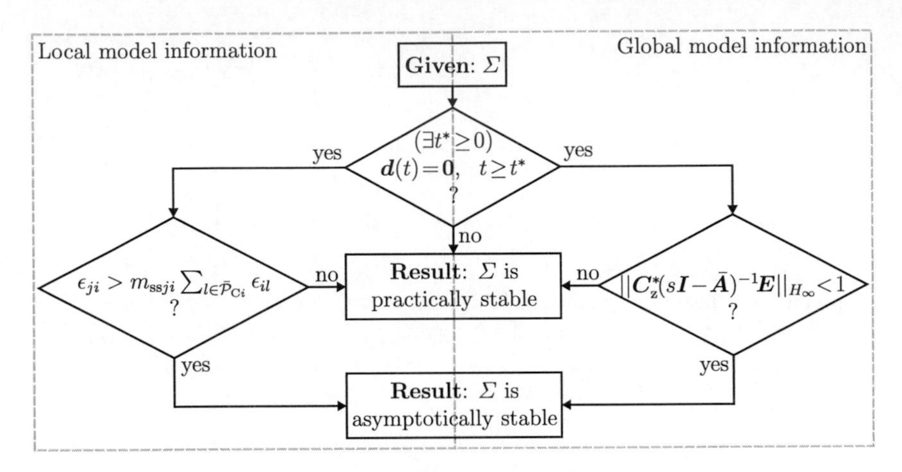

Figure 4.13: Structure of the stability analysis for Σ defined in (4.20).

Σ with local model information and Proposition 4.1 checks the asymptotic stability of Σ with global model information (Fig. 4.13). If the disturbances are not vanishing, Σ is still practically stable (cf. Theorem 4.2). Under the proposed assumptions the main result of the stability analysis can be summarized as follows:

> The self-organizing control system Σ defined in (4.20) is always practically stable.

In the following, the results on the stability analysis in this section are applied to the self-organizing controlled symmetrically interconnected subsystem from Example 4.1.

Example 4.1 (cont.) *Situation-dependent compensation of physical coupling inputs*

Figure 4.14 shows the phase portrait of the states $x_1(t)$ and $x_2(t)$ of the symmetrically interconnected subsystems P defined in (3.32) with $x_{01} = 1$ and $x_{02} = -1$ controlled by the local control units C_i defined in (4.14) with the parameters listed in Table 4.3 for the disturbances $d_i(t)$ depicted in the upper part of Fig. 4.4. Two different physical interconnection matrices L are concerned:

- **Scenario 1**: $L_{21} = L_{12} = 1$ (cf. Fig. 4.14 left)
- **Scenario 2**: $L_{21} = L_{12} = 3$ (cf. Fig. 4.14 right).

In Scenario 1 the decentralized control system Σ_{D} is asymptotically stable since the matrix $\bar{A}_{\mathrm{P}}$ is a

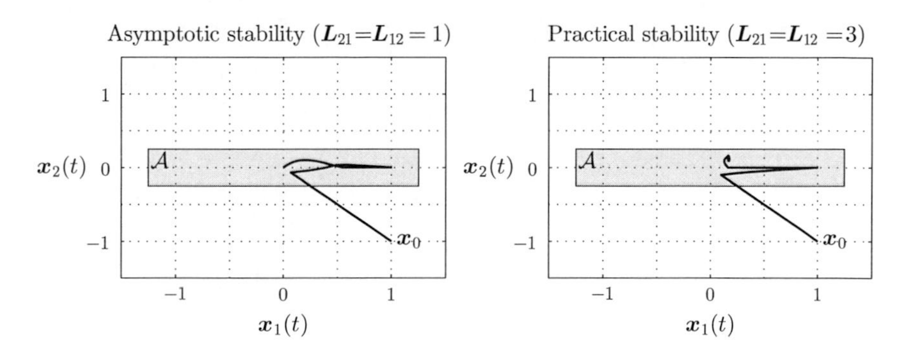

Figure 4.14: Phase portrait of the states $x_1(t)$ and $x_2(t)$ of the interconnected subsystems defined in (3.32) with a situation-dependent compensation of physical coupling inputs for two different physical interconnections.

Hurwitz matrix according to (4.8). In Scenario 2 the matrix $\bar{A}_P$ is not a Hurwitz matrix since

$$\bar{A}_P = \underbrace{\begin{pmatrix}1 & 0\\0 & 1\end{pmatrix}}_{A} + \underbrace{\begin{pmatrix}1 & 0\\0 & 1\end{pmatrix}}_{E}\underbrace{\begin{pmatrix}0 & 3\\3 & 0\end{pmatrix}}_{L}\underbrace{\begin{pmatrix}1 & 0\\0 & 1\end{pmatrix}}_{C_z} - \underbrace{\begin{pmatrix}1 & 0\\0 & 1\end{pmatrix}}_{B}\begin{pmatrix}\overbrace{3}^{K_{C1}} & 0\\0 & \underbrace{3}_{K_{C2}}\end{pmatrix} = \begin{pmatrix}-2 & 3\\3 & -2\end{pmatrix}.$$

From Theorem 4.3 and Proposition 4.1 it follows for the two scenarios

- **Scenario 1**: with $\|L\| = 1$

$$\text{Theorem 4.3} \rightarrow \begin{cases}\epsilon_{21} = 0.5 > m_{ss21}\epsilon_{12} = 0.25 \checkmark\\ \epsilon_{12} = 0.5 > m_{ss12}\epsilon_{21} = 0.25 \checkmark\end{cases}$$

$$\text{Proposition 4.1} \rightarrow \left\|C_z^*(sI - \bar{A})^{-1}E\right\|_{H_\infty} = 0.7071 < 1 \checkmark$$

- **Scenario 2**: with $\|L\| = 3$ and $m_{ss21} = m_{ss12} = 1.5$

$$\text{Theorem 4.3} \rightarrow \begin{cases}\epsilon_{21} = 0.5 \not> m_{ss21}\epsilon_{12} = 0.75 \lightning\\ \epsilon_{12} = 0.5 \not> m_{ss12}\epsilon_{21} = 0.75 \lightning\end{cases}$$

$$\text{Proposition 4.1} \rightarrow \left\|C_z^*(sI - \bar{A})^{-1}E\right\|_{H_\infty} = 2.1214 \not< 1 \lightning$$

For Scenario 1 both conditions are fulfilled, hence, Σ is asymptotically stable (cf. Fig. 4.14 left). For Scenario 2 both conditions are not fulfilled. Hence, Σ might not be asymptotically stable (cf. Fig. 4.14 right) but is still practically stable with respect to the set

$$\mathcal{A} = \left\{ \begin{pmatrix}x_1\\x_2\end{pmatrix} \in \mathbb{R}^2 \;\middle|\; \begin{pmatrix}x_1\\x_2\end{pmatrix} \leq \begin{pmatrix}\bar{x}_1 = 1.25\\ \bar{x}_2 = 0.25\end{pmatrix} \right\}.$$

□

4.3.3 Behavior compared to the corresponding centralized control system

In this section the behavior of the self-organizing control system Σ and the corresponding centralized control system Σ_{C} are compared. Therefore, the section is divided into three parts:

1. Considering the general model of the centralized control system Σ_{C} defined in (3.25), the specific model of Σ_{C} for a situation-dependent compensation of physical coupling inputs is derived (cf. Section 3.3.3).

2. A difference subsystem $\bar{\Sigma}_{\Delta i}$ is deduced which describes the difference between the behavior of the extended subsystem $\bar{\Sigma}_i^*$ and behavior of the controlled subsystems $\bar{\Sigma}_{\mathrm{C}i}$ within Σ_{C}.

3. An analysis of the behavior of the difference subsystem $\bar{\Sigma}_{\Delta i}$ shows that there exists an upper bound on the difference behavior of Σ compared to Σ_{C} which depends on the switching thresholds (Theorem 4.4).

Centralized control system Σ_{C}. According to (3.25) and (4.20), the centralized control systems Σ_{C} presented in Fig. 3.6 with a compensation of physical coupling inputs is defined by

$$\Sigma_{\mathrm{C}} : \begin{cases} \dot{\boldsymbol{x}}_{\mathrm{C}}(t) = \bar{\boldsymbol{A}}\boldsymbol{x}_{\mathrm{C}}(t) + \boldsymbol{G}\boldsymbol{d}(t), & \boldsymbol{x}_{\mathrm{C}}(0) = \boldsymbol{x}_{\mathrm{C}0} \\ \boldsymbol{v}_{\mathrm{C}}(t) = \boldsymbol{C}_{\mathrm{v}}\boldsymbol{x}_{\mathrm{C}}(t) \end{cases} \tag{4.45}$$

with the overall state $\boldsymbol{x}_{\mathrm{C}}(t) \in \mathbb{R}^n$ of Σ_{C}, the overall initial state $\boldsymbol{x}_{\mathrm{C}0} \in \mathbb{R}^n$ of Σ_{C} and the overall performance output $\boldsymbol{v}_{\mathrm{C}}(t) \in \mathbb{R}^{r_{\mathrm{v}}}$ of Σ_{C}. Consider the modeling of the self-organizing control system Σ as an uncertain system in (4.43), then Σ_{C} defined in (4.45) results from $\boldsymbol{L}_{\mathrm{u}}(t) = \boldsymbol{O}$ for all $t \geq 0$, which means that the physical coupling is permanently compensated. Hence, the matrices $\bar{\boldsymbol{A}}$, $\boldsymbol{G}$ and $\boldsymbol{C}_{\mathrm{v}}$ are all block-diagonal, which leads completely decoupled subsystems. With this, the controlled subsystems $\Sigma_{\mathrm{C}i}$, $(\forall i \in \mathcal{N})$, of the centralized control system Σ_{C} result to

$$\bar{\Sigma}_{\mathrm{C}i} : \begin{cases} \dot{\boldsymbol{x}}_{\mathrm{C}i}(t) = \bar{\boldsymbol{A}}_i\boldsymbol{x}_{\mathrm{C}i}(t) + \boldsymbol{G}_i\boldsymbol{d}_i(t), & \boldsymbol{x}_{\mathrm{C}i}(0) = \boldsymbol{x}_{\mathrm{C}0i} \\ \boldsymbol{v}_{\mathrm{C}i}(t) = \boldsymbol{C}_{\mathrm{v}i}\boldsymbol{x}_{\mathrm{C}i}(t), \end{cases} \tag{4.46}$$

with the state $\boldsymbol{x}_{\mathrm{C}i}(t) \in \mathbb{R}^{n_i}$ of $\bar{\Sigma}_{\mathrm{C}i}$, the initial state $\boldsymbol{x}_{\mathrm{C}0i} \in \mathbb{R}^{n_i}$ of $\bar{\Sigma}_{\mathrm{C}i}$ and the performance output $\boldsymbol{v}_{\mathrm{C}i}(t) \in \mathbb{R}^{r_{\mathrm{v}i}}$ of $\bar{\Sigma}_{\mathrm{C}i}$, building

$$\boldsymbol{x}_{\mathrm{C}}(t) = \begin{pmatrix} \boldsymbol{x}_{\mathrm{C}1}(t) \\ \vdots \\ \boldsymbol{x}_{\mathrm{C}N}(t) \end{pmatrix}, \quad \boldsymbol{x}_{\mathrm{C}0} = \begin{pmatrix} \boldsymbol{x}_{\mathrm{C}01} \\ \vdots \\ \boldsymbol{x}_{\mathrm{C}0N} \end{pmatrix}, \quad \boldsymbol{v}_{\mathrm{C}}(t) = \begin{pmatrix} \boldsymbol{v}_{\mathrm{C}1}(t) \\ \vdots \\ \boldsymbol{v}_{\mathrm{C}N}(t) \end{pmatrix}. \tag{4.47}$$

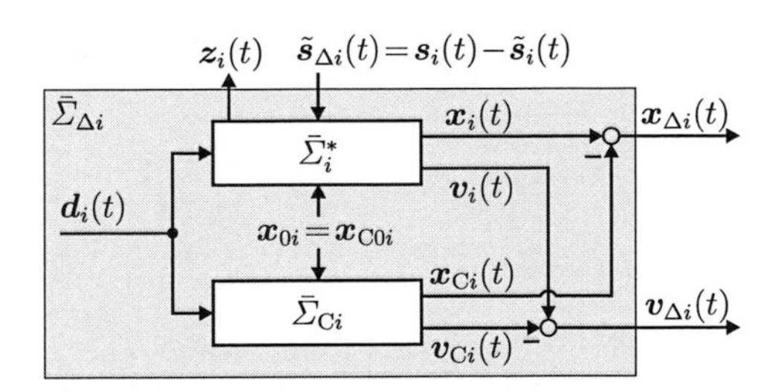

Figure 4.15: Difference subsystem $\bar{\Sigma}_{\Delta i}$ defined in (4.48).

Difference subsystem $\bar{\Sigma}_{\Delta i}$. The difference between the behavior of the extended subsystem $\bar{\Sigma}_i^*$ defined in (4.18) and the behavior of the controlled subsystem $\bar{\Sigma}_{Ci}$ defined in (4.46) belonging to the centralized control system Σ_C is described by the *difference subsystem*

$$\bar{\Sigma}_{\Delta i} : \begin{cases} \dot{x}_{\Delta i}(t) = \bar{A}_i x_{\Delta i}(t) + E_i \tilde{s}_{\Delta i}(t), \quad x_{\Delta i}(0) = x_{\Delta 0i} \\ v_{\Delta i}(t) = C_{vi} x_{\Delta i}(t) \end{cases} \tag{4.48}$$

with the difference state $x_{\Delta i}(t) = x_i(t) - x_{Ci}(t)$, the difference initial state $x_{\Delta 0i} = x_{0i} - x_{C0i}$, and the difference performance output $v_{\Delta i}(t) = v_i(t) - v_{Ci}(t)$ (cf. Fig. 4.15). Note that the difference dynamics in $\bar{\Sigma}_{\Delta i}$ is independent from the disturbance $d_i(t)$ affecting the subsystems since the impact of $d_i(t)$ is identical in $\bar{\Sigma}_i^*$ and $\bar{\Sigma}_{Ci}$ and is canceled out throughout the construction of $\bar{\Sigma}_{\Delta i}$.

Boundedness of the difference behavior. The following theorem gives an upper bound on the difference state $x_{\Delta i}(t)$ and the difference performance output $v_{\Delta i}(t)$, where for the analysis the initial states of $\bar{\Sigma}_i$ and $\bar{\Sigma}_{Ci}$ are set to be identical, i.e., $x_{C0i} = x_{0i}$. Hence, $x_{\Delta 0i} = 0$ holds true.

Theorem 4.4 (Individual boundedness of the local difference behavior) *For $x_{C0i} = x_{0i}$ the state $x_{\Delta i}(t)$ and the performance output $v_{\Delta i}(t)$ of the difference subsystem $\bar{\Sigma}_{\Delta i}$ defined in (4.48) are bounded for arbitrary disturbances as follows:*

$$\max_{t \geq 0} \|x_{\Delta i}(t)\| \leq \bar{x}_{\Delta i} := m_{xsi} \sum_{j \in \bar{\mathcal{P}}_{Ci}} \epsilon_{ij}, \quad \forall i \in \mathcal{N}$$

$$\max_{t \geq 0} \|v_{\Delta i}(t)\| \leq \bar{v}_{\Delta i} := m_{vsi} \sum_{j \in \bar{\mathcal{P}}_{Ci}} \epsilon_{ij}, \quad \forall i \in \mathcal{N}$$

with m_{xsi} defined in Lemma 4.2 and m_{vsi} defined in Lemma 4.3.

Proof. From $\bar{\Sigma}_{\Delta i}$ in (4.48) it follows that

$$\boldsymbol{x}_{\Delta i}(t) = \mathrm{e}^{\bar{\boldsymbol{A}}_i t}\boldsymbol{x}_{\Delta 0i} + \int_0^t \mathrm{e}^{\bar{\boldsymbol{A}}_i(t-\tau)}\boldsymbol{E}_i\tilde{\boldsymbol{s}}_{\Delta i}(\tau)\mathrm{d}\tau, \quad \forall i\in\mathcal{N}, \quad \forall t\geq 0. \tag{4.49}$$

From (4.49), Lemma 4.1 and $\boldsymbol{x}_{\Delta 0i} = \boldsymbol{0}$ according to $\boldsymbol{x}_{\mathrm{C}0i} = \boldsymbol{x}_{0i}$, the following holds true

$$\begin{aligned}\max_{t\geq 0}\|\boldsymbol{x}_{\Delta i}(t)\| &\leq \int_0^\infty \left\|\mathrm{e}^{\bar{\boldsymbol{A}}_i\tau}\boldsymbol{E}_i\right\|\mathrm{d}\tau \max_{t\geq 0}\|\tilde{\boldsymbol{s}}_{\Delta i}(t)\|, \quad \forall i\in\mathcal{N}\\ &\leq m_{\mathrm{xs}i}\sum\nolimits_{j\in\bar{\mathcal{P}}_{\mathrm{C}i}}\epsilon_{ij}, \quad \forall i\in\mathcal{N}.\end{aligned}$$

Due to Assumption 4.1, the matrices $\bar{\boldsymbol{A}}_i$, $(\forall i\in\mathcal{N})$ are Hurwitz and, therefore, the scalars $m_{\mathrm{xs}i}$ are finite. The proof of the boundedness of $\boldsymbol{v}_{\Delta i}(t)$ is similar to the one for $\boldsymbol{x}_{\Delta i}(t)$, where

$$\boldsymbol{v}_{\Delta i}(t) = \boldsymbol{C}_{\mathrm{v}i}\mathrm{e}^{\bar{\boldsymbol{A}}_i t}\boldsymbol{x}_{\Delta 0i} + \int_0^t \boldsymbol{C}_{\mathrm{v}i}\mathrm{e}^{\bar{\boldsymbol{A}}_i\bar{\tau}}\boldsymbol{E}_i\tilde{\boldsymbol{s}}_{\Delta i}(\tau)\mathrm{d}\tau, \quad \forall i\in\mathcal{N}, \quad \forall t\geq 0. \tag{4.50}$$

According to $\boldsymbol{x}_{\Delta 0i} = \boldsymbol{0}$ and Lemma 4.1, the relation in (4.50) leads to

$$\max_{t\geq 0}\|\boldsymbol{v}_{\Delta i}(t)\| \leq \int_0^\infty \left\|\boldsymbol{C}_{\mathrm{v}i}\mathrm{e}^{\bar{\boldsymbol{A}}_i\tau}\boldsymbol{E}_i\right\|\mathrm{d}\tau \max_{t\geq 0}\|\tilde{\boldsymbol{s}}_{\Delta i}(t)\|, \quad \forall i\in\mathcal{N},$$

which completes the proof. □

Theorem 4.4 shows that the difference between the behavior of the self-organizing control system Σ and the behavior of the corresponding centralized control system Σ_{C} only depends on the switching thresholds ϵ_{ij}. If, e.g., all thresholds are set to be zero the behavior of Σ and the behavior of Σ_{C} would be identical, i.e., $\boldsymbol{x}_{\Delta i}(t) = \boldsymbol{0}$ for all $t\geq 0$. Furthermore, Theorem 4.4 shows that the difference behavior can be analyzed for each subsystem individually. In summary, the following result shall be highlighted:

> The difference between the behavior of the self-organizing control system Σ and the behavior of the corresponding centralized control system Σ_{C} can be adjusted with arbitrary precision for each subsystem individually by the choice of the switching thresholds.

In Section 4.4.1 the analysis results in Theorem 4.4 are used to develop a method for choosing the switching thresholds ϵ_{ij} to fulfill Control aim 3.4.

Example 4.1 (cont.) *Situation-dependent compensation of physical coupling inputs*

Figure 4.16 shows the behavior of the symmetrically interconnected subsystems P defined in (3.32) with $\boldsymbol{x}_{01} = 1$ and $\boldsymbol{x}_{02} = 0$ controlled by the local control units C_i defined in (4.14) with the parameters

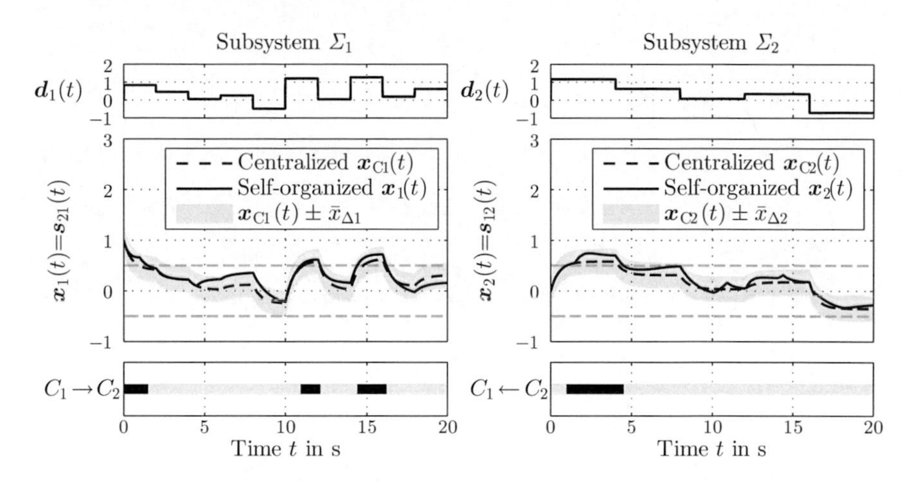

Figure 4.16: Analysis of the difference between the behavior of the symmetrically interconnected subsystems defined in (3.32) controlled by the self-organizing controller C defined in 4.15 and the behavior of the subsystems controlled by the corresponding centralized controller C_C.

listed in Table 4.3 (solid line) compared to the behavior of the subsystems controlled by the local control units C_i with permanent communication (centralized controller; dashed line). The gray area depicts the bound $\bar{x}_{\Delta 1} = \bar{x}_{\Delta 2} = 0.25$ on the state $\boldsymbol{x}_{\Delta i}$, $(\forall i \in \{1, 2\})$, of the difference subsystem $\bar{\Sigma}_{\Delta_i}$ according to Theorem 4.4. The states $\boldsymbol{x}_i(t)$ never leave these gray area, no matter what disturbances affect the subsystems. □

4.3.4 Effects of disturbances on the behavior of the self-organizing control system

This section investigates the behavior of the self-organizing control system Σ defined in (4.20) in case of disturbances affecting the subsystems. Therefore, the subsystems can be divided into two groups:

- The *disturbed subsystems* Σ_j , $(\forall j \in \mathcal{D})$, which are affected by a disturbance $\boldsymbol{d}_j(t)$, where

$$\mathcal{D} = \{j \in \mathcal{N} \mid \exists t \geq 0 \text{ s.t. } \boldsymbol{d}_j(t) \neq \boldsymbol{0}\} \tag{4.51}$$

  is the *set of disturbed subsystems.*

- The *undisturbed subsystems* Σ_p , $(\forall p \in \mathcal{U})$, which are not disturbed but might be influenced over the physical coupling. Recall that $\mathcal{U}$ is the set of undisturbed subsystems defined in (3.27).

Three main questions arise:

1. How does a disturbance affect the behavior of the undisturbed subsystems (Fig. 4.17 left)?
2. Which communication is necessary, to attenuate the effect of the disturbances within the self-organizing control system Σ (Fig. 4.17 right)?
3. Will the communication be deactivated when the disturbances vanish?

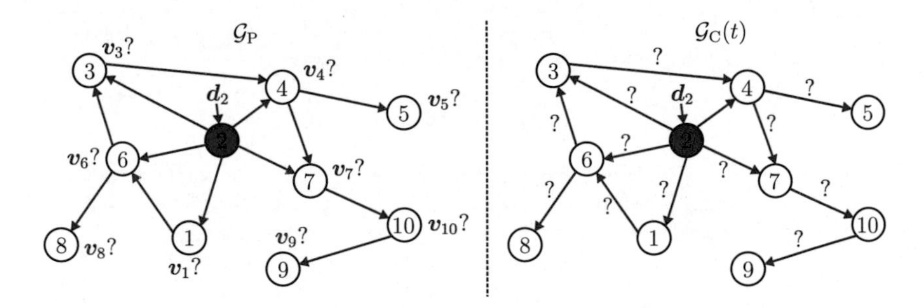

Figure 4.17: Disturbance propagation (left) and communication activation (right) in case of a disturbance for a disturbance affecting subsystem Σ_2.

Disturbance propagation. To answer the first question consider the analysis of the controlled subsystems $\bar{\Sigma}_i$ in Lemma 4.3, which gives bounds on the performance outputs $\boldsymbol{v}_i(t)$. With this, the following theorem gives bounds on the performance outputs $\boldsymbol{v}_i(t)$ of the undisturbed subsystems Σ_i, $(\forall i \in \mathcal{U})$, in the case that the subsystems Σ_j, $(\forall j \in \mathcal{D})$, are disturbed.

Theorem 4.5 (Disturbance propagation by compensating physical coupling inputs) *Consider the self-organizing control system Σ defined in (4.20) with $\boldsymbol{x}_0 = \boldsymbol{0}$ in which the subsystems Σ_j, $(\forall j \in \mathcal{D})$, are disturbed. Then the performance outputs $\boldsymbol{v}_i(t)$ of the undisturbed subsystems Σ_i, $(\forall i \in \mathcal{U})$, are upper bounded by the maximal performance output $\bar{v}_i$ with*

$$\max_{t \geq 0} \|\boldsymbol{v}_i(t)\| \leq \bar{v}_i := m_{\mathrm{vs}i} \sum_{l \in \bar{\mathcal{P}}_{\mathrm{C}i}} \epsilon_{il}, \quad \forall i \in \mathcal{U}. \tag{4.52}$$

Proof. According to (4.26) in Lemma 4.3, $\boldsymbol{x}_0 = \boldsymbol{0}$ and $\bar{d}_i = 0$, $(\forall i \in \mathcal{U})$, the relation in (4.52) holds true. □

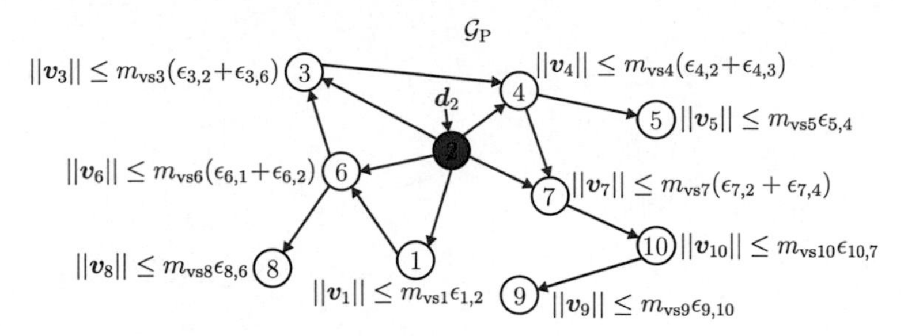

Figure 4.18: Bounds on the performance outputs of the subsystems in case of a disturbance affecting subsystem Σ_2 according to Theorem 4.5.

Figure 4.18 illustrates Theorem 4.5. The bounds in (4.52) give a measure on how strong a disturbance propagates through the closed-loop system. Theorem 4.5 shows that $\max_{t\geq 0}\|\boldsymbol{v}_i(t)\|$ of the undisturbed subsystems Σ_i , $(\forall i \in \mathcal{U})$, depend on the choice of the switching thresholds ϵ_{il} only. Hence, the disturbance propagation can be arbitrarily adjusted by the choice of the switching thresholds.

Communication activation. To answer the second question consider the analysis of the communication activation in Lemma 4.4. The following proposition defines the maximal communication graph $\bar{\mathcal{G}}_{\mathrm{CD}}(\mathcal{D})$ which can occur in case that the subsystems Σ_j, $(\forall j \in \mathcal{D})$, are disturbed.

Proposition 4.2 (Communication for bounding the disturbance propagation) *Consider the self-organizing control system Σ defined in (4.20) with $\boldsymbol{x}_0 = \boldsymbol{0}$. If the subsystems Σ_j, $(\forall j \in \mathcal{D})$, are disturbed, there might be communication from the local control unit C_i, $(i \in \mathcal{N})$ to the local control unit C_h, $(h \in \bar{\mathcal{F}}_{\mathrm{CD}i}(\mathcal{D}))$, with*

$$\bar{\mathcal{F}}_{\mathrm{CD}i}(\mathcal{D}) = \{h \in \bar{\mathcal{F}}_{\mathrm{C}i} \mid \bar{s}_{\mathrm{D}hi}(\mathcal{D}) \geq \epsilon_{hi}\}, \qquad \forall i \in \mathcal{N} \tag{4.53}$$

and

$$\bar{s}_{\mathrm{D}hi}(\mathcal{D}) := \begin{cases} m_{\mathrm{ss}hi} \sum_{l \in \bar{\mathcal{P}}_{\mathrm{C}i}} \epsilon_{il} & \text{if } i \in \mathcal{U} = \mathcal{N}\backslash\mathcal{D} \\ m_{\mathrm{sd}hi}\bar{d}_i + m_{\mathrm{ss}hi} \sum_{l \in \bar{\mathcal{P}}_{\mathrm{C}i}} \epsilon_{il} & \text{if } i \in \mathcal{D}, \end{cases} \qquad \forall h \in \bar{\mathcal{F}}_{\mathrm{C}i}, \quad \forall i \in \mathcal{N}$$

but not among any other local control units. With this, the set $\mathcal{E}_\mathrm{C}(t)$ of communication links within the communication graph $\mathcal{G}_\mathrm{C}(t) = (\mathcal{V}_\mathrm{C}, \mathcal{E}_\mathrm{C}(t))$ of Σ is bounded by the set $\bar{\mathcal{E}}_{\mathrm{CD}}(\mathcal{D})$ of maximal communication links

$$\emptyset \subseteq \mathcal{E}_\mathrm{C}(t) \subseteq \bar{\mathcal{E}}_{\mathrm{CD}}(\mathcal{D}) = \{(i \rightarrow j) \mid j \in \bar{\mathcal{F}}_{\mathrm{CD}i}(\mathcal{D}), i \in \mathcal{N}\}, \quad \forall t \geq 0. \tag{4.54}$$

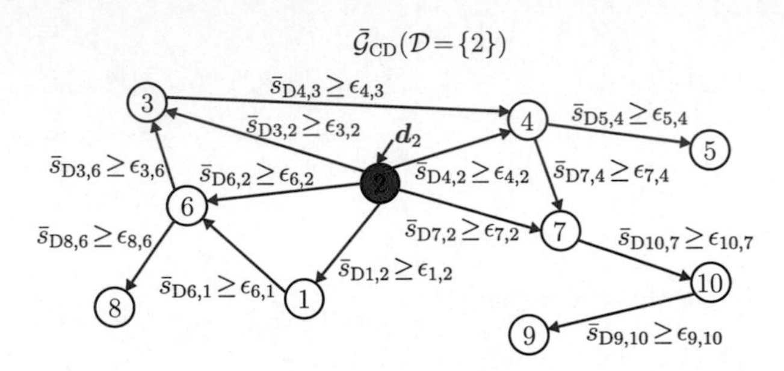

Figure 4.19: Illustration of the determination of the maximal communication graph $\bar{\mathcal{G}}_{\mathrm{CD}}(\mathcal{D})$ for the disturbance situation $\mathcal{D} = \{2\}$ according to Proposition 4.2.

The set $\bar{\mathcal{E}}_{\mathrm{CD}}(\mathcal{D})$ defines the resulting maximal communication graph

$$\bar{\mathcal{G}}_{\mathrm{CD}}(\mathcal{D}) = (\mathcal{V}_{\mathrm{C}}, \bar{\mathcal{E}}_{\mathrm{CD}}(\mathcal{D})). \tag{4.55}$$

Proof. The proof is given in Appendix A.3. □

The graph $\bar{\mathcal{G}}_{\mathrm{CD}}(\mathcal{D})$ in Proposition 4.2 defines the maximal possible communication among the local control units that can occur if the subsystems Σ_j, $(\forall j \in \mathcal{D})$, are disturbed. Figure 4.19 illustrates Proposition 4.2 for the set $\mathcal{D} = \{2\}$. According to the definition of $\bar{\mathcal{F}}_{\mathrm{CD}i}(\mathcal{D})$, $(\forall i \in \mathcal{N})$, in (4.53), the edges $(i \rightarrow h)$ only exists if the relation

$$\bar{s}_{\mathrm{D}hi}(\mathcal{D}) \geq \epsilon_{hi}, \quad h \in \bar{\mathcal{F}}_{\mathrm{CD}i}, \quad i \in \mathcal{N}$$

holds true. Therefore, $\bar{\mathcal{G}}_{\mathrm{CD}}(\mathcal{D})$ gives a measure on how a disturbance influences the communication among the local control units and defines what communication is necessary to attenuate the effect of a disturbance on the subsystems. Hence, Proposition 4.2 shows that the occurring communication depends on the acting disturbances.

Deactivation of the commmuniation for vanishing disturbances. A condition for the deactivation of certain communication links for disturbed subsystems has been given in Lemma 4.5 in Section 4.3.1. Furthermore, the investigation of the asymptotic stability of the self-organizing control system Σ in Section 4.3.2 showed that Σ is asymptotically stable if the decentralized control system Σ_{D} is asymptotically stable and the communication is eventually permanently deactivated for an undisturbed Σ. The following corollary uses these results to derive two conditions for which the communication will be deactivated for vanishing disturbances as well.

Corollary 4.1 (Communication deactivation for vanishing disturbances while compensating physical coupling inputs) *Consider the self-organizing control system Σ defined in* (4.20) *with $\boldsymbol{x}_0 = \mathbf{0}$ and $\boldsymbol{d}(t) = \mathbf{0}$ for $t \geq t^* \geq 0$, ($t^* \in \mathbb{R}_+$). The communication among all local control units C_i stays deactivated after a time $t_\mathrm{D} \in \mathbb{R}_+$, i.e.,*

$$\mathcal{E}_\mathrm{C}(t) = \emptyset, \quad t \geq t_\mathrm{D} \geq 0$$

if Σ is asymptotically stable according to Definition 2.2 which is implied by (4.42) *(cf. Theorem 4.3).*

Proof. If the self-organizing control system Σ defined in (4.20) is asymptotically stable, there always exists a deactivation time $t_{\mathrm{D}ji}$ that deactivates the communication link $(i \rightarrow j) \in \mathcal{E}_\mathrm{C}(t)$ according to (4.33), since $\lim_{t\rightarrow\infty} |\hat{\boldsymbol{s}}_{ji}(t)| = \mathbf{0}$ holds true . □

Corollary 4.1 derives that for an asymptotically stable self-organizing control system Σ the communication among the local control units is only necessary if a disturbance is present.

Summary. With the analysis of the effects of disturbances on the self-organizing control system defined in (4.20) in Theorem 4.5, Proposition 4.2 and Corollary 4.1 the three initial questions can be answered as follows:

1. The situation-dependent compensation of the physical coupling inputs bounds the influence of the disturbances on the undisturbed subsystems, where the resulting impact can be made arbitrarily small by the choice of the switching thresholds (Theorem 4.5).
2. The maximal possible communication graph can be predicted for given maximal values on the disturbances by analyzing each communication link individually (Proposition 4.2).
3. The communication among the local control units will be deactivated for vanishing disturbances if the closed-loop system Σ is asymptotically stable (Corollary 4.1).

Furthermore, the overall analysis scales linearly with the number of subsystems since local model information is used in Theorem 4.5 and Proposition 4.2.

Proposition 4.2 is used in Section 4.4.2 to choose the communication structure among the local control units. Since there are only two subsystems in Example 4.1 the results in Theorem 4.5 and Proposition 4.2 are illustrated in more detail at the example of the water supply system with 23 subsystems in Section 4.5.

4.4 Design of the local control units

In this section methods for designing the switching thresholds and the communication structure among the local control units are presented. First, the switching thresholds are chosen such that the resulting self-organizing control system Σ guarantees a desired disturbance attenuation as required in Control aim 3.3 (Section 4.4.1). Second, based on the designed switching thresholds a method for reducing the maximal communication graph $\bar{\mathcal{G}}_{\mathrm{C}}$ is presented which cancels out unnecessary communication links (Section 4.4.2). Furthermore, the overall design of the local control units is summarized in Section 4.4.3.

4.4.1 Design of the switching thresholds

For the following design procedure the switching thresholds ϵ_{ij}, $(j \in \bar{\mathcal{P}}_{\mathrm{C}i}, \forall i \in \mathcal{N})$ are set to be identical with the threshold $\bar{\epsilon}_i$, i.e.,

$$\epsilon_{ij} = \bar{\epsilon}_i, \quad \forall j \in \bar{\mathcal{P}}_{\mathrm{C}i}, \quad \forall i \in \mathcal{N}. \tag{4.56}$$

With this, for the design procedure each coupling input $\boldsymbol{s}_{ij}(t)$ on Σ_i is weighted identically

$$\sum_{j \in \bar{\mathcal{P}}_{\mathrm{C}i}} \epsilon_{ij} = \left|\bar{\mathcal{P}}_{\mathrm{C}i}\right| \bar{\epsilon}_i.$$

Note that the thresholds could also be weighted differently, e.g., to penalize the usage of specific communication links by using a larger weight compared to the other thresholds.

The following theorem uses the results in Lemma 4.3 and Theorem 4.5 to give a method for designing the thresholds $\bar{\epsilon}_i$, $(\forall i \in \mathcal{N})$, and, therefore, implicitly design the switching threshold ϵ_{ij}, $(j \in \bar{\mathcal{P}}_{\mathrm{C}i}, \forall i \in \mathcal{N})$.

Theorem 4.6 (Choose switching thresholds for compensating physical coupling inputs) *Consider the self-organizing control system Σ defined in (4.20) with $\boldsymbol{x}_0 = \boldsymbol{0}$. The performance output $\boldsymbol{v}_i(t)$, $(\forall i \in \mathcal{U})$, of the undisturbed subsystems Σ_i is bounded by the desired maximal performance output $\bar{v}_i^*$ as required in Control aim 3.3 if the switching thresholds ϵ_{ij} are chosen such that the relation*

$$\bar{\epsilon}_i \leq \frac{\bar{v}_i^*}{\left|\bar{\mathcal{P}}_{\mathrm{C}i}\right| \cdot m_{\mathrm{vs}i}}, \quad \forall i \in \mathcal{N} \tag{4.57}$$

holds true.

Proof. According to Theorem 4.5, $\boldsymbol{x}_0 = \boldsymbol{0}$ and (4.56), the relations

$$m_{\mathrm{vsi}} \cdot \bar{\epsilon}_i \cdot |\bar{\mathcal{P}}_{\mathrm{C}i}| \leq \bar{v}_i^*, \quad \forall i \in \mathcal{N} \tag{4.58}$$

have to be fulfilled to satisfy Control aim 3.3. Solving (4.58) for $\bar{\epsilon}_i$ leads to (4.57). □

Theorem 4.6 describes an appropriate choice of the switching thresholds ϵ_{ji}, $(j \in \bar{\mathcal{P}}_{\mathrm{C}i}, \forall i \in \mathcal{N})$, for each local control unit C_i to guarantee the desired maximal performance output $\bar{v}_i^*$ for disturbance attenuation according to (3.26). For the choice of the switching thresholds in Theorem 4.6 only local model information of the isolated subsystems Σ_i is necessary to apply (4.57). Note that the switching thresholds ϵ_{ij}, $(j \in \bar{\mathcal{P}}_{\mathrm{C}i}, \forall i \in \mathcal{N})$, designed by model information of Σ_i have to be used by all local control units C_j that can send information to C_i. Therefore, if the thresholds are designed locally at the local control units, they have to be exchanged initially.

Remark 4.2 *The switching thresholds can also be designed to guarantee the asymptotic stability of the self-organizing control system Σ by using linear programming [32, 37]. For this, the inequalities in* (4.42) *have to be summarized and a cost function to maximize the switching thresholds has to be defined. Furthermore, also the condition* (4.57) *in Theorem 4.6 can be integrated into the overall optimization problem to guarantee a desired disturbance attenuation. Compared to Theorem 4.6 the linear program would use global model information.*

4.4.2 Reduction of the communication structure

In this section the communication among the local control units is designed such that unnecessary communication links in $\bar{\mathcal{E}}_{\mathrm{C}}$ are removed to reduce the communication effort among the local control units. The set $\bar{\mathcal{E}}_{\mathrm{C}}$ of maximal communication links results from the structure of the central feedback gain matrix according to (2.21) and (4.9). An unnecessary communication link $(i \to j)$ is a link that will never be activated while controlling the plant P since the corresponding coupling signal $\hat{\boldsymbol{s}}_{ji}(t)$ determined in C_i never exceeds the threshold ϵ_{ji} (cf. (4.13b)).

Recalling the analysis of the disturbance propagation in Section 4.3.4, Proposition 4.2 already defines a communication structure $\mathcal{G}_{\mathrm{CD}}(\mathcal{D}) = (\mathcal{V}_{\mathrm{C}}, \mathcal{E}_{\mathrm{CD}}(\mathcal{D}))$ that will be used during the control of the plant P, for the case that the subsystems Σ_j, $(\forall j \in \mathcal{D})$, are disturbed. In the following the construction of $\mathcal{G}_{\mathrm{CD}}(\mathcal{D})$ is used to define a reduced maximal communication structure $\bar{\mathcal{G}}_{\mathrm{C}} = (\mathcal{V}_{\mathrm{C}}, \bar{\mathcal{E}}_{\mathrm{C}})$ by considering that all subsystems could be disturbed, i.e.,

$$\bar{\mathcal{G}}_{\mathrm{C}} = \mathcal{G}_{\mathrm{CD}}(\mathcal{D} = \mathcal{N}).$$

The self-organizing control system Σ^* with the reduced maximal communication $\bar{\mathcal{G}}_{\mathrm{C}}^*$ is defined by

$$\Sigma^*: \begin{cases} \Sigma \text{ defined in (4.20) with } \boldsymbol{x}_0 = \boldsymbol{x}_0^* \text{ and} \\ \mathcal{E}_{\mathrm{C}}(t) \text{ bounded by } \mathcal{E}_{\mathrm{CD}}(\mathcal{D}{=}\mathcal{N}) \text{ according to (4.54) in Proposition 4.2} \end{cases} \tag{4.59}$$

where $\boldsymbol{x}_0^*$ is the initial state of Σ^* and $\mathcal{E}_{\mathrm{CD}}(\mathcal{D}{=}\mathcal{N})$ is the set of maximal communication links within $\mathcal{G}_{\mathrm{CD}}(\mathcal{D}=\mathcal{N}) = (\mathcal{V}_{\mathrm{C}}, \mathcal{E}_{\mathrm{CD}}(\mathcal{D}=\mathcal{N}))$. To build the reduced maximal communication graph $\bar{\mathcal{G}}_{\mathrm{C}}^*$ the sets $\bar{\mathcal{F}}_{\mathrm{C}i}$ of maximal followers in Σ are replaced by sets $\bar{\mathcal{F}}_{\mathrm{CD}i}(\mathcal{D}=\mathcal{N})$ according to Proposition 4.2

$$\begin{aligned} \bar{\mathcal{F}}_{\mathrm{C}i} &= \bar{\mathcal{F}}_{\mathrm{CD}i}(\mathcal{D}=\mathcal{N}), \quad \forall i \in \mathcal{N} \\ \bar{\mathcal{P}}_{\mathrm{C}i} &= \{j \in \mathcal{N} \mid i \in \bar{\mathcal{F}}_{\mathrm{CD}j}(\mathcal{D}=\mathcal{N})\}, \quad \forall i \in \mathcal{N}, \end{aligned}$$

where the sets $\bar{\mathcal{P}}_{\mathrm{C}i}$ of maximal predecessors follow from $\bar{\mathcal{F}}_{\mathrm{CD}i}(\mathcal{D}=\mathcal{N})$.

The following theorem uses the results in Proposition 4.2 to state that the reduction of the communication within Σ^* defined in (4.59) compared to Σ defined in (4.20) leads to an identical behavior of Σ and Σ^*.

Theorem 4.7 (Reduction of the communication structure for Σ defined in (4.20)**)** *For $\boldsymbol{x}_0 = \boldsymbol{x}_0^* = \boldsymbol{0}$ the behavior of the self-organizing control system Σ defined in* (4.20) *with maximal communication graph $\bar{\mathcal{G}}_{\mathrm{C}}$ and the behavior of the self-organizing control system Σ^* defined in* (4.59) *with the reduced maximal communication graph $\bar{\mathcal{G}}_{\mathrm{C}}^*$ is identical.*

Proof. Based on Proposition 4.2 the sets $\bar{\mathcal{F}}_{\mathrm{C}i}^* = \mathcal{F}_{\mathrm{CD}i}(\mathcal{D}=\mathcal{N})$, $(\forall i \in \mathcal{N})$, includes all local control units C_k, $(\forall k \in \mathcal{F}_{\mathrm{CD}i}(\mathcal{D}=\mathcal{N}))$ to which the local control units C_i send information in the case that all subsystems are disturbed and $\boldsymbol{x}_0 = \boldsymbol{0}$. Hence, if the there is no communication from C_i, $(i \in \mathcal{N})$, to C_j, $(\forall j \in \bar{\mathcal{F}}_{\mathrm{C}i} \backslash \bar{\mathcal{F}}_{\mathrm{C}i}^*)$ within the system Σ anyway, then the behavior of Σ and the behavior of Σ^* is identical. □

Theorem 4.7 shows that not all communication links within $\bar{\mathcal{G}}_{\mathrm{C}}$ are necessary since some of them are not used anyway. The benefit of the communication reduction is that in the real control process there is no need to enable a connection among the local control units if the corresponding communication link is canceled out.

4.4.3 Overall design algorithm

In this section an algorithm for choosing the parameters of the local control units C_i, $(\forall i \in \mathcal{N})$, to guarantee a desired disturbance attenuation (Control aim 3.3) are presented.

The following local model information is used in Algorithm 4.1 to choose the parameters of the local control units C_i:

- The model of the subsystem Σ_i defined in (3.1).
- The set $\mathcal{P}_{\mathrm{P}i}$ of physical predecessors and the set $\mathcal{F}_{\mathrm{P}i}$ of physical followers of Σ_i.
- The matrices $\boldsymbol{L}_{ji}$ from (3.3) describing The physical impact from Σ_i to Σ_j, $(\forall j \in \mathcal{F}_{\mathrm{P}i})$.
- The desired maximal performance output $\bar{v}_i^*$, $(\forall i \in \mathcal{U})$, for the undisturbed subsystems.

Algorithm 4.1 is divided into three steps.

Initially, the feedback gain matrix $\boldsymbol{K}_{\mathrm{C}i}$ for the feedback unit F_i defined in (4.11) is designed such that Assumption 4.1 is fulfilled. Note that due to Assumption 3.2 there always exists a feedback gain matrix $\boldsymbol{K}_{\mathrm{C}i}$ that guarantees Assumption 4.1 (cf. Theorem 2.2). The matrix $\boldsymbol{K}_{\mathrm{C}i}$ can be determined by classical state feedback design methods (cf. Section 2.2 or methods presented in [22, 36, 94, 122]). Of course, Step 1 can be neglected if a stabilizing matrix $\boldsymbol{K}_{\mathrm{C}i}$ is already given.

In Step 2 the parameters for the observation unit O_i defined in (4.12) are chosen: the set $\bar{\mathcal{P}}_{\mathrm{C}i}$ of maximal communicational predecessors is specified by the set $\mathcal{P}_{\mathrm{P}i}$ of physical predecessors and the set $\bar{\mathcal{F}}_{\mathrm{C}i}$ of maximal communicational followers by the set $\mathcal{F}_{\mathrm{P}i}$ of physical followers. The matrix $\boldsymbol{C}_{\mathrm{z}i}$ is taken from the model of the subsystem Σ_i and, furthermore, the matrices $\boldsymbol{L}_{ji}$, $(\forall j \in \mathcal{F}_{\mathrm{P}i})$, are added.

Algorithm 4.1: Design local control units C_i, $(\forall i \in \mathcal{N})$, defined in (4.14) which compensate physical coupling inputs to bound the disturbance propagation according to Control aim 3.3.

Given: $\bar{v}_i^*$, Σ_i, $\mathcal{P}_{\mathrm{P}i}$, $\mathcal{F}_{\mathrm{P}i}$, $\boldsymbol{L}_{ji}$, $(\forall i \in \mathcal{N}, \forall j \in \mathcal{F}_{\mathrm{P}i})$

for i to N **do**

1 Construct F_i defined in (4.11) by determining $\boldsymbol{K}_{\mathrm{C}i}$ such that $\bar{\boldsymbol{A}}_i = \boldsymbol{A}_i - \boldsymbol{B}_i \boldsymbol{K}_{\mathrm{C}i}$ is a Hurwitz matrix.

2 Construct O_i defined in (4.12) by
- setting $\bar{\mathcal{P}}_{\mathrm{C}i} = \mathcal{P}_{\mathrm{P}i}$ and $\bar{\mathcal{F}}_{\mathrm{C}i} = \mathcal{F}_{\mathrm{P}i}$,
- extracting the matrix $\boldsymbol{C}_{\mathrm{z}i}$ from Σ_i and
- adding the matrices $\boldsymbol{L}_{ji}$, $(\forall j \in \bar{\mathcal{F}}_{\mathrm{C}i})$.

3 Construct D_i defined in (4.13) by
- setting $\bar{\mathcal{P}}_{\mathrm{C}i} = \mathcal{P}_{\mathrm{P}i}$ and $\bar{\mathcal{F}}_{\mathrm{C}i} = \mathcal{F}_{\mathrm{P}i}$,
- calculating the decoupling matrix $\boldsymbol{K}_{\mathrm{D}i}$ with (4.17) and
- choosing the switching thresholds $\bar{\epsilon}_i = \epsilon_{ij}$ for C_j, $(\forall j \in \bar{\mathcal{P}}_{\mathrm{C}i})$, according to (4.57).

end

Result: C_i consisting of F_i, O_i and D_i guaranteeing that the performance output $\boldsymbol{v}_i(t)$ of the undisturbed subsystems is upper bounded by $\bar{v}_i^*$ as claimed in Control aim 3.3.

The decision unit D_i defined in (4.13) is parametrized in Step 3: the local communication structure is defined by setting $\bar{\mathcal{P}}_{\mathrm{C}i} = \mathcal{P}_{\mathrm{P}i}$ and $\bar{\mathcal{F}}_{\mathrm{C}i} = \mathcal{F}_{\mathrm{P}i}$, the decoupling matrix $\boldsymbol{K}_{\mathrm{D}i}$ is determined with (4.17) and the switching thresholds ϵ_{ji} are calculated with (4.57) in Theorem 4.6 to fulfill Control aim 3.3.

The result of Algorithm 4.1 are the parametrized components F_i, O_i and D_i of all local control units C_i, $(\forall i \in \mathcal{N})$ such that the resulting self-organizing control system Σ defined in (4.20) is practically stable according Theorem 4.2. Furthermore, Σ guarantees that the maximal value of the performance output $\boldsymbol{v}_j(t)$ of the undisturbed subsystems Σ_j, $(\forall j \in \mathcal{U})$ is upper bounded by the desired maximal performance output $\bar{v}_j^*$ as claimed in Control aim 3.3.

Summary. The most interesting aspect of the presented algorithms is:

> The local control units C_i, $(\forall i \in \mathcal{N})$, are designed independently while using local model information of the subsystems Σ_i only.

With this, two properties for the controller design can be derived:

- The complexity of the overall design of the self-organizing controller C defined in (4.15) scales linearly with the number N of subsystems of the plant P.
- New subsystems can be easily integrated into an already existing self-organizing control system Σ. The integration of new subsystems investigated in Section 4.7.

Due to these properties, the proposed self-organizing control structure is very useful for controlling complex physically interconnected systems.

Example 4.1 (cont.) *Situation-dependent compensation of physical coupling inputs*

In the following the local control units C_i, $(\forall i = \{1, 2\})$, defined in (4.14) are designed for controlling the symmetrically interconnected subsystems P defined in (3.32). The control aim is that the resulting self-organizing control system defined in (4.20) guarantees that performance output $\boldsymbol{v}_1(t)$ of Σ_1 is bounded by $\bar{v}_1^* = 0.4$ if Σ_2 is disturbed and that performance output $\boldsymbol{v}_2(t)$ of Σ_2 is bounded by $\bar{v}_2^* = 0.2$ if Σ_1 is disturbed (cf. Control aim 3.3).

Table 4.2 illustrates the execution of Algorithm 4.1 for the design of the local control units C_1 and C_2 to guarantee the desired disturbance attenuation. In Step 1 of Algorithm 4.1 the feedback gain matrices $\boldsymbol{K}_{\mathrm{C}i}$ for the feedback units F_i are designed. For subsystem Σ_1 the eigenvalue of the system matrix $\bar{\boldsymbol{A}}_1$ of the controlled subsystem $\bar{\Sigma}_1$ is placed at $\lambda(\bar{\boldsymbol{A}}_1) = -1$. The feedback gain matrix $\boldsymbol{K}_{\mathrm{C}2}$ is the solution of the optimization problem in (2.11) with the weighting matrices $\boldsymbol{Q} = 1$ and $\boldsymbol{R} = 1$. Hence, the feedback units are designed independently for different requirements. The observation units O_i, $(\forall i = \{1, 2\})$, are constructed by defining the predecessors and followers, by using $\boldsymbol{C}_{\mathrm{z}i}$ from the model of the subsystems Σ_i and adding $\boldsymbol{L}_{ji}$, $(\forall j \in \mathcal{F}_{\mathrm{P}i})$, from the physical coupling structure (Step 2). The calculation of the decoupling matrices $\boldsymbol{K}_{\mathrm{D}i}$ for the decision units D_1 and D_2 with (4.17) in Step 3 is straight-forward and fulfills Assumption 4.2. Furthermore, the different desired maximal performance outputs $\bar{v}_i^*$ lead to different thresholds $\bar{\epsilon}_i$. For Σ_2 a much smaller deviation is tolerated than for Σ_1, therefore, $\epsilon_{21} = \bar{\epsilon}_2 = 0.28$ is smaller than $\epsilon_{12} = \bar{\epsilon}_1 = 0.4$ (Step 3).

Table 4.2: Design local control units for the symmetrically interconnected subsystems P in (3.32) by Algorithm 4.1 to guarantee $\bar{v}_1^* = 0.4$ and $\bar{v}_2^* = 0.2$.

	Local control unit C_1		Local control unit C_2	
Step	Execution	Result	Execution	Result
1	Place $\lambda(\bar{\boldsymbol{A}}_1) = -1$	$\boldsymbol{K}_{C1} = 2$	LQ-Design: $\boldsymbol{Q} = \boldsymbol{R} = 1$	$\boldsymbol{K}_{C2} = 2.41$
2	$\bar{\mathcal{P}}_{C1} = \mathcal{P}_{P1}$ $\bar{\mathcal{F}}_{C1} = \mathcal{F}_{P1}$	$\bar{\mathcal{P}}_{C1} = \{2\}$ $\bar{\mathcal{F}}_{C1} = \{2\}$	$\bar{\mathcal{P}}_{C2} = \mathcal{P}_{P2}$ $\bar{\mathcal{F}}_{C2} = \mathcal{F}_{P2}$	$\bar{\mathcal{P}}_{C2} = \{1\}$ $\bar{\mathcal{F}}_{C2} = \{1\}$
	extract $\boldsymbol{C}_{z1}$ from Σ_1	$\boldsymbol{C}_{z1} = 1$	extract $\boldsymbol{C}_{z2}$ from Σ_2	$\boldsymbol{C}_{z2} = 1$
	add $\boldsymbol{L}_{21}$	$\boldsymbol{L}_{21} = 1$	add $\boldsymbol{L}_{12}$	$\boldsymbol{L}_{12} = 1$
3	$\bar{\mathcal{P}}_{C1} = \mathcal{P}_{P1}$ $\bar{\mathcal{F}}_{C1} = \mathcal{F}_{P1}$	$\bar{\mathcal{P}}_{C1} = \{2\}$ $\bar{\mathcal{F}}_{C1} = \{2\}$	$\bar{\mathcal{P}}_{C2} = \mathcal{P}_{P2}$ $\bar{\mathcal{F}}_{C2} = \mathcal{F}_{P2}$	$\bar{\mathcal{P}}_{C2} = \{1\}$ $\bar{\mathcal{F}}_{C2} = \{1\}$
	$\boldsymbol{K}_{D1} = \underbrace{1^{-1}}_{\boldsymbol{B}_1^+} \cdot \underbrace{1}_{\boldsymbol{E}_1}$	$\boldsymbol{K}_{D1} = 1$	$\boldsymbol{K}_{D2} = \underbrace{1^{-1}}_{\boldsymbol{B}_2^+} \cdot \underbrace{1}_{\boldsymbol{E}_2}$	$\boldsymbol{K}_{D2} = 1$
	$\bar{\epsilon}_1 = \frac{1}{\underbrace{1}_{\lvert\bar{\mathcal{P}}_{C1}\rvert} \cdot \underbrace{1}_{m_{vs1}}} \underbrace{0.4}_{\bar{v}_{\Delta 1}^*}$	$\epsilon_{12} = \bar{\epsilon}_1 = 0.4$	$\bar{\epsilon}_2 = \frac{1}{\underbrace{1}_{\lvert\bar{\mathcal{P}}_{C2}\rvert} \cdot \underbrace{0.71}_{m_{vs2}}} \underbrace{0.2}_{\bar{v}_{\Delta 2}^*}$	$\epsilon_{21} = \bar{\epsilon}_2 = 0.28$

Figure 4.20 shows that the desired maximal performance outputs $\bar{v}_i^*$, ($\forall i = \{1, 2\}$), are satisfied by Σ with the designed local control units. If Σ_1 is disturbed, $\boldsymbol{v}_2(t)$ stays below $v_2^* = 0.2$ ($t = [0\,\text{s}, 20\,\text{s}]$). Furthermore, if Σ_2 is disturbed, $\boldsymbol{v}_1(t)$ stays below $\bar{v}_1^* = 0.4$ ($t = [20\,\text{s}, 40\,\text{s}]$). The stability analysis is similar to the one presented in Section 4.3.2. □

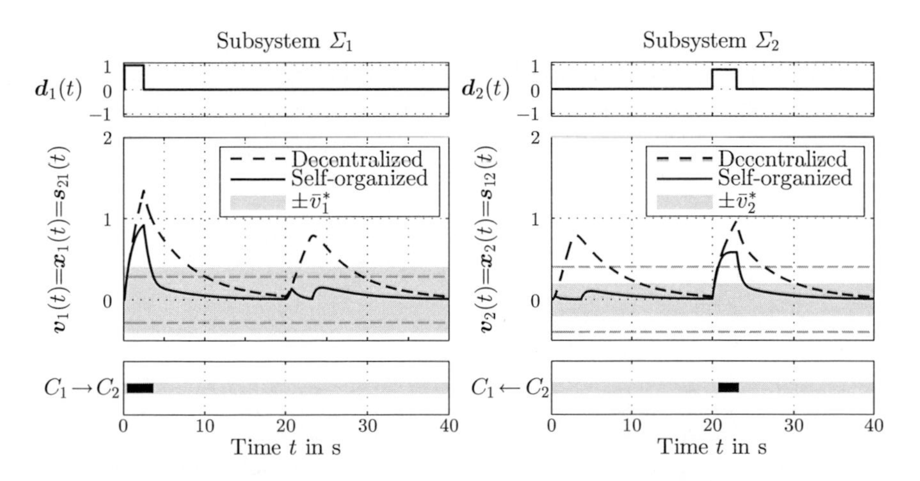

Figure 4.20: Behavior of the symmetrically interconnected subsystems defined in (3.32) controlled by the local control units designed by Algorithm 4.1 presented in Table 4.2 (switching threshold - gray dashed line).

4.5 Application example: Water supply system

In this section the proposed self-organizing control concept with a situation-dependent compensation of physical coupling inputs is applied to the water supply system introduced in Section 3.5.1. Section 4.5.1 presents the control aim and the resulting parameters of the local control units. With this, a stability analysis of the self-organizing control system Σ is performed and the disturbance propagation within Σ is investigated (Section 4.5.2). These analysis results are compared with the actual behavior of the closed-loop system (Section 4.5.3). Section 4.5.4 illustrates the relation between the performance and the communication effort of the self-organizing control system Σ.

4.5.1 Performance criteria and design of the local control units

The disturbance propagation shall be bounded such that the level $t_{\mathrm{TB}i}(t) = \boldsymbol{v}_i(t)$, of the undisturbed subsystems Σ_i, $(\forall i \in \mathcal{U})$, is upper bounded by 1 cm, i.e.,

$$\|\boldsymbol{v}_i(t)\| = |l_{\mathrm{TB}i}(t)| \leq \bar{v}_i^* = 1\,\mathrm{cm}, \quad \forall i \in \mathcal{U}. \tag{4.60}$$

Recall that the subsystems Σ_6, Σ_9 and Σ_{17} are assumed to be disturbed, i.e., $\mathcal{D} = \{6, 9, 17\}$.

With this, the local control units can be designed with Algorithm 4.1 which guarantees the desired disturbance attenuation. The steps of Algorithm 4.1 are as follows, where the resulting parameters are listed in Table 4.3:

1. The solutions of the optimization problem in (2.11) with the weighting matrices $\boldsymbol{Q}_i$ and $\boldsymbol{R}_i$ in (3.29) lead to the feedback gain matrices $\boldsymbol{K}_{\mathrm{C}i}$, which define the feedback units F_i and stabilize the isolated subsystems since $\max_{i \in \mathcal{N}, j \in \{1,2\}} \mathrm{Re}(\lambda_j(\bar{\boldsymbol{A}}_i)) = -0.12$.
2. The observation units O_i, $(\forall i \in \mathcal{N})$, are constructed by
   - defining the sets $\bar{\mathcal{P}}_{\mathrm{C}i} = \mathcal{P}_{\mathrm{P}i}$, $(\forall i \in \mathcal{N})$, of predecessors and sets $\bar{\mathcal{F}}_{\mathrm{C}i} = \mathcal{F}_{\mathrm{P}i}$ of followers resulting from $\mathcal{P}_{\mathrm{P}i}$ and $\mathcal{F}_{\mathrm{P}i}$ listed in Table 3.2,
   - using $\boldsymbol{C}_{\mathrm{z}i}$ from the model of the subsystems Σ_i defined in Section 3.5.1 and
   - adding $\boldsymbol{L}_{ji}$, $(\forall j \in \mathcal{F}_{\mathrm{P}i})$, listed in Table 3.1.
3. The decision units D_i, $(\forall i \in \mathcal{N})$, are constructed by
   - defining the sets $\bar{\mathcal{P}}_{\mathrm{C}i} = \mathcal{P}_{\mathrm{P}i}$, $(\forall i \in \mathcal{N})$, and $\bar{\mathcal{F}}_{\mathrm{C}i} = \mathcal{F}_{\mathrm{P}i}$,
   - determining the decoupling matrices $\boldsymbol{K}_{\mathrm{D}i}$, $(\forall i \in \mathcal{N})$, resulting from the matrices $\boldsymbol{B}_i$ and $\boldsymbol{E}_i$ of subsystems Σ_i in Section 3.5 by (4.17) to fulfill Assumption 4.2 and
   - choosing the thresholds $\bar{\epsilon}_i$ according to (4.57), where their values are identical for subsystem with the same dynamics since their number of predecessors is identical.

Table 4.3: Parameters of the local control units C_i defined in (4.14) for the water supply system.

	$\forall i \in \mathcal{N}_{\text{large}}$	$\forall i \in \mathcal{N}_{\text{middle}}$	$\forall i \in \mathcal{N}_{\text{small}}$
$\bar{\mathcal{P}}_{\text{C}i}$		$\mathcal{P}_{\text{P}i}$ from Table 3.2	
$\bar{\mathcal{F}}_{\text{C}i}$		$\mathcal{F}_{\text{P}i}$ from Table 3.2	
$\boldsymbol{K}_{\text{C}i}$	$\begin{pmatrix} 4.008 & 0.017 \\ 0.057 & 9.945 \end{pmatrix}$	$\begin{pmatrix} 3.841 & 0.061 \\ 0.188 & 9.761 \end{pmatrix}$	$\begin{pmatrix} 3.500 & 0.130 \\ 0.456 & 9.402 \end{pmatrix}$
$\boldsymbol{C}_{\text{z}i}$	$\begin{pmatrix} 0 & 0.4 \end{pmatrix}$	$\begin{pmatrix} 0 & 0.5 \end{pmatrix}$	$\begin{pmatrix} 0 & 0.8 \end{pmatrix}$
$\boldsymbol{L}_{ji}$	$5.5, \ \forall j \in \mathcal{F}_{\text{P}i}$	$4.4, \ \forall j \in \mathcal{F}_{\text{P}i}$	$4.0, \ \forall j \in \mathcal{F}_{\text{P}i}$
$\boldsymbol{K}_{\text{D}i}$		$\begin{pmatrix} 0 \\ 1 \end{pmatrix}$	
$\bar{\epsilon}_i$	0.0492	0.0499	0.1001
ϵ_{ij}		$\bar{\epsilon}_i, \ \forall j \in \bar{\mathcal{P}}_{\text{C}i}$	

With this, the local control units C_i, $(\forall i \in \mathcal{N})$, for the subsystems of the water supply system are completely defined. In the following, the water supply system controlled by the local control units C_i with the parameters listed in Table 4.3 are denoted as self-organizing controlled tanks.

4.5.2 Analysis of the behavior of the self-organizing controlled tanks

In the following the practical stability, the asymptotic stability, the disturbance propagation and the communication reduction of the self-organizing controlled tanks is analyzed.

Practical stability. The controlled tanks are practically stable (cf. Theorem 4.2). The bounds $\bar{x}_i$, $(\forall i \in \mathcal{N})$, which define the area $\mathcal{A}$ in (4.37) to which the states $\boldsymbol{x}_i(t)$ converge result to

$$\bar{x}_6 = 7.03\,\text{cm} \tag{4.61a}$$

$$\bar{x}_9 = 10.15\,\text{cm} \tag{4.61b}$$

$$\bar{x}_{17} = 8.03\,\text{cm} \tag{4.61c}$$

$$\bar{x}_j = 1.00\,\text{cm}, \quad \forall j \in \mathcal{N}\backslash\{6, 9, 17\}. \tag{4.61d}$$

Hence, for $\boldsymbol{x}_0 = \boldsymbol{0}$ the states $\boldsymbol{x}_i(t)$, $(\forall i \in \mathcal{N})$, will not leave $\mathcal{A}$.

Asymptotic stability. The analysis of the asymptotic stability listed in Table B.1 in the appendix Using condition (4.42) in Theorem 4.3 yields that the self-organizing controlled tanks are asymptotically stable. A part of this analysis for subsystem Σ_9 is presented in Table 4.4 which shows that all conditions in (4.42) for Σ_9 hold true which means that all communication links going out of C_9 will be deactivated after a certain time for vanishing disturbances.

Table 4.4: Part of the analysis of the asymptotic stability of the self-organizing controlled tanks with the evaluation of communication of C_9 by Theorem 4.3.

j	i	ϵ_{ji}	$m_{\mathrm{ss}ji} \sum_{l \in \bar{\mathcal{P}}_{\mathrm{C}i}} \epsilon_{il}$	$\epsilon_{ji} > m_{\mathrm{ss}ji} \sum_{l \in \bar{\mathcal{P}}_{\mathrm{C}i}} \epsilon_{il}$
7	9	0.0499	0.0176	True
8	9	0.0499	0.0176	True
10	9	0.0492	0.0176	True
11	9	0.0492	0.0176	True

If the self-organizing controlled tanks are considered as an uncertain system according to (4.43), the test of condition (4.44) in Proposition 4.1 with global model information leads to

$$\left\| \boldsymbol{C}_{\mathrm{z}}^{*} (s\boldsymbol{I} - \bar{\boldsymbol{A}})^{-1} \boldsymbol{E} \right\|_{H_\infty} = 3.9997 \not< 1,$$

which states that the self-organizing controlled tanks are perhaps not asymptotically stable. However, as Theorem 4.3 showed, Σ is asymptotically stable.

This shows that condition (4.42) in Theorem 4.3 yields a more precise analysis of the overall system since compared to Proposition 4.1 every physical coupling is analyzed individually.

Disturbance behavior for a single disturbance. Consider that only subsystem Σ_{17} is disturbed, i.e., $\mathcal{D} = \{17\}$. Then the propagation of $\boldsymbol{d}_{17}(t)$ is bounded according to (4.60), i.e., $\max_{t \geq 0} |l_{\mathrm{TB}i}| \leq 1\,\mathrm{cm}$ for all $i \in \mathcal{U} = \mathcal{N} \backslash \{17\}$.

The analysis of the communication activation with Proposition 4.2 for $\mathcal{D} = \{17\}$ listed in Table B.2 in the appendix shows that in the worst case at some time the local control unit C_{17} transmits information to C_{14}, C_{16}, C_{18} and C_{21}. A part of that analysis is listed in Table 4.5, which shows that only the thresholds ϵ_{14}, ϵ_{16}, ϵ_{18} or ϵ_{21} can be exceeded. The resulting maximal communication graph $\bar{\mathcal{G}}_{\mathrm{CD}}(\mathcal{D} = \{17\})$ is depicted in Fig. 4.23 (c).

Table 4.5: Part of the analysis of the communication activation of the self-organizing controlled tanks for $\mathcal{D} = \{17\}$ with the evaluation of communication of C_{16} and C_{17} by Proposition 4.2.

i	$\bar{\mathcal{F}}_{\mathrm{C}i}$	$\bar{\mathcal{F}}_{\mathrm{CD}i}(\mathcal{D}=\{17\})$	h	ϵ_{hi}	$\bar{s}_{\mathrm{D}hi}$	$\epsilon_{hi} \leq \bar{s}_{\mathrm{D}hi}$
16	$\{11, 17\}$	$\emptyset$	11	0.0492	0.0220	False
			17	0.0499	0.0220	False
17	$\{14, 16, 18, 21\}$	$\{14, 16, 18, 21\}$	14	0.0499	0.1765	True
			16	0.0499	0.1765	True
			18	0.1001	0.1765	True
			21	0.1001	0.1765	True

Reduction of the communication structure. If all subsystems are considered to be disturbed $\mathcal{D} = \mathcal{N}$, the determination of $\bar{\mathcal{G}}_{\mathrm{CD}}(\mathcal{D} = \mathcal{N})$ with Proposition 4.2 leads to the graph in Fig. 4.21 (cf. Table B.3). According to Theorem 4.7, the self-organizing controlled tanks with this reduced maximal communication structure $\bar{\mathcal{G}}_{\mathrm{C}}^{*} = \bar{\mathcal{G}}_{\mathrm{CD}}(\mathcal{D} = \mathcal{N})$ have the same performance as the closed-loop system with the original maximal communication structure $\bar{\mathcal{G}}_{\mathrm{C}} = \mathcal{G}_{\mathrm{P}}$ for $\boldsymbol{x}_{0i} = \mathbf{0}$, $(\forall i \in \mathcal{N})$. Figure 4.21 shows that only the depicted communication links $\bar{\mathcal{G}}_{\mathrm{CD}}(\mathcal{D} = \mathcal{N})$ are necessary to guarantee the required control aim in (4.60) for $\boldsymbol{x}_0 = \mathbf{0}$. Only the local control units C_6, C_9 and C_{17} have to transmit information and will transmit information. Note that in the following section the original maximal communication structure $\bar{\mathcal{G}}_{\mathrm{C}} = \mathcal{G}_{\mathrm{P}}$ and not the reduced one is used.

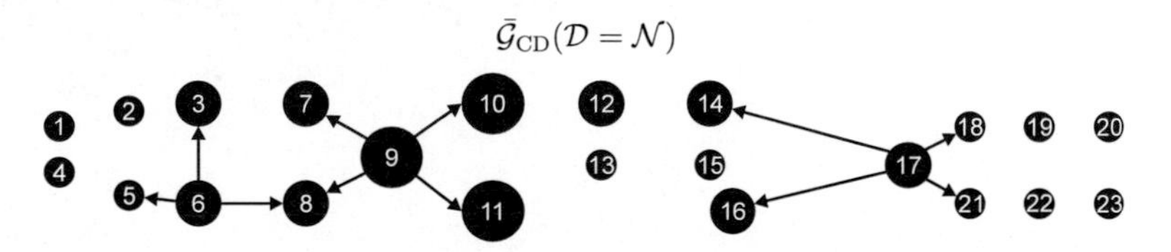

Figure 4.21: Maximal communication graph $\bar{\mathcal{G}}_{\mathrm{CD}}(\mathcal{D} = \mathcal{N})$ for the self-organizing controlled tanks in case that all subsystems are disturbed.

4.5.3 Behavior of the self-organizing controlled tanks

This section verifies the analysis results in the previous section by evaluating a simulation of the self-organizing controlled tanks. At first, the general behavior of the closed-loop system is presented and discussed.

Behavior of the self-organizing controlled tanks. In Fig. 4.22 the behavior of the water supply system for $\boldsymbol{x}_{0i} = \mathbf{0}$, $(\forall i \in \mathcal{N})$, using the local control units C_i designed in Section 4.5.1 is depicted. The topmost plot shows the disturbances $\boldsymbol{d}_i(t) = d_{\mathrm{TB}i}(t)$ affecting the subsystems Σ_i, where the disturbances affecting Σ_6, Σ_9 and Σ_{17} are colored. The disturbances affecting the other subsystems are all black. The second and third plot display the level $l_{\mathrm{TB}i}$ of tank TB while using the self-organizing controller C (solid line) and the decentralized controller C_{D} (dashed line). The transmission of information among the local control units is shown in the lowest plot. The black bars indicate at which time the communication links $(i \rightarrow j)$ are in the set $\mathcal{E}_{\mathrm{C}}(t)$ of communicational edges which means that C_i transmits information to C_j. The thin red vertical lines refer to the corresponding communication graphs $\mathcal{G}_{\mathrm{C}}(t)$ for different times shown in Fig. 4.23.

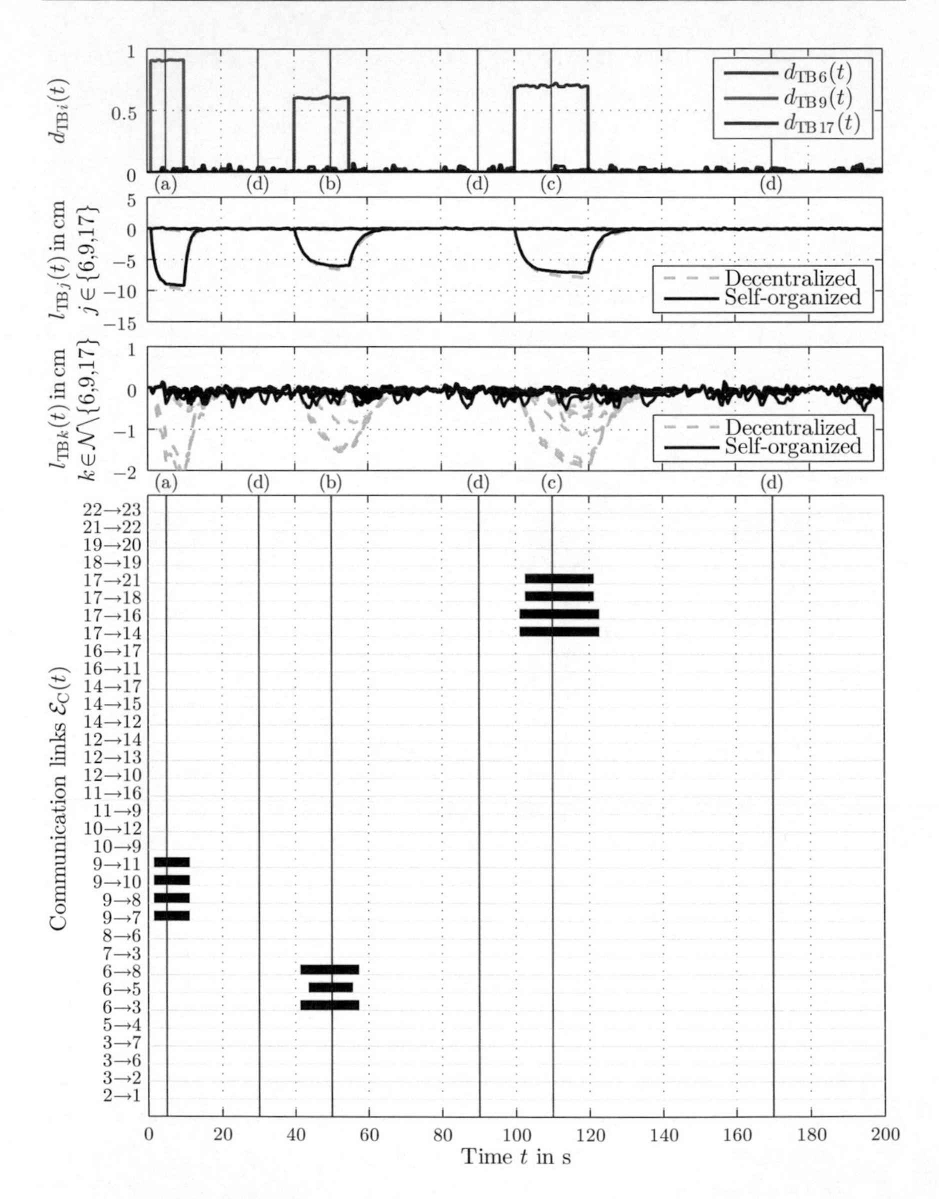

Figure 4.22: Behavior water supply system controlled by the self-organizing controller C defined in (4.15).

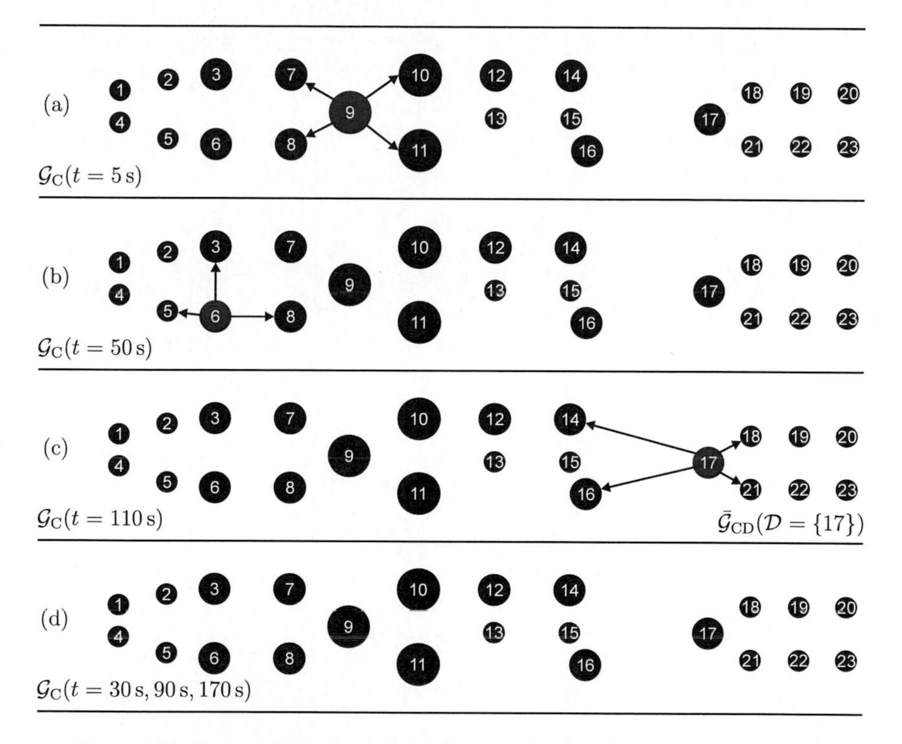

Figure 4.23: Communication graph $\mathcal{G}_C(t)$ at certain times for different disturbance situations in Fig. 4.22 (disturbed subsystems are red).

Figure 4.22 shows that the self-organizing controller C leads to better performance compared to the usage of a decentralized controller C_D. In particular the deviation of the maximal level $l_{TBk}(t)$, $(\forall k \in \mathcal{N} \backslash \{6, 9, 17\})$, of the subsystems with a small perturbation is much smaller while using C (third plot). The performance of the disturbed subsystems (Σ_6, Σ_9 and Σ_{17}) is similar for both control concepts since the effect of a disturbance at Σ_i can not be compensated by sending information to C_i or sending information from C_i to other local control units. Hence, the main benefit of the self-organizing control concept is the reduction of the disturbance propagation within the overall system.

At most times there is no communication among the local control units C_i (lower plot of Fig. 4.22 and Fig. 4.23 (d)). The communication is only activated if a subsystem is disturbed significantly. For example, if subsystem Σ_9 is disturbed, the corresponding local control unit C_9 sends information to C_7, C_8, C_{10} and C_{11} (Fig. 4.23 (a)). The communication structures in Fig. 4.23 show that it is sufficient to send information from the local control units of the disturbed subsystems to all local control units corresponding to subsystems on which the disturbed subsystems have a physical impact. With this, a further propagation of the disturbance is eliminated.

In summary, the self-organizing controlled tanks show characteristic properties of self-organization since the communication graph $\mathcal{G}_{\mathrm{C}}(t)$ adapts to the current acting disturbances to guarantee a desired disturbance attenuation.

Comparison to decentralized control and centralized control. Figure 4.24 depicts the physical coupling graph $\mathcal{G}_{\mathrm{C}}$ and the maximal values of the tanks levels in TBi for all $t \geq 0$ ($\max_{t\geq 0} |l_{\mathrm{TB}i}(t)|$) while using the self-organizing controller C (black bar), the decentralized controller C_{D} (gray bar) and the centralized controller C_{C} (blue bar) without giving the actual amount of the values. This depiction indicates the benefit of the self-organizing control concept. The performance with the self-organizing controller C is similar to the performance with the centralized controller C_{C}. The propagation of a disturbance is prevented by the self-organizing controller, where the communication is only activated if a subsystem is disturbed. Hence, the self-organizing controller C combines the benefit of the decentralized controller – a low communication effort – and the benefit of the centralized controller – a good disturbance attenuation.

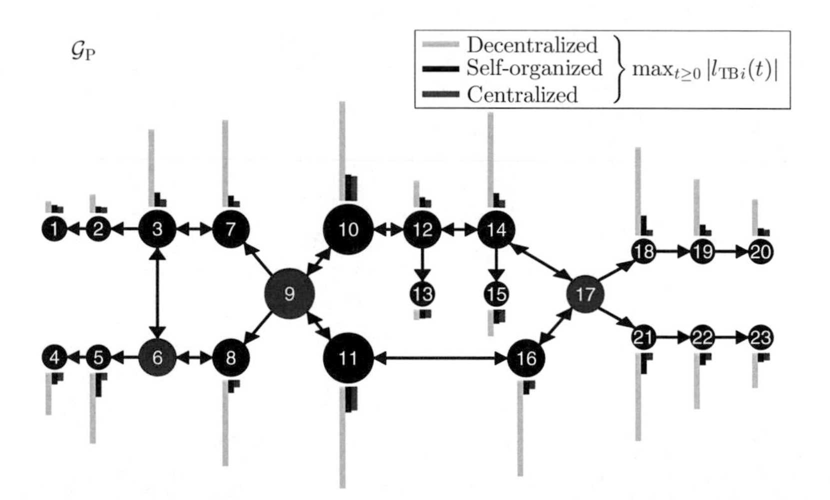

Figure 4.24: Physical coupling graph $\mathcal{G}_{\mathrm{P}}$ with the maximal values of the tank levels $\max_{t\geq 0} |l_{\mathrm{TB}i}|$, $(i = 1, \ldots, 23)$, for decentralized control, self-organized control and centralized control w.r.t the behavior depicted in Fig. 4.22.

Fulfillment of desired disturbance attenuation (control aim). The fulfillment of the control aim in (4.60) is guaranteed for all undisturbed subsystems Σ_i, $(\forall i \in \mathcal{U} = \mathcal{N} \backslash \{6, 9, 17\})$, (cf. Fig. 4.25). The levels $l_{\mathrm{TB}i}(t) = \boldsymbol{v}_i(t)$, $(\forall i \in \mathcal{U})$, are smaller than the desired maximal performance output of $\bar{v}_i^* = 1\,\mathrm{cm}$ for all $t \geq 0$.

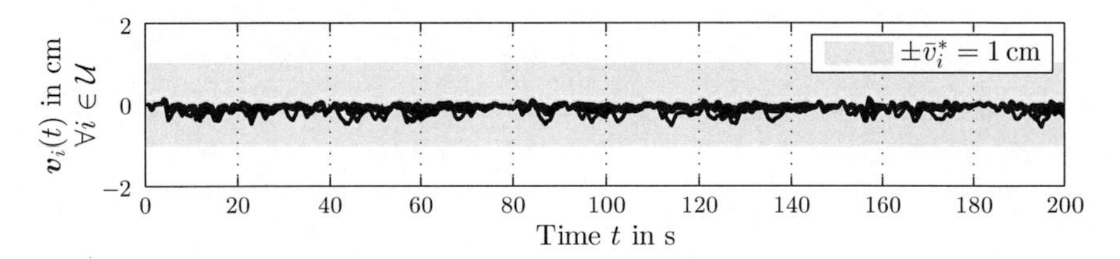

Figure 4.25: Evaluation of the control aim in (4.60) claiming a desired disturbance attenuation.

Verification of the analysis results. In the following the analysis results for the behavior of the self-organizing controlled tanks in Section 4.5.2 are compared with the actual behavior:

- **Practical stability**: Since the initial states are zero, all states $\boldsymbol{x}_i(t)$ are expected to stay in the set $\mathcal{A}$ defined by (4.37) and (4.61). Figure 4.26 shows a part of the overall verification of the practical stability presented in Fig. B.1 in the appendix. The gray area represents the state bound $\bar{x}_i$ defined in (4.61). The solid lines are the phase portrait of $x_{i,1}(t)$ and $x_{i,2}(t)$. The gray areas are not left for any $t \geq 0$. Hence, the closed-loop system is practically stable w.r.t the compact set $\mathcal{A}$ with $\bar{x}_i$ defined in (4.61).

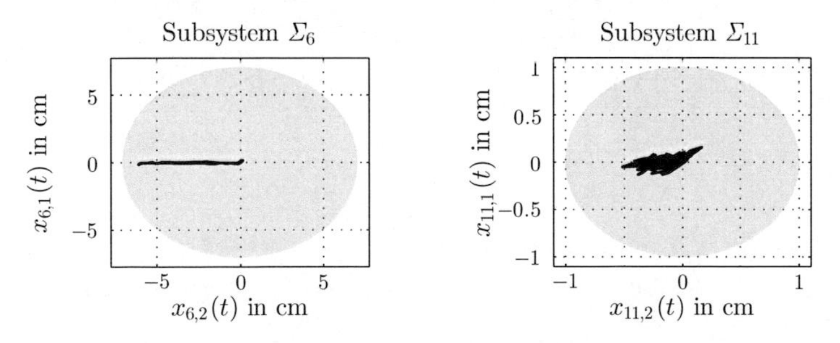

Figure 4.26: Verification of the practical stability of two tanks.

- **Asymptotic stability**: For vanishing disturbances the levels $l_{\mathrm{TB}i}$ converge to zero (cf. Fig. 4.22). The same holds for the levels $l_{\mathrm{TA}i}$ which are not plotted in Fig. 4.22. Hence, Σ is asymptotically stable for $\boldsymbol{d}_i(t) = \boldsymbol{0}$, $(\forall i \in \mathcal{N})$. This approves the result in Table B.1.

- **Disturbance behavior for a single disturbance**: For the evaluation of the disturbance behavior of the closed-loop system for a disturbance affecting Σ_{17}, the behavior of the tanks after 100 s is investigated in more detail since in that time interval only subsystem Σ_{17} is disturbed.

  The disturbance propagation of $\boldsymbol{d}_{17}(t)$ is illustrated in Fig. 4.27, which depicts the physical coupling graph $\mathcal{G}_\text{P}$ and the maximal values of the tank levels $l_{\text{TB}i}$ for all $t \geq 100\,\text{s}$ ($\max_{t\geq 100} |t_{\text{TB}i}(t)|$) from the behavior of the self-organizing controlled tanks shown in Fig. 4.22.

  With a decentralized controller C_D the disturbance $\boldsymbol{d}_{17}(t)$ has a great influence on subsystems that are physically coupled with Σ_{17}. With the self-organizing controller C this influence is eliminated.

  The analysis result of the communication activation in Table B.2 matches with the behavior of the self-organizing controlled tanks. There is only communication from C_{17} to C_{14}, C_{16}, C_{18} or C_{21} (cf. Fig. 4.23 (c)).

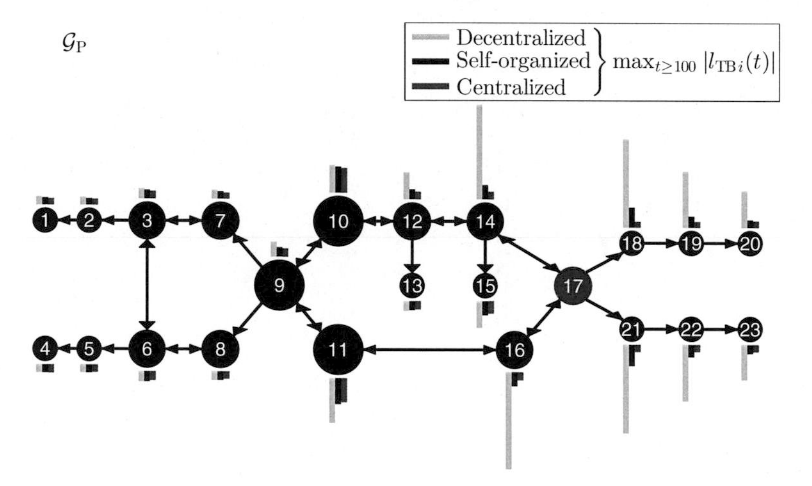

Figure 4.27: Physical coupling graph $\mathcal{G}_\text{P}$ with the maximal values of the tank levels $\max_{t\geq 0} |l_{\text{TB}i}|$, $(i = 1, \ldots, 23)$, for $t \geq 100\,\text{s}$.

- **Reduction of the communication structure**: Figure 4.23 approves the analysis results illustrated in Fig. 4.21, only the determined communication links within $\bar{\mathcal{G}}_\text{CD}(\mathcal{D} = \mathcal{N})$ are used by the local control units. Hence, the other communication links in $\bar{\mathcal{G}}_\text{C}$ are not necessary and can be canceled out.

4.5.4 Trade-off between performance and communication effort

Consider different self-organizing controllers C designed for certain desired maximal performance outputs $\bar{v}_i^*$, $(\forall i \in \mathcal{N})$, equal for every subsystem that lead to different self-organizing control systems Σ. Figure 4.28 compares the performance of the different Σ indicated by $\bar{v}_i^*$, $(\forall i \in \mathcal{N})$, and the communication effort indicated by the overall communication time t_Σ for the same disturbance situation as depicted in Fig. 4.22. The time t_Σ is the sum of all lengths of the communication intervals for each communication link $(i \to j) \in \mathcal{E}_\text{C}$. Roughly speaking, t_Σ is the length of all black bars in the lower plot of Fig. 4.22 in a row. For example, for $\bar{v}_i^* = 1\,\text{cm}$ the overall communication time is $t_\Sigma \approx 4$ min (cf. Fig. 4.22). Figure 4.28 shows that even for small values of $\bar{v}_i^*$ there is a huge reduction of the communication effort compared to a permanent communication within the centralized controller C_C. Hence, the desired maximal performance outputs $\bar{v}_i^* = 1\,\text{cm}$, $(\forall i \in \mathcal{N})$, claimed in (4.60) ensures a good compromise between a low communication effort and a reasonable disturbance attenuation.

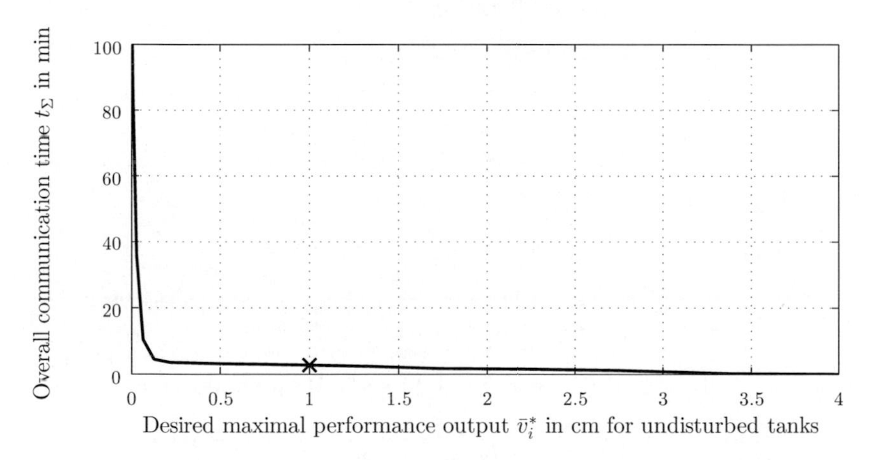

Figure 4.28: Communication time for certain self-organizing systems Σ of the form (4.20) resulting from different desired maximal performance outputs $\bar{v}_i^*$.

4.6 Extension of the concept to output feedback control

In this section Assumption 3.3 is neglected which means that the full subsystem state can not be measured anymore. The basic approach to adapt the proposed self-organizing control concept to output feedback control is presented in Section 4.6.1. The structure of the resulting self-organizing controller with output feedback is introduced in Section 4.6.2. In Section 4.6.3 it is shown that the analysis results and the design methods for the state feedback approach can be transfered to the output feedback approach. Section 4.6.4 illustrates the output feedback approach by its application to the water supply system. The main difference between the state feedback approach and the output feedback approach are discussed in Section 4.6.5. Finally, another control structure which uses output feedback is briefly discussed in Section 4.6.6.

4.6.1 Realization of a situation-dependent compensation of coupling signals by using output feedback

The main challenge is to realize a situation-dependent compensation coupling inputs through the local control units C_i, ($\forall i \in \mathcal{N}$), with the local measurement of the outputs $\boldsymbol{y}_i(t)$ only since the overall subsystem state $\boldsymbol{x}_i(t)$ is not accessible (Fig. 4.29). Considering the structure of the local control units C_i that measure the state $\boldsymbol{x}_i(t)$ illustrated in Fig. 4.7, the following two problems arise for output feedback control:

1. Stabilization of the isolated subsystems by local output feedback.

2. Local determination of the coupling signals $\hat{\boldsymbol{s}}_{ji}(t)$ using the measured output $\boldsymbol{y}_i(t)$.

The first problem is solved by using a local observer in the feedback units F_i that reconstructs the state $\boldsymbol{x}_i(t)$ of the subsystem Σ_i to generate an input $\boldsymbol{u}_i(t)$ similar to the state feedback control

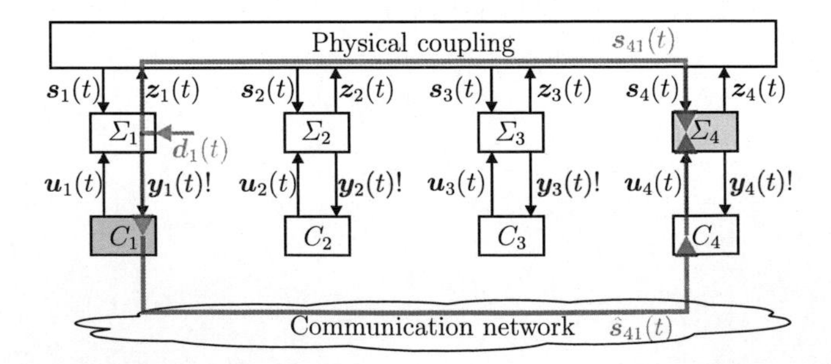

Figure 4.29: Realization of the self-organizing control concept by transmitting physical coupling signals and measuring the output $\boldsymbol{y}_i(t)$.

in (4.11). The question is, whether the state feedback in (4.11) and the observers for the different local control units can be designed independently from each other to guarantee a stable overall closed loop-system.

Considering the second problem, for the determination of $\hat{\boldsymbol{s}}_{ji}(t)$ by C_i it is not necessary to know the entire state $\boldsymbol{x}_i(t)$ since the coupling signals $\hat{\boldsymbol{s}}_{ji}(t)$ only depend on the coupling output $\boldsymbol{z}_i(t)$. The coupling output $\boldsymbol{z}_i(t)$ can be determined by using the measured output $\boldsymbol{y}_i(t)$ if there exists a matrix $\boldsymbol{C}_{\mathrm{zy}i} \in \mathbb{R}^{r_{\mathrm{z}i} \times r_i}$ for which

$$\boldsymbol{z}_i(t) \stackrel{!}{=} \hat{\boldsymbol{z}}_i(t) = \boldsymbol{C}_{\mathrm{zy}i} \boldsymbol{y}_i(t), \quad \forall i \in \mathcal{N} \tag{4.62}$$

holds true, where $\hat{\boldsymbol{z}}_i(t)$ is the determined coupling output. The implementation of these basic approaches within the structure of the local control units is presented in the following section.

4.6.2 Structure of the self-organizing controller with output feedback

In the following, the adaption of the structure of the local control units to output feedback and the resulting model of the closed-loop system are presented.

Communication structure and structure of the local control units. The maximal communication graph $\bar{\mathcal{G}}_{\mathrm{C}}$ resulting from the sets $\bar{\mathcal{P}}_{\mathrm{C}i}$, $(\forall i \in \mathcal{N})$, of maximal communicational predecessors defined in (4.9) does not change compared to the state feedback approach since the physical coupling among the subsystems and the general approach of a situation-dependent compensation of these coupling inputs stays the same (c.f. Fig. 4.6). Hence, the set $\mathcal{E}_{\mathrm{C}}(t)$ is bounded by the same set $\bar{\mathcal{E}}_{\mathrm{C}}$ of maximal communicational edges according to (4.10). However, this does not imply that the behavior of $\mathcal{G}_{\mathrm{C}}(t)$ is identical to the one in the state feedback approach.

As described in Section 4.2, the local control units C_i consist of a feedback unit F_i, an observation unit O_i and a decision unit D_i (Fig. 4.30). The differences in the structure of C_i compared to the local control units measuring the state (cf. Fig. 4.7) are highlighted in red in Fig. 4.30. Due to the output feedback control, the feedback unit F_i includes a state observer $\Sigma_{\mathrm{O}i}$ to reconstruct the subsystem state $\boldsymbol{x}_i(t)$. Compared to Fig. 4.7 the coupling output $\boldsymbol{z}_i(t)$ is determined in O_i to calculate the coupling signals $\hat{\boldsymbol{s}}_{ji}(t)$. The decision unit D_i is identical to the one for the state feedback control defined in (4.13). In the following, the feedback unit F_i and the observation unit O_i are described in more detail

The feedback unit F_i, $(i \in \mathcal{N})$, generates the input $\boldsymbol{u}_i(t)$ for the subsystem Σ_i

$$F_i : \begin{cases} \boldsymbol{u}_i(t) = -\boldsymbol{K}_{\mathrm{C}i} \hat{\boldsymbol{x}}_i(t) - \tilde{\boldsymbol{u}}_i(t) \\ \Sigma_{\mathrm{O}i} \text{ defined in (4.64)}, \end{cases} \quad \forall i \in \mathcal{N}, \tag{4.63}$$

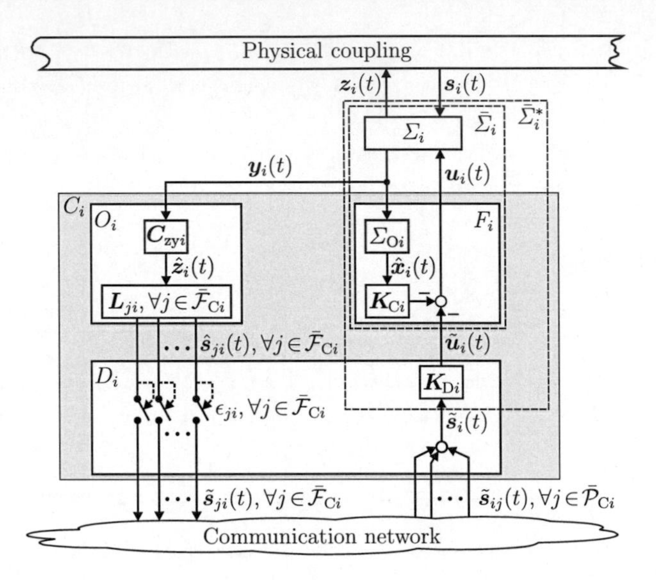

Figure 4.30: Local control unit C_i for situation-dependent compensation of the coupling inputs with an output feedback.

using a linear observer

$$\Sigma_{Oi}: \frac{\mathrm{d}}{\mathrm{d}t}\hat{\boldsymbol{x}}_i(t) = \bar{\boldsymbol{A}}_{Oi}\hat{\boldsymbol{x}}_i(t) - \boldsymbol{B}_i\boldsymbol{K}_{Ci}\hat{\boldsymbol{x}}_i(t) + \boldsymbol{K}_{Oi}\boldsymbol{y}_i(t), \quad \hat{\boldsymbol{x}}_i(0) = \hat{\boldsymbol{x}}_{0i}, \quad \forall i \in \mathcal{N} \tag{4.64}$$

that determines the *estimated subsystem state* $\hat{\boldsymbol{x}}_i(t)$. The matrix $\bar{\boldsymbol{A}}_{Oi} := \boldsymbol{A}_i - \boldsymbol{K}_{Oi}\boldsymbol{C}_i$ resulting from the decentralized observer feedback includes the observer gain matrix $\boldsymbol{K}_{Oi}$. Since the pairs $(\boldsymbol{A}_i, \boldsymbol{C}_i)$, $(\forall i \in \mathcal{N})$, are assumed to be controllable (Assumption 3.2), the observer gain matrix $\boldsymbol{K}_{Oi}$ can be chosen to arbitrarily place the eigenvalues of the matrix $\bar{\boldsymbol{A}}_{Oi}$ (cf. Theorem 2.3). Hence, the following assumption can always be fulfilled.

Assumption 4.3 *The observer gain matrix $\boldsymbol{K}_{Oi}$ is chosen to make $\bar{\boldsymbol{A}}_{Oi} = \boldsymbol{A}_i - \boldsymbol{K}_{Oi}\boldsymbol{C}_i$ a Hurwitz matrix.*

The observation unit O_i uses the measured output $\boldsymbol{y}_i(t)$ to calculate the coupling inputs $\hat{\boldsymbol{s}}_{ji}(t)$

$$O_i: \ \hat{\boldsymbol{s}}_{ji}(t) = \boldsymbol{L}_{ji}\hat{\boldsymbol{z}}_i(t) = \boldsymbol{L}_{ji}\boldsymbol{C}_{zyi}\boldsymbol{y}_i(t), \quad \forall j \in \bar{\mathcal{F}}_{Ci}, \quad i \in \mathcal{N}. \tag{4.65}$$

To guarantee the relation in (4.62),

$$\boldsymbol{C}_{zyi}\boldsymbol{C}_i = \boldsymbol{C}_{zi}, \quad \forall i \in \mathcal{N} \tag{4.66}$$

has to be fulfilled. The relation (4.66) can be fulfilled if and only if $\boldsymbol{C}_{\mathrm{z}i}\boldsymbol{C}_i^+\boldsymbol{C}_i = \boldsymbol{C}_{\mathrm{z}i}$ holds true (cf. Theorem 6.11 in [84]), where $\boldsymbol{C}_i^+$ denotes the pseudoinverse of $\boldsymbol{C}_i$ (cf. (2.1)). Then, the matrix $\boldsymbol{C}_{\mathrm{zy}i}$ can be determined by

$$\boldsymbol{C}_{\mathrm{zy}i} = \boldsymbol{C}_{\mathrm{z}i}\boldsymbol{C}_i^+, \quad i \in \mathcal{N}. \tag{4.67}$$

Therefore, the following assumption is made.

Assumption 4.4 *The matrices $\boldsymbol{C}_{\mathrm{zy}i}$, $(\forall i \in \mathcal{N})$, are chosen such that the relation (4.66) is satisfied.*

In summary, the local control units C_i, $(\forall i \in \mathcal{N})$ with output feedback are defined by

$$C_i : \begin{cases} F_i \text{ defined in (4.63)} \\ O_i \text{ defined in (4.65)} \\ D_i \text{ defined in (4.13)}, \end{cases} \qquad \forall i \in \mathcal{N}, \tag{4.68}$$

where the decision units D_i are identical to the state feedback approach. This leads to the overall definition of the self-organizing controller with output feedback

$$C: \; C_i, (\forall i \in \mathcal{N}), \text{ defined in (4.68) with } \mathcal{E}_{\mathrm{C}}(t) \text{ bounded according to (4.10).} \tag{4.69}$$

Controlled subsystems and self-organizing control system. The extended subsystems $\bar{\Sigma}_i^*$, $(\forall i \in \mathcal{N})$ result from (3.1), (4.13), (4.63) and Assumption 4.2 to

$$\bar{\Sigma}_i^* : \begin{cases} \frac{\mathrm{d}}{\mathrm{d}t}\tilde{\boldsymbol{x}}_i(t) = \tilde{\boldsymbol{A}}_i\tilde{\boldsymbol{x}}_i(t) + \tilde{\boldsymbol{E}}_i\tilde{\boldsymbol{s}}_{\Delta i}(t) + \tilde{\boldsymbol{G}}_i\boldsymbol{d}_i(t), \quad \tilde{\boldsymbol{x}}_i(0) = \tilde{\boldsymbol{x}}_{0i} \\ \boldsymbol{v}_i(t) = \tilde{\boldsymbol{C}}_{\mathrm{v}i}\tilde{\boldsymbol{x}}_i(t) \\ \boldsymbol{z}_i(t) = \tilde{\boldsymbol{C}}_{\mathrm{z}i}\tilde{\boldsymbol{x}}_i(t) \end{cases} \tag{4.70}$$

with

$$\tilde{\boldsymbol{A}}_i := \begin{pmatrix} \boldsymbol{A}_i - \boldsymbol{B}_i\boldsymbol{K}_{\mathrm{C}i} & \boldsymbol{B}_i\boldsymbol{K}_i \\ \boldsymbol{O} & \boldsymbol{A}_i - \boldsymbol{K}_{\mathrm{O}i}\boldsymbol{C}_i \end{pmatrix}, \tag{4.71}$$

$$\tilde{\boldsymbol{E}}_i := \begin{pmatrix} \boldsymbol{E}_i \\ \boldsymbol{E}_i \end{pmatrix}, \quad \tilde{\boldsymbol{G}}_i := \begin{pmatrix} \boldsymbol{G}_i \\ \boldsymbol{G}_i \end{pmatrix}, \quad \tilde{\boldsymbol{x}}_i(t) := \begin{pmatrix} \boldsymbol{x}_i(t) \\ \hat{\boldsymbol{e}}_i(t) \end{pmatrix}, \quad \tilde{\boldsymbol{x}}_{0i} := \begin{pmatrix} \boldsymbol{x}_{0i} \\ \hat{\boldsymbol{e}}_{0i} \end{pmatrix},$$

$$\tilde{\boldsymbol{C}}_{\mathrm{v}i} := \begin{pmatrix} \boldsymbol{C}_{\mathrm{v}i} & \boldsymbol{O} \end{pmatrix}, \quad \tilde{\boldsymbol{C}}_{\mathrm{z}i} := \begin{pmatrix} \boldsymbol{C}_{\mathrm{z}i} & \boldsymbol{O} \end{pmatrix},$$

where $\hat{e}_i(t) := \boldsymbol{x}_i(t) - \hat{\boldsymbol{x}}_i(t)$ is the observation error and $\hat{e}_{0i} := \boldsymbol{x}_{0i} - \hat{\boldsymbol{x}}_{0i}$ the corresponding initial state. Since the system matrix $\tilde{\boldsymbol{A}}_i$ defined in (4.71) is upper block-triangular, the separation theorem (Theorem 2.4) holds similarly for the extended subsystems $\bar{\Sigma}_i^*$ defined in (4.70).

Lemma 4.6 (Separation principle for the extended subsystems) *The spectrum $\sigma(\tilde{\boldsymbol{A}}_i)$ of all system matrices $\tilde{\boldsymbol{A}}_i$, $(\forall i \in \mathcal{N})$, of the extended subsystems $\bar{\Sigma}_i^*$ defined in* (4.70) *consists of the spectrum $\sigma(\boldsymbol{A}_i - \boldsymbol{B}_i\boldsymbol{K}_{\mathrm{C}i})$ of the system matrix $\boldsymbol{A}_i - \boldsymbol{B}_i\boldsymbol{K}_{\mathrm{C}i}$ of the extended subsystem $\bar{\Sigma}_i^*$ defined in* (4.70) *with state feedback and the spectrum $\sigma(\boldsymbol{A}_i - \boldsymbol{K}_{\mathrm{O}i}\boldsymbol{C}_i)$ of the system matrix $\boldsymbol{A}_i - \boldsymbol{K}_{\mathrm{O}i}\boldsymbol{C}_i$ of the linear observer $\Sigma_{\mathrm{O}i}$ defined in* (4.64)*:*

$$\sigma(\tilde{\boldsymbol{A}}_i) = \sigma(\boldsymbol{A}_i - \boldsymbol{B}_i\boldsymbol{K}_{\mathrm{C}i}) \cup \sigma(\boldsymbol{A}_i - \boldsymbol{K}_{\mathrm{O}i}\boldsymbol{C}_i), \quad \forall i \in \mathcal{N}.$$

Proof. Since the system matrix $\tilde{\boldsymbol{A}}_i$ defined in (4.71) of the extended subsystem $\bar{\Sigma}_i^*$ is a block upper triangular matrix, its eigenvalues are equal to the eigenvalues of the matrices $\boldsymbol{A}_i - \boldsymbol{B}_i\boldsymbol{K}_{\mathrm{C}i}$ and $\boldsymbol{A}_i - \boldsymbol{K}_{\mathrm{O}i}\boldsymbol{C}_i$. □

This result is not surprising but interesting. Since the difference signal $\tilde{\boldsymbol{s}}_{\Delta i}(t)$ which is an input for $\bar{\Sigma}_i^*$ is bounded according to Lemma 4.1, the stability of the closed-loop system Σ only depends on the spectrum $\sigma(\boldsymbol{A}_i - \boldsymbol{B}_i\boldsymbol{K}_{\mathrm{C}i})$, $(\forall i \in \mathcal{N})$ and the spectrum $\sigma(\boldsymbol{A}_i - \boldsymbol{K}_{\mathrm{O}i}\boldsymbol{C}_i)$. This analysis is performed in the following section.

Due to Assumption 4.1 and Assumption 4.3, and since the separation principle holds for all $\tilde{\boldsymbol{A}}_i$, $(\forall i \in \mathcal{N})$, defined in (4.71) according to Lemma 4.6, all $\tilde{\boldsymbol{A}}_i$ are Hurwitz matrices as well.

The self-organizing control system Σ with output feedback follows from (3.3), (4.70), (4.3) and (4.69) to

$$\Sigma : \begin{cases} \frac{\mathrm{d}}{\mathrm{d}t}\tilde{\boldsymbol{x}}(t) = \tilde{\boldsymbol{A}}\tilde{\boldsymbol{x}}(t) + \tilde{\boldsymbol{E}}\tilde{\boldsymbol{s}}_{\Delta}(t) + \tilde{\boldsymbol{G}}\boldsymbol{d}(t), \; \tilde{\boldsymbol{x}}(0) = \tilde{\boldsymbol{x}}_0 \\ \boldsymbol{v}(t) = \tilde{\boldsymbol{C}}_{\mathrm{v}}\tilde{\boldsymbol{x}}(t) \\ \boldsymbol{z}(t) = \tilde{\boldsymbol{C}}_{\mathrm{z}}\tilde{\boldsymbol{x}}(t) \\ \boldsymbol{s}(t) = \boldsymbol{L}\boldsymbol{z}(t) \text{ according to (3.3)} \\ O_i, \; (\forall i \in \mathcal{N}) \text{ defined in (4.65)} \\ D_i, \; (\forall i \in \mathcal{N}) \text{ defined in (4.13),} \end{cases} \tag{4.72}$$

with

$$\tilde{\boldsymbol{A}} = \mathrm{diag}(\tilde{\boldsymbol{A}}_1 \dots \tilde{\boldsymbol{A}}_N), \quad \tilde{\boldsymbol{E}} = \mathrm{diag}(\tilde{\boldsymbol{E}}_1 \dots \tilde{\boldsymbol{E}}_N), \quad \tilde{\boldsymbol{G}} = \mathrm{diag}(\tilde{\boldsymbol{G}}_1 \dots \tilde{\boldsymbol{G}}_N),$$
$$\tilde{\boldsymbol{C}}_{\mathrm{v}} = \mathrm{diag}(\tilde{\boldsymbol{C}}_{\mathrm{v}1} \dots \tilde{\boldsymbol{C}}_{\mathrm{v}N}), \quad \tilde{\boldsymbol{C}}_{\mathrm{z}} = \mathrm{diag}(\tilde{\boldsymbol{C}}_{\mathrm{z}1} \dots \tilde{\boldsymbol{C}}_{\mathrm{z}N})$$

and the overall state $\tilde{\boldsymbol{x}}(t)$ and the overall initial state $\tilde{\boldsymbol{x}}_0$ defined by

$$\tilde{\boldsymbol{x}}(t) = \begin{pmatrix} \tilde{\boldsymbol{x}}_1(t) \\ \vdots \\ \tilde{\boldsymbol{x}}_N(t) \end{pmatrix}, \quad \tilde{\boldsymbol{x}}_0 = \begin{pmatrix} \tilde{\boldsymbol{x}}_{01} \\ \vdots \\ \tilde{\boldsymbol{x}}_{0N} \end{pmatrix}.$$

Since the matrices $\tilde{\boldsymbol{A}}_i$, $(\forall i \in \mathcal{N})$, are Hurwitz matrices, the overall system matrix $\tilde{\boldsymbol{A}}$ is a Hurwitz matrix.

4.6.3 Transformation of the analysis results and design methods from the state feedback approach to the output feedback approach

Comparing the extended subsystem $\bar{\Sigma}_i^*$ with state feedback defined in (4.18) and the extended subsystem $\bar{\Sigma}_i^*$ with output feedback defined in (4.70) shows that the structure of both models is identical. The state $\boldsymbol{x}_i(t)$ is replaced by the state $\tilde{\boldsymbol{x}}_i(t)$ which includes the observation error $\hat{\mathrm{e}}_i(t)$ (cf. Table 4.6). Hence, also the corresponding initial states differ in the two models. Furthermore, the related matrices listed in Table 4.6 have a different structure for the output feedback. This analysis is summarized in the following corollary which shows that the results for the state feedback approach can be adjusted to use them for the output feedback approach as well.

Corollary 4.2 *The results in Lemma 4.1–4.5, Theorem 4.2–4.7 and Proposition 4.1–4.2, are stated for the self-organizing control system Σ defined in* (4.20) *with a state feedback. These results are obtained for the self-organizing control system Σ defined in* (4.72) *with an output feedback if the signal, vector and matrices in the first line of Table 4.6 are replaced by the signal and matrices in the second line of Table 4.6.*

Proof. The proof is obvious and is not given here. □

Corollary 4.2 shows that the self-organizing control system Σ with output feedback defined in (4.72) can be analyzed in the same way as the self-organizing control system Σ with state feedback defined in (4.20). However, the behavior of both closed-loop systems is different

Table 4.6: Comparison between the extended subsystems $\bar{\Sigma}_i^*$, $(\forall i \in \mathcal{N})$, defined in (4.18) using state feedback and the extended subsystems $\bar{\Sigma}_i^*$ defined in (4.70) using output feedback.

$\bar{\Sigma}_i^*$ in (4.18)	$\boldsymbol{x}_i(t)$	$\boldsymbol{x}_{0i}$	$\bar{\boldsymbol{A}}_i$	$\boldsymbol{E}_i$	$\boldsymbol{G}_i$	$\boldsymbol{C}_{\mathrm{v}i}$	$\boldsymbol{C}_{\mathrm{z}i}$
$\bar{\Sigma}_i^*$ in (4.70)	$\tilde{\boldsymbol{x}}_i(t)$	$\tilde{\boldsymbol{x}}_{0i}$	$\tilde{\boldsymbol{A}}_i$	$\tilde{\boldsymbol{E}}_i$	$\tilde{\boldsymbol{G}}_i$	$\tilde{\boldsymbol{C}}_{\mathrm{v}i}$	$\tilde{\boldsymbol{C}}_{\mathrm{z}i}$

since due to the observer Σ_{Oi} the output feedback approach includes additional dynamics in the extended subsystems $\bar{\Sigma}_i^*$. Hence, the choice of the dynamical behavior of the observer Σ_{Oi} specifies the deviation between the behavior of Σ with output feedback and Σ with state feedback. This deviation is investigated in more detail in the following section.

Stabilizability with output feedback. Theorem 4.1 is not listed in Corollary 4.2 since the conditions on the properties of the plant P, which guarantee that the self-organizing controller C with output feedback can stabilize the plant P within the overall closed-loop system Σ are different from the ones for the state feedback approach. To derive these conditions, the previous assumptions are neglected for the following theorem.

Theorem 4.8 (Practical stabilizability for output feedback) *The plant P defined in (3.10) can be practically stabilized w.r.t. to a compact set $\mathcal{A}$ by the self-organizing controller C with output feedback defined in (4.69) if all the following five conditions are satisfied:*

1. *The subsystems Σ_i can be physically decoupled, i.e., $\boldsymbol{B}_i\boldsymbol{B}_i^+\boldsymbol{E}_i = \boldsymbol{E}_i$, $(\forall i \in \mathcal{N})$.*
2. *The coupling output $\boldsymbol{z}_i(t)$ can be determined by the measured output $\boldsymbol{y}_i(t)$, i.e., $\boldsymbol{C}_{zi}\boldsymbol{C}_i^+\boldsymbol{C}_i = \boldsymbol{C}_{zi}$, $(\forall i \in \mathcal{N})$.*
3. *The subsystems Σ_i are completely controllable, i.e., the pairs $(\boldsymbol{A}_i, \boldsymbol{B}_i)$ are controllable, $(\forall i \in \mathcal{N})$.*
4. *The subsystems Σ_i are completely observable, i.e., the pairs $(\boldsymbol{A}_i, \boldsymbol{C}_i)$ are observable, $(\forall i \in \mathcal{N})$.*
5. *The disturbances $\boldsymbol{d}_i(t)$ are bounded, i.e., $\|\boldsymbol{d}_i(t)\| < \infty$, $(\forall t \geq 0, \forall i \in \mathcal{N})$.*

Proof. The extended subsystem $\bar{\Sigma}_i^*$ defined in (4.70) can be built if and only if relation (4.3) holds true. Therefore, according to Theorem 6.2 in [84], relation $\boldsymbol{B}_i\boldsymbol{B}_i^+\boldsymbol{E}_i = \boldsymbol{E}_i$ has to be fulfilled (Condition 1). The coupling output $\boldsymbol{z}_i(t)$ has to be determined with $\boldsymbol{y}_i(t)$ as in (4.62). According to Theorem 6.2 in [84], the relation $\boldsymbol{C}_{zi}\boldsymbol{C}_i^+\boldsymbol{C}_i = \boldsymbol{C}_{zi}$ has to hold, to guarantee that Assumption 4.4 can be fulfilled (Condition 2). Consider the bound for $\boldsymbol{x}_i(t)$ in (4.23) by replacing $\boldsymbol{x}_i(t)$, $\boldsymbol{x}_{0i}$, $\bar{\boldsymbol{A}}_i$, $\boldsymbol{E}_i$ and $\boldsymbol{G}_i$ by $\tilde{\boldsymbol{x}}_i(t)$, $\tilde{\boldsymbol{x}}_{0i}$, $\tilde{\boldsymbol{A}}_i$, $\tilde{\boldsymbol{E}}_i$ and $\tilde{\boldsymbol{G}}_i$, respectively. To fulfill the first requirement in Definition 2.3, $\lim_{t\to\infty} \|\mathrm{e}^{\tilde{\boldsymbol{A}}_i t}\tilde{\boldsymbol{x}}_{0i}\| = 0$ has to be satisfied. Therefore, $\tilde{\boldsymbol{A}}_i$ has to be a Hurwitz matrix. If $\tilde{\boldsymbol{A}}_i$ is a Hurwitz matrix, then $\|\mathrm{e}^{\tilde{\boldsymbol{A}}_i t}\|$ is bounded from above. Hence, there always exists an initial state $\|\tilde{\boldsymbol{x}}_{0i}\| \leq \delta$ that fulfills the second requirement in Definition 2.3. According to Lemma 4.6, the matrix $\tilde{\boldsymbol{A}}_i$ is a Hurwitz matrix if and only if $\boldsymbol{A}_i - \boldsymbol{B}_i\boldsymbol{K}_{Ci}$ and $\boldsymbol{A}_i - \boldsymbol{K}_{Oi}\boldsymbol{C}_i$ are Hurwitz matrices. According to Theorem 2.2, if the pair $(\boldsymbol{A}_i, \boldsymbol{B}_i)$ is controllable, then the feedback gain matrix $\boldsymbol{K}_{Ci}$ can be chosen to make $\boldsymbol{A}_i - \boldsymbol{B}_i\boldsymbol{K}_{Ci}$

a Hurwitz matrix (Condition 3). According to Theorem 2.3, if the pair $(\boldsymbol{A}_i, \boldsymbol{C}_i)$ is observable, an observer gain matrix $\boldsymbol{K}_{\mathrm{O}i}$ can be chosen to make $\boldsymbol{A}-\boldsymbol{K}_{\mathrm{O}i}\boldsymbol{C}$ a Hurwitz matrix (Condition 4). Furthermore, according to Lemma 4.2 the value for $\max_{t\geq 0} \|\boldsymbol{d}_i(t)\|$ has to be bounded which is fulfilled for $\|\boldsymbol{d}_i(t)\| < \infty$ (Condition 5). □

Theorem 4.8 states five sufficient conditions on the plant P defined in (3.10) for which the plant P can be practically stabilized with the proposed self-organizing controller C with output feedback. The relaxation of the first condition in Theorem 4.1, that the subsystem state $\boldsymbol{x}_i$ can be measured, leads to two weaker conditions in Theorem 4.8:

- The coupling outputs $\boldsymbol{z}_i(t)$, $(\forall i \in \mathcal{N})$, have to be calculable by the measured outputs $\boldsymbol{y}_i(t)$ according to (4.62) (Condition 2 in Theorem 4.8) which is always fulfilled when the state is measurable since $\boldsymbol{C}_i = \boldsymbol{I}_{n_i}$ holds true for $\boldsymbol{y}_i(t) = \boldsymbol{x}_i(t)$, i.e., $\boldsymbol{C}_{\mathrm{zy}i} = \boldsymbol{C}_{\mathrm{z}i}$.

- The subsystems Σ_i, $(\forall i \in \mathcal{N})$, have to be completely observable (Condition 4 in Theorem 4.8) to reconstruct the state $\boldsymbol{x}_i(t)$ which is not necessary if the state can be measured.

As the Theorem 4.1, all conditions depend on the dynamics of the isolated subsystems Σ_i, $(\forall i \in \mathcal{N})$ only. This leads to a similar property of the proposed self-organizing control concept with output feedback:

> The practical stabilizability of the plant P to be controlled by the self-organizing controller C with output feedback can be checked by using local model information of the subsystems only.

Overall design of the local control units. Algorithm 4.1 presented in Section 4.4.3 can be adapted to design the local control units C_i with output feedback defined in (4.68). Therefore, Step 1 and Step 2 in Algorithm 4.1 are changed which leads to Algorithm 4.2. The differences to Algorithm 4.1 are highlighted in red. Note that only the design of the feedback unit F_i and the observation unit O_i varies, whereas the design of the decision units D_i, $(\forall i \in \mathcal{N})$, stays the same.

Algorithm 4.2: Design local control units C_i, $(\forall i \in \mathcal{N})$, defined in (4.68) with output feedback which compensate physical coupling inputs to bound the disturbance propagation according to Control aim 3.3.

Given: $\bar{v}_i^*$, Σ_i, $\mathcal{P}_{\mathrm{P}i}$, $\mathcal{F}_{\mathrm{P}i}$, $\boldsymbol{L}_{ji}$, $(\forall i \in \mathcal{N}, \forall j \in \mathcal{F}_{\mathrm{P}i})$

for i **to** N **do**

1 Construct F_i defined in (4.63) by
- determining $\boldsymbol{K}_{\mathrm{C}i}$ such that $\bar{\boldsymbol{A}}_i = \boldsymbol{A}_i - \boldsymbol{B}_i \boldsymbol{K}_{\mathrm{C}i}$ is a Hurwitz matrix and
- determining $\boldsymbol{K}_{\mathrm{O}i}$ such that $\bar{\boldsymbol{A}}_{\mathrm{O}i} = \boldsymbol{A}_i - \boldsymbol{K}_{\mathrm{O}i} \boldsymbol{C}_i$ is a Hurwitz matrix.

2 Construct O_i defined in (4.65) by
- setting $\bar{\mathcal{P}}_{\mathrm{C}i} = \mathcal{P}_{\mathrm{P}i}$ and $\bar{\mathcal{F}}_{\mathrm{C}i} = \mathcal{F}_{\mathrm{P}i}$,
- extracting the matrix $\boldsymbol{C}_{\mathrm{z}i}$ from Σ_i,
- determining the matrix $\boldsymbol{C}_{\mathrm{zy}i}$ with (4.67) and
- adding the matrices $\boldsymbol{L}_{ji}$, $(\forall j \in \bar{\mathcal{F}}_{\mathrm{C}i})$.

3 Construct D_i defined in (4.13) by
- setting $\bar{\mathcal{P}}_{\mathrm{C}i} = \mathcal{P}_{\mathrm{P}i}$ and $\bar{\mathcal{F}}_{\mathrm{C}i} = \mathcal{F}_{\mathrm{P}i}$,
- calculating the decoupling matrix $\boldsymbol{K}_{\mathrm{D}i}$ with (4.17) and
- choosing the switching thresholds $\bar{\epsilon}_i = \epsilon_{ij}$ for C_j, $(\forall j \in \bar{\mathcal{P}}_{\mathrm{C}i})$, according to (4.57).

end

Result: C_i with output feedback consisting of F_i, O_i and D_i guaranteeing that the performance output $\boldsymbol{v}_i(t)$ of the undisturbed subsystems is upper bounded by $\bar{v}_i^*$ as claimed in Control aim 3.3.

4.6.4 Application example: Water supply system

For the water supply system described in Section 3.5.1 it is considered that there is no level sensor at tank TAi, i.e., $\boldsymbol{C}_i = \begin{pmatrix} 0 & 1 \end{pmatrix}$, for all $i \in \mathcal{N}$. The control aim is identical to the one for controlling the tanks by a state feedback presented in Section 4.5, i.e., the disturbance propagation should be bounded such that the level $t_{\mathrm{TB}i}(t) = \boldsymbol{v}_i(t)$, of the undisturbed subsystems Σ_i, $(\forall i \in \mathcal{U})$, are upper bounded by 1 cm according to (4.60).

Design of the local control units using output feedback. Algorithm 4.2 leads to the parameters of local control units C_i defined in (4.68) listed in Table 4.7. In the following the additional steps in Algorithm 4.2 compared to Algorithm 4.1 are presented:

1. The observer gain matrices $\boldsymbol{K}_{\mathrm{O}i}$, $(\forall i \in \mathcal{N})$, are designed such that the eigenvalues of $\boldsymbol{A}_i - \boldsymbol{K}_{\mathrm{O}i} \boldsymbol{C}_i$ are smaller than the dominant eigenvalues of $\bar{\boldsymbol{A}}_i$.

2. The matrix $\boldsymbol{C}_{\mathrm{zy}i}$, $(i \in \mathcal{N})$, for the observation unit O_i is determined by (4.67).

Table 4.7: Parameters of the local control units C_i using output feedback defined in (4.68) for the water supply system.

	$\forall i \in \mathcal{N}_{\text{large}}$	$\forall i \in \mathcal{N}_{\text{middle}}$	$\forall i \in \mathcal{N}_{\text{small}}$
$\bar{\mathcal{P}}_{\text{C}i}$, $\bar{\mathcal{F}}_{\text{C}i}$, $\boldsymbol{K}_{\text{C}i}$, $\boldsymbol{L}_{ji}$ and $\boldsymbol{K}_{\text{D}i}$ from Table 4.3			
$\boldsymbol{C}_{\text{zy}i}$	0.4	0.5	0.8
$\boldsymbol{K}_{\text{O}i}$	$\begin{pmatrix} 59.021 \\ 0.8426 \end{pmatrix}$	$\begin{pmatrix} 8.312 \\ 0.408 \end{pmatrix}$	$\begin{pmatrix} 2.061 \\ 0.283 \end{pmatrix}$
$\bar{\epsilon}_i$	0.020	0.021	0.042
ϵ_{ij}	$\bar{\epsilon}_i, \ \forall j \in \bar{\mathcal{P}}_{\text{C}i}$		

Behavior of the self-organizing controlled tanks while using output feedback. Figure 4.32 shows that the behavior of the self-organizing controlled tanks with output feedback is similar to the behavior with state feedback (cf. Fig. 4.22). The maximal value of the level $l_{\text{TB}i}(t)$, $(\forall i \in \{6, 9, 17\})$, of the disturbed subsystems (Σ_6, Σ_9 and Σ_{17}) is a little bit larger with output feedback. Furthermore, due to the smaller switching thresholds ϵ_{ji} and the overshoot of the levels, the communication intervals are larger than with the state feedback approach. However, the aim of disturbance attenuation is achieved since the effect of the situation-dependent compensation of the coupling inputs leads to a much better performance of the undisturbed subsystems Σ_i, $(\forall i \in \mathcal{N}\backslash\{6, 9, 17\})$, compared to the use of a decentralized controller C_{D}.

Fulfillment of desired difference behavior (control aim). Figure 4.31 shows that the control aim claimed in (4.60) is also guaranteed for all undisturbed subsystems Σ_i, $(\forall i \in \mathcal{U} = \mathcal{N}\backslash\{6, 9, 17\})$, while using an output feedback. The levels $l_{\text{TB}i}(t) = \boldsymbol{v}_i(t)$, $(\forall i \in \mathcal{U})$, are smaller than the desired maximal performance output of $\bar{v}_i^* = 1\,\text{cm}$ for all $t \geq 0$.

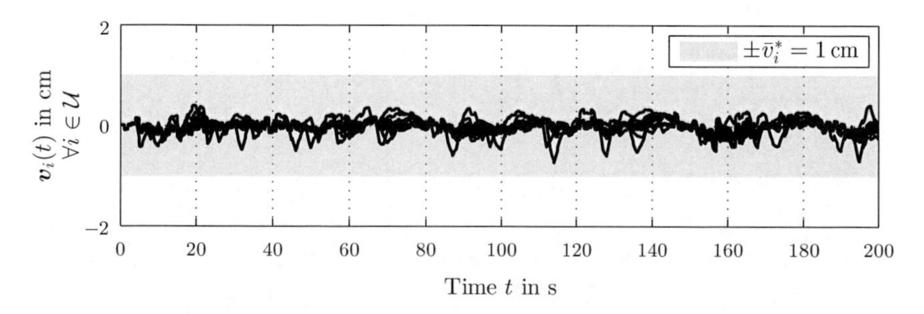

Figure 4.31: Verification of the control aim (4.60); Disturbance attenuation of the self-organizing controlled water supply system while using output feedback.

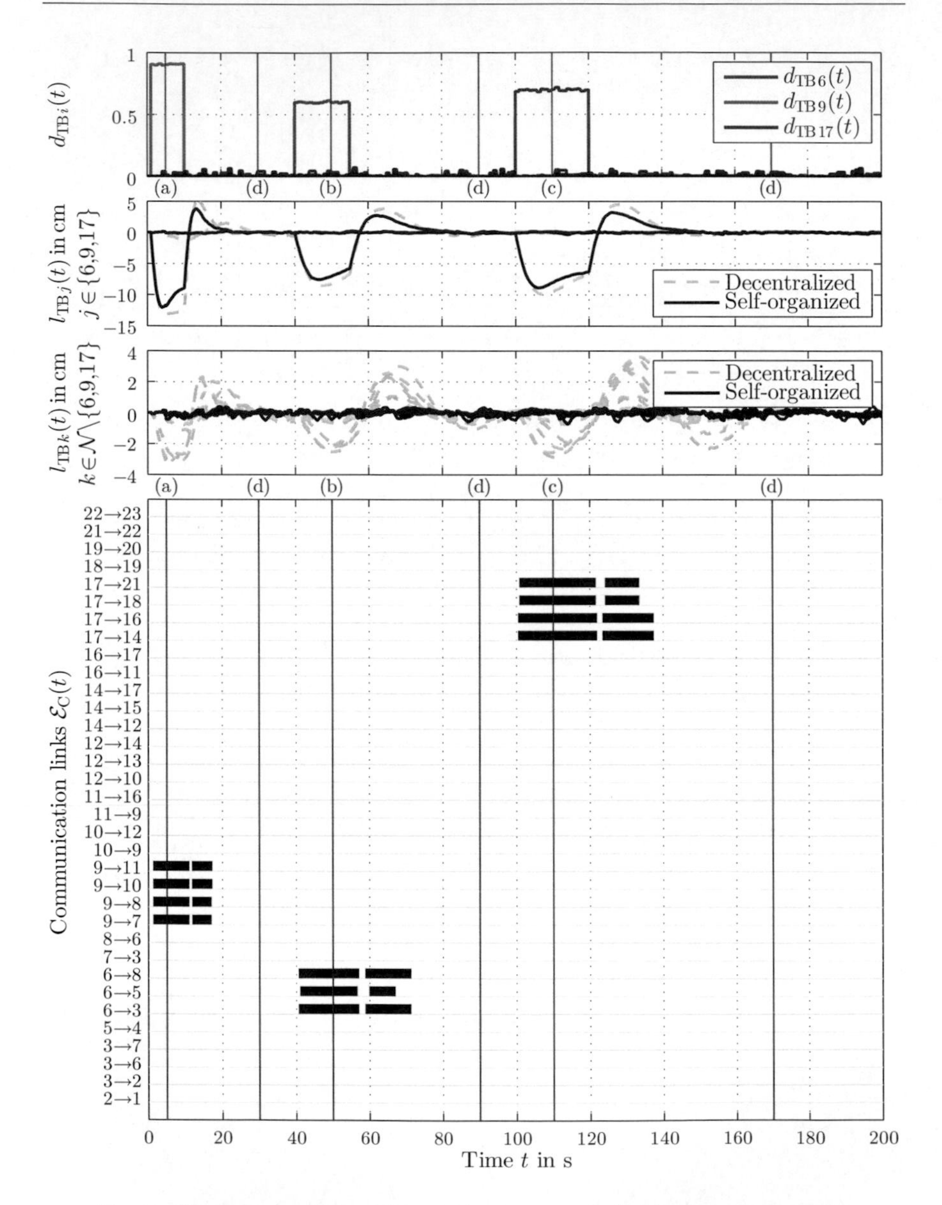

Figure 4.32: Behavior of the water supply system controlled by the self-organizing controller C with output feedback defined in (4.69).

4.6.5 Comparison with state feedback approach

The main similarities and differences of the output feedback control approach and the state feedback control approach can be summarized as follows:

- The structure of the local control units C_i is similar (compare Fig. 4.7 and Fig. 4.30). But there is an additional dynamics due to the linear observer within C_i using output feedback.
- For guaranteeing the practical stabilizability of the plant P to be controlled by the self-organizing controller with output feedback, the subsystems additionally have to be completely controllable and the coupling outputs $\boldsymbol{z}_i(t)$, $(\forall i \in \mathcal{N})$, have to be calculable from the corresponding measured outputs $\boldsymbol{y}_i(t)$ (cf. (4.62) and Assumption 4.4).
- The observer and the state feedback can be designed independently to guarantee the practical stability of the self-organizing control system with output feedback (cf. Lemma 4.6 and Theorem 4.8).
- The analysis results for the self-organizing control system and the design methods for the local control units are similar to the ones for the state feedback approach (Corollary 4.2).

In summary, the basic properties of the self-organizing control system Σ with state feedback maintain for the output feedback approach.

4.6.6 Extension to output feedback without observer

Consider that the first problem stated in Section 4.6.1 - the stabilization of the isolated subsystems - is solved by local control units C_i including a feedback unit F_i with a permanent local output feedback

$$F_i : \boldsymbol{u}_i(t) = -\boldsymbol{K}_{\mathrm{Y}i}\boldsymbol{y}_i(t) - \tilde{\boldsymbol{u}}_i(t), \quad \forall i \in \mathcal{N}$$

while using the measured output $\boldsymbol{y}_i(t)$ instead of the observed subsystem state $\hat{\boldsymbol{x}}_i(t)$ and the feedback gain matrix $\boldsymbol{K}_{\mathrm{Y}i} \in \mathbb{R}^{m_i \times r_i}$ instead of $\boldsymbol{K}_{\mathrm{C}i}$ (Fig. 4.33). The difference in the structure of C_i compared to the local control using an observer (cf. Fig. 4.30) is highlighted in red in Fig. 4.33. With this local feedback the extended subsystem $\bar{\Sigma}_i^*$ is given by

$$\bar{\Sigma}_i^* : \begin{cases} \dot{\boldsymbol{x}}_i(t) = \bar{\boldsymbol{A}}_{\mathrm{Y}i}\boldsymbol{x}_i(t) + \boldsymbol{E}_i\tilde{\boldsymbol{s}}_{\Delta i}(t) + \boldsymbol{G}_i\boldsymbol{d}_i(t), & \boldsymbol{x}_i(0) = \boldsymbol{x}_{0i} \\ \boldsymbol{v}_i(t) = \boldsymbol{C}_{\mathrm{v}i}\boldsymbol{x}_i(t) \\ \boldsymbol{z}_i(t) = \boldsymbol{C}_{\mathrm{z}i}\boldsymbol{x}_i(t) \end{cases}$$

resulting from (3.1), (4.13) and (4.63) and Assumption 4.2, where the corresponding system matrix is given by

$$\bar{\boldsymbol{A}}_{\mathrm{Y}i} = \boldsymbol{A}_i - \boldsymbol{B}_i \boldsymbol{K}_{\mathrm{Y}i}, \quad \forall i \in \mathcal{N}.$$

Hence, if $\boldsymbol{K}_{\mathrm{Y}i}$ is chosen such that $\bar{\boldsymbol{A}}_{\mathrm{Y}i}$ is a Hurwitz matrix, the analysis results and the design methods for the state feedback approach presented in Sections 4.3 and 4.4 generally hold for the self-organizing control system C with the local control units C_i illustrated in Fig. 4.33 as well. The only difference is that the matrices $\bar{\boldsymbol{A}}_i = \boldsymbol{A}_i - \boldsymbol{B}_i \boldsymbol{K}_{\mathrm{C}i}$, $(\forall i \in \mathcal{N})$, are replaced by the matrices $\bar{\boldsymbol{A}}_{\mathrm{Y}i} = \boldsymbol{A}_i - \boldsymbol{B}_i \boldsymbol{K}_{\mathrm{Y}i}$. However, it is well known that it is not guaranteed that there exists a feedback gain matrix $\boldsymbol{K}_{\mathrm{Y}i}$ which leads to an Hurwitz matrix $\bar{\boldsymbol{A}}_{\mathrm{Y}i}$. Nevertheless, if such a matrix $\boldsymbol{K}_{\mathrm{Y}i}$ exists the proposed self-organizing control concept can be applied.

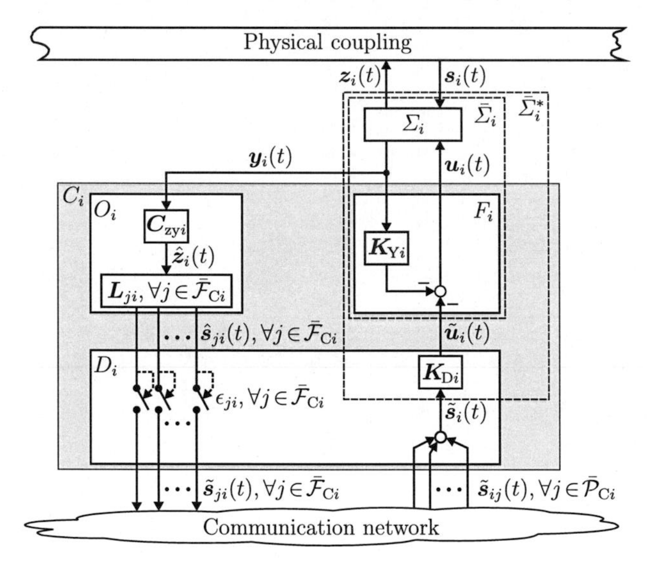

Figure 4.33: Local control unit C_i for situation-dependent compensation of the coupling inputs with an output feedback without observer.

4.7 Integration of a new subsystem

In the following a procedure for integrating a new subsystem into an already present self-organizing control system of the form (4.20) is presented (Fig. 4.34) which is related to the desired property of scalability. The overall initial situation is described by:

- There is a present self-organizing control system Σ^{old} of the form (4.20) consisting of the plant P^{old} with $N-1$ subsystems Σ_i, $(\forall i \in \mathcal{N}^{\mathrm{old}} := \{1, 2 \ldots N-1\})$, and the self-organizing controller C^{old} of the form (4.15).
- There is a new subsystem Σ_N controlled by the local control unit C_N of the form (4.14). The parameters of C_N are not yet chosen.
- Each local control unit C_i, $(\forall i \in \mathcal{N})$, has a design agent A_i which can adapt (design) the parameters of C_i.
- The design agents A_i, $(\forall i \in \mathcal{N})$, know the local control aim, have local model information of the corresponding subsystem Σ_i and can exchange model information and controller parameters over the communication network which is indicated by the thick arrows in Fig. 4.34.

The overall aim is to integrate subsystem Σ_N with its corresponding local control unit C_N into the present self-organizing control system Σ^{old} to build a new self-organizing control system Σ^{new} which is practically stable and guarantees a desired disturbance attenuation according to Control aim 3.3. Hence, there are three main questions to be answered to fulfill the overall aim:

1. How should the design agents A_N design the parameters of the local control unit C_N?

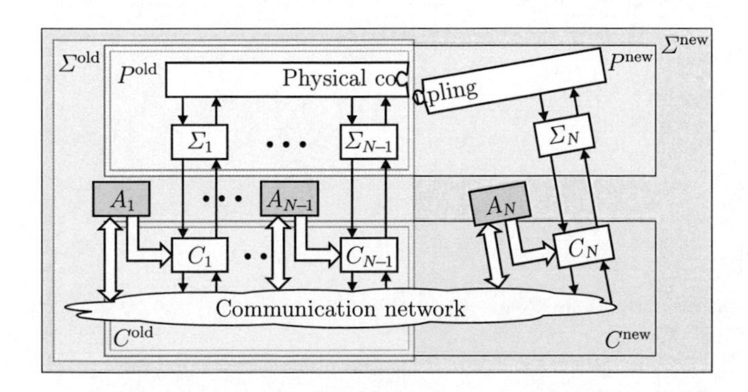

Figure 4.34: Integration of a new subsystem Σ_N into a present self-organizing control system Σ^{old} of the form (4.20).

2. Which local control units C_i, $(i \in \mathcal{N}^{\text{old}})$, have to be updated in which way by the design agents A_i?

3. Which information have to be exchanged among the design agents A_i, $(\forall i \in \mathcal{N})$?

Procedure to integrate a new subsystem. A procedure for the integration of the new subsystem Σ_N is given in Table 4.8 which lists the executed steps of design agents A_i, $(\forall i \in \mathcal{N})$. It is distinguished between the design agent A_N of the new subsystem Σ_N and the design agents A_i, $(\forall i \in \mathcal{N}^{\text{old}})$, of the old subsystems Σ_i since the steps of all A_i are similar. All parameters of C_N have to be designed by A_N, whereas only the parameters of some C_i, $(i \in \mathcal{N}^{\text{old}})$, have to be updated by A_i.

The design agents A_i, $(\forall i \in \mathcal{N}^{\text{old}})$ have the same given information as in Algorithm 4.1. Design agent A_N additionally has information about the coupling matrices $\boldsymbol{L}_{Nk}$, $(\forall k \in \mathcal{P}_{\text{P}N})$, which describe the physical coupling impact on Σ_N by the old subsystems Σ_i, $(\forall i \in \mathcal{N}^{\text{old}})$.

In Step 1 design agent A_N sends the physical coupling matrices $\boldsymbol{L}_{Nk}$, $(\forall k \in \mathcal{P}_{\text{P}N})$, to the design agents A_k which need the actual value of $\boldsymbol{L}_{Nk}$ and use $\boldsymbol{L}_{Nk}$ to update their sets $\mathcal{F}_{\text{P}k}$ of physical followers. The design agent A_N sends a message to the design agents A_j, $(\forall j \in \mathcal{F}_{\text{P}N})$, to inform A_j that the new subsystem Σ_N has a physical impact on the subsystems Σ_j (Step 2). What kind of message is sent from A_N to A_j is not specified in more detail since the implementation does not matter for understanding the procedure. If A_i, $(i \in \mathcal{N}^{\text{old}})$, receives a

Table 4.8: Overall procedure of the design agents A_i, $(\forall i \in \mathcal{N})$, to integrate a new subsystem Σ_N into an existing self-organizing control system Σ^{old}.

	Design agent A_N	Design agents A_i, $(\forall i \in \mathcal{N}^{\text{old}})$
Given:	$\bar{v}_N^*$, Σ_N, $\mathcal{P}_{\text{P}N}$, $\mathcal{F}_{\text{P}N}$, $\boldsymbol{L}_{jN}$, $(\forall j \in \mathcal{F}_{\text{P}N})$, $\boldsymbol{L}_{Nk}$, $(\forall k \in \mathcal{P}_{\text{P}N})$	$\bar{v}_i^*$, Σ_i, $\mathcal{P}_{\text{P}i}^{\text{old}}$, $\mathcal{F}_{\text{P}i}^{\text{old}}$, $\boldsymbol{L}_{ji}$, $(\forall j \in \mathcal{F}_{\text{P}i}^{\text{old}})$
Step 1	Send $\boldsymbol{L}_{N,k}$, $(\forall k \in \mathcal{P}_{\text{P}N})$ to A_k	Update physical followers $\mathcal{F}_{\text{P}i} = \begin{cases} \mathcal{F}_{\text{P}i}^{\text{old}} \cup \{N\} & \text{if } \boldsymbol{L}_{Ni} \text{ was send to } A_i \\ \mathcal{F}_{\text{P}i}^{\text{old}} & \text{else} \end{cases}$
Step 2	Send message A_j, $(\forall j \in \mathcal{F}_{\text{P}N})$	Update physical predecessors $\mathcal{P}_{\text{P}i} = \begin{cases} \mathcal{P}_{\text{P}i}^{\text{old}} \cup \{N\} & \text{if } A_N \text{ gets message from } A_i \\ \mathcal{P}_{\text{P}i}^{\text{old}} & \text{else} \end{cases}$
Step 3	Perform Algorithm 4.1 for designing C_N	- Step 1&3 of Alg. 4.1 if $\mathcal{F}_{\text{P}i} \neq \mathcal{F}_{\text{P}i}^{\text{old}}$ or $\mathcal{P}_{\text{P}i} \neq \mathcal{P}_{\text{P}i}^{\text{old}}$ - Step 2 of Alg. 4.1 if $\mathcal{F}_{\text{P}i} \neq \mathcal{F}_{\text{P}i}^{\text{old}}$ - Step 3 of Alg. 4.1 if $\mathcal{P}_{\text{P}i} \neq \mathcal{P}_{\text{P}i}^{\text{old}}$
Step 4	Integrate/Update $\epsilon_{ij} = \bar{\epsilon}_i$ at C_j, $(\forall j \in \bar{\mathcal{P}}_{\text{C}i})$, if $\bar{\epsilon}_i$ is sent to A_j	
Result:	Σ^{new} is practically stable and guarantees Control aim 3.3.	

message from A_N, it updates its set $\mathcal{P}_{\mathrm{P}i}$ of physical followers (Step 2). In Step 3 A_N performs Algorithm 4.1 to design the local control unit C_N. The design agents A_i, $(\forall i \in \mathcal{N}^{\mathrm{old}})$, of the subsystems Σ_i with a changed physical coupling structure, i.e., $\mathcal{F}_{\mathrm{P}i} \neq \mathcal{F}_{\mathrm{P}i}^{\mathrm{old}}$ or $\mathcal{P}_{\mathrm{P}i} \neq \mathcal{P}_{\mathrm{P}i}^{\mathrm{old}}$, have to update the local communication structure of C_i defined by $\bar{\mathcal{F}}_{\mathrm{C}i}$ and $\bar{\mathcal{P}}_{\mathrm{C}i}$ by performing Step 1 and Step 3 of Algorithm 4.1. Furthermore, if the set of physical followers has changed, i.e., $\mathcal{F}_{\mathrm{P}i} \neq \mathcal{F}_{\mathrm{P}i}^{\mathrm{old}}$, then the observation unit O_i of C_i has to be updated by performing Step 2 of Algorithm 4.1. Moreover, if the set of physical predecessors has changed, i.e., $\mathcal{P}_{\mathrm{P}i} \neq \mathcal{P}_{\mathrm{P}i}^{\mathrm{old}}$, then the threshold $\bar{\epsilon}_i = \epsilon_{ij}$, $(\forall j \in \mathcal{P}_{\mathrm{P}i})$ has to be redesigned and has to be sent to the corresponding design agents A_j by performing Step 3 of Algorithm 4.1.

The redesign of the feedback-gain matrix $\boldsymbol{K}_{\mathrm{C}i}$, $(i \in \mathcal{N}^{\mathrm{old}})$, and the decoupling matrix $\boldsymbol{K}_{\mathrm{D}i}$ is not necessary since they depend on the dynamics of Σ_i only. Note that the Step 1-2 are not performed in parallel. Agent A_N first sends information which is then handled in A_i, $(\forall i \in \mathcal{N}^{\mathrm{old}})$.

The result of the procedure in Table 4.8 is that the new self-organizing control system Σ^{new} is practically stable according to Theorem 4.2 and that Σ^{new} guarantees the desired disturbance attenuation claimed in Control aim 3.3.

Summary. With the proposed procedure in Table 4.8 the three mentioned questions can be answered as follows:

1. The design agent A_N has to perform Algorithm 4.1 to design the parameters of C_N.

2. All local control units for which the physical coupling structure changes due to the new subsystem have to be updated, where the observation unit and/or the switching thresholds have to be adapted.

3. A_N has to inform the other design agents about the new coupling structure. Redesigned switching threshold have to be exchanged.

The design and update of the local control units is performed in a distributed manner with local model information only. The update of the present local control units is initialized be the design agent C_N of the new subsystems Σ_N by informing the present design agents about the new physical coupling structure among the subsystems.

In summary, the following property of the proposed self-organizing control concept with a situation-dependent compensation of physical coupling inputs is highlighted:

> New subsystems can be easily integrated into a self-organizing control system Σ of the form (4.20), where the update of the local control units is limited to the part of the physical coupling graph $\mathcal{G}_{\mathrm{P}}$ connected with the new subsystem (Fig. 4.35).

This property is illustrated in the following example.

Example 4.2 *Integration of a new subsystem*

Consider the physical coupling graph in Fig. 4.35. The subsystems Σ_i, ($\forall i \in \{1, \dots, 6\}$), belong to the present self-organizing control system Σ^{old}. Subsystem Σ_7 is the new subsystem which shall be integrated into Σ^{old} to build a new self-organizing control system Σ^{new}. According to the overall procedure in Table 4.8, the steps of the design agents A_i are as follows

- **Design agent** A_7**:**
  1. Send $\boldsymbol{L}_{74}$ to A_4.
  2. Send message to A_5 that Σ_7 has a physical impact on Σ_5.
  3. Design C_7 by performing Algorithm 4.1.
  4. Integrate $\epsilon_{57} = \bar{\epsilon}_5$ in D_7.
- **Design agent** A_4**:**
  1. Update $\mathcal{F}_{\text{P4}} = \mathcal{F}_{\text{P4}}^{\text{old}} \cup \{7\} = \{2, 3, 5, 7\}$.
  2. -
  3. Update $\bar{\mathcal{F}}_{\text{C4}} = \mathcal{F}_{\text{P4}} = \{2, 3, 5, 7\}$. Update O_4 with $\boldsymbol{L}_{74}$ and $\bar{\mathcal{F}}_{\text{C4}}$.
  4. Update $\epsilon_{54} = \bar{\epsilon}_5$ and integrate $\epsilon_{74} = \bar{\epsilon}_7$ in D_4.
- **Design agent** A_5**:**
  1. -
  2. Update $\mathcal{P}_{\text{P5}} = \mathcal{P}_{\text{P5}}^{\text{old}} \cup \{7\} = \{2, 4, 7\}$.
  3. Update $\bar{\mathcal{P}}_{\text{C5}} = \mathcal{P}_{\text{P5}} = \{2, 4, 7\}$. Redesign $\bar{\epsilon}_5$ and send $\bar{\epsilon}_5$ to A_2, A_4 and A_7.
  4. -
- **Design agent** A_2**:**
  1. -
  2. -
  3. -
  4. Update $\epsilon_{52} = \bar{\epsilon}_5$ in D_2.

Only the local control units C_2, C_4 and C_5 have do be updated, whereas the other local control units remain unchanged (Fig. 4.35). Hence, the update of the local control units is limited to a certain part of the physical coupling graph $\mathcal{G}_{\text{P}}$. □

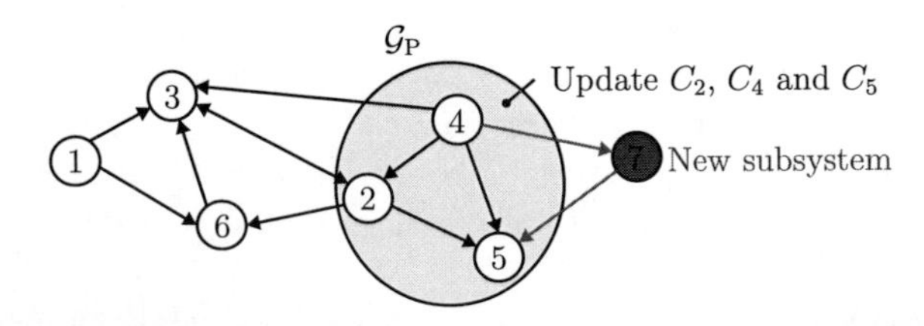

Figure 4.35: Example for the update of the local control units in the case of a new subsystem.

4.8 Fault tolerance of the self-organizing control system

This section investigates how the self-organizing control system tolerates possible faults within the subsystems or the local control units. In Section 4.8.2 the behavior of the closed-loop system in case of a subprocess fault is analyzed. A discussion of the propagation of other faults within the self-organizing control system is presented in Section 4.8.3. The behavior of the self-organizing controlled coupled tanks affected by different faults is illustrated in Section 4.8.4.

4.8.1 Effects of faults on the closed-loop system

As for disturbances, the effect of a fault at a single subsystem spreads out on other subsystems due to the physical coupling or the communication among the local control units (Fig. 4.36). The analysis of the disturbance propagation in Section 4.3.4 illustrates that the proposed self-organizing controller C defined in (4.15) bounds the influence of a disturbance on the other subsystems through the situation-dependent compensation of the physical coupling inputs. In the following it is investigated whether Σ has a similar property in case of faulty subsystems or not. Therefore, three main questions are considered:

1. How does a fault at one subsystem influence the behavior of the other subsystem within the closed-loop system Σ?

2. Which communication is necessary to suppress the fault propagation within the self-organizing control system?

3. Which local control units have to be reconfigured and how do they have to be reconfigured to guarantee the desired performance of the overall system?

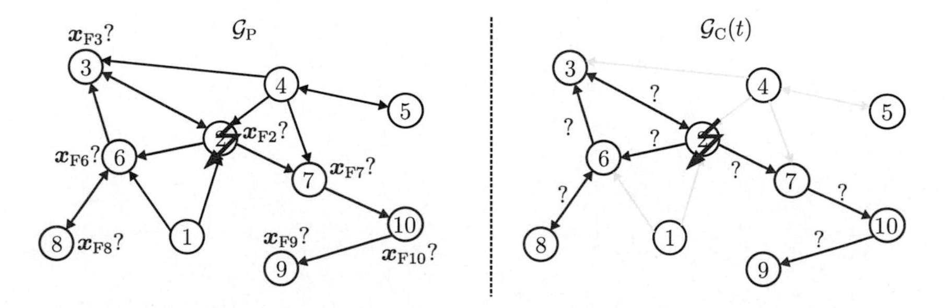

Figure 4.36: Possible fault propagation (left) and communication activation in case of a fault (right) with $\mathcal{N}_F = \{2\}$.

The first two questions consider the passive fault tolerance of the self-organizing control system Σ defined in (4.20). The third question concerns active fault tolerance of Σ with a reconfiguration of the local control units C_i.

In the following all three questions are investigated for subprocess faults (Section 4.8.2). Furthermore, the first two questions are briefly discussed in Section 4.8.3 for sensor faults, actuator faults and communication faults.

4.8.2 Impact of a subprocess fault

This section investigates the three mentioned questions for subprocess faults. Recall that condition (3.14) has to be fulfilled for subprocess fault in a subsystem $\Sigma_{\mathrm{F}i}$, $(i \in \mathcal{N})$, (cf. Section 3.2.4). With this, it is considered that

$$\boldsymbol{A}_{\mathrm{F}k} \neq \boldsymbol{A}_k, \quad \forall k \in \mathcal{N}_{\mathrm{F}} \tag{4.73a}$$

$$\boldsymbol{A}_{\mathrm{F}j} = \boldsymbol{A}_j, \quad \forall j \in \mathcal{N} \backslash \mathcal{N}_{\mathrm{F}} \tag{4.73b}$$

hold true for the subsystems $\Sigma_{\mathrm{F}k}$, $(\forall k \in \mathcal{N}_{\mathrm{F}})$. Recall that $\mathcal{N}_{\mathrm{F}}$ is the set of faulty subsystems. For the other matrices within $\Sigma_{\mathrm{F}i}$ the following holds true

$$\boldsymbol{E}_{\mathrm{F}i} = \boldsymbol{E}_i, \quad \boldsymbol{G}_{\mathrm{F}i} = \boldsymbol{G}_i, \quad \boldsymbol{C}_{\mathrm{Fz}i} = \boldsymbol{C}_{\mathrm{z}i}, \quad \boldsymbol{B}_{\mathrm{F}i} = \boldsymbol{B}_i, \quad \boldsymbol{C}_{\mathrm{F}i} = \boldsymbol{C}_i, \quad \forall i \in \mathcal{N}. \tag{4.74}$$

Hence, the fault only affects a part of the subprocess of the subsystem. The local control units are assumed to be fault-free, i.e.,

$$C_{\mathrm{F}i} = C_i \text{ defined in (4.14)}, \quad \forall i \in \mathcal{N}. \tag{4.75}$$

According to the nominal extended subsystems $\bar{\Sigma}_i^*$ defined in (4.18), the model of the extended subsystems $\bar{\Sigma}_{\mathrm{F}i}^*$, $(\forall i \in \mathcal{N})$, that might be faulty follows from (3.13), (4.14), (4.73) and (4.74) to

$$\bar{\Sigma}_{\mathrm{F}i}^* : \begin{cases} \dot{\boldsymbol{x}}_{\mathrm{F}i}(t) = \bar{\boldsymbol{A}}_{\mathrm{F}i}\boldsymbol{x}_{\mathrm{F}i}(t) + \boldsymbol{E}_i\tilde{\boldsymbol{s}}_{\mathrm{F}\Delta i}(t) + \boldsymbol{G}_i\boldsymbol{d}_i(t), \quad \boldsymbol{x}_{\mathrm{F}i}(0) = \boldsymbol{x}_{\mathrm{F}0i} \\ \boldsymbol{v}_{\mathrm{F}i}(t) = \boldsymbol{C}_{\mathrm{v}i}\boldsymbol{x}_{\mathrm{F}i}(t) \\ \boldsymbol{z}_{\mathrm{F}i}(t) = \boldsymbol{C}_{\mathrm{z}i}\boldsymbol{x}_{\mathrm{F}i}(t) \end{cases} \tag{4.76}$$

with the system matrix $\bar{\boldsymbol{A}}_{\mathrm{F}i} = \boldsymbol{A}_{\mathrm{F}i} - \boldsymbol{B}_i\boldsymbol{K}_{\mathrm{C}i}$ and $\tilde{\boldsymbol{s}}_{\mathrm{F}\Delta i}(t) = \boldsymbol{s}_{\mathrm{F}i}(t) - \tilde{\boldsymbol{s}}_{\mathrm{F}i}(t)$, where $\tilde{\boldsymbol{s}}_{\mathrm{F}i}(t)$ is the overall transmitted coupling input in the faulty case determined by the local control units $C_{\mathrm{F}i}$.

According to (3.17), (4.14), (4.75) and (4.76), the faulty self-organizing control system Σ_{F}

results to

$$\Sigma_{\mathrm{F}}: \begin{cases} \bar{\Sigma}^*_{\mathrm{F}i}, (\forall i \in \mathcal{N}), \text{ defined in (4.76)} \\ \boldsymbol{s}_{\mathrm{F}}(t) = \boldsymbol{L}\boldsymbol{z}_{\mathrm{F}}(t) \text{ according to (3.17)} \\ O_i, (\forall i \in \mathcal{N}), \text{ defined in (4.12)} \\ D_i, (\forall i \in \mathcal{N}), \text{ defined in (4.13)} \end{cases} \tag{4.77}$$

which includes the faulty subsystems $\Sigma_{\mathrm{F}k}$, $(\forall k \in \mathcal{N}_{\mathrm{F}})$, and the fault-free subsystems $\Sigma_{\mathrm{F}j}$, $(\forall j \in \mathcal{N} \backslash \mathcal{N}_{\mathrm{F}})$ defined by (3.13), (4.73) and (4.74). Note that the feedback units F_i in $C_{\mathrm{F}i}$ are included in the model of the extended subsystems $\bar{\Sigma}^*_{\mathrm{F}i}$ defined in (4.76).

Behavior of the faulty self-organizing control system. Consider that the feedback gain matrices $\boldsymbol{K}_{\mathrm{C}i}$, $(\forall i \in \mathcal{N})$, within the feedback unit F_i defined in (4.11) maintain. Then the resulting extended subsystem $\bar{\Sigma}^*_{\mathrm{F}i}$, $(\forall i \in \mathcal{N})$, that might be faulty can be divided into three groups (Fig. 4.37):

1. The fault-free extended subsystems $\bar{\Sigma}^*_{\mathrm{F}i}$, $(\forall i \in \mathcal{N} \backslash \mathcal{N}_{\mathrm{F}})$, are asymptotically stable and have the same dynamics as the nominal isolated extended subsystems $\bar{\Sigma}^*_i$ since their system matrices are identical

$$\bar{\boldsymbol{A}}_{\mathrm{F}i} = \bar{\boldsymbol{A}}_i = \boldsymbol{A}_i - \boldsymbol{B}_i \boldsymbol{K}_{\mathrm{C}i}, \quad \forall i \in \mathcal{N} \backslash \mathcal{N}_{\mathrm{F}}.$$

2. The faulty extended subsystems $\bar{\Sigma}^*_{\mathrm{F}i}$, $(\forall i \in \mathcal{H}_{\mathrm{F}})$, are **not** asymptotically stable if their faulty system matrices $\bar{\boldsymbol{A}}_{\mathrm{F}i}$ are **not** Hurwitz matrices, where $\mathcal{H}_{\mathrm{F}}$ is the *set of non Hurwitz matrices* defined by

$$\mathcal{H}_{\mathrm{F}} = \{i \in \mathcal{N}_{\mathrm{F}} \mid \exists j \in \{1, \ldots, n_h\} \text{ s.t. } \mathrm{Re}(\lambda_j(\bar{\boldsymbol{A}}_{\mathrm{F}i})) \geq 0\}.$$

3. Generally subprocess faults do not imply that the system matrix $\bar{\boldsymbol{A}}_{\mathrm{F}i}$ is not a Hurwitz matrix. The matrices $\bar{\boldsymbol{A}}_{\mathrm{F}i}$ are often only different to the nominal ones, but are still Hurwitz matrices. In this case the faulty extended subsystems $\bar{\Sigma}^*_{\mathrm{F}i}$, $(\forall i \in \mathcal{N}_{\mathrm{F}} \backslash \mathcal{H}_{\mathrm{F}})$, are asymptotically stable but have different dynamics than the nominal isolated extended subsystems $\bar{\Sigma}^*_i$.

The classification of these three groups can transfered to the analysis of the behavior of the faulty self-organizing control system Σ_{F}. It is assumed that all faults are present initially and are persistent. With this, the following proposition gives bounds on the states $\boldsymbol{x}_{\mathrm{F}i}(t)$, $(\forall i \in \mathcal{N})$, in the case of a process fault defined in (4.73) while using the results in Lemma 4.2.

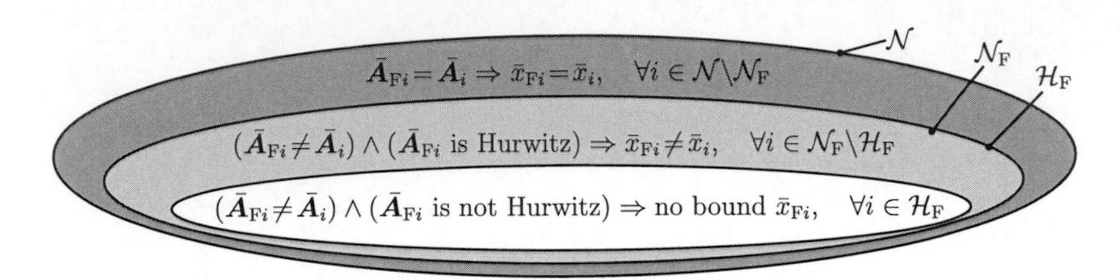

Figure 4.37: Boundedness of the subsystem states in the case of a subprocess fault.

Proposition 4.3 (Fault propagation) *Consider the faulty self-organizing control system Σ_F defined in (4.77). Then the states $\boldsymbol{x}_{\mathrm{F}i}(t)$, $(\forall i \in \mathcal{N}\backslash\mathcal{H}_\mathrm{F})$, of the subsystems $\Sigma_{\mathrm{F}i}$ are bounded by*

$$\|\boldsymbol{x}_{\mathrm{F}i}(t)\| \leq \left\| \mathrm{e}^{\bar{\boldsymbol{A}}_{\mathrm{F}i}t} \boldsymbol{x}_{\mathrm{F}0i} \right\| + \bar{x}_{\mathrm{F}i}, \quad \forall i \in \mathcal{N}\backslash\mathcal{H}_\mathrm{F}, \quad \forall t \geq 0 \tag{4.78}$$

with

$$\begin{aligned} \bar{x}_{\mathrm{F}p} &= m_{\mathrm{Fxd}p} \bar{d}_p + m_{\mathrm{Fxs}p} \sum\nolimits_{l \in \bar{\mathcal{P}}_{\mathrm{C}p}} \epsilon_{pl}, \quad \forall p \in \mathcal{N}_\mathrm{F}\backslash\mathcal{H}_\mathrm{F} \\ \bar{x}_{\mathrm{F}j} &= \bar{x}_j, \quad \forall j \in \mathcal{N}\backslash\mathcal{N}_\mathrm{F} \\ m_{\mathrm{Fxd}p} &:= \int_0^\infty \left\| \mathrm{e}^{\bar{\boldsymbol{A}}_{\mathrm{F}p}\tau} \boldsymbol{G}_p \right\| \mathrm{d}\tau, \quad \forall p \in \mathcal{N}_\mathrm{F}\backslash\mathcal{H}_\mathrm{F} \\ m_{\mathrm{Fxs}p} &:= \int_0^\infty \left\| \mathrm{e}^{\bar{\boldsymbol{A}}_{\mathrm{F}p}\tau} \boldsymbol{E}_p \right\| \mathrm{d}\tau, \quad \forall p \in \mathcal{N}_\mathrm{F}\backslash\mathcal{H}_\mathrm{F}. \end{aligned} \tag{4.79}$$

Proof. The proof is given in Appendix A.4. □

Proposition 4.3 shows that there exist three different situations concerning the boundedness of the subsystem states (cf. Fig. 4.37):

1. The states of the fault-free subsystems $\Sigma_{\mathrm{F}i}$, $(\forall i \in \mathcal{N}\backslash\mathcal{N}_\mathrm{F})$, have the same bounds as in the fault-free case.

2. The faulty subsystems $\Sigma_{\mathrm{F}i}$, $(\forall i \in \mathcal{N}_\mathrm{F}\backslash\mathcal{H}_\mathrm{F})$ that are still asymptotically stable within the decentralized feedback have modified bounds according to (4.79).

3. If the isolated extended subsystems $\bar{\Sigma}^*_{\mathrm{F}i}$, $(\forall i \in \mathcal{H}_\mathrm{F})$ are not stable, the corresponding states $\boldsymbol{x}_{\mathrm{F}i}(t)$ are not bounded.

Hence, the performance of the controlled fault-free subsystems is similar as in the fault-free case and the fault propagation is bounded (cf. Fig. 4.38 left).

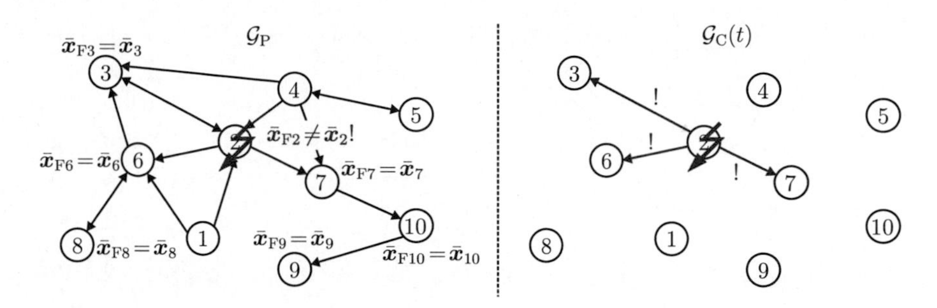

Figure 4.38: Actual fault propagation (left) and communication activation in case of a fault (right) with $\mathcal{N}_{\mathrm{F}} = \{2\}$.

Discussion of the realization in technical systems. Consider that $\mathcal{N}_{\mathrm{F}} = \mathcal{H}_{\mathrm{F}} \neq \emptyset$ holds true, then the physical impact of the subsystems $\Sigma_{\mathrm{F}i}$, $(\forall i \in \mathcal{H}_{\mathrm{F}})$, on their neighboring subsystems could become infinite. Therefore, the local control units whose corresponding subsystems are affected by the subsystems $\Sigma_{\mathrm{F}i}$, $(\forall i \in \mathcal{H}_{\mathrm{F}})$, would need a infinite control input to compensate the physical coupling. In real systems this is not possible. However, in real system also the states of the faulty subsystems would be bounded and, therefore, also the required control input to the fault-free subsystems is bounded. The question is whether the maximal possible input is sufficient to compensate the physical impact. Furthermore, there also is the opportunity to reconfigure the feedback gain matrix $\boldsymbol{K}_{\mathrm{C}i}$ within the feedback unit F_i to stabilize the corresponding faulty subsystems, where in the meantime the other local control units try to compensate the physical coupling inputs as good as possible.

Stability of the faulty self-organizing control system. Roughly speaking, there are two parts within the self-organizing control system for $\mathcal{H}_{\mathrm{F}} \neq \emptyset$. The part that is stable and includes the subsystems $\Sigma_{\mathrm{F}i}$, $(\forall i \in \mathcal{N}\backslash\mathcal{H}_{\mathrm{F}})$ and the part that is unstable that includes the subsystems $\Sigma_{\mathrm{F}j}$, $(\forall j \in \mathcal{H}_{\mathrm{F}})$. But obviously, according to Definition 2.2 the faulty self-organizing control system Σ_{F} is not stable for $\mathcal{H}_{\mathrm{F}} \neq \emptyset$. If $\mathcal{H}_{\mathrm{F}} = \emptyset$ holds true, then according to Theorem 4.2 Σ_{F} is practically stable w.r.t the compact set

$$\mathcal{A} = \{\boldsymbol{x} = (\boldsymbol{x}_1^{\mathrm{T}} \ldots \boldsymbol{x}_N^{\mathrm{T}})^{\mathrm{T}} \in \mathbb{R}^n \mid \|\boldsymbol{x}_i\| \leq \bar{x}_{\mathrm{F}i}, i \in \mathcal{N}\}.$$

with the bounds $\bar{x}_{\mathrm{F}i}$, $(\forall i \in \mathcal{N})$, from Proposition 4.3.

Communication for the fault attenuation. To guarantee that the faults do not spread out on other subsystems, the local control units of the faulty subsystems have to send information to the other local control units if necessary, to compensate the effect of the coupling input of the faulty subsystems. Since the states $\boldsymbol{x}_{\mathrm{F}i}(t)$, $(\forall i \in \mathcal{H}_{\mathrm{F}})$, of subsystems $\Sigma_{\mathrm{F}i}$ are not bounded according to Proposition 4.3, also the physical coupling inputs $\boldsymbol{s}_{\mathrm{F}ji}(t) = \boldsymbol{L}_{ji}\boldsymbol{C}_{\mathrm{z}i}\boldsymbol{x}_{\mathrm{F}i}(t)$ to Σ_j, $(\forall j \in \mathcal{F}_{\mathrm{P}i})$, are not bounded. Hence, these coupling inputs have to be compensated by sending the determined coupling inputs $\hat{\boldsymbol{s}}_{\mathrm{F}ji}(t) = \boldsymbol{L}_{ji}\boldsymbol{C}_{\mathrm{z}i}\boldsymbol{x}_{\mathrm{F}i}(t)$ from C_i to C_j (cf. Fig 4.38 right). This might be a drawback with respect to a small communication effort, but otherwise all subsystems would be affected by a fault in a single subsystem.

Reconfiguration of the self-organizing controller. Consider that the faults are known. For the adaption of the local control units $C_{\mathrm{F}i}$ to guarantee that the reconfigured faulty self-organizing control system guarantees the practical stability and a desired disturbance attenuation as claimed in Control aim 3.3 two main problems have to be solved:

- The **reconfigurability** of the overall faulty plant P_{F} can be tested by using local model information of the faulty subsystem $\Sigma_{\mathrm{F}i}$, $(\forall i \in \mathcal{N}_{\mathrm{F}})$, only. Recalling Condition 3 in Theorem 4.1, the faulty subsystems $\Sigma_{\mathrm{F}i}$, $(\forall i \in \mathcal{N}_{\mathrm{F}})$, have to be completely controllable, i.e.,

  $$\text{the pair } (\boldsymbol{A}_{\mathrm{F}i}, \boldsymbol{B}_i) \text{ has to be controllable for all } i \in \mathcal{N}_{\mathrm{F}}.$$

  Then the feedback gain matrix $\boldsymbol{K}_{\mathrm{C}i}$ within the local control units $C_{\mathrm{F}i}$, $(\forall i \in \mathcal{N}_{\mathrm{F}})$, of the corresponding faulty subsystems $\Sigma_{\mathrm{F}i}$ can be redesigned to make $\bar{\boldsymbol{A}}_{\mathrm{F}i} = \boldsymbol{A}_{\mathrm{F}i} - \boldsymbol{B}_i\boldsymbol{K}_{\mathrm{C}i}$ Hurwitz matrices and, therefore, practically stabilize the faulty self-organizing control system Σ_{F} defined in (4.77).

- For the **reconfiguration** of the local control units $C_{\mathrm{F}i}$, $(\forall i \in \mathcal{N}_{\mathrm{F}})$, of the faulty subsystems $\Sigma_{\mathrm{F}i}$ to guarantee practical stability and the desired disturbance attenuation claimed in Control aim 3.3, Algorithm 4.1 can be used. To use this algorithm the model of the faulty subsystem $\Sigma_{\mathrm{F}i}$, $(\forall i \in \mathcal{N}_{\mathrm{F}})$, has to be given instead of the model of the corresponding nominal subsystem Σ_i which only differ in the system matrix.

Note that only the local control units of the faulty subsystems have to be reconfigured.

Summary. The overall investigation on the self-organizing control system Σ_{F} affected by subprocess faults leads to the following answers for the three initially asked questions:

1. The behavior of the controlled fault-free subsystems $\Sigma_{\mathrm{F}i}$, $(\forall i \in \mathcal{N}\backslash\mathcal{N}_{\mathrm{F}})$, is similar to the behavior of the corresponding controlled nominal subsystems Σ_i.

2. The communication time of the local control units $C_{\mathrm{F}i}$, $(\forall i \in \mathcal{N}\backslash\mathcal{N}_{\mathrm{F}})$, of the faulty subsystems $\Sigma_{\mathrm{F}i}$ changes and might be activated all of the time to compensate the coupling input if $\Sigma_{\mathrm{F}i}$ is unstable.

3. The reconfigurability of the faulty plant P_{F} can be tested by using local model information of the faulty subsystem $\Sigma_{\mathrm{F}i}$, $(\forall i \in \mathcal{N}_{\mathrm{F}})$ and, furthermore, only the local control units $C_{\mathrm{F}i}$ of the faulty subsystems $\Sigma_{\mathrm{F}i}$ have to be reconfigured to guarantee a desired overall performance.

In summary, the proposed self-organizing control structure with the situation-dependent compensation of the coupling inputs ensures that the subprocess faults do not propagate on other subsystems within the closed-loop system.

Example 4.1 (cont.) *Situation-dependent compensation of physical coupling inputs*

In the following it is assumed that subsystem Σ_1 of the symmetrically interconnected subsystems P defined in (3.32) has a process fault, i.e., $\boldsymbol{A}_{\mathrm{F}1} = 1.8 \neq \boldsymbol{A}_1 = 1$. Consider that the local control units C_i, $(\forall i \in \{1, 2\})$, include the parameters $\boldsymbol{K}_{\mathrm{C}i}$, $\boldsymbol{K}_{\mathrm{D}i}$, $\boldsymbol{C}_{zi}$, $\boldsymbol{L}_{21}$ and $\boldsymbol{L}_{12}$ listed in Table 4.2. The switching thresholds $\epsilon_{12} = 0.4$ and $\epsilon_{21} = 0.28$ follow from Step 3 in Algorithm 4.1 such that the performance output $\boldsymbol{v}_1(t)$ is bounded by $\bar{v}_1^* = 0.4$ if Σ_2 is disturbed and $\boldsymbol{v}_2(t)$ is bounded by $\bar{v}_1^* = 0.2$ if Σ_1 is disturbed according to Control aim 3.3. Hence, if subsystem Σ_1 is faulty, the system matrix $\bar{\boldsymbol{A}}_{\mathrm{F}1}$ of the extended subsystem $\bar{\Sigma}_{\mathrm{F}1}^*$ results to $\bar{\boldsymbol{A}}_{\mathrm{F}1} = -0.2 \neq \bar{\boldsymbol{A}}_1 = -1$. which is different from the system matrix $\bar{\boldsymbol{A}}_1$ of the nominal extended subsystem $\bar{\Sigma}_1^*$. To guarantee the same performance as in the nominal case, the feedback gain matrix $\boldsymbol{K}_{\mathrm{C}1}$ can be determined with Step 2 in Algorithm 4.1 by considering $\bar{\boldsymbol{A}}_{\mathrm{F}1}$

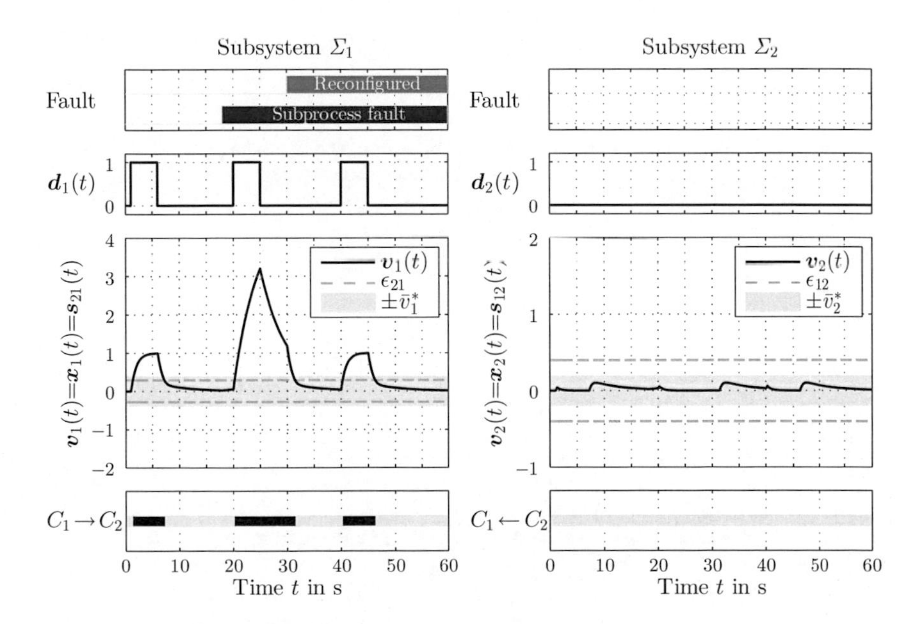

Figure 4.39: Behavior of the symmetrically interconnected subsystems defined in (3.32) in case of a subprocess fault in Σ_1.

instead of $\boldsymbol{A}_1$ to reconfigure C_1, i.e.,

$$\boldsymbol{A}_{\mathrm{F}1} = 1.8 \quad \Rightarrow \quad \boldsymbol{K}_{\mathrm{C}1} = 2.8.$$

The other parameters of C_1 maintain.

Figure 4.39 depicts the behavior of the self-organizing controlled subsystems in case of a persistent process fault in Σ_1 at $t = 18\,\mathrm{s}$. The local control unit C_1 is reconfigured by adapting the feedback gain matrix from $\boldsymbol{K}_{\mathrm{C}1} = 2$ to $\boldsymbol{K}_{\mathrm{C}1} = 2.8$ at $t = 8\,\mathrm{s}$, where the diagnosis of the process fault is assumed to be given. Two interesting properties can be observed: First, the control aim for subsystem Σ_2 is fulfilled at any time, no matter if Σ_1 is faulty or not. Therefore, C_1 has to send $\hat{\boldsymbol{s}}_{21}(t)$ to C_2 for a much longer time interval than in the nominal case. Second, after the local reconfiguration the performance of Σ_2 is similar as in the fault-free case. In summary, the communication time increases with a subprocess fault, but after the local reconfiguration a similar performance as in the fault-free case can be achieved. □

4.8.3 Propagation of actuator faults, sensor faults and communication faults

In this section the passive fault tolerance of the self-organizing control system Σ is investigated for actuator faults, sensor faults and communication faults (Fig. 4.40). In particular, the first two questions given in Section 4.8.1 are discussed for these kinds of faults. The gray area in Fig. 4.40 indicates the propagation of the specific faults to other subsystem (Question 1), where blue arrows show which communication is necessary to attenuation the propagation of the fault (Question 2).

Actuator fault. An actuator fault in subsystem $\Sigma_{\mathrm{F}i}$, $(i \in \mathcal{N}_{\mathrm{F}})$, is modeled by

$$\boldsymbol{B}_{\mathrm{F}i} \neq \boldsymbol{B}_i, \quad i \in \mathcal{N}.$$

This causes two main problems within the closed-loop system Σ. First, the system matrix of the controlled faulty subsystem $\Sigma_{\mathrm{F}i}$ changes to $\bar{\boldsymbol{A}}_{\mathrm{F}i} = \boldsymbol{A}_i - \boldsymbol{B}_{\mathrm{F}i}\boldsymbol{K}_{\mathrm{C}i}$. Second, the condition in (4.3) to compensate the coupling inputs into $\Sigma_{\mathrm{F}i}$ by transmitting information might be violated, i.e.,

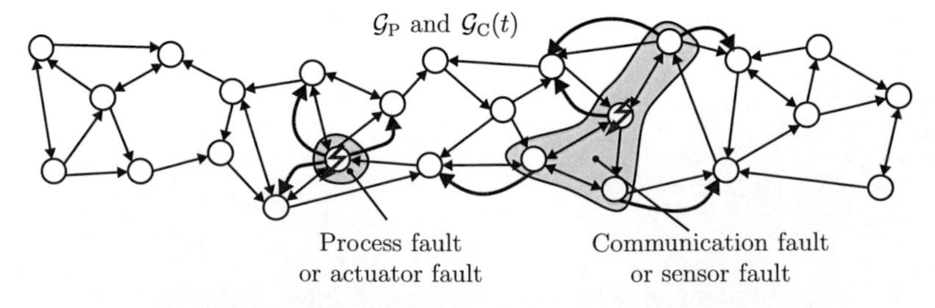

Figure 4.40: Propagation of a fault on other controlled subsystems.

$\boldsymbol{B}_{\mathrm{F}i}\boldsymbol{K}_{\mathrm{D}i} \neq \boldsymbol{E}_i$ (cf. Lemma 4.1). This leads to the following answers to the first two questions from Section 4.8.1:

1. Despite all possible changes in the dynamics of the controlled faulty subsystem the correct coupling input $\hat{\boldsymbol{s}}_{ji}$, $(\forall j \in \bar{\mathcal{F}}_{\mathrm{C}i})$, can be transmitted to the other local control units C_j. Hence, an actuator fault in $\Sigma_{\mathrm{F}i}$ does not propagate to other subsystems within the plant.

2. The communication time from C_i to C_j, $(\forall j \in \bar{\mathcal{F}}_{\mathrm{C}i})$, might change, where also a permanent activation is possible to compensate the coupling input of an unstable controlled faulty subsystem $\Sigma_{\mathrm{F}i}$.

In summary, an actuator fault has a similar impact on the behavior of Σ as a subprocess fault investigated in the previous section.

Sensor fault. If subsystem $\Sigma_{\mathrm{F}i}$, $(i \in \mathcal{N}_{\mathrm{F}})$, has a sensor fault, the relation

$$\boldsymbol{C}_{\mathrm{F}i} \neq \boldsymbol{C}_i = \boldsymbol{I}_{n_i}, \quad i \in \mathcal{N}_{\mathrm{F}}.$$

holds true which leads to two main problems. First, the system matrix of the controlled faulty subsystem $\Sigma_{\mathrm{F}i}$ changes to $\bar{\boldsymbol{A}}_{\mathrm{F}i} = \boldsymbol{A}_i - \boldsymbol{B}_i\boldsymbol{K}_{\mathrm{C}i}\boldsymbol{C}_{\mathrm{F}i}$ since the complete subsystem state can not be measured anymore. Second, the determined coupling signal $\hat{\boldsymbol{s}}_{\mathrm{F}ji}(t)$ is not identical to the real coupling input $\boldsymbol{s}_{\mathrm{F}ji}(t)$

$$\hat{\boldsymbol{s}}_{\mathrm{F}ji}(t) = \boldsymbol{L}_{ji}\boldsymbol{C}_{\mathrm{z}i}\boldsymbol{C}_{\mathrm{F}i}\boldsymbol{x}_{\mathrm{F}i}(t) \neq \boldsymbol{s}_{\mathrm{F}ji}(t) = \boldsymbol{L}_{ji}\boldsymbol{C}_{\mathrm{z}i}\boldsymbol{x}_{\mathrm{F}i}(t), \quad i \in \mathcal{N}. \tag{4.80}$$

With this, the concerned questions can be answered as follows:

1. Since the neighboring local control units C_j, $(\forall j \in \bar{\mathcal{F}}_{\mathrm{C}i})$, might receive false information about the coupling input $\boldsymbol{s}_{\mathrm{F}ji}(t)$ into Σ_j, $(\forall j \in \bar{\mathcal{F}}_{\mathrm{C}i})$, according to (4.80), the coupling inputs into Σ_j are usually not be compensated anymore (cf. Lemma 4.1). Hence, the behavior of the controlled subsystems Σ_j, $(\forall j \in \mathcal{F}_{\mathrm{P}i})$, might change.

2. The communication time from C_i to C_j, $(\forall j \in \bar{\mathcal{P}}_{\mathrm{C}i})$, and from C_j, $(\forall l \in \bar{\mathcal{P}}_{\mathrm{C}j})$, might change, where also a permanent activation is possible to compensate the coupling input of an unstable controlled subsystem.

Communication fault. In case of a communication fault in a local control unit $C_{\mathrm{F}i}$, $(i \in \mathcal{N})$, either the transmission unit or the receiver unit is faulty or both:

- **Faulty transmission unit:** The local control units $C_{\mathrm{F}j}$, $(\forall j \in \bar{\mathcal{F}}_{\mathrm{C}i})$, do not receive information from the faulty local control unit $C_{\mathrm{F}i}$, $(i \in \mathcal{N}_{\mathrm{F}})$. Hence, the coupling inputs to $\Sigma_{\mathrm{F}j}$, $(\forall j \in \bar{\mathcal{F}}_{\mathrm{C}i})$, are not bounded anymore (cf. Lemma 4.1).

- **Faulty receiver unit:** The faulty local control unit $C_{\mathrm{F}i}$, $(i \in \mathcal{N}_{\mathrm{F}})$, does not receive any information from the local control units $C_{\mathrm{F}j}$, $(\forall j \in \bar{\mathcal{P}}_{\mathrm{C}i})$, whose subsystems have a physical impact on $\Sigma_{\mathrm{F}i}$. With this, the coupling inputs into the subsystem $\Sigma_{\mathrm{F}i}$, $(i \in \mathcal{N})$, of the faulty local control unit C_i is not bounded anymore (cf. Lemma 4.1).

In the worst case that both, the transmission unit and receiver unit, are faulty the concerned questions can be answered as follows:

1. A communication fault does not only affect the faulty subsystem $\Sigma_{\mathrm{F}i}$, $(i \in \mathcal{N}_{\mathrm{F}})$, the fault propagates also to subsystems $\Sigma_{\mathrm{F}j}$, $(\forall j \in \bar{\mathcal{F}}_{\mathrm{C}i})$, on which $\Sigma_{\mathrm{F}i}$ has a physical impact (Fig. 4.40).

2. The communication time from C_i to C_j, $(\forall j \in \bar{\mathcal{P}}_{\mathrm{C}i})$, and from C_j, $(\forall l \in \bar{\mathcal{P}}_{\mathrm{C}j})$, might change.

In summary, a communication fault has a similar impact on the behavior of Σ as a sensor fault.

Boundedness of the fault propagation with situation-dependent compensation of physical coupling inputs. The analysis of subprocess faults in Section 4.8.2 and the discussion of the impact of other faults on other subsystems show that the proposed self-organizing controller C defined in (4.15) leads to a self-organizing control system Σ which bounds the propagation of a fault. Concerning the boundedness of the fault propagation, the considered faults can be divided into two groups (Fig. 4.40):

- A subprocess fault and an actuator fault only affect the desired performance of the faulty subsystems.

- A sensor fault and a communication fault do not only affect the desired performance of the faulty subsystem $\Sigma_{\mathrm{F}i}$, $(i \in \mathcal{N}_{\mathrm{F}})$, the fault also propagates to all subsystems $\Sigma_{\mathrm{F}j}$, $(\forall j \in \bar{\mathcal{F}}_{\mathrm{C}i})$, on which $\Sigma_{\mathrm{F}i}$ has a physical impact.

The situation-dependent compensation of the coupling inputs isolates the part of the closed-loop system affected by the fault. If a fault vanishes, the communication for this compensation will be deactivated since the isolation of the faulty subsystem is not necessary anymore. In summary, the self-organizing control system shows the desired property of fault-tolerance of self-organizing systems.

4.8.4 Application example: Water supply system

This section investigates the passive fault-tolerance of the self-organizing system Σ defined in (4.20) consisting of the water supply system introduced in Section 3.5.1 which is controlled by the local control units defined in (4.14) with the parameters listed in Table 4.3 to fulfill the control aim in (4.60). Note that there is no reconfiguration of the local control units. The behavior of this overall system in a fault-free case is described in Section 4.5. The considered sensor fault and actuator fault of the controlled tanks are described in Section 3.5.1.

Fault-tolerant behavior of the self-organizing control system. Figure 4.41 shows the behavior of the water supply system for $\boldsymbol{x}_{0i} = \mathbf{0}$, $(\forall i \in \mathcal{N})$, using the local control units C_i designed in Section 4.5.1. Compared to Fig. 4.22 the upper plot in Fig. 4.41 specifies the times at which subsystem Σ_{10} has a sensor fault (cf. (3.30)) or subsystem Σ_{17} has an actuator fault (cf. (3.31)). The subsystems are affected by small disturbances as depicted in Fig. 4.22, where the larger disturbances at Σ_6, Σ_9 and Σ_{17} are not present.

The actuator fault at Σ_{17} leads to a large value of the performance output $l_{\mathrm{TB}17}(t)$. But due to the communication from C_{17} to C_{14}, C_{16}, C_{18} and C_{21}, the performance of the fault-free subsystems is not affected compared to decentralized control (Fig. 4.42 (a)).

Subsystem Σ_{10} is affected by a temporary sensor fault. Due to the faulty measurement, the coupling input from Σ_{10} on Σ_9 and Σ_{12} can not be determined correctly. Hence, there is no communication from C_{10} to C_9 and C_{12} and, therefore, the sensor fault also affects the performance of Σ_9 and Σ_{10} indicated by the pink color of the circles in Fig. 4.42. Due to this deviation of $l_{\mathrm{TB}9}(t)$ and $l_{\mathrm{TB}12}(t)$ from the operating point, C_9 and C_{12} have to transmit information, to prevent a further propagation of the fault (Fig. 4.42 (b)). This indicates that the propagation of a sensor fault is eliminated by communication of neighboring local control units.

Similar as for a disturbance, communication is only necessary if the fault is present, otherwise the communication among the local control units is deactivated (Fig. 4.42 (c)).

Comparison to decentralized control and centralized control. Figure 4.43 depicts the physical coupling graph $\mathcal{G}_{\mathrm{P}}$ and the maximal values of the tank levels in TBi ($\max_{t\geq 0} |l_{\mathrm{TB}i}(t)|$) while using the self-organizing controller C (black bar), the decentralized controller C_{D} (gray bar) and the centralized controller C_{C} (blue bar) (without giving the actual amount of the values). The values for the subsystems that are faulty or affected by the faults are not given since they are much larger than the values of the fault-free subsystems. This depiction shows that the self-organizing controller leads to a much better performance of the fault-free subsystems compared to decentralized control. Rather, the performance is similar to centralized control. Hence, the propagation of the fault can be bounded with the self-organizing controller.

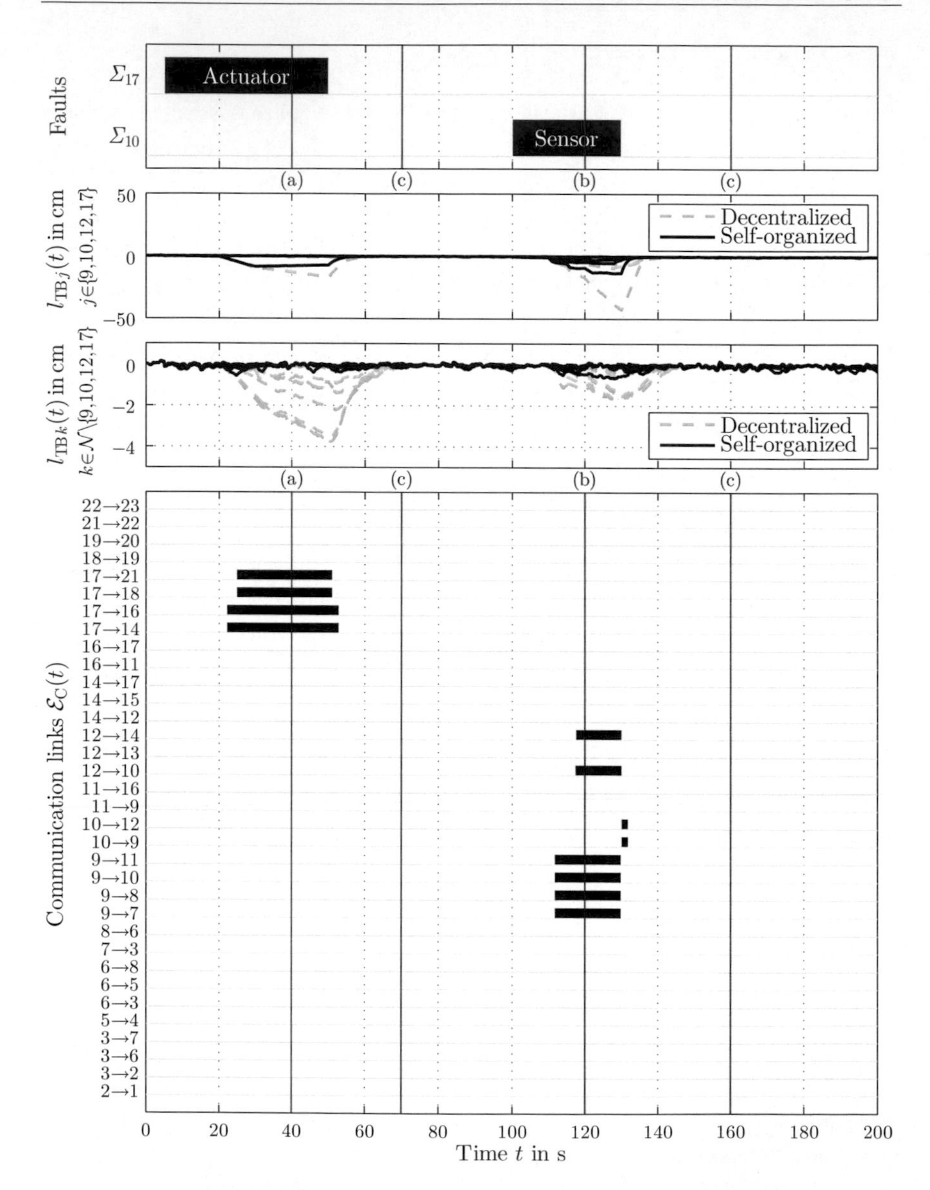

Figure 4.41: Fault-tolerant behavior of the water supply system controlled by the self-organizing controller C defined in (4.15).

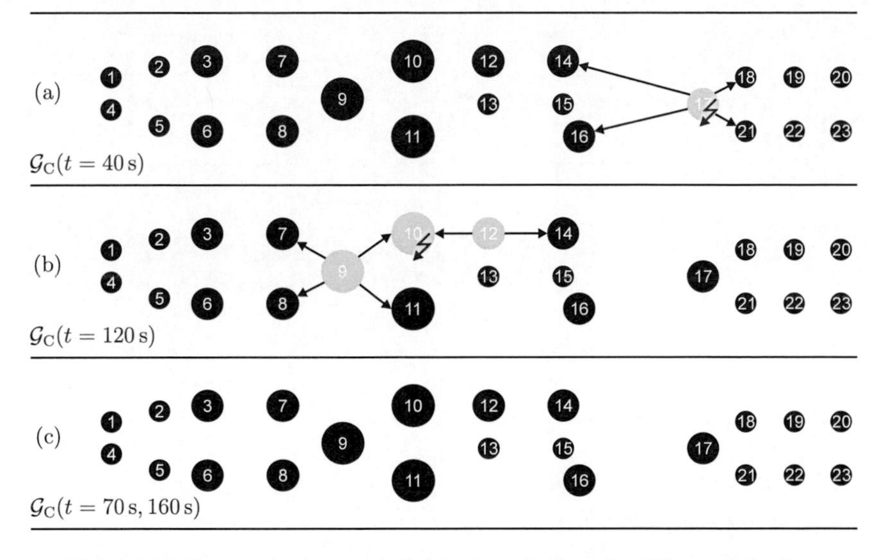

Figure 4.42: Communication graph $\mathcal{G}_C(t)$ at certain times for different fault situations in Fig. 4.41.

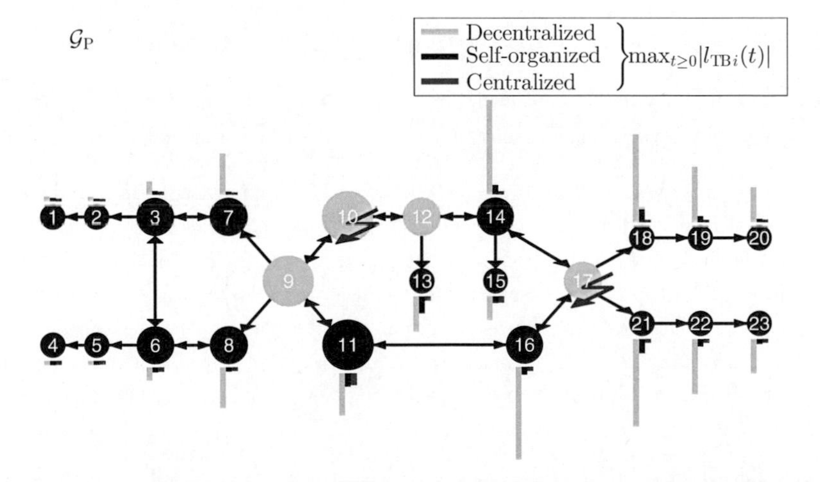

Figure 4.43: Physical coupling graph $\mathcal{G}_P$ with the maximal values of the tank levels $\max_{t\in[0,120]} |l_{TBi}|$, $(i = 1, \ldots, 23)$, for a decentralized control, self-organized control and centralized control w.r.t the fault-tolerant behavior depicted in Fig. 4.41.

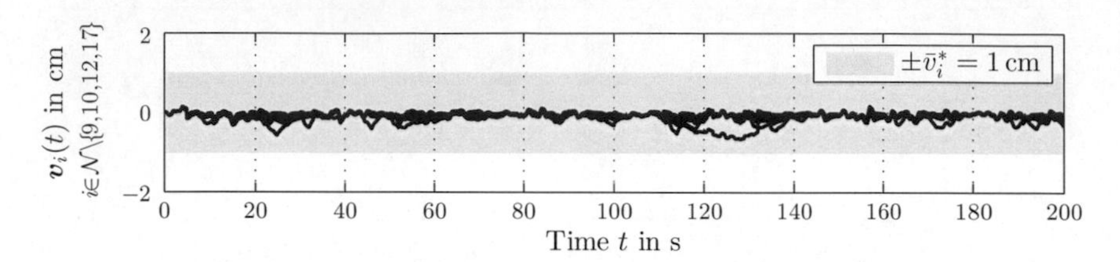

Figure 4.44: Verification of the control aim in (4.60) in case of faults affecting the subsystems.

Fulfillment of desired difference behavior (control aim). The control aim in (4.60) is guaranteed for the fault-free subsystem except subsystems Σ_9 and Σ_{12} which are affected by the sensor fault in Σ_{10} (cf. Fig. 4.44). The performance output $\boldsymbol{v}_i(t) = l_{\mathrm{TB}i}(t)$, $(\forall i \in \mathcal{N}\backslash\{9,10,12,17\})$, is smaller than the desired maximal performance output of $\bar{v}_i^* = 1\,\mathrm{cm}$ for all $t \geq 0$. The same fact also holds for the practical stability of the subsystems Σ_i, $(\forall i \in \mathcal{N}\backslash\{9,10,12,17\})$, illustrated in Fig. B.2 in the appendix.

In summary, the expected passive fault-tolerant behavior of the self-organizing control system is verified by results of the simulation.

4.9 Summary and evaluation of the proposed control concept

The main properties of the self-organizing control concept with situation-dependent compensation of physical coupling inputs are illustrated in Fig. 4.45 and can be summarized as follows:

- **Communication for disturbance attenuation**: The communication among the local control units is adapted to the currently acting disturbances, where the communication activation depends on the magnitude of the disturbance (cf. Proposition 4.2). Due to this situation-dependent communication, the disturbance propagation can be arbitrarily adjusted by the choice of the switching thresholds (cf. Theorem 4.5 and Theorem 4.6). Furthermore, unnecessary communication links resulting from the structure of the physical coupling graph $\mathcal{G}_{\mathrm{P}}$, can be canceled out of the self-organizing controller C (cf. Theorem 4.7).

- **Analysis and design by local model information**: The practical stability and the asymptotic stability of the resulting self-organizing control system can be checked by local model information (cf. Theorem 4.2 and Theorem 4.3). If the isolated subsystems are stabilizable, then a self-organizing controller C defined in (4.15) can always be designed such that the self-organizing control system is practically stable (cf. Theorem 4.1). There

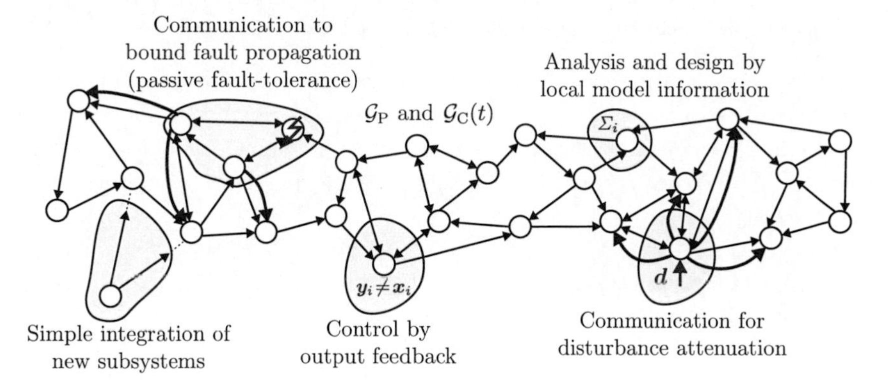

Figure 4.45: Main properties of the self-organizing control concept with a situation-dependent compensation of the physical coupling inputs.

is no restriction on the properties of the physical coupling among the subsystems and the properties of the overall plant P. The local control units C_i can be designed individually by using local model information to guarantee an overall control aim (cf. Algorithm 4.1).

- **Control by output feedback**: The basic local control units need to measure the overall subsystem state to stabilize the isolated subsystems and to determine the coupling inputs to the other subsystem. To eliminate that restriction, a new structure for the local control units is presented which uses the measured output of the subsystem to fulfill the same tasks (cf. Theorem 4.8). The analysis and the design methods for the state feedback approach can also be used for the output feedback approach (cf. Corollary 4.2).

- **Simple integration of new subsystems**: Since the analysis and design methods need only local model information, new subsystems can be easily integrated into a present self-organizing control system without a complete new design of all local control units (cf. Table 4.8). The local feedbacks for stabilizing the isolated subsystems maintain. Some of the switching thresholds have to be redesigned for local control units whose corresponding subsystems are physically connected with the new subsystem.

- **Communication to bound fault propagation**: Due to the situation-dependent compensation of the physical coupling inputs, the proposed self-organizing controller also eliminates the propagation of a local fault within the overall system. Depending on the type of the fault the fault propagation is different but limited. Similar to the behavior in case of a disturbance, the communication among the local control units is adapted to the currently acting fault, where the communication is deactivated if the fault vanishes.

The main restriction of the proposed self-organizing control concept is that relation (4.3) has to be fulfilled which guarantees that the effect of the coupling inputs on the subsystems can be completely compensated by an applicable control input. Hence, only subsystems for which relation (4.3) is fulfilled can be controlled such that the closed-loop system has the presented self-organizing behavior.

Properties of self-organization. The claimed properties of a self-organizing control system in Section1.2 for the proposed control concept are verified as follows:

- **Flexibility**: The communication structure is adjusted to currently acting disturbances, certain initial states or faults within the subsystems. The examples have shown that the communication among the local control units is only activated if it is necessary, e.g., if a disturbance occurs. If the disturbance vanishes, the communication is deactivated. Furthermore, in most cases the activation of the communication is restricted to an area around the disturbed subsystem or faulty subsystem. Hence, a local change of the situation of the subsystem (disturbance, fault) is attenuated close to that subsystem and does not affect the overall system.

- **Scalability**: The analysis of the overall system and the design of the local control units can be made by using local model information of the subsystem only. Therefore, the analysis and the design scales linearly with the number of subsystem which is a desired property for the control of complex systems. Due to these properties, new subsystems can be easily integrated into or removed from a present self-organizing control system. In the case of a new subsystem most of the local control units can keep their parameters and only local control units whose corresponding subsystems are influenced by the changed physical coupling structure have to be adapted (cf. Section 4.7).

- **Fault tolerance**: The self-organizing controller leads to a closed-loop system which prevents the propagation of local faults within the overall system. If a subsystem is faulty its physical impact on the other subsystems is eliminated by transmitting information to the neighboring local control units. With this, the faulty part of the overall system is isolated from the healthy part in a natural way.

In summary, the proposed self-organizing control concept shows all desired self-organizing properties.

5 Mimicry of a centralized controller

This chapter presents a control concept which mimics the behavior of a centralized controller. The locally determined control signals are exchanged among the local control units, whenever their values have a significant amount on the overall local control signal. Therewith, the difference between the behavior of the self-organizing control system and the centralized control system can be arbitrarily adjusted. The control concept is applied to a water supply system.

Chapter contents

5.1 Situation-dependent transmission of control signals

The control of physically coupled subsystems defined in (3.10) by a centralized controller

$$C_\mathrm{C}: \ \boldsymbol{u}_\mathrm{C}(t) = -\boldsymbol{K}_\mathrm{C}\boldsymbol{x}_\mathrm{C}(t) \tag{5.1}$$

leads to the centralized control system

$$\Sigma_\mathrm{C}: \ \dot{\boldsymbol{x}}_\mathrm{C}(t) = \underbrace{(\boldsymbol{A}_\mathrm{P} - \boldsymbol{B}\boldsymbol{K}_\mathrm{C})}_{\boldsymbol{A}_\mathrm{C}}\boldsymbol{x}_\mathrm{C}(t) + \boldsymbol{G}\boldsymbol{d}(t), \quad \boldsymbol{x}_\mathrm{C}(0) = \boldsymbol{x}_\mathrm{C0} \tag{5.2}$$

depicted in the left part of Fig. 5.1. The matrix $\boldsymbol{K}_\mathrm{C} \in \mathbb{R}^{m\times n}$ is the central feedback gain matrix, $\boldsymbol{u}_\mathrm{C}(t) \in \mathbb{R}^m$ is the central control input, $\boldsymbol{x}_\mathrm{C}(t) \in \mathbb{R}^n$ is the state of Σ_C and $\boldsymbol{x}_\mathrm{C0} \in \mathbb{R}^n$ is the initial state of Σ_C. With this, the main **aim** of this chapter can be summarized as follows:

> Develop a self-organizing controller C that leads to a self-organizing control system Σ that mimics the behavior of the centralized control system Σ_C with a central controller C_C (Control aim 3.4).

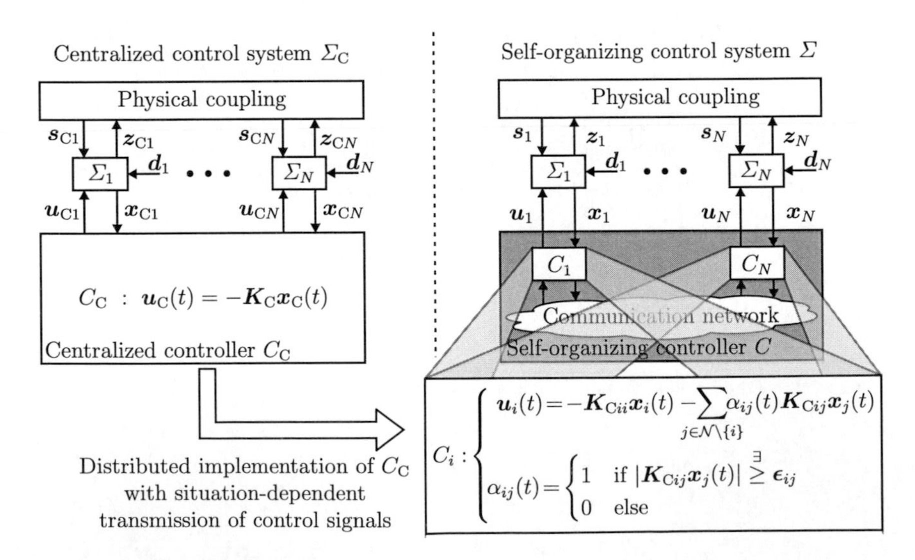

Figure 5.1: Idea of the self-organizing control concept which mimics a centralized controller C_C.

For this purpose, consider the decomposition of the central control input $\boldsymbol{u}_{\mathrm{C}}(t)$ into subsystem control inputs

$$\boldsymbol{u}_{\mathrm{C}i}(t) = \sum_{j \in \mathcal{N}} \boldsymbol{K}_{\mathrm{C}ij} \boldsymbol{x}_{\mathrm{C}j}(t), \quad \forall i \in \mathcal{N},$$

where the matrices $\boldsymbol{K}_{\mathrm{C}ij} \in \mathbb{R}^{m_i \times n_i}$, $(\forall i, j \in \mathcal{N})$, build the central feedback gain matrix $\boldsymbol{K}_{\mathrm{C}} = (\boldsymbol{K}_{\mathrm{C}ij})$ and $\boldsymbol{x}_{\mathrm{C}i}(t) \in \mathbb{R}^{n_i}$ are the subsystem states of Σ_{C} with

$$\boldsymbol{u}_{\mathrm{C}}(t) = \begin{pmatrix} \boldsymbol{u}_{\mathrm{C}1}(t) \\ \vdots \\ \boldsymbol{u}_{\mathrm{C}N}(t) \end{pmatrix}, \quad \boldsymbol{x}_{\mathrm{C}}(t) = \begin{pmatrix} \boldsymbol{x}_{\mathrm{C}1}(t) \\ \vdots \\ \boldsymbol{x}_{\mathrm{C}N}(t) \end{pmatrix}.$$

The right part of Fig. 5.1 shows the self-organizing decomposition of the control inputs

$$\boldsymbol{u}_i(t) = \underbrace{-\boldsymbol{K}_{\mathrm{C}ii} \boldsymbol{x}_i(t)}_{\text{permanent}} - \underbrace{\sum\nolimits_{j \in \mathcal{N} \setminus \{i\}} \alpha_{ij}(t) \boldsymbol{K}_{\mathrm{C}ij} \boldsymbol{x}_j(t)}_{\text{situation-dependent}}, \quad \forall i \in \mathcal{N}, \tag{5.3}$$

into a permanent control input and a situation-dependent control input, where $\alpha_{ij}(t) \in \{0, 1\}$ indicates whether the control signal $\hat{\boldsymbol{u}}_{ij}(t) := \boldsymbol{K}_{\mathrm{C}ij} \boldsymbol{x}_j(t)$ is transmitted from C_j to C_i $(\alpha_{ij}(t) = 1)$ or not $(\alpha_{ij}(t) = 0)$. With this implementation, there exist two extremes: full communication $(\alpha_{ij}(t) = 1, \forall i, j \in \mathcal{N}, \forall t \geq 0)$, which would be identical to the centralized control system Σ_{C}, and no communication $(\alpha_{ij}(t) = 0, \forall i, j \in \mathcal{N}, \forall t \geq 0)$, which would result in a decentralized control system Σ_{D} according to (3.24). On the one hand, full communication would induce the same system performance as the centralized control system Σ_{C} but causes a high communication effort among the local control units. On the other hand, no communication can worsen the system performance compared to full communication, but does not load the communication network. Hence, a rule has to be defined in which situations a transmission of $\hat{\boldsymbol{u}}_{ij}(t)$ from C_j to C_i is appropriate.

If, for example, due to a disturbance an element of the local control signal $\hat{\boldsymbol{u}}_{ij}(t) = \boldsymbol{K}_{\mathrm{C}ij} \boldsymbol{x}_j(t)$ exceeds a given switching threshold $\boldsymbol{\epsilon}_{ij} \in \mathbb{R}^{m_i}$, the value of $\boldsymbol{K}_{\mathrm{C}ij} \boldsymbol{x}_j(t)$ is sent from C_j to C_i $(\alpha_{ij}(t) = 1)$ to compensate the effect of the disturbance. If all elements of $\boldsymbol{K}_{\mathrm{C}ij} \boldsymbol{x}_j(t)$ are below the threshold $\boldsymbol{\epsilon}_{ij}$, its value is not sent $(\alpha_{ij}(t) = 0)$, i.e.,

$$\alpha_{ij}(t) = \begin{cases} 1 & \text{if } |\boldsymbol{K}_{\mathrm{C}ij} \boldsymbol{x}_j(t)| \overset{\exists}{\geq} \boldsymbol{\epsilon}_{ij} \\ 0 & \text{else,} \end{cases} \quad \forall j, i \in \mathcal{N}, \quad j \neq i. \tag{5.4}$$

Hence, if no subsystem is disturbed and $\boldsymbol{x}_0 = \boldsymbol{0}$, then no communication is necessary to guarantee a desired system performance. Roughly speaking, C_j decides whether $\hat{\boldsymbol{u}}_{ij}(t) = \boldsymbol{K}_{\mathrm{C}ij}\boldsymbol{x}_j(t)$ is so large that the local control unit C_i of the neighboring subsystem Σ_i needs to know $\hat{\boldsymbol{u}}_{ij}(t)$ to react to the effect of the disturbance and, therefore, improve its control performance. Recall that the symbol $\overset{\exists}{\geq}$ indicates that the relation $|\boldsymbol{K}_{\mathrm{C}ij}\boldsymbol{x}_j(t)| \overset{\exists}{\geq} \boldsymbol{\epsilon}_{ij}$ holds true if there exists some element of the vector $\boldsymbol{K}_{\mathrm{C}ij}\boldsymbol{x}_j(t)$ which is greater or equal to the corresponding element of the vector $\boldsymbol{\epsilon}_{ij}$ (cf. Section 2.1).

Based on these considerations, the main **idea** of the self-organizing control concept proposed in this chapter can be summarized as follows:

> The local control signals $\hat{\boldsymbol{u}}_{ij}(t) = \boldsymbol{K}_{\mathrm{C}ij}\boldsymbol{x}_j(t)$ are transmitted from the local control units C_j to the local control units C_i whenever $\hat{\boldsymbol{u}}_{ij}(t)$ exceeds the switching threshold $\boldsymbol{\epsilon}_{ij}$ in order to obtain a similar performance as with the centralized controller C_{C}.

Equations (5.3) and (5.4) characterize the principle of operation of the local control units C_i, $(\forall i \in \mathcal{N})$, (Fig. 5.1), whereas their actual realization is different from this operation. The decision of transmitting the local control signals $\hat{\boldsymbol{u}}_{ij}(t)$, $(\forall j \in \mathcal{N}\backslash\{i\})$, is made at C_j and the composition of all transmitted local control signals $\boldsymbol{K}_{\mathrm{C}ij}\boldsymbol{x}_j(t)$ is performed by C_i, see Section 5.2. Nevertheless, the basic tasks of the of the local control units C_i, $(\forall i \in \mathcal{N})$, can be defined as follows:

1. Perform a permanent local state feedback according to (5.3).
2. Determine the local control signal $\hat{\boldsymbol{u}}_{ji}(t) = \boldsymbol{K}_{\mathrm{C}ji}\boldsymbol{x}_i(t)$ for the other local control units C_j, $(\forall j \in \mathcal{N}\backslash\{i\})$.
3. Decide whether it is beneficial to send the local control signals $\hat{\boldsymbol{u}}_{ji}(t)$ to other local control units C_j, $(\forall j \in \mathcal{N}\backslash\{i\})$, according to (5.4)
4. Use the incoming local control signals $\hat{\boldsymbol{u}}_{ij}(t)$ from C_j to mimic the behavior of the centralized controller C_{C} by adapting the local control input $\boldsymbol{u}_i(t)$ for subsystem Σ_i.

In the following, the main idea of the concept is illustrated for concerning two subsystems only. Therefore, the structure of the local control units which performs these four tasks is presented.

Structure of the self-organizing controller for two subsystems. For two subsystems the centralized controller C_{C} defined in (5.1) has the form

$$\underbrace{\begin{pmatrix} \boldsymbol{u}_{\mathrm{C1}}(t) \\ \boldsymbol{u}_{\mathrm{C2}}(t) \end{pmatrix}}_{\boldsymbol{u}_{\mathrm{C}}(t)} = \underbrace{\begin{pmatrix} \boldsymbol{K}_{\mathrm{C11}} & \boldsymbol{K}_{\mathrm{C12}} \\ \boldsymbol{K}_{\mathrm{C21}} & \boldsymbol{K}_{\mathrm{C22}} \end{pmatrix}}_{\boldsymbol{K}_{\mathrm{C}}} \underbrace{\begin{pmatrix} \boldsymbol{x}_{\mathrm{C1}}(t) \\ \boldsymbol{x}_{\mathrm{C2}}(t) \end{pmatrix}}_{\boldsymbol{x}_{\mathrm{C}}(t)} \tag{5.5}$$

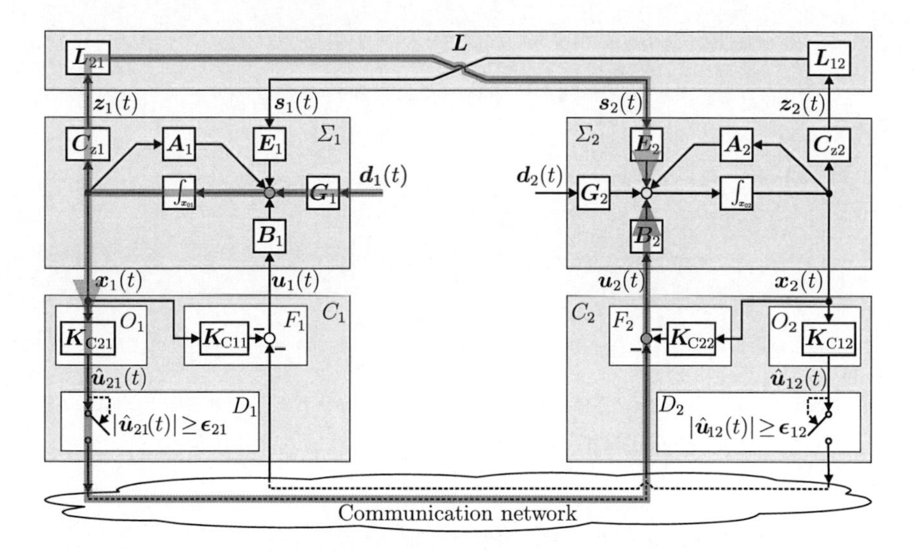

Figure 5.2: Mimicry of a centralized controller C_{C} for two subsystems.

Figure 5.2 depicts the resulting structure of the self-organizing control concept with transmission of control signals for two subsystems. In the case of a disturbance affecting subsystem Σ_1, the components of the local control units C_1 and C_2 work as follows:

- The **feedback unit** F_1 performs a local state feedback $-\boldsymbol{K}_{\mathrm{C}11}\boldsymbol{x}_1(t)$ to generate a permanent input to subsystem Σ_1 (1. Task).
- The **observation unit** O_1 calculates the local control signal by $\hat{\boldsymbol{u}}_{21}(t) = \boldsymbol{K}_{\mathrm{C}21}\boldsymbol{x}_1(t)$ for C_2 by using the state $\boldsymbol{x}_1(t)$ (2. Task).
- The **decision unit** D_1 sends the local control signal $\hat{\boldsymbol{u}}_{21}(t) = \boldsymbol{K}_{\mathrm{C}21}\boldsymbol{x}_1(t)$ to the local control unit C_2 if $\hat{\boldsymbol{u}}_{21}(t)$ exceeds the switching threshold $\boldsymbol{\epsilon}_{21}$ (3. Task), i.e.,

$$|\hat{\boldsymbol{u}}_{21}(t)| = |\boldsymbol{K}_{\mathrm{C}21}\boldsymbol{x}_1(t)| \overset{\exists}{\geq} \boldsymbol{\epsilon}_{21} \quad \Leftrightarrow \quad C_1 \text{ sends } \hat{\boldsymbol{u}}_{21}(t) \text{ to } C_2. \tag{5.6}$$

- The **decision unit** D_2 in the local control unit C_2 uses $\hat{\boldsymbol{u}}_{21}(t) = \boldsymbol{K}_{\mathrm{C}21}\boldsymbol{x}_1(t)$ to adapt its input $\boldsymbol{u}_2(t)$ such that

$$\boldsymbol{u}_2(t) = \underbrace{-\boldsymbol{K}_{\mathrm{C}22}\boldsymbol{x}_2(t)}_{\text{permanent}} - \underbrace{\boldsymbol{K}_{\mathrm{C}21}\boldsymbol{x}_1(t),}_{\text{situation-dependent}} \tag{5.7}$$

  which matches with the input of the central controller.

This control strategy tries to mimic the behavior of the centralized controller C_{C} defined in (5.5).

To illustrate the deviation between the two control concepts, consider the behavior of subsystem Σ_1 controlled by C_{C}

$$\dot{\boldsymbol{x}}_{\mathrm{C1}}(t) = (\boldsymbol{A}_1 - \boldsymbol{B}_1\boldsymbol{K}_{\mathrm{C11}})\,\boldsymbol{x}_{\mathrm{C1}}(t) - \boldsymbol{B}_1\boldsymbol{K}_{\mathrm{C12}}\boldsymbol{x}_{\mathrm{C2}}(t) + \boldsymbol{E}_1\boldsymbol{s}_{\mathrm{C1}}(t) + \boldsymbol{G}_1\boldsymbol{d}_1(t), \quad \boldsymbol{x}_{\mathrm{C1}}(0) = \boldsymbol{x}_{\mathrm{C01}}$$

and the behavior of Σ_1 controlled by the local control unit C_1 with situation-dependent communication

$$\dot{\boldsymbol{x}}_1(t) = (\boldsymbol{A}_1 - \boldsymbol{B}_1\boldsymbol{K}_{11})\,\boldsymbol{x}_1(t) - \alpha_{12}(t)\boldsymbol{B}_1\boldsymbol{K}_{\mathrm{C12}}\boldsymbol{x}_2(t) + \boldsymbol{E}_1\boldsymbol{s}_1(t) + \boldsymbol{G}_1\boldsymbol{d}_1(t), \quad \boldsymbol{x}_1(0) = \boldsymbol{x}_{01}$$

The signal $\boldsymbol{s}_{\mathrm{C1}}(t)$ represents the coupling input to Σ_1 in the centralized control case. The difference dynamics between both subsystems results from

$$\begin{aligned}
\dot{\boldsymbol{x}}_{\Delta 1}(t) &= \dot{\boldsymbol{x}}_1(t) - \dot{\boldsymbol{x}}_{\mathrm{C1}}(t) \\
&= (\boldsymbol{A}_1 - \boldsymbol{B}_1\boldsymbol{K}_{\mathrm{C11}})\,\boldsymbol{x}_{\Delta 1}(t) - \boldsymbol{B}_1\boldsymbol{K}_{\mathrm{C12}}\,(\alpha_{12}(t)\boldsymbol{x}_2(t) - \boldsymbol{x}_{\mathrm{C2}}(t)) + \boldsymbol{E}_1\boldsymbol{s}_{\Delta 1}(t) \\
&= (\boldsymbol{A}_1 - \boldsymbol{B}_1\boldsymbol{K}_{\mathrm{C11}})\,\boldsymbol{x}_{\Delta 1}(t) - \boldsymbol{B}_1\boldsymbol{K}_{\mathrm{C12}}(\alpha_{12}(t)\boldsymbol{x}_2(t) \underbrace{- \boldsymbol{x}_{\mathrm{C2}}(t) + \boldsymbol{x}_2(t)}_{\boldsymbol{x}_{\Delta 2}(t)} - \boldsymbol{x}_2(t)) + \boldsymbol{E}_1\boldsymbol{s}_{\Delta 1}(t) \\
&= (\boldsymbol{A}_1 - \boldsymbol{B}_1\boldsymbol{K}_{\mathrm{C11}})\,\boldsymbol{x}_{\Delta 1}(t) - \boldsymbol{B}_1\boldsymbol{K}_{\mathrm{C12}}\boldsymbol{x}_{\Delta 2}(t) - \boldsymbol{B}_1 \underbrace{\boldsymbol{K}_{\mathrm{C12}}\boldsymbol{x}_2(t)(\alpha_{12}(t) - 1)}_{|\cdot| \le \boldsymbol{\epsilon}_{12}} + \boldsymbol{E}_1\boldsymbol{s}_{\Delta 1}(t)
\end{aligned}$$

with $\boldsymbol{x}_{\Delta 1}(t) = \boldsymbol{x}_1(t) - \boldsymbol{x}_{\mathrm{C1}}(t)$, $\boldsymbol{s}_{\Delta 1}(t) = \boldsymbol{s}_1(t) - \boldsymbol{s}_{\mathrm{C1}}(t)$ and the initial state $\boldsymbol{x}_{\Delta 1}(0) = \boldsymbol{x}_{\Delta 01} = \boldsymbol{x}_{01} - \boldsymbol{x}_{\mathrm{C01}}$. According to the transmission condition (5.4), the absolute value of $\boldsymbol{K}_{\mathrm{C12}}\boldsymbol{x}_2(t)(\alpha_{12}(t) - 1)$ is bounded by the switching threshold $\boldsymbol{\epsilon}_{12}$, i.e., $|\boldsymbol{K}_{\mathrm{C12}}\boldsymbol{x}_2(t)(\alpha_{12}(t) - 1)| \le \boldsymbol{\epsilon}_{12}$. With this, the overall difference behavior of the interconnected subsystems results to

$$\begin{aligned}
\begin{pmatrix} \dot{\boldsymbol{x}}_{\Delta 1}(t) \\ \dot{\boldsymbol{x}}_{\Delta 2}(t) \end{pmatrix} &= \underbrace{\left(\underbrace{\begin{pmatrix} \boldsymbol{A}_1 & \boldsymbol{E}_1\boldsymbol{L}_{21}\boldsymbol{C}_{\mathrm{z2}} \\ \boldsymbol{E}_2\boldsymbol{L}_{12}\boldsymbol{C}_{\mathrm{z1}} & \boldsymbol{A}_2 \end{pmatrix}}_{\boldsymbol{A}_{\mathrm{P}}} - \underbrace{\begin{pmatrix} \boldsymbol{B}_1 \\ \boldsymbol{B}_2 \end{pmatrix}}_{\boldsymbol{B}} \underbrace{\begin{pmatrix} \boldsymbol{K}_{\mathrm{C11}} & \boldsymbol{K}_{\mathrm{C12}} \\ \boldsymbol{K}_{\mathrm{C21}} & \boldsymbol{K}_{\mathrm{C22}} \end{pmatrix}}_{\boldsymbol{K}_{\mathrm{C}}} \right)}_{\boldsymbol{A}_{\mathrm{C}}} \begin{pmatrix} \boldsymbol{x}_{\Delta 1}(t) \\ \boldsymbol{x}_{\Delta 2}(t) \end{pmatrix} \\
&\quad + \underbrace{\begin{pmatrix} \boldsymbol{B}_1 \\ \boldsymbol{B}_2 \end{pmatrix}}_{\boldsymbol{B}} \underbrace{\begin{pmatrix} \boldsymbol{K}_{\mathrm{C12}}\boldsymbol{x}_2(t)\,(\alpha_{12} - 1) \\ \boldsymbol{K}_{\mathrm{C21}}\boldsymbol{x}_1(t)\,(\alpha_{21} - 1) \end{pmatrix}}_{|\cdot| \le \begin{pmatrix} \boldsymbol{\epsilon}_{12} \\ \boldsymbol{\epsilon}_{21} \end{pmatrix}}, \qquad \begin{pmatrix} \boldsymbol{x}_{\Delta 1}(0) \\ \boldsymbol{x}_{\Delta 2}(0) \end{pmatrix} = \begin{pmatrix} \boldsymbol{x}_{\Delta 01} \\ \boldsymbol{x}_{\Delta 02} \end{pmatrix}
\end{aligned} \tag{5.8}$$

where $\boldsymbol{A}_{\mathrm{C}}$ is the system matrix of the centralized control system Σ_{C}. Hence, equation (5.8) shows that for $\boldsymbol{x}_{\Delta 0i} = \boldsymbol{0}$, $(\forall i \in \{1,2\})$, which holds true for $\boldsymbol{x}_{0i} = \boldsymbol{x}_{\mathrm{C}0i}$, the difference between the behavior of the centralized control system Σ_{C} and the behavior of the self-organizing control system Σ can be made arbitrarily small by the choice of the switching thresholds ϵ_{ij} no matter if the subsystems are disturbed or not. The following example illustrates these properties.

Example 5.1 *Situation-dependent transmission of control signals*

In this example the self-organizing control concept with a situation-dependent transmission of control signals illustrated in Fig. 5.2 is used to control the symmetrically interconnected subsystems defined in (3.32). The example is divided into two parts: First, the construction of the local control units C_1 and C_2 based on a centralized controller C_{C} is presented. Second, the behavior of the resulting self-organizing control system Σ is investigated.

Local control units. The parameters of the components of the two identical local control units C_1 and C_2 are listed in Table 5.1. The central feedback gain matrix $\boldsymbol{K}_{\mathrm{C}}$ of the centralized controller C_{C} defined in (5.1) results from the solutions of the optimization problem in (2.11) w.r.t the plant P defined in (3.32) while using the weighting matrices $\boldsymbol{Q} = 3\boldsymbol{I}_2$ and $\boldsymbol{R} = \boldsymbol{I}_2$ to

$$\boldsymbol{K}_{\mathrm{C}} = \begin{pmatrix} 3.19 & 1.46 \\ 1.46 & 3.19 \end{pmatrix}. \tag{5.9}$$

This centralized feedback stabilizes the centralized control system Σ_{C} defined in (5.2) since the overall system matrix

$$\boldsymbol{A}_{\mathrm{C}} = \underbrace{\begin{pmatrix} 1 & 1 \\ 1 & 1 \end{pmatrix}}_{\boldsymbol{A}_{\mathrm{P}}} - \underbrace{\begin{pmatrix} 1 & 0 \\ 0 & 1 \end{pmatrix}}_{\boldsymbol{B}} \underbrace{\begin{pmatrix} 3.19 & 1.46 \\ 1.46 & 3.19 \end{pmatrix}}_{\boldsymbol{K}_{\mathrm{C}}} = \begin{pmatrix} -2.19 & -0.46 \\ -0.46 & -2.19 \end{pmatrix} \tag{5.10}$$

is a Hurwitz matrix ($\lambda_1(\boldsymbol{A}_{\mathrm{C}}) = -2.65$ and $\lambda_2(\boldsymbol{A}_{\mathrm{C}}) = -1.73$). The central feedback gain matrix $\boldsymbol{K}_{\mathrm{C}}$ leads to the local feedback gain matrices $\boldsymbol{K}_{\mathrm{C}11}$ and $\boldsymbol{K}_{\mathrm{C}22}$ for the feedback units F_i as well as to the gain matrices $\boldsymbol{K}_{\mathrm{C}21}$ and $\boldsymbol{K}_{\mathrm{C}12}$ for the observation units O_i. The switching thresholds are chosen to be identical, i.e., $\epsilon_{21} = \epsilon_{12} = 0.2$. The links within the communication graph $\mathcal{G}_{\mathrm{C}}(t)$ depicted in Fig. 5.3 result from the structure of $\boldsymbol{K}_{\mathrm{C}}$. For example, the communication link $(1 \rightarrow 2)$ is active if the relation $|\hat{\boldsymbol{u}}_{21}(t)| = |\boldsymbol{K}_{\mathrm{C}21}\boldsymbol{x}_1(t)| \overset{\exists}{\geq} \epsilon_{21} = 0.2$ holds true.

Behavior of the self-organizing control system Σ. The behavior of Σ with $\boldsymbol{x}_{0i} = 0$, $(\forall i \in \{1,2\})$, and a disturbance affecting subsystem Σ_1 is depicted in Fig. 5.4 (black solid line). Due to the physical coupling, the disturbance $\boldsymbol{d}_1(t) \neq 0$ forces the states $\boldsymbol{x}_1(t)$ and $\boldsymbol{x}_2(t)$ out of the equilibrium point. Therefore, the control signals $\hat{\boldsymbol{u}}_{21}(t) = \boldsymbol{K}_{\mathrm{C}21}\boldsymbol{x}_1(t)$ and $\hat{\boldsymbol{u}}_{12}(t) = \boldsymbol{K}_{\mathrm{C}12}\boldsymbol{x}_2(t)$ exceed the switching thresholds ϵ_{12} and ϵ_{21}, respectively. When $\hat{\boldsymbol{u}}_{21}(t)$ exceeds the threshold ϵ_{21}, the local control unit C_1 sends $\hat{\boldsymbol{u}}_{21}(t)$ to the local control unit C_2 (indicated by the black bar in the lowest plot). If $\hat{\boldsymbol{u}}_{12}(t)$ exceeds ϵ_{12}, then C_2 sends $\hat{\boldsymbol{u}}_{12}(t)$ to C_1. After the disturbance $\boldsymbol{d}_1(t)$ goes to zero, the communication from C_1

Table 5.1: Parameters of the local control units C_1 and C_2 presented in Fig. 5.2 for the control of the symmetrically interconnected subsystems.

Feedback units F_1 and F_2	Observation units O_1 and O_2	Decision units D_1 and D_2
$\boldsymbol{K}_{\mathrm{C}11} = \boldsymbol{K}_{\mathrm{C}22} = 3.19$	$\boldsymbol{K}_{\mathrm{C}21} = \boldsymbol{K}_{\mathrm{C}12} = 1.46$	$\epsilon_{21} = \epsilon_{12} = 0.2$

$\exists$
$|\hat{\boldsymbol{u}}_{12}(t)| \geq \boldsymbol{\epsilon}_{12}$

1 2 $\mathcal{G}_{\mathrm{C}}(t)$

$\exists$
$|\hat{\boldsymbol{u}}_{21}(t)| \geq \boldsymbol{\epsilon}_{21}$

Figure 5.3: Communication graph $\mathcal{G}_{\mathrm{C}}(t)$ for the situation-dependent transmission of local control signals of the symmetrically interconnected subsystems.

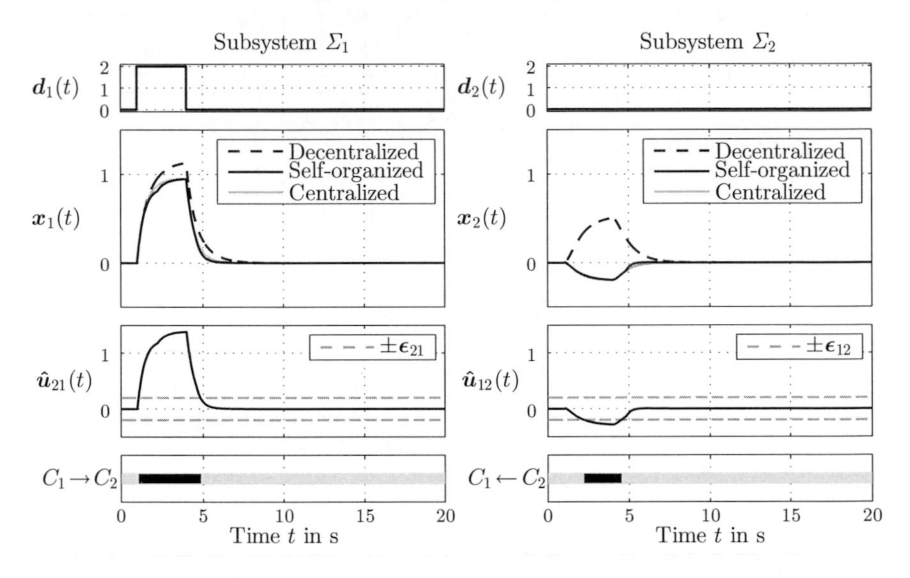

Figure 5.4: Behavior of the disturbed symmetrically interconnected subsystems defined (3.32) controlled be the local control units C_i specified in Fig. 5.2 and Table 5.1.

to C_2 and from C_2 to C_1 is deactivated. This limited communication generates a similar behavior of the self-organizing control system Σ compared to the centralized control system Σ_{C} (gray solid line in Fig. 5.4). In comparison to the decentralized control system Σ_{D} (black dashed line in Fig. 5.4) especially the deviation of $\boldsymbol{x}_2(t)$ from the equilibrium point can be reduced with the self-organizing controller using a situation-dependent communication. But Fig. 5.4 also shows that generally the performance of Σ can of course only be as good the performance of Σ_{C}.

Figure 5.5 shows interesting properties of the self-organizing control system Σ. The left plot depicts the maximal value $\max_{t\geq 0} |\boldsymbol{x}_{\Delta 2}(t)| = \max_{t\geq 0} |\boldsymbol{x}_2(t) - \boldsymbol{x}_{\mathrm{C}2}(t)|$ of the difference state $\boldsymbol{x}_{\Delta 2}(t)$ for different maximal values of the disturbance $\boldsymbol{d}_1(t)$ as in Fig. 5.4 with the thresholds $\boldsymbol{\epsilon}_{21} = \boldsymbol{\epsilon}_{12} = 0.2$ as stated above. The difference value $\max_{t\geq 0} |\boldsymbol{x}_{\Delta 2}(t)|$ with the self-organizing controller C is bounded (solid line), whereas for the decentralized controller C_{D} $\max_{t\geq 0} |\boldsymbol{x}_{\Delta 2}(t)|$ (dashed line) increases linearly with $\max_{t\geq 0} |\boldsymbol{d}_1(t)|$. To achieve this boundedness the communication time increases with $\max_{t\geq 0} |\boldsymbol{d}_1(t)|$ (dot-dashed line). The *communication time* is the length of the time interval in which C_1 transmits information to C_2 which is the length of the black bar in the lower left plot of Fig. 5.4.

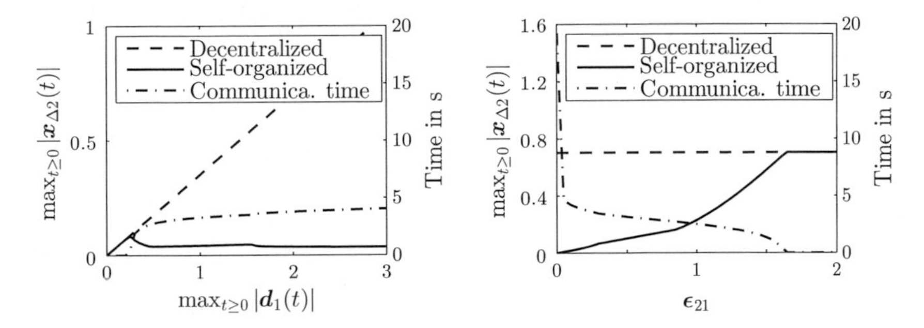

Figure 5.5: Relation between the maximal disturbance $\boldsymbol{d}_1(t)$ at Σ_1 and maximal value of the difference state $\boldsymbol{x}_{\Delta2}(t) = \boldsymbol{x}_2(t) - \boldsymbol{x}_{C2}(t)$ at Σ_2 (left); Relation between the switching threshold ϵ_{21} in C_1 and maximal value of $\boldsymbol{x}_{\Delta2}(t)$ of Σ_2 (right).

The relation between $\max_{t\geq0}|\boldsymbol{x}_{\Delta2}(t)|$ and the switching threshold ϵ_{21} is depicted in the right part of Fig. 5.5. The value for $\max_{t\geq0}|\boldsymbol{x}_{\Delta2}(t)|$ increases with the switching threshold ϵ_{21} until $\epsilon_{21} > 1.6$. Thereafter, $\max_{t\geq0}|\boldsymbol{x}_{\Delta2}(t)|$ is equal to the value while using a decentralized controller C_D since the large threshold ϵ_{21} is never exceeded by $\hat{\boldsymbol{u}}_{21}(t)$. If $\epsilon_{21} = 0$ holds, then C_1 sends $\hat{\boldsymbol{u}}_{21}(t)$ to C_2 at all times. Even for small values of the threshold ϵ_{21} there is huge reduction of the communication time but only a small deviation from the behavior of the subsystems while using the centralized controller C_C. Hence, information has to be exchanged only for a small time interval to mimic the behavior of the centralized control system Σ_C in an adequate way. This illustrates that there is a trade-off between $\max_{t\geq0}|\boldsymbol{x}_{\Delta2}(t)|$ and the communication effort.

In summary, this example shows that the proposed self-organizing control concept with transmission of local control signals has the following interesting properties:

- The local control units only exchange information if a disturbance affects a subsystem. Hence, the communication is situation-dependent.
- The difference between the behavior of the self-organizing control system Σ and the behavior of the centralized control system Σ_C can be adjusted by the choice of the switching thresholds, where the difference can be made arbitrarily small no matter of the amount of the disturbances.
- There is a trade-off between the communication effort and the imitation of the performance of the centralized control system Σ_C.

□

Overview of the chapter. In the following sections it is shown that the mentioned properties of the proposed control concept also exist if it is applied to an arbitrary number of subsystem. Therefore, the content of the next sections is as follows:

- The structure of the self-organizing controller C is extended to an arbitrary number of subsystems (Section 5.2).

- An analysis of the self-organizing control system Σ yields bounds on the difference behavior between Σ and Σ_{C}, conditions for the stability of Σ as well as estimations on the behavior of Σ in case of disturbances (Section 5.3).
- Methods for designing the switching thresholds and the communication structure to guarantee the control aims claimed in Section 3.4 are given (Section 5.4).
- The self-organizing control strategy is demonstrated by its application to a water supply system (Section 5.5).
- The self-organizing properties of the closed-loop system with the situation-dependent transmission of local control signals are evaluated (Section 5.6).

5.2 Structure of the self-organizing controller and model of the overall closed-loop system

This section presents the structure of the self-organizing networked controller C which transmits control signals illustrated for two subsystems in the previous section. The properties of the communication structure are introduced in Section 5.2.1 and the structure of the local control units C_i is derived in Section 5.2.2. Section 5.2.3 presents the model of the resulting self-organizing control system Σ.

5.2.1 Communication structure depending on the centralized controller

To improve the control performance compared to decentralized control, the local control units C_i, $(\forall i \in \mathcal{N})$, need the opportunity to receive the local control signal $\hat{\boldsymbol{u}}_{ij}(t)$ from all other local control units if the overall feedback gain matrix $\boldsymbol{K}_{\mathrm{C}}$ is fully populated, i.e., $\boldsymbol{K}_{\mathrm{C}ji} \neq \boldsymbol{O}$, $(\forall i, j \in \mathcal{N})$. Otherwise, the local control units C_i, $(\forall i \in \mathcal{N})$, receive the local control signals $\hat{\boldsymbol{u}}_{ij}(t)$ from a selection of local control units. With this, the sets $\bar{\mathcal{P}}_{\mathrm{C}i}$, $(\forall i \in \mathcal{N})$, of maximal communicational predecessors introduced in Section 2.3 result to

$$\bar{\mathcal{P}}_{\mathrm{C}i} = \{j \in \mathcal{N} \backslash \{i\} \mid \|\boldsymbol{K}_{\mathrm{C}ij}\| \neq 0\}, \quad \forall i \in \mathcal{N}, \tag{5.11}$$

which lead to the set $\bar{\mathcal{E}}_{\mathrm{C}}$ of maximal communicational edges defined in (2.21). The set $\bar{\mathcal{E}}_{\mathrm{C}}$ defines the maximal communication graph $\bar{\mathcal{G}}_{\mathrm{C}} = (\mathcal{V}_{\mathrm{C}}, \bar{\mathcal{E}}_{\mathrm{C}})$ according to (2.19) and bounds the set $\mathcal{E}_{\mathrm{C}}(t)$ of communication links within the communication graph $\mathcal{G}_{\mathrm{C}}(t)$ as stated in (3.19) (cf. Fig. 3.4). Recall that the sets $\mathcal{P}_{\mathrm{P}i}$, $(\forall i \in \mathcal{N})$, also define the sets $\bar{\mathcal{F}}_{\mathrm{C}i}$ of maximal com-

$$\boldsymbol{K}_{\mathrm{C}} = \begin{pmatrix} \boldsymbol{K}_{\mathrm{C}11} & \boldsymbol{K}_{\mathrm{C}12} & \boldsymbol{K}_{\mathrm{C}13} & \boldsymbol{K}_{\mathrm{C}14} \\ \boldsymbol{K}_{\mathrm{C}21} & \boldsymbol{K}_{\mathrm{C}22} & \boldsymbol{O} & \boldsymbol{O} \\ \boldsymbol{K}_{\mathrm{C}31} & \boldsymbol{K}_{\mathrm{C}32} & \boldsymbol{K}_{\mathrm{C}33} & \boldsymbol{O} \\ \boldsymbol{O} & \boldsymbol{K}_{\mathrm{C}42} & \boldsymbol{K}_{\mathrm{C}43} & \boldsymbol{K}_{\mathrm{C}44} \end{pmatrix} \Rightarrow \bar{\mathcal{G}}_{\mathrm{C}} : \left\{ \right.$$

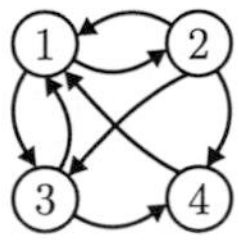

Figure 5.6: Construction of the maximal communication graph $\bar{\mathcal{G}}_{\mathrm{C}}$ for the situation-dependent transmission of the control signals by $\boldsymbol{K}_{\mathrm{C}}$, where $\boldsymbol{K}_{\mathrm{C}12}$, $\boldsymbol{K}_{\mathrm{C}13}$, $\boldsymbol{K}_{\mathrm{C}14}$, $\boldsymbol{K}_{\mathrm{C}21}$, $\boldsymbol{K}_{\mathrm{C}31}$, $\boldsymbol{K}_{\mathrm{C}32}$, $\boldsymbol{K}_{\mathrm{C}42}$ and $\boldsymbol{K}_{\mathrm{C}43}$ are non-zero matrices.

municational followers according to (2.22), i.e., $\bar{\mathcal{F}}_{\mathrm{C}i} = \{j \in \mathcal{N}\backslash\{i\} \mid \|\boldsymbol{K}_{\mathrm{C}ji}\| \neq 0\}$ for all $i \in \mathcal{N}$.

In summary, the maximal communication graph $\bar{\mathcal{G}}_{\mathrm{C}}$ results from the structure of the central feedback gain matrix $\boldsymbol{K}_{\mathrm{C}}$ which is illustrated in Fig. 5.6. Furthermore, the set $\mathcal{E}_{\mathrm{C}}(t)$ of communication links is bounded according to (2.25), (3.19) and (5.11), i.e.,

$$\emptyset \subseteq \mathcal{E}_{\mathrm{C}}(t) \subseteq \bar{\mathcal{E}}_{\mathrm{C}} = \{(j \rightarrow i) \mid \|\boldsymbol{K}_{\mathrm{C}ij}\| \neq 0,\ i,j \in \mathcal{V}_{\mathrm{C}},\ i \neq j\}, \quad \forall t \geq 0. \tag{5.12}$$

5.2.2 Structure of the local control units

The local control units C_i consist of a feedback unit F_i, an observation unit O_i and a decision unit D_i (Fig. 5.7). The feedback unit F_i generates the local feedback which not necessarily stabilizes the isolated subsystem Σ_i. It is assumed that the complete state $\boldsymbol{x}_i(t)$ can be measured (Assumption 3.3). The observation unit O_i determines the current situation of the agents Σ_i indicated by the magnitude of the calculated control signals $\hat{\boldsymbol{u}}_{ji}(t)$ for the local control units C_j, $(\forall j \in \bar{\mathcal{F}}_{\mathrm{C}i})$. The decision unit D_i sends the calculated control signals $\hat{\boldsymbol{u}}_{ji}(t)$ to C_j, $(\forall j \in \bar{\mathcal{F}}_{\mathrm{C}i})$, within the *transmitted control signal* $\tilde{\boldsymbol{u}}_{ji}(t)$ if $\|\hat{\boldsymbol{u}}_{ji}(t)\|$ exceeds the switching threshold $\boldsymbol{\epsilon}_{ji}$. Furthermore, D_i uses $\tilde{\boldsymbol{u}}_{ij}(t)$, $(\forall j \in \bar{\mathcal{P}}_{\mathrm{C}i})$, from the local control units C_j to generate the *communicated control input* $\tilde{\boldsymbol{u}}_i(t)$ which supports the local feedback by using information from the other subsystem Σ_j to mimic the behavior of the centralized control system Σ_{C}. The following paragraphs describe these components in a formal way.

Feedback unit F_i. The feedback unit F_i generates the input $\boldsymbol{u}_i(t)$ for the subsystem Σ_i:

$$F_i : \boldsymbol{u}_i(t) = -\boldsymbol{K}_{\mathrm{C}ii}\boldsymbol{x}_i(t) - \tilde{\boldsymbol{u}}_i(t), \quad i \in \mathcal{N}, \tag{5.13}$$

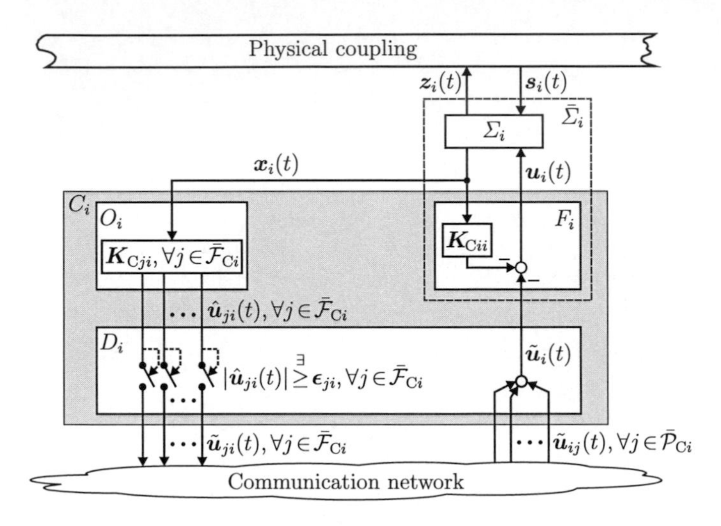

Figure 5.7: Local control unit C_i defined in (5.16) for situation-dependent transmission of control signals.

using the block-diagonal matrices $\boldsymbol{K}_{Cii}$ of the central feedback gain matrix $\boldsymbol{K}_C$ and the communicated control input $\tilde{\boldsymbol{u}}_i(t)$ determined by the decision unit D_i according to (5.15c).

Observation unit O_i**.** The observation unit O_i provides the *calculated control signals* $\hat{\boldsymbol{u}}_{ji}(t)$ for the decision unit D_i defined in (5.15):

$$O_i : \hat{\boldsymbol{u}}_{ji}(t) = \boldsymbol{K}_{Cji}\boldsymbol{x}_i(t), \quad \forall j \in \bar{\mathcal{F}}_{Ci}, \quad i \in \mathcal{N}, \tag{5.14}$$

where $\hat{\boldsymbol{u}}_{ji}(t)$ describes how great the impact of the state $\boldsymbol{x}_i(t)$ over the central feedback gain matrix $\boldsymbol{K}_C$ on the input $\boldsymbol{u}_j(t)$ of Σ_j is.

Decision unit D_i**.** The decision unit D_i, $(i \in \mathcal{N})$, has two tasks and is defined as follows:

$$D_i : \begin{cases} \tilde{\boldsymbol{u}}_{ji}(t) = \begin{cases} \hat{\boldsymbol{u}}_{ji}(t) & \text{if } j \in \mathcal{F}_{Ci}(t) \\ \boldsymbol{\eta} & \text{else,} \end{cases} \quad \forall j \in \mathcal{N} \backslash \{i\} & \text{(5.15a)} \\ \mathcal{F}_{Ci}(t) = \{ j \in \bar{\mathcal{F}}_{Ci} \mid |\hat{\boldsymbol{u}}_{ji}(t)| \overset{\exists}{\geq} \boldsymbol{\epsilon}_{ji} \} & \text{(5.15b)} \\ \tilde{\boldsymbol{u}}_i(t) = \sum_{j \in \mathcal{P}_{Ci}(t)} \tilde{\boldsymbol{u}}_{ij}(t) & \text{(5.15c)} \\ \mathcal{P}_i(t) = \{ j \in \bar{\mathcal{P}}_{Ci} \mid \tilde{\boldsymbol{u}}_{ij}(t) \neq \boldsymbol{\eta} \}. & \text{(5.15d)} \end{cases}$$

First (in (5.15a) and (5.15b)), the decision unit D_i decides whether the calculated control signals $\hat{\boldsymbol{u}}_{ji}(t)$ are transmitted from the local control unit C_i to the local control units C_j, $(j \in \bar{\mathcal{F}}_{\mathrm{C}i})$, through the transmitted control signals $\tilde{\boldsymbol{u}}_{ji}(t)$. The symbol $\boldsymbol{\eta} \in \mathbb{R}^{m_j}$ is a placeholder for the transmitted control signal $\tilde{\boldsymbol{u}}_{ji}(t)$ and means that $\hat{\boldsymbol{u}}_{ji}(t)$ is not transmitted from C_i to C_j. The calculated control signal $\hat{\boldsymbol{u}}_{ji}(t)$ is transmitted to C_j if at least on element of $\hat{\boldsymbol{u}}_{ji}(t)$ exceeds the corresponding element of the switching threshold $\boldsymbol{\epsilon}_{ji}$ and thus, the index j is in the current set $\mathcal{F}_{\mathrm{C}i}(t)$ of communicational followers. Hence, the determination of the set $\mathcal{F}_{\mathrm{C}i}(t)$ by the switching condition in (5.15b) defines the current communication graph $\mathcal{G}_{\mathrm{C}}(t)$ according to (2.16).

Second (in (5.15c) and (5.15d)), the decision unit D_i calculates the communicated control input $\tilde{\boldsymbol{u}}_i(t)$ for the feedback unit F_i based on the transmitted control signals $\tilde{\boldsymbol{u}}_{ij}(t)$, $(\forall j \in \bar{\mathcal{P}}_{\mathrm{C}i})$.

In summary, the local control units C_i, $(\forall i \in \mathcal{N})$, are defined by

$$C_i : \begin{cases} F_i \text{ defined in (5.13)} \\ O_i \text{ defined in (5.14)} \\ D_i \text{ defined in (5.15),} \end{cases} \qquad \forall i \in \mathcal{N}. \tag{5.16}$$

The corresponding self-organizing controller C with the situation-dependent communication graph $\mathcal{G}_{\mathrm{C}}(t) = (\mathcal{V}_{\mathrm{C}}, \mathcal{E}_{\mathrm{C}}(t))$ results to

$$C : C_i, (\forall i \in \mathcal{N}), \text{ defined in (5.16) with } \mathcal{E}_{\mathrm{C}}(t) \text{ bounded according to (5.12).} \tag{5.17}$$

5.2.3 Overall model of the self-organizing control system

The controlled subsystems $\bar{\Sigma}_i$, $(\forall i \in \mathcal{N})$, depicted in Fig. 5.7 result from (3.1) and (5.13) to

$$\bar{\Sigma}_i : \begin{cases} \dot{\boldsymbol{x}}_i(t) = \bar{\boldsymbol{A}}_i \boldsymbol{x}_i(t) - \boldsymbol{B}_i \tilde{\boldsymbol{u}}_i(t) + \boldsymbol{E}_i \boldsymbol{s}_i(t) + \boldsymbol{G}_i \boldsymbol{d}_i(t), \quad \boldsymbol{x}_i(0) = \boldsymbol{x}_{0i} \\ \boldsymbol{v}_i(t) = \boldsymbol{C}_{\mathrm{v}i} \boldsymbol{x}_i(t) \\ \boldsymbol{z}_i(t) = \boldsymbol{C}_{\mathrm{z}i} \boldsymbol{x}_i(t) \end{cases} \tag{5.18}$$

with $\bar{\boldsymbol{A}}_i = \boldsymbol{A}_i - \boldsymbol{B}_i \boldsymbol{K}_{\mathrm{C}ii}$.

The resulting self-organizing control system Σ includes all observation units O_i , $(\forall i \in \mathcal{N})$, and decision units D_i and follows from (3.3), (5.17) and (5.18) to

$$\Sigma: \begin{cases} \dot{\boldsymbol{x}}(t) = \bar{\boldsymbol{A}}_{\mathrm{P}}\boldsymbol{x}(t) - \boldsymbol{B}\tilde{\boldsymbol{u}}(t) + \boldsymbol{G}\boldsymbol{d}(t), \quad \boldsymbol{x}(0) = \boldsymbol{x}_0 \\ \boldsymbol{v}(t) = \boldsymbol{C}_{\mathrm{v}}\boldsymbol{x}(t) \\ \boldsymbol{z}(t) = \boldsymbol{C}_{\mathrm{z}}\boldsymbol{x}(t) \\ \boldsymbol{s}(t) = \boldsymbol{L}\boldsymbol{z}(t) \text{ according to (3.3)} \\ O_i, \ (\forall i \in \mathcal{N}), \text{ defined in (5.14)} \\ D_i, \ (\forall i \in \mathcal{N}), \text{ defined in (5.15)} \end{cases} \tag{5.19}$$

with $\bar{\boldsymbol{A}}_{\mathrm{P}} = \bar{\boldsymbol{A}} + \boldsymbol{E}\boldsymbol{L}\boldsymbol{C}_{\mathrm{z}} \in \mathbb{R}^{n \times n}$ and $\bar{\boldsymbol{A}} = \mathrm{diag}(\bar{\boldsymbol{A}}_1 \ldots \bar{\boldsymbol{A}}_N) \in \mathbb{R}^{n \times n}$. The overall vector of communicated control inputs $\tilde{\boldsymbol{u}}(t) \in \mathbb{R}^m$ determined by the decision units D_i is defined by

$$\begin{aligned} \tilde{\boldsymbol{u}}(t) &= \begin{pmatrix} \tilde{\boldsymbol{u}}_1(t) \\ \vdots \\ \tilde{\boldsymbol{u}}_N(t) \end{pmatrix} \\ \tilde{\boldsymbol{u}}_i(t) &= \sum_{j \in \mathcal{P}_{\mathrm{C}i}(t)} \boldsymbol{K}_{\mathrm{C}ij}\boldsymbol{x}_j(t) \\ &= \sum_{j \in \bar{\mathcal{P}}_{\mathrm{C}i}} \boldsymbol{K}_{\mathrm{C}ij}\boldsymbol{x}_j(t) - \sum_{l \in \bar{\mathcal{P}}_{\mathrm{C}i} \setminus \mathcal{P}_{\mathrm{C}i}(t)} \boldsymbol{K}_{\mathrm{C}il}\boldsymbol{x}_l(t), \quad \forall i \in \mathcal{N}. \end{aligned} \tag{5.20}$$

5.3 Analysis of the self-organizing control system

In this section the self-organizing control system Σ is analyzed with respect to its behavior compared to the behavior of the corresponding centralized control system Σ_{C} (Section 5.3.1), the behavior of the individual subsystems (Section 5.3.2), its stability (Section 5.3.3), and its behavior in case of disturbances affecting the subsystems (Section 5.3.4).

5.3.1 Behavior compared to the corresponding centralized control system

In this section the behavior of the self-organizing control system Σ and the corresponding centralized control system Σ_{C} are compared. Therefore, the section is divided into three parts:

1. The model of the centralized control system Σ_{C} defined in (5.2) is extended by considering disturbances and performance outputs.

2. A difference system Σ_{Δ} is derived which describes the difference between the behavior of the self-organizing control system Σ and behavior of the centralized control system Σ_{C}.

3. An analysis of the behavior of Σ_Δ shows that there exists an upper bound on the difference behavior of Σ compared to Σ_C (Theorem 5.1).

Extended model of the centralized controlled system. The model of the centralized control system Σ_C while considering disturbance and performance outputs follows from (3.10) and (5.1) to

$$\Sigma_\mathrm{C}: \begin{cases} \dot{\boldsymbol{x}}_\mathrm{C}(t) = \boldsymbol{A}_\mathrm{C}\boldsymbol{x}_\mathrm{C}(t) + \boldsymbol{G}\boldsymbol{d}(t), & \boldsymbol{x}_\mathrm{C}(0) = \boldsymbol{x}_{\mathrm{C}0} \\ \boldsymbol{v}_\mathrm{C}(t) = \boldsymbol{C}_\mathrm{v}\boldsymbol{x}_\mathrm{C}(t) \end{cases} \tag{5.21}$$

with

$$\boldsymbol{A}_\mathrm{C} = \boldsymbol{A}_\mathrm{P} - \boldsymbol{B}\boldsymbol{K}_\mathrm{C}. \tag{5.22}$$

The overall performance output

$$\boldsymbol{v}_\mathrm{C}(t) = \begin{pmatrix} \boldsymbol{v}_{\mathrm{C}1}(t) \\ \vdots \\ \boldsymbol{v}_{\mathrm{C}N}(t) \end{pmatrix} \in \mathbb{R}^{r_\mathrm{v}}$$

of Σ_C is assembled by the corresponding performance outputs $\boldsymbol{v}_i(t) \in \mathbb{R}^{r_{\mathrm{v}i}}$, $(\forall i \in \mathcal{N})$, of the subsystems. Due to Assumption 3.5, Theorem 2.2 states that there exists a feedback gain matrix $\boldsymbol{K}_\mathrm{C}$ such that the system matrix $\boldsymbol{A}_\mathrm{C}$ defined in (5.22) is a Hurwitz matrix. Therefore, the following assumption is weak.

Assumption 5.1 *The feedback gain matrix $\boldsymbol{K}_\mathrm{C}$ is chosen such that the matrix $\boldsymbol{A}_\mathrm{C}$ defined in (5.22) is a Hurwitz matrix.*

Difference behavior between the self-organizing control system Σ and the centralized control system Σ_C. The difference behavior between Σ defined in (5.19) and Σ_C defined in (5.21) is described by the overall difference state $\boldsymbol{x}_\Delta(t) = \boldsymbol{x}(t) - \boldsymbol{x}_\mathrm{C}(t)$. Equations (5.19), (5.21) and $\dot{\boldsymbol{x}}_\Delta(t) = \dot{\boldsymbol{x}}(t) - \dot{\boldsymbol{x}}_\mathrm{C}(t)$ yield

$$\dot{\boldsymbol{x}}_\Delta(t) = \bar{\boldsymbol{A}}_\mathrm{P}\boldsymbol{x}_\Delta(t) - \boldsymbol{B}(\tilde{\boldsymbol{u}}(t) - \tilde{\boldsymbol{u}}_\mathrm{C}(t)), \quad \boldsymbol{x}_\Delta(0) = \boldsymbol{x}_{\Delta 0} \tag{5.23}$$

with the difference initial state $\boldsymbol{x}_{\Delta 0} = \boldsymbol{x}_0 - \boldsymbol{x}_{\mathrm{C}0}$ and

$$\boldsymbol{x}_{\Delta}(t) = \begin{pmatrix} \boldsymbol{x}_{\Delta 1}(t) \\ \vdots \\ \boldsymbol{x}_{\Delta N}(t) \end{pmatrix}, \qquad \tilde{\boldsymbol{u}}_{\mathrm{C}}(t) := \begin{pmatrix} \tilde{\boldsymbol{u}}_{\mathrm{C}1}(t) \\ \vdots \\ \tilde{\boldsymbol{u}}_{\mathrm{C}N}(t) \end{pmatrix}$$

$$\tilde{\boldsymbol{u}}_{\mathrm{C}i}(t) := \sum_{j \in \bar{\mathcal{P}}_{\mathrm{C}i}} \boldsymbol{K}_{\mathrm{C}ij} \boldsymbol{x}_{\mathrm{C}j}(t) = \sum_{j \in \bar{\mathcal{P}}_{\mathrm{C}i}} \hat{\boldsymbol{u}}_{\mathrm{C}ij}(t), \quad \forall i \in \mathcal{N} \tag{5.24}$$

according to (5.1). The difference states $\boldsymbol{x}_{\Delta i}(t) = \boldsymbol{x}_i(t) - \boldsymbol{x}_{\mathrm{C}i}(t)$, $(\forall i \in \mathcal{N})$ of the subsystems assemble the overall difference state $\boldsymbol{x}_{\Delta}(t)$. The calculated control signal $\hat{\boldsymbol{u}}_{\mathrm{C}ij}(t) = \boldsymbol{K}_{\mathrm{C}ij}\boldsymbol{x}_{\mathrm{C}j}(t)$ of Σ_{C} corresponds to the calculated control signal $\hat{\boldsymbol{u}}_{ij}(t) = \boldsymbol{K}_{\mathrm{C}ij}\boldsymbol{x}_j(t)$ of Σ determined by the observation unit O_i defined in (5.14). With this, the model of the overall difference system Σ_{Δ} for the self-organizing control system Σ defined in (5.19) and the centralized control system Σ_{C} defined in (5.21) follows from (5.20), (5.23) and (5.24) to

$$\Sigma_{\Delta} : \begin{cases} \dot{\boldsymbol{x}}_{\Delta}(t) = \boldsymbol{A}_{\mathrm{C}} \boldsymbol{x}_{\Delta}(t) + \boldsymbol{B} \boldsymbol{u}_{\Delta}(t), & \boldsymbol{x}_{\Delta}(0) = \boldsymbol{x}_{\Delta 0} \\ \boldsymbol{v}_{\Delta}(t) = \boldsymbol{C}_{\mathrm{v}} \boldsymbol{x}_{\Delta}(t) \end{cases} \tag{5.25}$$

with $\boldsymbol{v}_{\Delta}(t) := \boldsymbol{v}(t) - \boldsymbol{v}_{\mathrm{C}}(t)$ and

$$\boldsymbol{v}_{\Delta}(t) = \begin{pmatrix} \boldsymbol{v}_{\Delta 1}(t) \\ \vdots \\ \boldsymbol{v}_{\Delta N}(t) \end{pmatrix}, \qquad \boldsymbol{u}_{\Delta}(t) := \begin{pmatrix} \boldsymbol{u}_{\Delta 1}(t) \\ \vdots \\ \boldsymbol{u}_{\Delta N}(t) \end{pmatrix}$$

$$\boldsymbol{u}_{\Delta i}(t) := \sum_{j \in \bar{\mathcal{P}}_{\mathrm{C}i} \setminus \mathcal{P}_{\mathrm{C}i}(t)} \boldsymbol{K}_{\mathrm{C}ij} \boldsymbol{x}_j(t) = \sum_{j \in \bar{\mathcal{P}}_{\mathrm{C}i} \setminus \mathcal{P}_{\mathrm{C}i}(t)} \hat{\boldsymbol{u}}_{ij}(t), \quad \forall i \in \mathcal{N}. \tag{5.26}$$

The difference performance outputs $\boldsymbol{v}_{\Delta i}(t) = \boldsymbol{v}_i(t) - \boldsymbol{v}_{\mathrm{C}i}(t)$, $(\forall i \in \mathcal{N})$ of the subsystems assemble the overall difference performance output $\boldsymbol{v}_{\Delta}(t)$. The signal $\boldsymbol{u}_{\Delta i}(t)$, $(i \in \mathcal{N})$ describes the deviation between the control input with permanent communication ($\mathcal{E}_{\mathrm{C}}(t) = \bar{\mathcal{E}}_{\mathrm{C}}$) and the control input with situation-dependent communication ($\mathcal{E}_{\mathrm{C}}(t) \subseteq \bar{\mathcal{E}}_{\mathrm{C}}$). The change of the set $\mathcal{E}_{\mathrm{C}}(t)$ of communication links within the communication graph $\mathcal{G}_{\mathrm{C}}(t) = (\mathcal{V}_{\mathrm{C}}, \mathcal{E}_{\mathrm{C}}(t))$ is indicated in (5.26) by the time-varying set $\mathcal{P}_{\mathrm{C}i}(t)$ of communicational predecessors (cf. (5.15d)). The more local control units C_j, $(j \in \mathcal{P}_{\mathrm{C}i})$ are sending the calculated control signals $\hat{\boldsymbol{u}}_{ij}(t)$, the less signals $\hat{\boldsymbol{u}}_{ij}(t)$ are in $\boldsymbol{u}_{\Delta i}(t)$ and, therefore, the smaller is the deviation between the behavior of Σ and Σ_{C}. The following lemma states an upper bound for $\boldsymbol{u}_{\Delta i}(t)$, $(\forall i \in \mathcal{N})$, depending on the switching thresholds $\boldsymbol{\epsilon}_{ij}$, $(\forall j \in \bar{\mathcal{P}}_{\mathrm{C}i})$.

Lemma 5.1 (Boundedness of the difference signals $\boldsymbol{u}_{\Delta i}(t)$) *The difference signals $\boldsymbol{u}_{\Delta i}$, ($\forall i \in \mathcal{N}$), within the overall difference system Σ_Δ defined in* (5.25) *are bounded by*

$$\max_{t\geq 0} |\boldsymbol{u}_{\Delta i}(t)| \leq \sum\nolimits_{j\in\bar{\mathcal{P}}_{\mathrm{C}i}} \boldsymbol{\epsilon}_{ij} \quad \forall i \in \mathcal{N}. \tag{5.27}$$

Proof. According to (5.15), the index j is added to the set $\mathcal{P}_{\mathrm{C}i}(t)$ if $|\hat{\boldsymbol{u}}_{ij}(t)| \overset{\exists}{\geq} \boldsymbol{\epsilon}_{ij}$ holds true for $i \in \mathcal{N}$ and $j \in \bar{\mathcal{P}}_{\mathrm{C}i}$, i.e., $j \notin \bar{\mathcal{P}}_{\mathrm{C}i} \backslash \mathcal{P}_{\mathrm{C}i}(t)$. Hence, $\hat{\boldsymbol{u}}_{ij}(t)$ vanishes from the sum in (5.26) for $|\hat{\boldsymbol{u}}_{ij}(t)| \overset{\exists}{\geq} \boldsymbol{\epsilon}_{ij}$. Therefore, the sum $\sum_{j\in\bar{\mathcal{P}}_{\mathrm{C}i}\backslash\mathcal{P}_{\mathrm{C}i}(t)} \hat{\boldsymbol{u}}_{ij}(t)$ in (5.26) is bounded by the sum $\sum_{j\in\bar{\mathcal{P}}_{\mathrm{C}i}} \boldsymbol{\epsilon}_{ij}$ of switching thresholds $\boldsymbol{\epsilon}_{ij}$ which completes the proof. □

With this, the overall signal $\boldsymbol{u}_\Delta(t)$, which is the only input of the overall difference system Σ_Δ, is bounded by the overall vector $\boldsymbol{\epsilon}$ of switching thresholds

$$\max_{t\geq 0} |\boldsymbol{u}_\Delta(t)| \leq \boldsymbol{\epsilon}$$

with

$$\boldsymbol{\epsilon} := \begin{pmatrix} \sum_{j\in\bar{\mathcal{P}}_{\mathrm{C}1}} \boldsymbol{\epsilon}_{1j} \\ \vdots \\ \sum_{j\in\bar{\mathcal{P}}_{\mathrm{C}N}} \boldsymbol{\epsilon}_{Nj} \end{pmatrix}.$$

With $\epsilon_{\max} := \|\boldsymbol{\epsilon}\|_\infty$, the following holds true

$$\max_{t\geq 0} \|\boldsymbol{u}_\Delta(t)\|_\infty < \epsilon_{\max}. \tag{5.28}$$

Analysis of the difference behavior. From the model of the overall difference system Σ_Δ defined in (5.25) it follows that

$$\boldsymbol{x}_\Delta(t) = \mathrm{e}^{\boldsymbol{A}_\mathrm{C} t}\boldsymbol{x}_{\Delta 0} + \int_0^t \mathrm{e}^{\boldsymbol{A}_\mathrm{C}(t-\tau)}\boldsymbol{B}\boldsymbol{u}_\Delta(\tau)\mathrm{d}\tau, \quad \forall t \geq 0. \tag{5.29}$$

Since the signal $\boldsymbol{u}_\Delta(t)$ is bounded according Lemma 5.1 and $\boldsymbol{A}_\mathrm{C}$ is a Hurwitz matrix according to Assumption 5.1, apparently the difference state $\boldsymbol{x}_\Delta(t)$ is also bounded. The following theorem gives an upper bound for the overall difference state $\boldsymbol{x}_\Delta(t)$ and the overall difference performance output $\boldsymbol{v}_\Delta(t)$ of Σ_Δ, where for the analysis the initial states of Σ and Σ_C are set to be identical, i.e., $\boldsymbol{x}_{\mathrm{C}0} = \boldsymbol{x}_0$. Hence, $\boldsymbol{x}_{\Delta 0} = \boldsymbol{0}$ holds true.

Theorem 5.1 (Boundedness of the overall difference behavior) *For $\boldsymbol{x}_{\mathrm{C}0} = \boldsymbol{x}_0$ the state $\boldsymbol{x}_\Delta(t)$ and the performance output $\boldsymbol{v}_\Delta(t)$ of the overall difference system Σ_Δ defined in (5.25) are bounded as follows:*

$$\max_{t\geq 0} |\boldsymbol{x}_\Delta(t)| \leq \bar{\boldsymbol{x}}_\Delta := \boldsymbol{M}_{\mathrm{xu}}\boldsymbol{\epsilon}$$
$$\max_{t\geq 0} |\boldsymbol{v}_\Delta(t)| \leq \bar{\boldsymbol{v}}_\Delta := \boldsymbol{M}_{\mathrm{vu}}\boldsymbol{\epsilon}$$

with

$$\boldsymbol{M}_{\mathrm{xu}} := \int_0^\infty |\mathrm{e}^{\boldsymbol{A}_{\mathrm{C}}\tau}\boldsymbol{B}|\mathrm{d}\tau$$
$$\boldsymbol{M}_{\mathrm{vu}} := \int_0^\infty |\boldsymbol{C}_{\mathrm{v}}\mathrm{e}^{\boldsymbol{A}_{\mathrm{C}}\tau}\boldsymbol{B}|\mathrm{d}\tau. \quad (5.30)$$

Proof. From (5.29), Lemma 5.1, and $\boldsymbol{x}_{\Delta 0} = \boldsymbol{0}$ it follows that

$$\max_{t\geq 0} |\boldsymbol{x}_\Delta(t)| \leq \int_0^\infty \left|\mathrm{e}^{\boldsymbol{A}_{\mathrm{C}}\tau}\boldsymbol{B}\right| \mathrm{d}\tau \max_{t\geq 0} |\boldsymbol{u}_\Delta(t)|$$
$$\leq \int_0^\infty \left|\mathrm{e}^{\boldsymbol{A}_{\mathrm{C}}\tau}\boldsymbol{B}\right| \mathrm{d}\tau \cdot \boldsymbol{\epsilon}.$$

Since $\boldsymbol{A}_{\mathrm{C}}$ is a Hurwitz matrix according to Assumption 5.1, the matrix $\boldsymbol{M}_{\mathrm{xu}}$ is finite.

From the model Σ_Δ in (5.25) it follows that

$$\boldsymbol{v}_\Delta(t) = \boldsymbol{C}_{\mathrm{v}}\mathrm{e}^{\boldsymbol{A}_{\mathrm{C}}t}\boldsymbol{x}_{\Delta 0} + \int_0^t \boldsymbol{C}_{\mathrm{v}}\mathrm{e}^{\boldsymbol{A}_{\mathrm{C}}(t-\tau)}\boldsymbol{B}\boldsymbol{u}_\Delta(\tau)\mathrm{d}\tau, \quad \forall t \geq 0. \quad (5.31)$$

With this, from Lemma 5.1 and $\boldsymbol{x}_{\Delta 0} = \boldsymbol{0}$ it follows that

$$\max_{t\geq 0} |\boldsymbol{v}_\Delta(t)| \leq \int_0^\infty \left|\boldsymbol{C}_{\mathrm{v}}\mathrm{e}^{\boldsymbol{A}_{\mathrm{C}}\tau}\boldsymbol{B}\right| \mathrm{d}\tau \max_{t\geq 0} |\boldsymbol{u}_\Delta(t)|$$
$$\leq \int_0^\infty \left|\boldsymbol{C}_{\mathrm{v}}\mathrm{e}^{\boldsymbol{A}_{\mathrm{C}}\tau}\boldsymbol{B}\right| \mathrm{d}\tau \cdot \boldsymbol{\epsilon}$$

Since $\boldsymbol{A}_{\mathrm{C}}$ is Hurwitz matrix according to Assumption 5.1, the matrix $\boldsymbol{M}_{\mathrm{vu}}$ is finite which completes the proof. □

Theorem 5.1 shows that the state $\boldsymbol{x}_\Delta(t)$ of the overall difference system Σ_Δ is bounded by $\bar{\boldsymbol{x}}_\Delta$, where $\bar{\boldsymbol{x}}_\Delta$ only depends on the choice of the switching thresholds $\boldsymbol{\epsilon}_{ij}$, $(\forall j \in \bar{\mathcal{P}}_{\mathrm{C}i}, i \in \mathcal{N})$, but not on the disturbances $\bar{\boldsymbol{d}}_i$. If all switching thresholds $\boldsymbol{\epsilon}_{ij}$ are chosen to be zero, then $\max_{t\geq 0} |\boldsymbol{x}_\Delta(t)| = \boldsymbol{0}$ holds, meaning that the behavior of Σ and Σ_{C} is identical.

From Theorem 5.1 the bounds on the difference states $\boldsymbol{x}_{\Delta i}(t)$, $(\forall i \in \mathcal{N})$, the performance outputs $\boldsymbol{v}_{\Delta i}(t)$ and the calculated control signals $\hat{\boldsymbol{u}}_{\Delta ji}(t) := \hat{\boldsymbol{u}}_{ji}(t) - \hat{\boldsymbol{u}}_{\mathrm{C}ji}(t)$ of a subsystem of the overall difference system Σ_Δ directly follow.

Corollary 5.1 (Boundedness of the signals of the difference subsystem) *For $\boldsymbol{x}_{\mathrm{C}0} = \boldsymbol{x}_0$, the states $\boldsymbol{x}_{\Delta i}(t) = \boldsymbol{x}_i(t) - \boldsymbol{x}_{\mathrm{C}i}(t)$, ($\forall i \in \mathcal{N}$), the performance outputs $\boldsymbol{v}_{\Delta i}(t) = \boldsymbol{v}_i(t) - \boldsymbol{v}_{\mathrm{C}i}(t)$ and the calculated control signals $\hat{\boldsymbol{u}}_{\Delta ji}(t) = \hat{\boldsymbol{u}}_{ji}(t) - \hat{\boldsymbol{u}}_{\mathrm{C}ji}(t)$ of a difference subsystem of the overall difference system Σ_Δ defined in (5.25) are bounded as follows:*

$$\begin{aligned}
&\max_{t\geq 0} |\boldsymbol{x}_{\Delta i}(t)| \leq \bar{\boldsymbol{x}}_{\Delta i} := \boldsymbol{M}_{\mathrm{xu}i}\boldsymbol{\epsilon}, && \forall i \in \mathcal{N} \\
&\max_{t\geq 0} |\boldsymbol{v}_{\Delta i}(t)| \leq \bar{\boldsymbol{v}}_{\Delta i} := \boldsymbol{M}_{\mathrm{vu}i}\boldsymbol{\epsilon}, && \forall i \in \mathcal{N} \\
&\max_{t\geq 0} |\hat{\boldsymbol{u}}_{\Delta ji}(t)| \leq \bar{\boldsymbol{u}}_{\Delta ji} := \boldsymbol{M}_{\mathrm{uu}ji}\boldsymbol{\epsilon}, && \forall i \in \mathcal{N}, \qquad \forall j \in \bar{\mathcal{F}}_{\mathrm{C}i}
\end{aligned}$$

with

$$\boldsymbol{M}_{\mathrm{xu}i} := \int_0^\infty |\check{\boldsymbol{C}}_{\mathrm{x}i}\mathrm{e}^{\boldsymbol{A}_{\mathrm{C}}\tau}\boldsymbol{B}|\mathrm{d}\tau, \qquad \forall i \in \mathcal{N} \tag{5.32}$$

$$\boldsymbol{M}_{\mathrm{vu}i} := \int_0^\infty |\check{\boldsymbol{C}}_{\mathrm{v}i}\mathrm{e}^{\boldsymbol{A}_{\mathrm{C}}\tau}\boldsymbol{B}|\mathrm{d}\tau, \qquad \forall i \in \mathcal{N} \tag{5.33}$$

$$\boldsymbol{M}_{\mathrm{uu}ji} := \int_0^\infty |\check{\boldsymbol{C}}_{\mathrm{u}ji}\mathrm{e}^{\boldsymbol{A}_{\mathrm{C}}\tau}\boldsymbol{B}|\mathrm{d}\tau, \qquad \forall i \in \mathcal{N}, \quad \forall j \in \bar{\mathcal{F}}_{\mathrm{C}i} \tag{5.34}$$

$$\begin{aligned}
\check{\boldsymbol{C}}_{\mathrm{x}i} &:= \mathrm{diag}(\boldsymbol{O}_{n_j \times \sum_{j=1}^{i-1} n_j} \;\; \boldsymbol{I}_{n_i \times n_i} \;\; \boldsymbol{O}_{n_j \times \sum_{j=i+1}^{N} n_j}), && \forall i \in \mathcal{N} \\
\check{\boldsymbol{C}}_{\mathrm{v}i} &:= \mathrm{diag}(\boldsymbol{O}_{r_{\mathrm{v}j} \times \sum_{j=1}^{i-1} n_j} \;\; \boldsymbol{C}_{\mathrm{v}i} \;\; \boldsymbol{O}_{r_{\mathrm{v}j} \times \sum_{j=i+1}^{N} n_j}), && \forall i \in \mathcal{N} \\
\check{\boldsymbol{C}}_{\mathrm{u}ji} &:= \mathrm{diag}(\boldsymbol{O}_{m_j \times \sum_{j=1}^{i-1} n_j} \;\; \boldsymbol{K}_{\mathrm{C}ji} \;\; \boldsymbol{O}_{m_j \times \sum_{j=i+1}^{N} n_j}), && \forall i \in \mathcal{N}, \quad \forall j \in \bar{\mathcal{F}}_{\mathrm{C}i}.
\end{aligned}$$

Proof. The proof is given in Appendix A.5. □

With the results from Corollary 5.1 the influence of the change of the switching thresholds ϵ_{ij} can be analyzed individually for each difference subsystem. In the next section, the results from Corollary 5.1 are used to analyze the controlled subsystems $\bar{\Sigma}_i$ defined in (5.18).

Example 5.1 (cont.) *Situation-dependent transmission of control signals*

Figure 5.8 shows the behavior of the symmetrically interconnected subsystems P defined in (3.32) with $\boldsymbol{x}_{01} = 1$ and $\boldsymbol{x}_{02} = 0.1$, controlled by the local control units C_i defined in (5.16) with the parameters listed in Table 5.1 (solid line) compared to the behavior of the same symmetrically interconnected subsystems controlled by the centralized controller C_{C} defined in (5.1) with $\boldsymbol{K}_{\mathrm{C}}$ from (5.9) (dashed line). The gray areas indicate the bounds $\bar{\boldsymbol{x}}_{\Delta i}$, ($\forall i \in \{1, 2\}$), on the subsystem states $\boldsymbol{x}_{\Delta i}(t)$ of the difference system Σ_Δ determined according to Theorem 5.1 with

$$\bar{\boldsymbol{x}}_\Delta = \begin{pmatrix} \bar{\boldsymbol{x}}_{\Delta 1} \\ \bar{\boldsymbol{x}}_{\Delta 2} \end{pmatrix} = \underbrace{\begin{pmatrix} 0.479 & 0.099 \\ 0.099 & 0.479 \end{pmatrix}}_{\boldsymbol{M}_{\mathrm{xu}}} \underbrace{\begin{pmatrix} 0.2 \\ 0.2 \end{pmatrix}}_{\boldsymbol{\epsilon}} = \begin{pmatrix} 0.116 \\ 0.116 \end{pmatrix}.$$

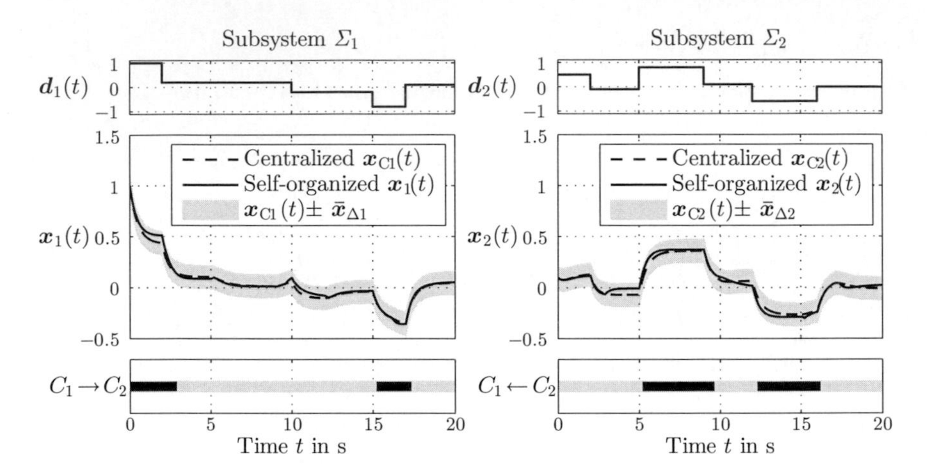

Figure 5.8: Analysis of the difference between the behavior of the symmetrically interconnected subsystems defined in (3.32) controlled by the self-organizing controller C defined in (5.17) and the behavior of the subsystems controlled by the corresponding centralized controller C_C defined in (5.1).

The bound $\bar{x}_\Delta$ on the difference behavior is valid since the states $x_i(t)$, $(\forall i \in \{1, 2\})$, never leave the gray areas, no matter what disturbances affect the subsystems. □

5.3.2 Behavior of a controlled subsystem

In this section the behavior of the controlled subsystems is analyzed. The investigations in the previous section show that the difference between the self-organizing control system Σ and the centralized control system Σ_C is bounded (cf. Theorem 5.1). Based on this boundedness and on the dynamics of Σ_C, the following two results are derived:

1. The states of the controlled subsystems are bounded (Lemma 5.2).
2. There exist conditions on the activation (Lemma 5.3) and the deactivation (Lemma 5.4) of the individual communication links depending on the disturbances and on the choice of the switching thresholds.

Behavior of the controlled subsystems. From $x_\Delta(t) = x(t) - x_C(t)$, $x_0 = x_{C0}$, (5.21) and (5.29) it follows that

$$x(t) = x_C(t) + x_\Delta(t)$$

$$
\begin{aligned}
&= \mathrm{e}^{\boldsymbol{A}_{\mathrm{C}}t}\boldsymbol{x}_{\mathrm{C}0} + \int_0^t \mathrm{e}^{\boldsymbol{A}_{\mathrm{C}}(t-\tau)}\boldsymbol{G}\boldsymbol{d}(\tau)\mathrm{d}\tau + \boldsymbol{x}_{\Delta}(t) \\
&= \mathrm{e}^{\boldsymbol{A}_{\mathrm{C}}t}\boldsymbol{x}_{0} + \int_0^t \mathrm{e}^{\boldsymbol{A}_{\mathrm{C}}(t-\tau)}\boldsymbol{G}\boldsymbol{d}(\tau)\mathrm{d}\tau + \int_0^t \mathrm{e}^{\boldsymbol{A}_{\mathrm{C}}(t-\tau)}\boldsymbol{B}\boldsymbol{u}_{\Delta}(\tau)\mathrm{d}\tau
\end{aligned}
\tag{5.35}
$$

holds for all $t \geq 0$. Since the disturbance $\boldsymbol{d}(t)$ is bounded according to Assumption 3.1 and the overall difference state $\boldsymbol{x}_{\Delta}(t)$ is bounded according to Theorem 5.1, obviously the overall state $\boldsymbol{x}(t)$ of the self-organizing control system Σ is also bounded. In the following the behavior of $\boldsymbol{x}(t)$ presented in (5.35) and the model of Σ defined in (5.19) is used to derive bounds on the signals of the controlled subsystems.

Lemma 5.2 (Boundedness of the local signals of the controlled subsystems) *The states $\boldsymbol{x}_i(t)$, $(i \in \mathcal{N})$, the performance outputs $\boldsymbol{v}_i(t)$ and the determined control signals $\hat{\boldsymbol{u}}_{ji}(t)$, $(\forall j \in \bar{\mathcal{F}}_{\mathrm{C}i})$ of the controlled subsystems within the self-organizing control system Σ defined in* (5.19) *are bounded by*

$$
|\boldsymbol{x}_i(t)| \leq \left|\check{\boldsymbol{C}}_{\mathrm{x}i}\mathrm{e}^{\boldsymbol{A}_{\mathrm{C}}t}\right| |\boldsymbol{x}_0| + \boldsymbol{M}_{\mathrm{xd}i}\bar{\boldsymbol{d}} + \boldsymbol{M}_{\mathrm{xu}i}\boldsymbol{\epsilon}, \qquad \forall t \geq 0, \quad \forall i \in \mathcal{N} \tag{5.36}
$$

$$
|\boldsymbol{v}_i(t)| \leq \left|\check{\boldsymbol{C}}_{\mathrm{v}i}\mathrm{e}^{\boldsymbol{A}_{\mathrm{C}}t}\right| |\boldsymbol{x}_0| + \boldsymbol{M}_{\mathrm{vd}i}\bar{\boldsymbol{d}} + \boldsymbol{M}_{\mathrm{vu}i}\boldsymbol{\epsilon}, \qquad \forall t \geq 0, \quad \forall i \in \mathcal{N} \tag{5.37}
$$

$$
|\hat{\boldsymbol{u}}_{ji}(t)| \leq \left|\check{\boldsymbol{C}}_{\mathrm{u}ji}\mathrm{e}^{\boldsymbol{A}_{\mathrm{C}}t}\right| |\boldsymbol{x}_0| + \boldsymbol{M}_{\mathrm{ud}ji}\bar{\boldsymbol{d}} + \boldsymbol{M}_{\mathrm{uu}ji}\boldsymbol{\epsilon}, \quad \forall t \geq 0, \quad \forall i \in \mathcal{N}, \quad \forall j \in \bar{\mathcal{F}}_{\mathrm{C}i} \tag{5.38}
$$

with $\boldsymbol{M}_{\mathrm{xu}i}$, $\boldsymbol{M}_{\mathrm{vu}i}$, $\boldsymbol{M}_{\mathrm{uu}ji}$ from (5.32)–(5.34) *and*

$$
\boldsymbol{M}_{\mathrm{xd}i} := \int_0^{\infty} \left|\check{\boldsymbol{C}}_{\mathrm{x}i}\mathrm{e}^{\boldsymbol{A}_{\mathrm{C}}\tau}\boldsymbol{G}\right| \mathrm{d}\tau, \quad \forall i \in \mathcal{N} \tag{5.39}
$$

$$
\boldsymbol{M}_{\mathrm{vd}i} := \int_0^{\infty} \left|\check{\boldsymbol{C}}_{\mathrm{v}i}\mathrm{e}^{\boldsymbol{A}_{\mathrm{C}}\tau}\boldsymbol{G}\right| \mathrm{d}\tau, \quad \forall i \in \mathcal{N} \tag{5.40}
$$

$$
\boldsymbol{M}_{\mathrm{ud}ji} := \int_0^{\infty} \left|\check{\boldsymbol{C}}_{\mathrm{u}ji}\mathrm{e}^{\boldsymbol{A}_{\mathrm{C}}\tau}\boldsymbol{G}\right| \mathrm{d}\tau, \quad \forall i \in \mathcal{N}, \quad \forall j \in \bar{\mathcal{F}}_{\mathrm{C}i}. \tag{5.41}
$$

Proof. The proof is given in Appendix A.6. □

According to Lemma 5.2, the states $\boldsymbol{x}_i(t)$, $(\forall i \in \mathcal{N})$, of Σ_i are bounded for a bounded initial state. This property is used to derive the practical stability of Σ in Section 5.3.3. Furthermore, in the following the bounds on $\boldsymbol{v}_i(t)$, $(\forall i \in \mathcal{N})$, are used to characterize the performance of the self-organizing control system Σ and the bounds on the determined control signal $\hat{\boldsymbol{u}}_{ji}(t)$, $(\forall j \in \bar{\mathcal{F}}_{\mathrm{C}i})$, are used to find conditions for the activation and the deactivation of the communication links within the communication graph $\mathcal{G}_{\mathrm{C}}(t)$.

Activition and deactivation of the communication links. To investigate the communication effort among the local control units three questions concerning the transmission of the determined control signal $\hat{\boldsymbol{u}}_{ji}(t)$ from the local control unit C_i to the local control unit C_j are considered:

1. Is there any communication from C_i to C_j?
2. If there is communication from C_i to C_j, will the communication be deactivated eventually?
3. At which times will the communication be activated and deactivated, respectively?

Recall that communication from C_i to C_j implies that there exists a communication link $(i \to j)$ within the set $\mathcal{E}_{\mathrm{C}}(t)$ of communicational edges, i.e., $(i \to j) \in \mathcal{E}_{\mathrm{C}}(t)$, which defines the time-varying part the communication graph $\mathcal{G}_{\mathrm{C}}(t) = (\mathcal{V}_{\mathrm{C}}, \mathcal{E}_{\mathrm{C}}(t))$.

The left part of Fig. 5.9 shows the situation in which C_i sends information to C_j, where in the right part no communication from C_i to C_j is present by subdividing the determined control signal and the switching thresholds as follows:

$$\hat{\boldsymbol{u}}_{ji}(t) = \begin{pmatrix} \hat{u}_{ji1}(t) \\ \vdots \\ \hat{u}_{jim_j}(t) \end{pmatrix}, \quad \boldsymbol{\epsilon}_{ji}(t) = \begin{pmatrix} \epsilon_{ji1}(t) \\ \vdots \\ \epsilon_{jim_j}(t) \end{pmatrix}, \quad i \in \mathcal{N}, \quad j \in \bar{\mathcal{F}}_{\mathrm{C}i}.$$

According to the definition of the decision unit D_i in (5.15), communication from C_i to C_j is activated at a time $t_{\mathrm{A}ji}$ if the relation

$$(\exists t_{\mathrm{A}ji} \geq 0) \quad t_{\mathrm{A}ji} = \min_{t \geq 0} t \ : \ |\hat{\boldsymbol{u}}_{ji}(t)| \overset{\exists}{\geq} \boldsymbol{\epsilon}_{ji}, \quad j \in \bar{\mathcal{F}}_{\mathrm{C}i}, \quad i \in \mathcal{N} \tag{5.42}$$

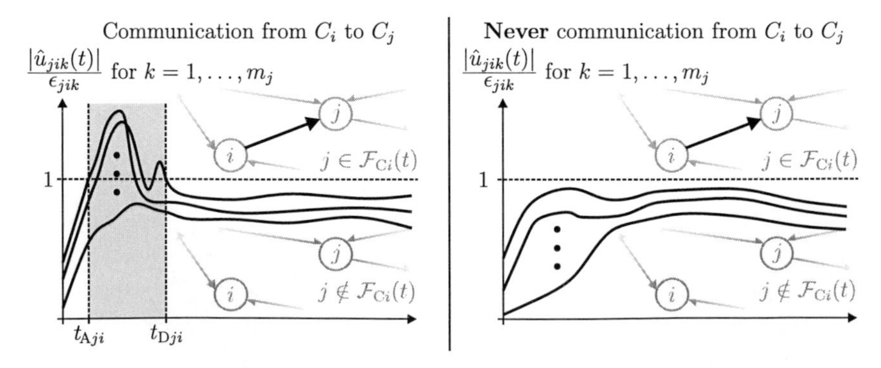

Figure 5.9: Activation and deactivation of the communication within Σ defined in (5.19).

holds true. The communication from C_i to C_j stays deactivated after a time $t_{\mathrm{D}ji}$ if the following relation holds true

$$(\exists t_{\mathrm{D}ji} \geq 0) \quad t_{\mathrm{D}ji} = \max_{t \geq 0} t : |\hat{\boldsymbol{u}}_{ji}(t)| \overset{\exists}{\geq} \boldsymbol{\epsilon}_{ji}, \quad j \in \bar{\mathcal{F}}_{\mathrm{C}i}, \quad i \in \mathcal{N}. \tag{5.43}$$

Therefore, the following lemma concerns the existence of the *activation time* $t_{\mathrm{A}ji}$ for the communication link $(i \to j) \in \bar{\mathcal{E}}_{\mathrm{C}}$.

Lemma 5.3 (Activation of the communication from C_i to C_j for Σ defined in (5.19)) *Consider the self-organizing control system Σ defined in* (5.19). *The communication from the local control unit C_i, $(i \in \mathcal{N})$, to the local control unit C_j, $(j \in \bar{\mathcal{F}}_{\mathrm{C}i})$ is activated only if there exists a time*

$$\bar{t}_{\mathrm{A}ji} = \min_{t \geq 0} t : \; \bar{\boldsymbol{u}}_{\mathrm{T}ji}(t) \overset{\exists}{\geq} \boldsymbol{\epsilon}_{ji}, \quad j \in \bar{\mathcal{F}}_{\mathrm{C}i}, \quad i \in \mathcal{N} \tag{5.44}$$

which is a lower bound on time activation time $t_{\mathrm{A}ji}$, i.e., $t_{\mathrm{A}ji} \geq \bar{t}_{\mathrm{A}ji}$, with

$$\bar{\boldsymbol{u}}_{\mathrm{T}ji}(t) := \left|\check{\boldsymbol{C}}_{\mathrm{u}ji}\mathrm{e}^{\boldsymbol{A}_{\mathrm{C}}t}\right| |\boldsymbol{x}_0| + \int_0^t \left|\check{\boldsymbol{C}}_{\mathrm{u}ji}\mathrm{e}^{\boldsymbol{A}_{\mathrm{C}}\tau}\boldsymbol{G}\right| \mathrm{d}\tau \cdot \bar{\boldsymbol{d}} + \int_0^t \left|\check{\boldsymbol{C}}_{\mathrm{u}ji}\mathrm{e}^{\boldsymbol{A}_{\mathrm{C}}\tau}\boldsymbol{B}\right| \mathrm{d}\tau \cdot \boldsymbol{\epsilon} \tag{5.45}$$

for all $i \in \mathcal{N}$, all $j \in \bar{\mathcal{F}}_{\mathrm{C}i}$ and $t \geq 0$.

Proof. The proof is given in Appendix A.7. □

Lemma 5.3 presents a necessary condition for the activation of the communication from C_i to C_j, where the lower bound $\bar{t}_{\mathrm{A}ji}$ on the activation time $t_{\mathrm{A}ji}$ depends on the initial state $\boldsymbol{x}_0$, the bound $\bar{\boldsymbol{d}}$ on the overall disturbance $\boldsymbol{d}(t)$ and on the choice of overall vector $\boldsymbol{\epsilon}$ of switching thresholds $\boldsymbol{\epsilon}_{ji}$. The condition is not sufficient, because the signal $\bar{\boldsymbol{u}}_{\mathrm{T}ji}(t)$ is an upper bound on the absolute value $|\hat{\boldsymbol{u}}_{ji}(t)|$ of the determined control signal. Hence, it is not guaranteed that $|\hat{\boldsymbol{u}}_{ji}(t)|$ exceeds $\boldsymbol{\epsilon}_{ji}$ if $\bar{\boldsymbol{u}}_{\mathrm{T}ji}(t)$ exceeds $\boldsymbol{\epsilon}_{ji}$. An interesting aspect of Lemma 5.3 is that the activation of each communication link can be checked separately.

Furthermore, Lemma 5.3 and (5.38) derive that for $\boldsymbol{x}_0 = \boldsymbol{0}$ the communication from C_i to C_j is activated only if the following relation is fulfilled

$$\boldsymbol{M}_{\mathrm{ud}ji}\bar{\boldsymbol{d}} + \boldsymbol{M}_{\mathrm{uu}ji}\boldsymbol{\epsilon} \overset{\exists}{\geq} \boldsymbol{\epsilon}_{ji}, \quad j \in \bar{\mathcal{F}}_{\mathrm{C}i}, \quad i \in \mathcal{N}. \tag{5.46}$$

The relation

$$\bar{\boldsymbol{d}} \overset{\exists}{\geq} \boldsymbol{M}_{\mathrm{ud}ji}^{+}\left(\boldsymbol{\epsilon}_{ji} - \boldsymbol{M}_{\mathrm{uu}ji}\boldsymbol{\epsilon}\right), \quad j \in \bar{\mathcal{F}}_{\mathrm{C}i}, \quad i \in \mathcal{N}$$

for $\boldsymbol{M}_{\mathrm{ud}ji}^{+}\boldsymbol{M}_{\mathrm{ud}ji}\,(\boldsymbol{\epsilon}_{ji} - \boldsymbol{M}_{\mathrm{uu}ji}\boldsymbol{\epsilon}) = \boldsymbol{\epsilon}_{ji} - \boldsymbol{M}_{\mathrm{uu}ji}\boldsymbol{\epsilon}$ (cf. (2.1)) gives a necessary condition on the bound $\bar{\boldsymbol{d}}$ of the overall disturbance $\boldsymbol{d}(t)$ for which the communication from C_i to C_j might be activated.

With this consideration, the following lemma concerns the existence of a *deactivation time* $t_{\mathrm{D}ji}$ according to (5.43).

Lemma 5.4 (Deactivation of the communication from C_i to C_j for Σ defined in (5.19)**)** *Consider the self-organizing control system Σ defined in* (5.19)*. The communication from the local control unit C_i, $(i \in \mathcal{N})$ to the local control unit C_j, $(j \in \bar{\mathcal{F}}_{\mathrm{C}i})$ stays deactivated after some deactivation time $t_{\mathrm{D}ji}$ if the relations*

$$\boldsymbol{M}_{\mathrm{ud}ji}\bar{\boldsymbol{d}} + \boldsymbol{M}_{\mathrm{uu}ji}\boldsymbol{\epsilon} < \boldsymbol{\epsilon}_{ji}, \quad j \in \bar{\mathcal{F}}_{\mathrm{C}i}, \quad i \in \mathcal{N} \tag{5.47}$$

holds true, i.e., $j \notin \mathcal{F}_{\mathrm{C}i}(t)$ for all $t > t_{\mathrm{D}ji} \geq 0$. The deactivation time $t_{\mathrm{D}ji}$ is upper bounded by the time $\bar{t}_{\mathrm{D}ji}$ which is defined by

$$t_{\mathrm{D}ji} \leq \bar{t}_{\mathrm{D}ji} := \max_{t \geq 0} t : \ \bar{\boldsymbol{u}}_{\mathrm{T}ji}(t) \overset{\exists}{\geq} \boldsymbol{\epsilon}_{ji}, \quad j \in \bar{\mathcal{F}}_{\mathrm{C}i}, \quad i \in \mathcal{N}. \tag{5.48}$$

Proof. The proof is given in Appendix A.8. □

Condition (5.47) in Lemma 5.4 is sufficient but not necessary for the deactivation of the communication from C_i to C_j since the left part of the inequality is an upper bound on the absolute value $\lim_{t\to\infty} |\hat{\boldsymbol{u}}_{ji}(t)|$ of the determined control signal. Hence, the communication from C_i to C_j might also be deactivated if the relation in (5.47) is not fulfilled.

The deactivation of the communication depends on the bound $\bar{\boldsymbol{d}}$ on the disturbance $\boldsymbol{d}(t)$ and the switching thresholds $\boldsymbol{\epsilon}_{ji}$, but not on the initial state $\boldsymbol{x}_0$ since the natural response of Σ vanishes for $t \to \infty$.

If there is no disturbance, i.e., $\boldsymbol{d}(t) = \boldsymbol{0}$ for all $t \geq 0$, the deactivation only depends on the switching thresholds. With this, the condition

$$\boldsymbol{M}_{\mathrm{uu}ji}\boldsymbol{\epsilon} < \boldsymbol{\epsilon}_{ji}, \quad j \in \bar{\mathcal{F}}_{\mathrm{C}i}, \quad i \in \mathcal{N}$$

has to be fulfilled to guarantee the deactivation of the communication from C_i to C_j.

In summary, for $\boldsymbol{x}_0 = \boldsymbol{0}$, the three initial questions concerning the communication from C_i to C_j can be answered as follows:

1. The communication is activated only if the relation in (5.46) holds true.

2. The communication will be deactivated forever if the relation in (5.47) holds true.

3. Lemma 5.3 and Lemma 5.4 derive a lower bound for the activation time $t_{\mathrm{A}ji}$ and an upper bound for the deactivation time $t_{\mathrm{D}ji}$. These bounds can be used to determine a communication interval for the link $(i \to j) \in \mathcal{E}_{\mathrm{C}}$ in the case a specific disturbance $\boldsymbol{d}(t)$ affecting the plant P which is illustrated in more detail in the following example.

The conditions for the activation and the deactivation of communication links in Lemma 5.3 and Lemma 5.4 are used in the following sections to derive a condition for the asymptotic stability of the self-organizing control system Σ (Section 5.3.3), to give a condition for activating the communication w.r.t. certain disturbance situations (Section 5.3.4) and to reduce links within the maximal communication graph $\bar{\mathcal{G}}_{\mathrm{C}}$ for guaranteeing a desired performance of Σ (Section 5.4.2).

Summary. The following result of this section is highlighted:

> The behavior of the controlled subsystems and the change of communication links can be analyzed individually.

In the following the analysis methods presented in this section are used to investigate the behavior of the self-organizing controlled symmetrically interconnected subsystems.

Example 5.1 (cont.) *Situation-dependent transmission of control signals*

Figure 5.10 depicts the behavior of the symmetrically interconnected subsystems P defined in (3.32) controlled by the local control units C_i defined in (5.16) with the parameters listed in Table 5.1 for the initial states $\boldsymbol{x}_{01} = 0$ and $\boldsymbol{x}_{02} = -1$. The gray area in the second and third row of plots shows the bounds on the state $\boldsymbol{x}_i(t)$ according to (5.36) in Lemma 5.2 and the signal $\pm\bar{\boldsymbol{u}}_{\mathrm{T}ji}(t)$ according to (5.45) in Lemma 5.3 which is an upper bound on the calculated control signal $\hat{\boldsymbol{u}}_{ji}(t)$. Due to the change of the disturbance $\boldsymbol{d}_1(t)$ at $t = t^* = 5\,\mathrm{s}$, the analysis of the behavior of Σ_1 is divided into two time intervals. In the first time interval $T_1 = [0\,\mathrm{s}, 5\,\mathrm{s}]$ the initial states are $\boldsymbol{x}_{01} = 0$ and $\boldsymbol{x}_{02} = -1$ and the maximal disturbance is $\bar{d}_1 = 2$. In the second time interval $T_2 = [5\,\mathrm{s}, 20\,\mathrm{s}]$ the initial states are the bounds $|\boldsymbol{x}_i(t)|$ at $t = t^* = 5\,\mathrm{s}$ from the first time interval, i.e., $\boldsymbol{x}_{01}^* = |\boldsymbol{x}_1(t = 5\,\mathrm{s})| = 1.072$ and $\boldsymbol{x}_{02}^* = |\boldsymbol{x}_2(t = 5\,\mathrm{s})| = 0.315$. Within T_2 the maximal disturbance is $\bar{d}_1^* = 0$.

Figure 5.10 shows that the states $\boldsymbol{x}_i(t)$ and the determined control signals $\hat{\boldsymbol{u}}_{ji}(t)$ never leave the corresponding gray areas. Hence, the upper bounds from Lemma 5.2 and Lemma 5.3 hold true. The bounds on the activation times and deactivation times of the communication resulting from Lemma 5.3 and Lemma 5.4 are

$$\bar{t}_{\mathrm{A}21} = 0.10\,\mathrm{s}, \quad \bar{t}_{\mathrm{A}12} = 0\,\mathrm{s}, \quad \bar{t}_{\mathrm{A}21}^* = 5\,\mathrm{s}, \quad \bar{t}_{\mathrm{A}12}^* = 5\,\mathrm{s},$$
$$\bar{t}_{\mathrm{D}21} = \infty, \quad \bar{t}_{\mathrm{D}12} = \infty, \quad \bar{t}_{\mathrm{D}21}^* = 6.35\,\mathrm{s}, \quad \bar{t}_{\mathrm{D}12}^* = 6.80\,\mathrm{s}.$$

Hence, $\bar{t}_{\mathrm{A}21}^*$ is the lower bound on the time at which the communication would be activated for the analysis of the time interval T_2, whereas $\bar{t}_{\mathrm{D}21}^*$ is the corresponding upper bound on the deactivation time $t_{\mathrm{D}21}^*$ in T_2. The result $\bar{t}_{\mathrm{D}21} = \infty$ occurs due to the fact that $\bar{\boldsymbol{u}}_{\mathrm{T}21}(t)$ always is larger than ϵ_{21} for $t \geq t_{\mathrm{A}21}$ (cf. Lemma 5.4). The determination of the bounds on the activation time and the deactivation time can be used to calculate the time interval in which the communication is active while using information about

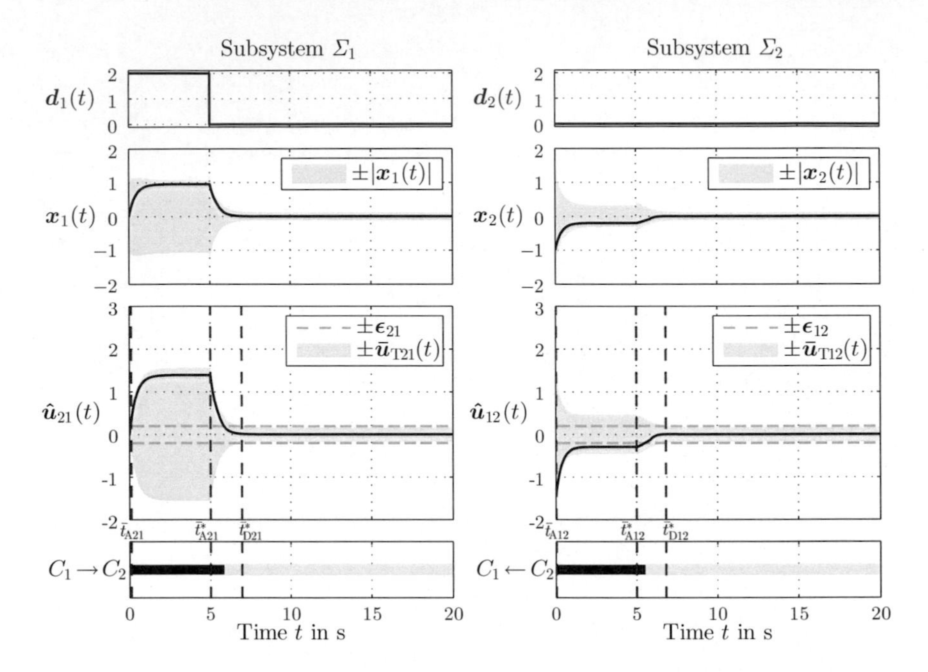

Figure 5.10: Analysis of the behavior of the interconnected subsystems defined in (3.32) controlled by the local control units C_i defined in (5.16) with the parameters listed in Table 5.1.

the course of the maximum values of the disturbances. For the considered course of the disturbances the communication from C_1 to C_2 is at most active for $\bar{t}^*_{D21} - \bar{t}_{A21} = 6.25\,\mathrm{s}$ and from C_2 to C_1 for $\bar{t}^*_{D12} - \bar{t}_{A12} = 6.80\,\mathrm{s}$, where the actual communication intervals have a length of $5.85\,\mathrm{s}$ and $5.52\,\mathrm{s}$, respectively.

In summary, the example shows that the results from Lemma 5.2, Lemma 5.3 and Lemma 5.4 provide a good approximation of the actual behavior of the interconnected subsystems. Furthermore, the analysis of the activation and the deactivation of the communication can be used to specify the communication effort for certain disturbance situations. □

5.3.3 Practical stability and asymptotic stability

This section investigates the first two control aims stated in Section 3.4. Therefore, the section is divided into two parts. The first part concerns the practical stability of the self-organizing control system Σ (Control aim 3.2). A condition on the properties of the plant P for practical stabilization using the proposed self-organizing controller C is given. Furthermore, it is shown that under the stated assumptions Σ always is practically stable. In the second part two sufficient conditions on the asymptotic stability of the undisturbed self-organizing control system

Σ are derived (Control aim 3.1). The first condition uses a distributed analysis and the second condition uses a central analysis, where both conditions use global model information.

Practical stability. In the following a condition on the properties of the plant P is derived, which guarantees that the self-organizing controller C with a situation-dependent transmission of control signals can stabilize the physically coupled subsystems. For this purpose, the previous assumptions are neglected.

Theorem 5.2 (Practical stabilizability through transmission of control signals) *The plant P defined in (3.10) can be practically stabilized w.r.t. to a compact set $\mathcal{A}$ by the self-organizing controller C defined in (5.17) if all the following three conditions hold true:*

1. *The relation $\boldsymbol{C} = \boldsymbol{I}_r$ is fulfilled.*
2. *The pair $(\boldsymbol{A}_{\mathrm{P}}, \boldsymbol{B})$ is controllable.*
3. *There exists an upper bound on $\boldsymbol{d}(t)$, i.e., $\|\boldsymbol{d}(t)\| < \infty$ for all $t \geq 0$.*

Proof. From (3.12), (5.28) and (5.35) it follows that the overall state $\boldsymbol{x}(t)$ is bounded according to

$$\|\boldsymbol{x}(t)\| \leq \left\|\mathrm{e}^{\boldsymbol{A}_{\mathrm{C}}t}\right\| \|\boldsymbol{x}_0\| + \int_0^\infty \left\|\mathrm{e}^{\boldsymbol{A}_{\mathrm{C}}\tau}\boldsymbol{G}\right\| \mathrm{d}\tau \max_{t\geq 0} \|\boldsymbol{d}(t)\|_\infty + \int_0^\infty \left\|\mathrm{e}^{\boldsymbol{A}_{\mathrm{C}}\tau}\boldsymbol{B}\right\| \mathrm{d}\tau \max_{t\geq 0} \|\boldsymbol{u}_\Delta(t)\|_\infty \tag{5.49}$$

for all $t \geq 0$. To fulfill the first requirement in Definition 2.3, the elements $\int_0^\infty \left\|\mathrm{e}^{\boldsymbol{A}_{\mathrm{C}}\tau}\boldsymbol{G}\right\| \mathrm{d}\tau$, $\left\|\mathrm{e}^{\boldsymbol{A}_{\mathrm{C}}t}\right\|$, $\max_{t\geq 0} \|\boldsymbol{d}(t)\|_\infty$, $\int_0^\infty \left\|\mathrm{e}^{\boldsymbol{A}_{\mathrm{C}}\tau}\boldsymbol{B}\right\|$ and $\max_{t\geq 0} \|\boldsymbol{u}_\Delta(t)\|_\infty$ have to be upper bounded which is investigated as follows:

- $\max_{t\geq 0} \|\boldsymbol{u}_\Delta(t)\|_\infty$ is upper bounded according to Lemma 5.1 and (5.28).
- $\left\|\mathrm{e}^{\boldsymbol{A}_{\mathrm{C}}t}\right\|$, $\int_0^\infty \left\|\mathrm{e}^{\boldsymbol{A}_{\mathrm{C}}\tau}\boldsymbol{G}\right\| \mathrm{d}\tau$ and $\int_0^\infty \left\|\mathrm{e}^{\boldsymbol{A}_{\mathrm{C}}\tau}\boldsymbol{B}\right\|$ are upper bounded if $\boldsymbol{A}_{\mathrm{C}}$ is a Hurwitz matrix. According to Theorem 2.2, the feedback gain matrix $\boldsymbol{K}_{\mathrm{C}}$ can be chosen to make $\boldsymbol{A}_{\mathrm{C}}$ a Hurwitz matrix if the pair $(\boldsymbol{A}_{\mathrm{P}}, \boldsymbol{B})$ is controllable and the full overall state $\boldsymbol{x}(t)$ can be measured which is claimed by the first and second condition, respectively.
- The third condition guarantees that $\max_{t\geq 0} \|\boldsymbol{d}(t)\|_\infty$ is upper bounded.

To fulfill the second requirement in Definition 2.3, $\lim_{t\to\infty} \left\|\mathrm{e}^{\boldsymbol{A}_{\mathrm{C}}t}\right\| = 0$ has to be satisfied. Therefore, $\boldsymbol{A}_{\mathrm{C}}$ has to be a Hurwitz matrix which is again possible if the first condition and the second condition are guaranteed which completes the proof. □

Theorem 5.2 states three conditions on the dynamics of the plant P defined in (3.10) such that P can be practically stabilized by the proposed self-organizing controller C. The first and the second condition are necessary but not sufficient since some part of the undisturbed plant which could not be influenced by the controller might be stable in advance. The third condition is necessary but not sufficient since the effect of the disturbance $\boldsymbol{d}(t)$ might be compensated within Σ even if $\boldsymbol{d}(t)$ is not upper bounded. In summary, Theorem 5.2 shows that the proposed self-organizing controller C can practically stabilize the undisturbed plant P if a centralized controller C defined in (5.1) can asymptotically stabilize the undisturbed plant P (cf. Theorem 2.1 and Theorem 2.2).

Remark 5.1 *The second condition in Theorem 5.2 can be relaxed by considering that the plant P is decomposed into a part with unstable eigenvalues and a part with stable eigenvalues. It is well known, that only the part with the unstable eigenvalues has to be controllable to guarantee that $\boldsymbol{A}_{\mathrm{C}} = \bar{\boldsymbol{A}}_{\mathrm{P}} - \boldsymbol{B}\boldsymbol{K}_{\mathrm{C}}$ can be made a Hurwitz matrix (cf. [20, 76, 94]) and, therefore, also the plant P controlled by the proposed self-organizing controller C can be practically stabilized.*

The following theorem shows that with the previous assumptions the self-organizing control system Σ is always practically stable and, therefore, always fulfills Control aim 3.2.

Theorem 5.3 (Practical stability of the self-organizing control system Σ in (5.19)) *The self-organizing control system Σ defined in (5.19) is practically stable with respect to the compact set*

$$\mathcal{A} = \{\boldsymbol{x} \in \mathbb{R}^n \mid ||\boldsymbol{x}|| \leq \bar{x} := m_{\mathrm{xd}} d_{\max} + m_{\mathrm{xu}} \epsilon_{\max}\}, \tag{5.50}$$

with

$$m_{\mathrm{xd}} := \int_0^\infty \left\| \mathrm{e}^{\boldsymbol{A}_{\mathrm{C}}\tau} \boldsymbol{G} \right\| \mathrm{d}\tau$$

$$m_{\mathrm{xu}} := \int_0^\infty \left\| \mathrm{e}^{\boldsymbol{A}_{\mathrm{C}}\tau} \boldsymbol{B} \right\| \mathrm{d}\tau.$$

Proof. Consider equation (5.49). The first requirement in Definition 2.3 is fulfilled since the elements $\left\|\mathrm{e}^{\boldsymbol{A}_{\mathrm{C}}t}\right\|$, $\int_0^\infty \left\|\mathrm{e}^{\boldsymbol{A}_{\mathrm{C}}\tau}\boldsymbol{G}\right\| \mathrm{d}\tau$, $\max_{t\geq 0} \|\boldsymbol{d}(t)\|_\infty$, $\int_0^\infty \left\|\mathrm{e}^{\boldsymbol{A}_{\mathrm{C}}\tau}\boldsymbol{B}\right\|$ and $\max_{t\geq 0} \|\boldsymbol{u}_\Delta(t)\|_\infty$ are upper bounded. $\left\|\mathrm{e}^{\boldsymbol{A}_{\mathrm{C}}t}\right\|$, $m_{\mathrm{xd}} = \int_0^\infty \left\|\mathrm{e}^{\boldsymbol{A}_{\mathrm{C}}\tau}\boldsymbol{G}\right\| \mathrm{d}\tau$ and $m_{\mathrm{xu}} = \int_0^\infty \left\|\mathrm{e}^{\boldsymbol{A}_{\mathrm{C}}\tau}\boldsymbol{B}\right\|$ are upper bounded since $\boldsymbol{A}_{\mathrm{C}}$ is a Hurwitz matrix according to Assumption 5.1. $\max_{t\geq 0} \|\boldsymbol{u}_\Delta(t)\|_\infty$ is upper bounded by $\epsilon_{\max}$ according to Lemma 5.1 and (5.28). $\max_{t\geq 0} \|\boldsymbol{d}(t)\|_\infty$ is upper bounded by $d_{\max}$ according to Assumption 3.1. The second requirement in Definition 2.3 is fulfilled since $\lim_{t\to\infty} \left\|\mathrm{e}^{\boldsymbol{A}_{\mathrm{C}}t}\right\| = 0$ holds true according to Assumption 5.1 which completes the proof. □

Theorem 5.3 shows that for a given centralized controller C_{C} which asymptotically stabilizes the undisturbed plant P (Assumption 5.1) the resulting self-organizing control system Σ is always practically stable. An interesting aspect is that the corresponding decentralized control system Σ_{D} defined in (3.24) does not have to be asymptotically stable since the situation-dependent communication of the control signals always keeps the state of Σ close to the equilibrium point $\boldsymbol{x} = \boldsymbol{0}$.

Asymptotic stability. In the following, two different conditions for testing the fulfillment of Control aim 3.1 by the self-organizing control system Σ defined in (5.19) are given:

1. The first condition (Theorem 5.4) checks whether the communication among the local control units stays deactivated after a certain time t_{D}. If there is no communication among the local control units, the self-organizing control system Σ behaves like the decentralized control system Σ_{D} defined in (3.24). Hence, if there exists such a deactivation time t_{D} and Σ_{D} is asymptotically stable, then Σ is also asymptotically stable.

2. For the second condition (Proposition 5.1) the model of the self-organizing control system Σ is rewritten as an uncertain system. The results on the asymptotic stability of such uncertain systems in Theorem 2.5 are used to derive a condition on the asymptotic stability of self-organizing control system Σ.

For both conditions it is assumed that the overall disturbance is zero, i.e., $\boldsymbol{d}(t) = \boldsymbol{0}$ for all $t \geq 0$.

Deactivation of all communication links (first condition). For the first condition consider the undisturbed decentralized control system

$$\Sigma_{\mathrm{D}}: \ \dot{\boldsymbol{x}}_{\mathrm{D}}(t) = \bar{\boldsymbol{A}}_{\mathrm{P}} \boldsymbol{x}_{\mathrm{D}}(t), \quad \boldsymbol{x}_{\mathrm{D}}(0) = \boldsymbol{x}_{\mathrm{D0}} \tag{5.51}$$

resulting from Σ defined in (5.19) with $\mathcal{E}_{\mathrm{C}}(t) = \emptyset$ for all $t \geq 0$ according to the definition of Σ_{D} in (3.24) with $\bar{\boldsymbol{A}}_{\mathrm{P}} = \bar{\boldsymbol{A}} + \boldsymbol{E}\boldsymbol{L}\boldsymbol{C}_{\mathrm{z}}$. Hence, $\tilde{\boldsymbol{u}}(t) = \boldsymbol{0}$ holds true for all $t \geq 0$. The signal $\boldsymbol{x}_{\mathrm{D}}(t) \in \mathbb{R}^n$ is the overall state of Σ_{D} and the vector $\boldsymbol{x}_{\mathrm{D0}} \in \mathbb{R}^n$ is the overall initial state of Σ_{D}. For simplicity reasons the performance output $\boldsymbol{v}_{\mathrm{D}}(t) \in \mathbb{R}^{r_{\mathrm{v}}}$ is not modeled within Σ_{D} defined in (5.51).

If the matrix $\bar{\boldsymbol{A}}_{\mathrm{P}} = \bar{\boldsymbol{A}} + \boldsymbol{E}\boldsymbol{L}\boldsymbol{C}_{\mathrm{z}}$ is a Hurwitz matrix, then the decentralized controlled system Σ_{D} defined in (5.51) is asymptotically stable (cf. Theorem 2.1).

The idea for a condition for the asymptotic stability of Σ is illustrated in Fig. 5.11. A fact resulting from the definition of Σ_{D} in (5.51) is: If all communication among the local control

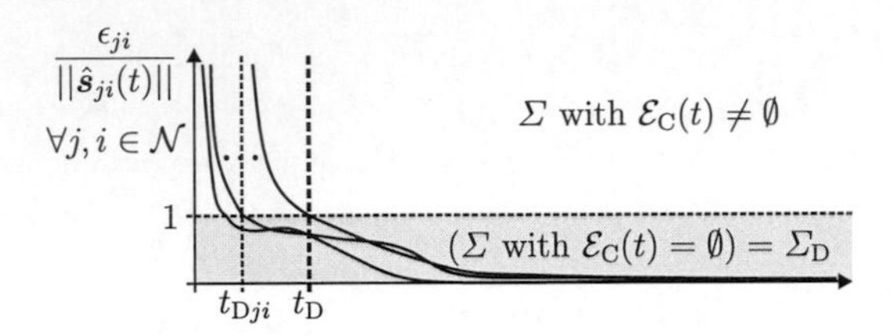

Figure 5.11: Idea for finding a condition to test the asymptotic stability of Σ defined in (5.19).

units C_i, ($\forall i \in \mathcal{N}$), stays deactivated after some time t_D, i.e., $\mathcal{E}_C(t) = \emptyset$ for all $t \geq t_D$, then Σ behaves like the decentralized controlled system Σ_D after the time t_D

$$(\Sigma \text{ with } \mathcal{E}_C(t) = \emptyset, \forall t \geq t_D) \Rightarrow (\Sigma = \Sigma_D, \forall t \geq t_D) \tag{5.52}$$

for $\boldsymbol{x}_D(t_D) = \boldsymbol{x}(t_D)$. Hence, if there exists a time $t_D \geq 0$ after which all communication among the local control units C_i stays deactivated, i.e., $\mathcal{E}_C(t) = \emptyset$ for all $t \geq t_D$, and Σ_D defined in (5.51) is asymptotically stable, the self-organizing control system Σ defined in (5.19) is asymptotically stable

$$(\Sigma_D \text{ asympt. stable}) \wedge (\Sigma \text{ with } \mathcal{E}_C(t) = \emptyset, \forall t \geq t_D) \Rightarrow (\Sigma \text{ asympt. stable}). \tag{5.53}$$

According to (5.43), the overall communication among the local control units stays deactivated, i.e. $\mathcal{E}_C(t) = \emptyset$, after the time t_D if the relation

$$(\exists t_D \geq 0) \quad |\hat{\boldsymbol{u}}_{ji}(t)| < \boldsymbol{\epsilon}_{ji}, \quad \forall t > t_D, \quad \forall j \in \bar{\mathcal{F}}_{Ci}, \quad \forall i \in \mathcal{N} \tag{5.54}$$

holds true, where t_D is the maximum of all deactivation times t_{Dji}, ($\forall j \in \bar{\mathcal{F}}_{Ci}, \forall i \in \mathcal{N}$) in (5.43).

In the following theorem these relations are used to derive a condition for asymptotic stability of the self-organizing control system Σ while using the results from Lemma 5.4 which investigates the deactivation of the communication links.

Theorem 5.4 (Asymptotic stability of Σ defined in (5.19)) *The self-organizing control system Σ defined in (5.19) with $\boldsymbol{d}(t) = \mathbf{0}$, $(\forall t \geq 0)$, is asymptotically stable if the following conditions are fulfilled*

1. *The system matrix $\bar{\boldsymbol{A}}_\mathrm{P} = \bar{\boldsymbol{A}} + \boldsymbol{ELC}_\mathrm{z}$ of Σ_D defined in (5.51) is a Hurwitz matrix.*
2. *The following relation holds true*

$$c_{ji} > \boldsymbol{M}_{\mathrm{uu}ji}c, \quad \forall j \in \bar{\mathcal{F}}_{\mathrm{C}i}, \quad \forall i \in \mathcal{N}. \tag{5.55}$$

Proof. According to (5.53), two requirements have to be fulfilled to guarantee asymptotic stability of Σ:

1. The decentralized control system Σ_D defined in (5.51) has to be asymptotically stable. According to Theorem 2.1, this requirement is covered by the first condition in the theorem.
2. The overall communication among the local control units has to stay deactivated after some time $t_\mathrm{D} \geq 0$. According to Lemma 5.4, this requirement is covered by the second condition in the theorem.

With this, the proof is completed. □

Theorem 5.4 gives a sufficient condition for the asymptotic stability of the self-organizing control system Σ defined in (5.19), where the deactivation of each communication link $(i \to j)$ is checked individually by the conditions in (5.55). The second part of the condition in Theorem 5.4 is a kind of small gain condition by checking the gains of each communication link individually.

Reformulation as an uncertain system (second condition). For the second condition the self-organizing control system Σ defined in (5.19) is rewritten as an uncertain system $\bar{\Sigma}_\mathrm{u}$ according to (2.15)

$$\Sigma: \ \dot{\boldsymbol{x}}(t) = \left(\bar{\boldsymbol{A}}_\mathrm{P} + \boldsymbol{B}^* \boldsymbol{K}_\mathrm{CU}(t)\right) \boldsymbol{x}(t) + \boldsymbol{Gd}(t), \quad \boldsymbol{x}(0) = \boldsymbol{x}_0 \tag{5.56}$$

with

$$\boldsymbol{K}_\mathrm{CU}(t) := -\frac{1}{\sqrt{n}\,\|\boldsymbol{K}^*\|} \boldsymbol{K}_\mathrm{COM}(t) \in \mathbb{R}^{m \times n}$$
$$\boldsymbol{K}_\mathrm{COM}(t) := (\boldsymbol{K}_{\mathrm{COM}ij}(t)) \in \mathbb{R}^{m \times n}$$

$$\begin{aligned}
\boldsymbol{K}_{\mathrm{COM}ij}(t) &:= \begin{cases} \boldsymbol{O}_{m_i \times n_j} & \text{if } |\hat{\boldsymbol{u}}_{ij}(t)| \overset{\exists}{\geq} \boldsymbol{\epsilon}_{ij} \\ \boldsymbol{K}^*_{ij} & \text{else} \end{cases}, \quad \forall i,j \in \mathcal{N} \\
\boldsymbol{K}^* &:= \left(\boldsymbol{K}^*_{ij}\right) \in \mathrm{I\!R}^{m \times n} \\
\boldsymbol{K}^*_{ij} &:= \begin{cases} \boldsymbol{O}_{m_i \times n_j} & \text{if } i = j \\ \boldsymbol{K}_{\mathrm{C}ij} & \text{else} \end{cases}, \quad \forall i,j \in \mathcal{N} \\
\boldsymbol{B}^* &:= \sqrt{n} \left\| \boldsymbol{K}^* \right\| \boldsymbol{B},
\end{aligned}$$

where the matrices $\boldsymbol{K}_{\mathrm{C}ij} \in \mathrm{I\!R}^{m_i \times n_j}$ build the feedback gain matrix $\boldsymbol{K}_{\mathrm{C}} \in \mathrm{I\!R}^{m \times n}$ according to (5.1). The matrix $\boldsymbol{K}_{\mathrm{COM}}(t)$ represents the current feedback gain matrix of the overall closed-loop system Σ defined in (5.19) generated by transmitting the control signals among the local control units C_i, $(\forall i \in \mathcal{N})$, but without the local feedbacks from the feedback units F_i. The definition of the local control units C_i in (5.16) shows that whenever $|\hat{\boldsymbol{u}}_{ij}(t)| \overset{\exists}{\geq} \boldsymbol{\epsilon}_{ij}$ holds true, the determined control signal $\hat{\boldsymbol{u}}_{ji}(t)$ is sent from C_i to C_j. Note that the relation

$$\boldsymbol{B}\boldsymbol{K}_{\mathrm{COM}}(t) = \underbrace{\sqrt{n} \left\| \boldsymbol{K}^* \right\| \boldsymbol{B}}_{\boldsymbol{B}^*} \underbrace{\frac{1}{\sqrt{n} \left\| \boldsymbol{K}^* \right\|} \boldsymbol{K}_{\mathrm{COM}}(t)}_{\boldsymbol{K}_{\mathrm{CU}}(t)}$$

holds. In the following, Theorem 2.5 is reformulated for the self-organizing control system Σ.

Proposition 5.1 (**Testing asymptotic stability by modeling Σ as an uncertain system**) *Consider the self-organizing control system Σ defined in* (5.19) *which can be reformulated as the uncertain system presented in* (5.56). *Then Σ with $\boldsymbol{d}(t) = \boldsymbol{0}$,* $(\forall t \geq 0)$, *is asymptotically stable if the following two conditions are fulfilled:*

1. *The system matrix $\bar{\boldsymbol{A}}_{\mathrm{P}} = \bar{\boldsymbol{A}} + \boldsymbol{E}\boldsymbol{L}\boldsymbol{C}_{\mathrm{z}}$ of Σ_{D} defined in* (5.51) *is a Hurwitz matrix.*
2. *The relation $\left\| (s\boldsymbol{I} - \bar{\boldsymbol{A}}_{\mathrm{P}})^{-1} \boldsymbol{B}^* \right\|_{H_\infty} < 1$ holds true.*

Proof. The relation

$$\left\| \boldsymbol{K}_{\mathrm{COM}}(t) \right\| \leq \sqrt{n} \left\| \boldsymbol{K}^* \right\|$$

holds true, since $\left\| \boldsymbol{K}^* \right\|_{\mathrm{F}} \leq \sqrt{n} \left\| \boldsymbol{K}^* \right\|$ holds true according to (2.3), $\left\| \boldsymbol{K}_{\mathrm{COM}}(t) \right\|_{\mathrm{F}} \leq \left\| \boldsymbol{K}^* \right\|_{\mathrm{F}}$ holds true for all $t \geq 0$ according to (2.2) and $\left\| \boldsymbol{K}_{\mathrm{COM}}(t) \right\| \leq \left\| \boldsymbol{K}_{\mathrm{COM}}(t) \right\|_{\mathrm{F}}$ holds true for all $t \geq 0$ according to (2.3). Hence, the relation $\left\| \boldsymbol{K}_{\mathrm{CU}}(t) \right\| \leq 1$, $(\forall t \geq 0)$, claimed for $\boldsymbol{K}_{\mathrm{u}}(t)$ in (2.14) is fulfilled. The proof is completed by replacing the matrices $\boldsymbol{A}$, $\boldsymbol{B}$, $\boldsymbol{K}_{\mathrm{u}}(t)$ and $\boldsymbol{C}$ in (2.15) by $\bar{\boldsymbol{A}}_{\mathrm{P}}$, $\boldsymbol{B}^*$, $\boldsymbol{K}_{\mathrm{CU}}(t)$ and $\boldsymbol{I}_n$ from (5.56), respectively, which leads to similar conditions as in Theorem 2.5. □

Proposition 5.1 presents a sufficient condition for the asymptotic stability of the self-organizing control system Σ. The first part of the condition is identical to the first part of the condition in Theorem 5.4. The second part of the condition is some kind of small gain condition (cf. [82]). A drawback of Proposition 5.1 is that compared to Theorem 5.4 there is no distinction between the individual communication links. The second condition in Proposition 5.1 is testing the worst case feedback gain of the situation-dependent communication. Hence, Proposition 5.1 usually is more conservative than Theorem 5.4.

Summary of the stability analysis. Figure 5.12 summarizes the results of the stability analysis of the self-organizing control system Σ defined in (5.19). Theorem 5.4 and Proposition 5.1 also hold if there exists a time $t^* \geq 0$ after which the overall disturbance vanishes, i.e., $\boldsymbol{d}(t) = \boldsymbol{0}$ for all $t \geq t^*$. For the overview in Fig. 5.12 the undisturbed closed-loop system Σ is considered. Then Theorem 5.4 checks the asymptotic stability of Σ by analyzing the gain of each communication link individually (left part of Fig. 5.12) and Proposition 5.1 checks the asymptotic stability of Σ with a worst case analysis of the overall communication (right part of Fig. 5.12). If the disturbances are not vanishing, Σ is still practically stable (cf. Theorem 5.2). Under the given assumptions the main result of the stability analysis is:

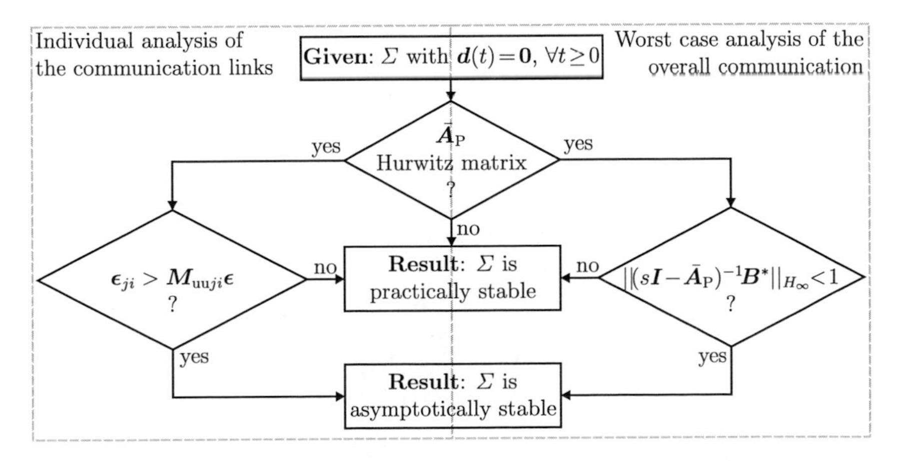

Figure 5.12: Structure of the stability analysis for Σ defined in (5.19).

The realization of a self-organizing controller C from a centralized controller C_C which leads to an asymptotically stable undisturbed centralized control system Σ_C also leads to a practically stable self-organizing control system Σ which fulfills Control aim 3.2.

In the following, the results on the stability analysis in this section are applied to the self-organizing controlled symmetrically interconnected subsystem from Example 5.1.

Example 5.1 (cont.) *Situation-dependent transmission of control signals*

Figure 5.13 shows the phase portrait of the states $\boldsymbol{x}_1(t)$ and $\boldsymbol{x}_2(t)$ of the symmetrically interconnected subsystems P defined in (3.32) with $\boldsymbol{x}_{01} = 3$ and $\boldsymbol{x}_{02} = -3$, controlled by the local control units C_i defined in (5.16) with the parameters listed in Table 5.1 for the disturbances $\boldsymbol{d}_i(t)$ depicted in the upper part of Fig 5.4 and for two different physical interconnection matrices $\boldsymbol{L}$:

- **Scenario 1** with $\boldsymbol{L}_{21} = \boldsymbol{L}_{12} = 1$ (cf. Fig. 5.13 left):
  - According to **Theorem 5.2**, the closed-loop system Σ is practically stable w.r.t. the compact set $\mathcal{A} = \{\boldsymbol{x} \in \mathbb{R}^2 \mid \|\boldsymbol{x}\| \leq \bar{x} = 1.27\}$, since

$$\boldsymbol{A}_\mathrm{C} = \begin{pmatrix} -2.19 & -0.46 \\ -0.46 & -2.19 \end{pmatrix} \quad \Rightarrow \quad \begin{cases} \lambda_1(\boldsymbol{A}_\mathrm{C}) = -2.65 < 0 \checkmark \\ \lambda_2(\boldsymbol{A}_\mathrm{C}) = -1.73 < 0 \checkmark . \end{cases}$$

  - According to **Theorem 5.4**, the closed-loop system Σ is **asymptotically stable**.
    1. The system matrix $\bar{\boldsymbol{A}}_\mathrm{P}$ of Σ_D defined in (5.51) is a Hurwitz matrix? **[yes]**

$$\bar{\boldsymbol{A}}_\mathrm{P} = \begin{pmatrix} -2.19 & 1 \\ 1 & -2.19 \end{pmatrix} \quad \Rightarrow \quad \begin{cases} \lambda_1(\bar{\boldsymbol{A}}_\mathrm{P}) = -3.19 < 0 \checkmark \\ \lambda_2(\bar{\boldsymbol{A}}_\mathrm{P}) = -1.19 < 0 \checkmark \end{cases}$$

    2. The relation $\boldsymbol{\epsilon}_{ji} > \boldsymbol{M}_{\mathrm{uu}ji}\boldsymbol{\epsilon}$ holds true for all $j \in \bar{\mathcal{F}}_{\mathrm{C}i}$ and all $i \in \mathcal{N}$? **[yes]**

$$\boldsymbol{\epsilon}_{21} = 0.2 > \boldsymbol{M}_{\mathrm{uu}21}\boldsymbol{\epsilon}_{12} = 0.168 \checkmark$$
$$\boldsymbol{\epsilon}_{12} = 0.2 > \boldsymbol{M}_{\mathrm{uu}12}\boldsymbol{\epsilon}_{21} = 0.168 \checkmark$$

  - According to **Proposition 5.1**, the closed-loop system Σ might **not** be asymptotically stable.
    1. The system matrix $\bar{\boldsymbol{A}}_\mathrm{P}$ of Σ_D defined in (5.51) is a Hurwitz matrix? **[yes]**
    2. The relation $\left\|(s\boldsymbol{I} - \bar{\boldsymbol{A}}_\mathrm{P})^{-1}\boldsymbol{B}^*\right\|_{H_\infty} < 1$ holds true? **[no]**

$$\sqrt{2}\,\|\boldsymbol{K}^*\| = 2.06 \quad \Rightarrow \quad \left\|(s\boldsymbol{I} - \bar{\boldsymbol{A}}_\mathrm{P})^{-1}\boldsymbol{B}^*\right\|_{H_\infty} = 1.73 \not< 1 \ ⚡$$

- **Scenario 2** with $\boldsymbol{L}_{21} = \boldsymbol{L}_{12} = 3$ (cf. Fig. 5.13 right):
  - According to **Theorem 5.2**, the closed-loop system Σ is practically stable w.r.t. the compact set $\mathcal{A} = \{\boldsymbol{x} \in \mathbb{R}^2 \mid \|\boldsymbol{x}\| \leq \bar{x} = 3.40\}$, since

$$\boldsymbol{A}_\mathrm{C} = \begin{pmatrix} -2.19 & 1.54 \\ 1.54 & -2.19 \end{pmatrix} \quad \Rightarrow \quad \begin{cases} \lambda_1(\boldsymbol{A}_\mathrm{C}) = -3.73 < 0 \checkmark \\ \lambda_2(\boldsymbol{A}_\mathrm{C}) = -0.65 < 0 \checkmark . \end{cases}$$

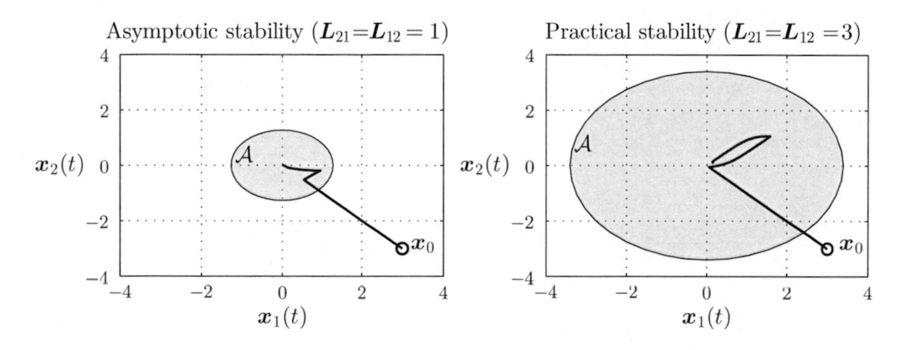

Figure 5.13: Stability analysis of the interconnected subsystems defined in (3.32) controlled by the local control units C_i defined in (5.16) with the parameters listed in Table 5.1.

- According to **Theorem 5.4**, the closed-loop system Σ might **not** be asymptotically stable.
  1. The system matrix $\bar{A}_{\mathrm{P}}$ of Σ_{D} defined in (5.51) is a Hurwitz matrix? [**no**]

$$\bar{A}_{\mathrm{P}} = \begin{pmatrix} -2.19 & 3 \\ 3 & -2.19 \end{pmatrix} \quad \Rightarrow \quad \begin{cases} \lambda_1(\bar{A}_{\mathrm{P}}) = -5.19 < 0 \checkmark \\ \lambda_2(\bar{A}_{\mathrm{P}}) = 0.81 \not< 0 \lightning \end{cases}$$

- According to **Proposition 5.1**, the closed-loop system Σ might **not** be asymptotically stable.
  1. The system matrix $\bar{A}_{\mathrm{P}}$ of Σ_{D} defined in (5.51) is a Hurwitz matrix? [**no**]

For Scenario 1, the second condition of Proposition 5.1 is not fulfilled. Hence, Proposition 5.1 states that the self-organizing control system Σ might not be asymptotically stable. But according to Theorem 5.4 and the left part of Fig. 5.13 Σ is asymptotically stable. This result shows that the individual analysis of the communication links in Theorem 5.4 is more precise then the worst case analysis of the overall communication in Proposition 5.1.

For Scenario 2, $\bar{A}_{\mathrm{P}}$ is no Hurwitz matrix. Hence, the first condition in Theorem 5.4 and in Proposition 5.1 is not fulfilled. The right part of Fig. 5.13 supports this result. But for both scenarios Fig. 5.13 shows that Σ is practically stable w.r.t. the corresponding compact sets $\mathcal{A}$. □

5.3.4 Effects of disturbances on the self-organizing control system

In this section the behavior of the self-organizing control system Σ defined in (5.19) is investigated for the case that some subsystems Σ_j, $(\forall j \in \mathcal{D} \subseteq \mathcal{N})$, are disturbed, where $\mathcal{D}$ is the set of disturbed subsystems defined by

$$\mathcal{D} = \{j \in \mathcal{N} \mid \exists t \geq 0 \text{ s.t. } \boldsymbol{d}_j(t) \neq \boldsymbol{0}\}. \tag{5.57}$$

Three main questions are concerned:

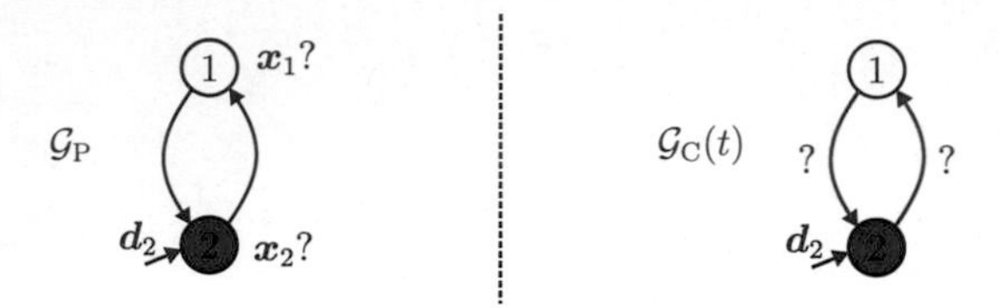

Figure 5.14: Behavior of the subsystems (left) and communication activation (right) for a disturbance affecting subsystem Σ_2 ($\mathcal{D} = \{2\}$).

1. How do disturbances affect the behavior of disturbed subsystems and the behavior of the undisturbed subsystems (Fig. 5.14 left)?
2. Which communication is necessary, to respond to the effect of the disturbances within the self-organizing control system Σ (Fig. 5.14 right)?
3. Will the communication be deactivated, when the disturbances vanish?

Behavior of the subsystems. To answer the first question, consider the analysis of signals the controlled subsystems in Lemma 5.2, which gives bounds on the states $\boldsymbol{x}_i(t)$, $(\forall i \in \mathcal{N})$ and performance outputs $\boldsymbol{v}_i(t)$. Since only some subsystems are disturbed the overall bound $\bar{\boldsymbol{d}}$ of the disturbances depends on the set $\mathcal{D}$ of disturbed subsystems

$$\bar{\boldsymbol{d}}^*(\mathcal{D}) := \begin{pmatrix} \bar{\boldsymbol{d}}_1^*(\mathcal{D}) \\ \vdots \\ \bar{\boldsymbol{d}}_N^*(\mathcal{D}) \end{pmatrix} \quad \text{with} \quad \bar{\boldsymbol{d}}_i^*(\mathcal{D}) := \begin{cases} \bar{\boldsymbol{d}}_i & \text{if } i \in \mathcal{D} \\ 0 & \text{else,} \end{cases} \quad \forall i \in \mathcal{N}. \tag{5.58}$$

With this, the following proposition gives bounds on the states $\boldsymbol{x}_i(t)$, $(\forall i \in \mathcal{N})$ and performance outputs $\boldsymbol{v}_i(t)$ in the case that the subsystems Σ_j, $(\forall j \in \mathcal{D})$, are disturbed.

Proposition 5.2 (Disturbance propagation within the overall system) *Consider the self-organizing control system Σ defined in (5.19) with $\boldsymbol{x}_0 = \boldsymbol{0}$ and that the subsystems Σ_j, $(\forall j \in \mathcal{D})$, are disturbed. Then the states $\boldsymbol{x}_i(t)$, $(\forall i \in \mathcal{N})$, and the performance outputs $\boldsymbol{v}_i(t)$ of the subsystems Σ_i are bounded by:*

$$\max_{t\geq 0} |\boldsymbol{x}_i(t)| \leq \bar{\boldsymbol{x}}_{\mathrm{D}i}(\mathcal{D}) := \boldsymbol{M}_{\mathrm{xd}i}\bar{\boldsymbol{d}}^*(\mathcal{D}) + \boldsymbol{M}_{\mathrm{xu}i}\boldsymbol{\epsilon}(\mathcal{D}), \quad \forall i \in \mathcal{N}$$
$$\max_{t\geq 0} |\boldsymbol{v}_i(t)| \leq \bar{\boldsymbol{v}}_{\mathrm{D}i}(\mathcal{D}) := \boldsymbol{M}_{\mathrm{vd}i}\bar{\boldsymbol{d}}^*(\mathcal{D}) + \boldsymbol{M}_{\mathrm{vu}i}\boldsymbol{\epsilon}(\mathcal{D}), \quad \forall i \in \mathcal{N}.$$

Proof. Due to the distinction between disturbed subsystems and undisturbed subsystems in

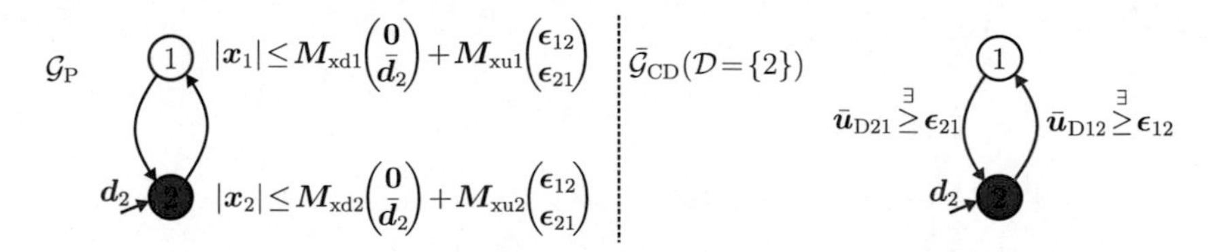

Figure 5.15: Bounds on the behavior of the subsystems in case of a disturbance affecting subsystem Σ_2 according to Proposition 5.2 (left); Resulting maximal communication graph $\bar{\mathcal{G}}_{\mathrm{CD}}(\mathcal{D})$ for the disturbance situation $\mathcal{D} = \{2\}$ according to Proposition 5.3 (right).

$\bar{\boldsymbol{d}}^*(\mathcal{D})$ defined in (5.58), Lemma 5.2 leads to the bounds $\bar{\boldsymbol{x}}_{\mathrm{D}i}(\mathcal{D})$, $(\forall i \in \mathcal{N})$, for all $\boldsymbol{x}_i(t)$ and the bounds $\bar{\boldsymbol{v}}_{\mathrm{D}i}(\mathcal{D})$ for all $\boldsymbol{v}_i(t)$. $\square$

The left part of Fig. 5.15 illustrates Proposition 5.2. The bounds in Proposition 5.2 give a measure on how strong a disturbance propagates through the overall system. The bounds in Proposition 5.2 for all subsystems Σ_i, $(\forall i \in \mathcal{N})$, depend on the bounds $\bar{\boldsymbol{d}}_j$ of the disturbed subsystems Σ_j, $(\forall j \in \mathcal{D})$, and on the choice of all switching thresholds, but the impact of the disturbance does not depend on the choice of the switching thresholds. Hence, all subsystem are influenced in the case of a disturbance independent from the choice of the switching thresholds. Furthermore, the effect of the disturbance on the undisturbed subsystems increases with the amount of the disturbances.

Communication activation for disturbance attenuation. To answer the second question, consider the analysis of the communication activation in Lemma 5.3. The following proposition defines the maximal communication graph $\bar{\mathcal{G}}_{\mathrm{CD}}(\mathcal{D})$ which can occur in the case that the subsystems Σ_j, $(\forall j \in \mathcal{D})$, are disturbed.

Proposition 5.3 (Necessary communication to mimic the centralized controller) *Consider the self-organizing control system Σ defined in (5.19) with $\boldsymbol{x}_0 = \mathbf{0}$. If the subsystems Σ_j, $(\forall j \in \mathcal{D})$, are disturbed, there might be communication from the local control unit C_i, $(i \in \mathcal{N})$ to the local control unit C_h, $(h \in \bar{\mathcal{F}}_{\mathrm{CD}i}(\mathcal{D}))$ with*

$$\bar{\mathcal{F}}_{\mathrm{CD}i}(\mathcal{D}) = \{h \in \bar{\mathcal{F}}_{\mathrm{C}i} \mid \bar{\boldsymbol{u}}_{\mathrm{D}hi}(\mathcal{D}) \overset{\exists}{\geq} \epsilon_{hi}\}, \quad \forall i \in \mathcal{N}$$
$$\text{and } \bar{\boldsymbol{u}}_{\mathrm{D}hi}(\mathcal{D}) := \boldsymbol{M}_{\mathrm{ud}hi}\bar{\boldsymbol{d}}^* + \boldsymbol{M}_{\mathrm{uu}hi}\boldsymbol{\epsilon}, \quad \forall h \in \bar{\mathcal{F}}_{\mathrm{C}i}, \quad \forall i \in \mathcal{N}$$

but not among any other local control units. Hence, the set $\mathcal{E}_{\mathrm{C}}(t)$ of communication links within the communication graph $\mathcal{G}_{\mathrm{C}}(t) = (\mathcal{V}_{\mathrm{C}}, \mathcal{E}_{\mathrm{C}}(t))$ of Σ is bounded by the set $\bar{\mathcal{E}}_{\mathrm{CD}}(\mathcal{D})$ of maximal *communication links*

$$\emptyset \subseteq \mathcal{E}_{\mathrm{C}}(t) \subseteq \bar{\mathcal{E}}_{\mathrm{CD}}(\mathcal{D}) = \{(i \to j) \mid j \in \bar{\mathcal{F}}_{\mathrm{CD}i}(\mathcal{D}), i \in \mathcal{N}\}, \quad \forall t \geq 0. \tag{5.59}$$

The set $\bar{\mathcal{E}}_{\mathrm{CD}}(\mathcal{D})$ defines the resulting maximal communication graph

$$\bar{\mathcal{G}}_{\mathrm{CD}}(\mathcal{D}) = (\mathcal{V}_{\mathrm{C}}, \bar{\mathcal{E}}_{\mathrm{CD}}(\mathcal{D})). \tag{5.60}$$

Proof. The proof is given in Appendix A.9. □

The graph $\bar{\mathcal{G}}_{\mathrm{CD}}(\mathcal{D})$ in Proposition 5.3 defines the maximal possible communication among the local control units that can occur if the subsystems Σ_j, $(\forall j \in \mathcal{D})$, are disturbed. The right part of Fig. 5.15 illustrates Proposition 5.3 for the set $\mathcal{D} = \{2\}$ which result to the maximal communication graph $\bar{\mathcal{G}}_{\mathrm{CD}}(\mathcal{D} = \{2\})$. According to the definition of $\bar{\mathcal{F}}_{\mathrm{CD}i}(\mathcal{D})$, $(\forall i \in \mathcal{N})$, the edges $(i \to h)$ only exists if the relation

$$\bar{\boldsymbol{u}}_{\mathrm{D}hi}(\mathcal{D}) \overset{\exists}{\geq} \boldsymbol{\epsilon}_{hi}, \quad h \in \bar{\mathcal{F}}_{\mathrm{CD}i}, \quad i \in \mathcal{N}$$

holds true. Therefore, $\bar{\mathcal{G}}_{\mathrm{CD}}(\mathcal{D} = \{2\})$ gives a measure on how a disturbance influences the communication among the local control units and defines what communication is necessary to mimic the behavior of the centralized control system Σ_{C} in a special disturbance situation. Hence, Proposition 5.3 shows that the occurring communication depends on the acting disturbances.

Deactivation of the communiation for vanishing disturbances. A condition for the deactivation of certain communication links for disturbed subsystems has been presented in Lemma 5.4 in Section 5.3.2. Furthermore, the investigation of the asymptotic stability of the self-organizing control system Σ in Section 5.3.3 showed that Σ is asymptotically stable if the decentralized controlled system Σ_{D} is asymptotically stable and the communication is deactivated for an undisturbed Σ. The following corollary uses these results to derive two alternative conditions for the deactivation of the communication for vanishing disturbances.

Corollary 5.2 (Communication deactivation for vanishing disturbances) *Consider the self-organizing control system Σ defined in (5.19) with $\boldsymbol{x}_0 = \boldsymbol{0}$ and $\boldsymbol{d}(t) = \boldsymbol{0}$ for $t \geq t_{\mathrm{D}} \geq 0$, ($t_{\mathrm{D}} \in \mathbb{R}_+$). The communication among all local control units C_i stays deactivated after a time $t_{\mathrm{D}} \in \mathbb{R}_+$, i.e.,*

$$\mathcal{E}_{\mathrm{C}}(t) = \emptyset, \quad t \geq t_{\mathrm{D}} \geq 0$$

if at least one of the following conditions is fulfilled:

1. *Σ is asymptotically stable according to Definition 2.2.*

2. *The relations in* (5.55) *hold true.*

Proof. If the self-organizing control system Σ defined in (5.19) is asymptotically stable, then there always exists a deactivation time $t_{\mathrm{D}ji}$ that deactivates the communication link $(i \to j) \in \mathcal{E}_{\mathrm{C}}(t)$ according to (5.43) since $\lim_{t\to\infty} |\hat{\boldsymbol{u}}_{ji}(t)| = \mathbf{0}$ holds true (Condition 1). Condition 2 follows from (5.47) in Lemma 5.4 for $\bar{\boldsymbol{d}} = \mathbf{0}$. □

The second condition in Corollary 5.2 directly follows from Lemma 5.4 which is also a sufficient condition for the asymptotic stability of Σ (cf. Theorem 5.4). The first condition in Corollary 5.2 derives that for an asymptotically stable self-organizing system Σ the communication among the local control units is only necessary if a disturbance is present. Both conditions are relevant since they can occur independently.

Summary. With the analysis of the effects of disturbances on the self-organizing control system Σ defined in (5.19) in Proposition 5.2, Proposition 5.3 and Corollary 5.2 the three initial questions can be answered as follows:

1. A single disturbance influences all other subsystems and the effect on the subsystems increases with amount of the disturbance.

2. A maximal possible communication graph can be predicted for certain disturbance situations by analyzing each communication link individually.

3. The communication among the local control units will be deactivated for vanishing disturbances if the closed-loop system Σ is asymptotically stable or the relations in (5.55) hold true.

Furthermore, Proposition 5.3 is used in Section 5.4.2 to design the communication among the local control units. Due to the small number of subsystems in Example 5.1, the results in Proposition 5.2 and Proposition 5.3 are illustrated in more detail at the application example of the water supply system with 23 subsystems in Section 5.5.

5.4 Design of the local control units

In this section methods for designing the switching thresholds and the communication structure among the local control units are presented. First, the switching thresholds are designed such that the resulting self-organizing control system Σ guarantees a desired difference behavior compared to the centralized control system Σ_{C} as required in Control aim 3.4 (Section 5.4.1). Second, based on the designed switching threshold a method for reducing the maximal communication graph $\bar{\mathcal{G}}_{\mathrm{C}}$ is presented which cancels out unnecessary communication links (Section 5.4.2). Furthermore, overall algorithms for designing the local control units are presented in Section 5.4.3.

5.4.1 Design of the switching thresholds

This section presents methods for designing the switching thresholds ϵ_{ij} of the local control units C_i defined in (5.16) to guarantee Control aim 3.4. Therefore, the switching thresholds ϵ_{ij}, $(j \in \bar{\mathcal{P}}_{\mathrm{C}i}, \forall i \in \mathcal{N})$ that bound the control signals according to (5.27) are weighted by the parameters c_{ij}, $(j \in \bar{\mathcal{P}}_{\mathrm{C}i}, \forall i \in \mathcal{N})$

$$\epsilon_{ij} = \frac{c_{ij}}{\sum_{l \in \bar{\mathcal{P}}_{\mathrm{C}i}} c_{il}} \cdot \left|\bar{\mathcal{P}}_{\mathrm{C}i}\right| \cdot \bar{\epsilon}_i, \quad \forall j \in \bar{\mathcal{P}}_{\mathrm{C}i}, \quad \forall i \in \mathcal{N}, \tag{5.61}$$

where $\bar{\epsilon}_i$ is an average switching threshold. The greater the weighting c_{ij} the greater is the penalization of the activation of the communication link $(j \to i) \in \mathcal{E}_{\mathrm{C}}(t)$. With this, certain communication links can be favored for being activated or not activated.

According to Theorem 5.1, and with $\bar{\epsilon}_i$ from (5.61) the relation

$$\boldsymbol{M}_{\mathrm{vu}} \boldsymbol{P} \bar{\boldsymbol{\epsilon}} = \bar{\boldsymbol{v}}_{\Delta}^{*} \tag{5.62}$$

has to be fulfilled to guarantee Control aim 3.4 with

$$\bar{\boldsymbol{v}}_{\Delta}^{*} := \begin{pmatrix} \bar{\boldsymbol{v}}_{\Delta 1}^{*} \\ \vdots \\ \bar{\boldsymbol{v}}_{\Delta N}^{*} \end{pmatrix}, \quad \bar{\boldsymbol{\epsilon}} := \begin{pmatrix} \bar{\epsilon}_1 \\ \vdots \\ \bar{\epsilon}_N \end{pmatrix},$$
$$\boldsymbol{P} := \mathrm{diag}(\boldsymbol{P}_1 \ldots \boldsymbol{P}_N),$$
$$\boldsymbol{P}_i := |\bar{\mathcal{P}}_{\mathrm{C}i}| \boldsymbol{I}_{m_i}, \quad \forall i \in \mathcal{N}.$$

Note that $\boldsymbol{P}\bar{\boldsymbol{\epsilon}} = \boldsymbol{\epsilon}$ since according to (5.61) the relation

$$\sum_{j \in \bar{\mathcal{P}}_{\mathrm{C}i}} \epsilon_{ij} = \frac{\sum_{l \in \bar{\mathcal{P}}_{\mathrm{C}i}} c_{il}}{\sum_{l \in \bar{\mathcal{P}}_{\mathrm{C}i}} c_{il}} \cdot \left|\bar{\mathcal{P}}_{\mathrm{C}i}\right| \cdot \bar{\epsilon}_i = \left|\bar{\mathcal{P}}_{\mathrm{C}i}\right| \cdot \bar{\epsilon}_i, \quad \forall i \in \mathcal{N}$$

holds true. The desired maximal difference performance outputs $\bar{\boldsymbol{v}}_{\Delta i}^{*} \in \mathbb{R}^{r_{\mathrm{v}i}}$, $(\forall i \in \mathcal{N})$, are an upper bound for the difference performance outputs $\boldsymbol{v}_{\Delta i}(t)$, i.e., $|\boldsymbol{v}_{\Delta i}(t)| \leq \bar{\boldsymbol{v}}_{\Delta i}^{*}$. The bound

$\bar{\boldsymbol{v}}^*_{\Delta i}$, $(\forall i \in \mathcal{N})$, is assumed to be chosen such that the relation $\|\bar{\boldsymbol{v}}^*_{\Delta i}\| \leq \bar{v}^*_{\Delta i}$ is fulfilled. With this, the relation

$$|\boldsymbol{v}_{\Delta i}(t)| \leq \bar{\boldsymbol{v}}^*_{\Delta i} \quad \Rightarrow \quad \|\boldsymbol{v}_{\Delta i}(t)\| \leq \bar{v}^*_{\Delta i}, \qquad \forall i \in \mathcal{N}, \quad \forall t \geq 0$$

holds true, where the right part is the requirement in Control aim 3.4.

The following theorem uses the relation in (5.62) to give a condition for designing the thresholds ϵ_{ij}, $(j \in \bar{\mathcal{P}}_{\mathrm{C}i}, \forall i \subset \mathcal{N})$ to guarantee Control aim 3.4.

Theorem 5.5 (Choose switching thresholds to guarantee desired difference behavior) *Consider the self-organizing control system Σ defined in (5.19) and the weighting of the switching thresholds ϵ_{ij}, $(j \in \bar{\mathcal{P}}_{\mathrm{C}i}, \forall i \in \mathcal{N})$, according to (5.61). The difference performance output $\boldsymbol{v}_{\Delta i}(t)$, $(\forall i \in \mathcal{N})$, for all subsystems Σ_i is bounded by the desired maximal difference performance output $\bar{v}^*_{\Delta i}$ as required in Control aim 3.4 if the switching thresholds ϵ_{ij} are chosen such that the relation*

$$\bar{\boldsymbol{\epsilon}} \leq \boldsymbol{M}^+_{\mathrm{Pvu}} \bar{\boldsymbol{v}}^*_{\Delta} \tag{5.63}$$

holds true and $\boldsymbol{M}^+_{\mathrm{Pvu}} \boldsymbol{M}_{\mathrm{Pvu}} \bar{\boldsymbol{\epsilon}} = \bar{\boldsymbol{\epsilon}}$. Note that $\boldsymbol{M}^+_{\mathrm{Pvu}}$ denotes the pseudoinverse of $\boldsymbol{M}_{\mathrm{Pvu}} := \boldsymbol{M}_{\mathrm{vu}} \boldsymbol{P}$ according to (2.1).

Proof. Solving (5.62) for $\bar{\boldsymbol{\epsilon}}$ leads to (5.63). Relation (5.62) can be fulfilled if $\boldsymbol{M}^+_{\mathrm{Pvu}} \boldsymbol{M}_{\mathrm{Pvu}} \bar{\boldsymbol{\epsilon}} = \bar{\boldsymbol{\epsilon}}$ holds true (cf. Theorem 6.2 in [84]). □

Theorem 5.5 presents an appropriate choice of the switching thresholds ϵ_{ji}, $(j \in \bar{\mathcal{P}}_{\mathrm{C}i}, \forall i \in \mathcal{N})$, for each local control unit C_i to guarantee the desired maximal difference performance output $\bar{v}^*_{\Delta i}$, $(\forall i \in \mathcal{N})$, according to Control aim 3.4. For the choice of the switching thresholds in Theorem 5.5 global model information of the plant P is necessary to apply (5.63).

Remark 5.2 *The switching thresholds can also be designed to guarantee the asymptotic stability of the self-organizing control system Σ by using linear programming [32, 37]. For this, the inequalities in (5.55) have to be assembled and a cost function to maximize the switching thresholds has to be defined. Furthermore, the conditions in Theorem 5.5 to guarantee a desired difference behavior can be integrated into the overall optimization problem as well.*

5.4.2 Reduction of the communication structure

In this section the communication within the self-organizing controller C defined in (5.17) is designed such that the unnecessary communication links in $\bar{\mathcal{E}}_{\mathrm{C}}$ are removed to reduce the communication effort among the local control units, where the set $\bar{\mathcal{E}}_{\mathrm{C}}$ of maximal communication links results from the structure of the central feedback gain matrix according to (2.21) and (5.11). An unnecessary communication link $(i \rightarrow j)$ is a link that will never be activated during the control of the plant P since the corresponding control signal $\hat{\boldsymbol{u}}_{ji}(t)$ determined in C_i does not exceed the threshold $\boldsymbol{\epsilon}_{ji}$ for all $t \geq 0$ (cf. (5.15b)).

Recalling the analysis of the disturbance propagation in Section 5.3.4, Proposition 5.3 already defines a communication graph $\mathcal{G}_{\mathrm{CD}}(\mathcal{D}) = (\mathcal{V}_{\mathrm{C}}, \mathcal{E}_{\mathrm{CD}}(\mathcal{D}))$ that will be used during the control of the plant P, for the case that the subsystems Σ_j, $(\forall j \in \mathcal{D})$, are disturbed. In the following the construction of $\mathcal{G}_{\mathrm{CD}}(\mathcal{D})$ is used to define a reduced maximal communication structure $\bar{\mathcal{G}}_{\mathrm{C}} = (\mathcal{V}_{\mathrm{C}}, \bar{\mathcal{E}}_{\mathrm{C}})$ by considering that all subsystems could be disturbed, i.e.,

$$\bar{\mathcal{G}}_{\mathrm{C}} = \mathcal{G}_{\mathrm{CD}}(\mathcal{D} = \mathcal{N}).$$

The self-organizing control system Σ^* with reduced communication is defined by

$$\Sigma^* : \begin{cases} \Sigma \text{ defined in (5.19) with } \boldsymbol{x}_0 = \boldsymbol{x}_0^* \text{ and} \\ \mathcal{E}_{\mathrm{C}}(t) \text{ bounded by } \mathcal{E}_{\mathrm{CD}}(\mathcal{D}{=}\mathcal{N}) \text{ according to (5.59) in Proposition 5.3} \end{cases} \tag{5.64}$$

where $\boldsymbol{x}_0^*$ is the initial state of Σ^* and $\mathcal{E}_{\mathrm{CD}}(\mathcal{D}{=}\mathcal{N})$ is the set of maximal communication links within $\mathcal{G}_{\mathrm{CD}}(\mathcal{D}{=}\mathcal{N}) = (\mathcal{V}_{\mathrm{C}}, \mathcal{E}_{\mathrm{CD}}(\mathcal{D}{=}\mathcal{N}))$. To build the reduced maximal communication graph $\bar{\mathcal{G}}_{\mathrm{C}} = \mathcal{G}_{\mathrm{CD}}(\mathcal{D} = \mathcal{N})$ the sets $\bar{\mathcal{F}}_{\mathrm{C}i}$ of maximal followers of Σ are replaced by sets $\bar{\mathcal{F}}_{\mathrm{CD}i}(\mathcal{D} = \mathcal{N})$ according to Proposition 5.3:

$$\begin{aligned} \bar{\mathcal{F}}_{\mathrm{C}i} &= \bar{\mathcal{F}}_{\mathrm{CD}i}(\mathcal{D}{=}\mathcal{N}), \quad \forall i{\in}\mathcal{N} \\ \bar{\mathcal{P}}_{\mathrm{C}i} &= \{j \in \mathcal{N} \mid i \in \bar{\mathcal{F}}_{\mathrm{CD}j}(\mathcal{D}{=}\mathcal{N})\}, \quad \forall i{\in}\mathcal{N}, \end{aligned}$$

where the sets $\bar{\mathcal{P}}_{\mathrm{C}i}$ of maximal predecessors also follow from $\bar{\mathcal{F}}_{\mathrm{CD}i}(\mathcal{D} = \mathcal{N})$.

The following theorem uses the results in Proposition 5.3 to state that the reduction of the communication within Σ^* defined in (5.64) compared to Σ defined in (5.19) leads to an identical behavior of Σ and Σ^*.

Theorem 5.6 (Reduction of the communication structure for Σ defined in (5.19) **)** *For $\boldsymbol{x}_0 = \boldsymbol{x}_0^* = \boldsymbol{0}$ the behavior of the self-organizing control system Σ defined in* (5.19) *and the behavior of the self-organizing control system Σ^* defined in* (5.64) *with the reduced communication is identical.*

Proof. Based on Proposition 5.3 the sets $\bar{\mathcal{F}}_{\mathrm{C}i} = \mathcal{F}_{\mathrm{CD}i}(\mathcal{D} = \mathcal{N})$, $(\forall i \in \mathcal{N})$, include all local control units C_k, $(\forall k \in \mathcal{F}_{\mathrm{CD}i}(\mathcal{D} = \mathcal{N}))$ to which the local control units C_i send information in the case that all subsystems are disturbed and $\boldsymbol{x}_0 = \boldsymbol{0}$. Hence, if there is no communication from C_i, $(i \in \mathcal{N})$, to C_j, $(\forall j \in \bar{\mathcal{F}}_{\mathrm{C}i} \backslash \bar{\mathcal{F}}_{\mathrm{C}i})$ within the system Σ anyway, then the behavior of Σ and the behavior of Σ^* is identical. □

Theorem 5.6 shows that not all communication links within self-organizing controller C defined in (5.17) are necessary since some of them are not used anyway. The benefit of the communication reduction is that in the real control process there is no need to enable a connection among the local control units if the corresponding communication link is canceled out.

5.4.3 Global design algorithm

In this section an algorithm for choosing all parameters of the local control units C_i, $(\forall i \in \mathcal{N})$, defined in (5.16) is presented. The algorithm guarantees a desired difference behavior between the self-organizing control system Σ defined in (5.19) and the centralized control system Σ_{C} defined in (5.21) according to Control aim 3.4.

Algorithm 5.1 uses global model information for designing the local control units C_i:

- The model of the plant P defined in (3.10).
- The overall desired maximal difference performance output $\bar{\boldsymbol{v}}^*_{\Delta}$, $(\forall i \in \mathcal{N})$, according to Control aim 3.4.
- The parameters c_{ij} for all $i, j \in \mathcal{N}$ and $i \neq j$ weight the corresponding communication links $(j \to i)$ according to (5.61).

Algorithm 5.1 is divided into four steps.

Algorithm 5.1: Design local control units C_i, $(\forall i \in \mathcal{N})$, in (5.16) to guarantee a desired difference behavior between Σ and Σ_{C} according to Control aim 3.4

Given: P, $\bar{\boldsymbol{v}}^*_{\Delta}$, c_{ij} for all $i, j \in \mathcal{N}$ and $i \neq j$

1. Determine $\boldsymbol{K}_{\mathrm{C}}$ such that $\boldsymbol{A}_{\mathrm{C}} = \boldsymbol{A}_{\mathrm{P}} - \boldsymbol{B}\boldsymbol{K}_{\mathrm{C}}$ is a Hurwitz matrix.
2. Construct F_i, $(\forall i \in \mathcal{N})$, defined in (5.13) with the block-diagonal matrices $\boldsymbol{K}_{\mathrm{C}ii}$ from $\boldsymbol{K}_{\mathrm{C}}$.
3. Construct O_i, $(\forall i \in \mathcal{N})$, defined in (5.14) by
   - determining $\bar{\mathcal{P}}_{\mathrm{C}i}$ with (5.11) as well as $\bar{\mathcal{F}}_{\mathrm{C}i}$ with (2.22) and
   - using the matrices $\boldsymbol{K}_{\mathrm{C}ji}$, $(\forall j \in \bar{\mathcal{F}}_{\mathrm{C}i})$, from $\boldsymbol{K}_{\mathrm{C}}$.
4. Construct D_i, $(\forall i \in \mathcal{N})$, defined in (5.15) by
   - determining $\bar{\mathcal{P}}_{\mathrm{C}i}$ with (5.11) as well as $\bar{\mathcal{F}}_{\mathrm{C}i}$ with (2.22) and
   - calculating the switching thresholds $\boldsymbol{\epsilon}_{ji}$, $(\forall j \in \bar{\mathcal{F}}_{\mathrm{C}i})$, with (5.63) in Theorem 5.5.

Result: C_i consisting of F_i, O_i and D_i guaranteeing that the difference performance output $\boldsymbol{v}_{\Delta i}(t)$ of all subsystems is upper bounded by $\bar{\boldsymbol{v}}^*_{\Delta i}$ as claimed in Control aim 3.4.

Initially, the feedback gain matrix $\boldsymbol{K}_{\mathrm{C}}$ of the centralized controller C_{C} defined in (5.1) has to be determined such that Assumption 5.1 is fulfilled. Note that due to Assumption 3.5 there always exists a feedback gain matrix $\boldsymbol{K}_{\mathrm{C}}$ that guarantees Assumption 5.1 (cf. Theorem 2.2). The matrix $\boldsymbol{K}_{\mathrm{C}}$ can be determined by classical state feedback design methods (cf. Section 2.2 or methods presented in [22, 36, 94, 122]). Of course, Step 1 can be neglected if already a stabilizing centralized controller C_{C} is given.

In Step 2 the feedback units F_i, $(\forall i \in \mathcal{N})$, defined in (5.13) are designed by integrating the block-diagonal matrices $\boldsymbol{K}_{\mathrm{C}ii}$ of the central feedback gain matrix $\boldsymbol{K}_{\mathrm{C}}$.

The observation units O_i, $(\forall i \in \mathcal{N})$, defined in (5.14) are constructed by defining the sets $\bar{\mathcal{P}}_{\mathrm{C}i}$, $(\forall i \in \mathcal{N})$, of maximal communicational predecessors with (5.11) as well as the sets $\bar{\mathcal{F}}_{\mathrm{C}i}$ of maximal communicational followers with (2.22) and by using the matrices $\boldsymbol{K}_{\mathrm{C}ji}$, $(\forall j \in \bar{\mathcal{F}}_{\mathrm{C}i})$, from $\boldsymbol{K}_{\mathrm{C}}$ (Step 3).

In Step 4 the decision units D_i, $(\forall i \in \mathcal{N})$, defined in (5.15) are constructed by defining $\bar{\mathcal{P}}_{\mathrm{C}i}$ with (5.11) as well as $\bar{\mathcal{F}}_{\mathrm{C}i}$ with (2.22) and calculating the switching thresholds $\boldsymbol{\epsilon}_{ji}$, $(\forall j \in \bar{\mathcal{F}}_{\mathrm{C}i})$, with (5.63) in Theorem 5.5 to fulfill Control aim 3.4.

The result of Algorithm 5.1 is that the self-organizing control system Σ defined in (5.19) is practically stable according to Theorem 5.2 and that Σ guarantees that the overall difference performance output $\boldsymbol{v}_{\Delta}(t)$ of the difference system Σ_{Δ} defined in (5.25) is bounded by the overall desired maximal difference performance output $\bar{\boldsymbol{v}}_{\Delta}^{*}$ as claimed in Control aim 3.4.

Example 5.1 (cont.) *Situation-dependent transmission of control signals*

In the following, the two local control units C_1 and C_2 defined in (5.16) are designed for controlling the symmetrically interconnected subsystems P defined in (3.32) such that the resulting self-organizing control system Σ defined in (5.19) guarantees that the difference performance outputs $\boldsymbol{v}_{\Delta 1}(t)$ and $\boldsymbol{v}_{\Delta 2}(t)$ are bounded by $\bar{v}_{\Delta 1}^{*} = \bar{\boldsymbol{v}}_{\Delta 1}^{*} = 0.4$ and $\bar{v}_{\Delta 2}^{*} = \bar{\boldsymbol{v}}_{\Delta 2}^{*} = 0.2$ according to Control aim 3.4. The weighting of the communication links is identical $c_{21} = c_{12} = 1$.

Table 5.2 illustrates the execution of Algorithm 5.1 for the design of the two local control units to guarantee the desired difference behavior. The feedback gain matrix $\boldsymbol{K}_{\mathrm{C}}$ is the solution of the optimization problem in (2.11) w.r.t. the symmetrically interconnected subsystems P defined in (3.32) and the weighting matrices $\boldsymbol{Q} = 3\boldsymbol{I}_2$ as well as $\boldsymbol{R} = \boldsymbol{I}_2$ (Step 1), where $\boldsymbol{A}_{\mathrm{C}}$ presented in (5.10) fulfills Assumption 5.1. The feedback units F_1 and F_2 are constructed by the block-diagonal matrices $\boldsymbol{K}_{\mathrm{C}11}$ and $\boldsymbol{K}_{\mathrm{C}22}$ from $\boldsymbol{K}_{\mathrm{C}}$ (Step 2).

The determination of the sets $\bar{\mathcal{P}}_{\mathrm{C}i}$ of maximal communicational predecessors and the sets $\bar{\mathcal{F}}_{\mathrm{C}i}$ of maximal communicational followers in Step 3 is straight-forward. The observation units O_1 and O_2 include the sets $\bar{\mathcal{P}}_{\mathrm{C}i}$ of maximal communicational predecessors and the sets $\bar{\mathcal{F}}_{\mathrm{C}i}$ of maximal communicational followers as well as the matrices $\boldsymbol{K}_{\mathrm{C}21}$ and $\boldsymbol{K}_{\mathrm{C}12}$ from $\boldsymbol{K}_{\mathrm{C}}$ (Step 3). Due to the different desired maximal difference performance outputs $\bar{\boldsymbol{v}}_{\Delta i}^{*}$, the designed thresholds $\boldsymbol{\epsilon}_{12}$ and $\boldsymbol{\epsilon}_{21}$ are also different. For $\boldsymbol{v}_{\Delta 2}(t)$ a much smaller deviation than for $\boldsymbol{v}_{\Delta 1}(t)$ is tolerated. Therefore, $\boldsymbol{\epsilon}_{21}$ is much smaller than $\boldsymbol{\epsilon}_{12}$ (Step 4).

Figure 5.16 shows that the desired maximal difference performance outputs $\bar{v}_{\Delta i}^{*} = \bar{\boldsymbol{v}}_{\Delta i}^{*}$, $(\forall i = \{1, 2\})$, are satisfied by Σ with the designed local control units. Due to the small switching threshold $\boldsymbol{\epsilon}_{21}$, there is more communication from C_1 to C_2 than from C_2 to C_1. The stability analysis would be similar to the one presented in Section 5.3.3. □

Table 5.2: Design local control units C_1 and C_2 for the symmetrically interconnected subsystems P in (3.32) by Algorithm 5.1 to guarantee $\bar{v}^*_{\Delta 1} = \bar{\boldsymbol{v}}^*_{\Delta 1} = 0.4$ and $\bar{v}^*_{\Delta 2} = \bar{\boldsymbol{v}}^*_{\Delta 2} = 0.2$.

Step	Execution	Result
1	LQ-Design: $\boldsymbol{Q} = 3\boldsymbol{I}_2$ and $\boldsymbol{R} = \boldsymbol{I}_2$	$\boldsymbol{K}_\mathrm{C} = \begin{pmatrix} 3.19 & 1.46 \\ 1.46 & 3.19 \end{pmatrix}$
2	$\boldsymbol{K}_\mathrm{C} \Rightarrow \boldsymbol{K}_{\mathrm{C}11}$ and $\boldsymbol{K}_{\mathrm{C}22}$	$\boldsymbol{K}_{\mathrm{C}11} = \boldsymbol{K}_{\mathrm{C}22} = 3.19$
3	$\lVert \boldsymbol{K}_{\mathrm{C}12} \rVert \neq 0 \Rightarrow \bar{\mathcal{P}}_{\mathrm{C}1} = \{2\}$ $\lVert \boldsymbol{K}_{\mathrm{C}21} \rVert \neq 0 \Rightarrow \bar{\mathcal{P}}_{\mathrm{C}2} = \{1\}$	$\bar{\mathcal{F}}_{\mathrm{C}1} = \bar{\mathcal{P}}_{\mathrm{C}1} = \{2\}$ $\bar{\mathcal{F}}_{\mathrm{C}2} = \bar{\mathcal{P}}_{\mathrm{C}2} = \{1\}$
	$\boldsymbol{K}_\mathrm{C} \Rightarrow \boldsymbol{K}_{\mathrm{C}21}$ and $\boldsymbol{K}_{\mathrm{C}12}$	$\boldsymbol{K}_{\mathrm{C}21} = \boldsymbol{K}_{\mathrm{C}12} = 1.46$
4	$\lVert \boldsymbol{K}_{\mathrm{C}12} \rVert \neq 0 \Rightarrow \bar{\mathcal{P}}_{\mathrm{C}1} = \{2\}$ $\lVert \boldsymbol{K}_{\mathrm{C}21} \rVert \neq 0 \Rightarrow \bar{\mathcal{P}}_{\mathrm{C}2} = \{1\}$	$\bar{\mathcal{F}}_{\mathrm{C}1} = \bar{\mathcal{P}}_{\mathrm{C}1} = \{2\}$ $\bar{\mathcal{F}}_{\mathrm{C}2} = \bar{\mathcal{P}}_{\mathrm{C}2} = \{1\}$
	$\underbrace{\begin{pmatrix} \bar{\epsilon}_1 \\ \bar{\epsilon}_2 \end{pmatrix}}_{\bar{\boldsymbol{\epsilon}}} = \underbrace{\begin{pmatrix} 1 & 0 \\ 0 & 1 \end{pmatrix}}_{\boldsymbol{P}^{-1}} \underbrace{\begin{pmatrix} 2.18 & -0.45 \\ -0.45 & 2.18 \end{pmatrix}}_{\boldsymbol{M}^+_{\mathrm{vu}}} \underbrace{\begin{pmatrix} 0.4 \\ 0.2 \end{pmatrix}}_{\bar{\boldsymbol{v}}^*_\Delta}$	$\boldsymbol{\epsilon}_{12} = \bar{\epsilon}_1 = 0.78$ $\boldsymbol{\epsilon}_{21} = \bar{\epsilon}_2 = 0.26$

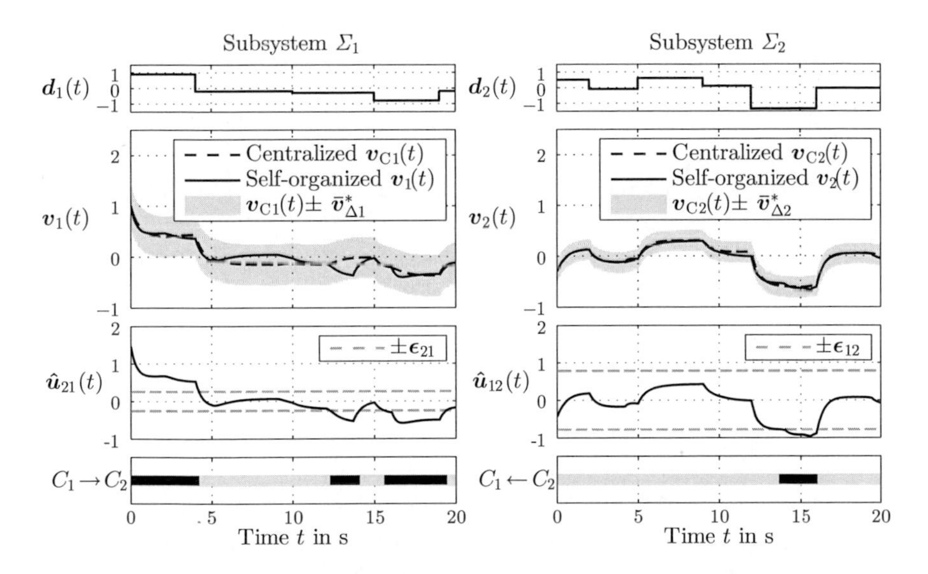

Figure 5.16: Behavior of the symmetrically interconnected subsystems defined in (3.32) controlled by the local control units designed by Algorithm 5.1 presented in Table 5.2.

5.5 Application example: Water supply system

In this section the proposed self-organizing control concept with a situation-dependent transmission of local control signals is applied to the water supply system introduced in Section 3.5.1. Section 5.5.1 presents the control aim and the resulting parameters of the local control units. With this, an analysis of the stability and the disturbance propagation of the resulting self-organizing control system is performed (Section 5.5.2), where these results are compared with actual behavior of the closed-loop system (Section 5.5.3). Section 5.5.4 illustrates the relation between the performance and the communication effort of the self-organizing control system Σ.

5.5.1 Performance criteria and design of the local control units

The deviation between the level $l_{\mathrm{TB}i}(t) = \boldsymbol{v}_i(t)$, $(\forall i \in \mathcal{N})$, with the self-organizing controller C defined in (5.17) and the corresponding level $l_{\mathrm{TBC}i}(t) = \boldsymbol{v}_{\mathrm{C}i}(t)$ in tank TBi controlled by the centralized controller C_{C} defined in (5.1) should be bounded

$$|\boldsymbol{v}_{\Delta i}(t)| = |l_{\mathrm{TB}i}(t) - l_{\mathrm{TBC}i}(t)| \leq \bar{\boldsymbol{v}}^{*}_{\Delta i} = 1\,\mathrm{cm}, \quad \forall i \in \mathcal{N} = \{1, \dots, 23\}. \tag{5.65}$$

Hence, the local control units C_i can be designed with Algorithm 5.1 which guarantees this desired difference behavior. The parameters c_{ij} for all $i, j \in \mathcal{N}$ and $i \neq j$ are listed in Table B.4. The resulting steps of the algorithm are as follows:

1. The solutions of the optimization problem in (2.11) w.r.t. the water supply system (plant P) defined in Section 3.5.1 using the weighting matrices $\boldsymbol{Q} = \mathrm{diag}(\boldsymbol{Q}_1 \dots \boldsymbol{Q}_N)$ and $\boldsymbol{R} = \mathrm{diag}(\boldsymbol{R}_1 \dots \boldsymbol{R}_N)$ leads to the central feedback gain matrix $\boldsymbol{K}_{\mathrm{C}}$ presented in Table B.5 and Table B.6 in the appendix, where the matrices $\boldsymbol{Q}_i$, $(\forall i \in \mathcal{N})$ and $\boldsymbol{R}_i$ are defined in (3.29). The resulting system matrix $\boldsymbol{A}_{\mathrm{C}} = \boldsymbol{A}_{\mathrm{P}} - \boldsymbol{B}\boldsymbol{K}_{\mathrm{C}}$ is a Hurwitz matrix since $\max_{k \in \{1,\dots,46\}} \mathrm{Re}(\lambda_k(\boldsymbol{A}_{\mathrm{C}})) = -0.124$.

2. The feedback units F_i, $(\forall i \in \mathcal{N})$, are constructed by the matrices $\boldsymbol{K}_{\mathrm{C}ii}$ from $\boldsymbol{K}_{\mathrm{C}}$ given in Table B.5 and Table B.6 in the appendix.

3. The observation units O_i, $(\forall i \in \mathcal{N})$, are constructed by

   - defining the sets $\bar{\mathcal{P}}_{\mathrm{C}i}$, $(\forall i \in \mathcal{N})$, of maximal communicational predecessors and sets $\bar{\mathcal{F}}_{\mathrm{C}i}$ of maximal communicational followers including the indices of all other subsystems, i.e.,

$$\bar{\mathcal{P}}_{\mathrm{C}i} = \bar{\mathcal{F}}_{\mathrm{C}i} = \mathcal{N} \backslash \{i\}, \quad \forall i \in \mathcal{N} \tag{5.66}$$

since the central feedback gain matrix $\boldsymbol{K}_{\mathrm{C}}$ is fully occupied and

- including the matrices $\boldsymbol{K}_{\mathrm{C}ji}$, $(\forall j \in \bar{\mathcal{F}}_{\mathrm{C}i})$, from $\boldsymbol{K}_{\mathrm{C}}$ given in Table B.5 and Table B.6 in the appendix.

4. The decision units D_i, $(\forall i \in \mathcal{N})$, are constructed by
   - defining $\bar{\mathcal{P}}_{\mathrm{C}i}$, $(\forall i \in \mathcal{N})$, and $\bar{\mathcal{F}}_{\mathrm{C}i}$ according to (5.66) and
   - choosing the switching thresholds ϵ_{ji}, $(\forall j \in \bar{\mathcal{F}}_{\mathrm{C}i}, \forall i \in \mathcal{N})$, according to (5.63) in Theorem 5.5 (cf. Tables B.7 - B.14 in the appendix).

With this, the local control units C_i, $(\forall i \in \mathcal{N})$, defined in (5.16) for controlling the subsystems of the water supply system are completely defined. In the following the water supply system controlled by these local control units C_i is denoted as self-organizing controlled tanks Σ.

5.5.2 Analysis of the behavior of the self-organizing controlled tanks

In the following the practical stability, the asymptotic stability, the disturbance propagation and the communication reduction of the self-organizing controlled tanks Σ is analyzed.

Practical stability. Since $\boldsymbol{A}_{\mathrm{C}}$ is a Hurwitz matrix, Theorem 5.2 states that the self organizing controlled tanks Σ are practically stable w.r.t. the compact set

$$\mathcal{A} = \{\boldsymbol{x} \in \mathbb{R}^{46} \mid ||\boldsymbol{x}|| \leq \bar{x} = 14.23\,\mathrm{cm}\}. \tag{5.67}$$

Hence, for $\boldsymbol{x}_0 = \boldsymbol{0}$ the overall states $\boldsymbol{x}(t)$ of Σ should not leave $\mathcal{A}$.

Asymptotic stability. According to Theorem 5.4, two conditions have to be fulfilled such that the undisturbed self-organizing controlled tanks Σ are asymptotically stable. The first condition is fulfilled since the system matrix $\bar{\boldsymbol{A}}_{\mathrm{P}}$ of the decentralized controlled tanks Σ_{D} defined in (5.51) is a Hurwitz matrix, where $\max_{k\in\{1,\dots,46\}} \mathrm{Re}(\lambda_k(\bar{\boldsymbol{A}}_{\mathrm{P}})) = -0.1241$. The fulfillment of the second condition is listed in Tables B.7 - B.14 in the appendix while using condition (5.55) in Theorem 5.4. A part of this analysis for subsystem Σ_9 is presented in Table 5.3 which shows that all conditions in (5.55) for Σ_9 hold true meaning that all communication links going out of C_9 will be deactivated after a certain time.

Consider the self-organizing controlled tanks Σ as an uncertain system defined in (5.56), then according to Proposition 5.1 two conditions have to be fulfilled to guarantee asymptotic stability of the undisturbed self-organizing controlled tanks Σ. The first condition is fulfilled since $\bar{\boldsymbol{A}}_{\mathrm{P}}$ is a Hurwitz matrix. But the second condition

$$\left\|(s\boldsymbol{I} - \bar{\boldsymbol{A}}_{\mathrm{P}})^{-1}\boldsymbol{B}^*\right\|_{H_\infty} = 8.34 \not< 1$$

Table 5.3: A part of the analysis of the asymptotic stability of the self-organizing controlled tanks Σ with the verification of communication of C_9 by Theorem 5.4.

j	i	$\boldsymbol{\epsilon}_{ji}^{\mathrm{T}}$	$(\boldsymbol{M}_{\mathrm{uu}ji}\boldsymbol{\epsilon})^{\mathrm{T}}$	$\boldsymbol{\epsilon}_{ji} > \boldsymbol{M}_{\mathrm{uu}ji}\boldsymbol{\epsilon}$
1	9	(0.000049 0.000835)	(0.000000 0.000003)	True
2	9	(0.000029 0.000758)	(0.000001 0.000040)	True
3	9	(0.000057 0.002501)	(0.000001 0.001210)	True
4	9	(0.000053 0.000918)	(0.000000 0.000003)	True
5	9	(0.000028 0.000710)	(0.000001 0.000040)	True
6	9	(0.000057 0.002501)	(0.000001 0.001210)	True
7	9	(0.000758 0.035198)	(0.000035 0.005249)	True
8	9	(0.000758 0.035198)	(0.000035 0.005249)	True
10	9	(0.000203 0.028367)	(0.000012 0.017536)	True
11	9	(0.000205 0.028327)	(0.000012 0.017536)	True
12	9	(0.000043 0.001957)	(0.000001 0.001144)	True
13	9	(0.000046 0.001003)	(0.000001 0.000032)	True
14	9	(0.000011 0.000504)	(0.000000 0.000005)	True
15	9	(0.000046 0.001001)	(0.000000 0.000000)	True
16	9	(0.000078 0.003511)	(0.000001 0.001150)	True
17	9	(0.000010 0.000434)	(0.000000 0.000005)	True
18	9	(0.000024 0.000633)	(0.000000 0.000000)	True
19	9	(0.000027 0.000556)	(0.000000 0.000000)	True
20	9	(0.000050 0.000941)	(0.000000 0.000000)	True
21	9	(0.000024 0.000611)	(0.000000 0.000000)	True
22	9	(0.000025 0.000517)	(0.000000 0.000000)	True
23	9	(0.000047 0.000886)	(0.000000 0.000000)	True

is not fulfilled. This result states that the self-organizing controlled tanks Σ might not be asymptotically stable. However, due to Theorem 5.4, Σ is known to be asymptotically stable.

This shows that condition (5.55) in Theorem 5.4 yields a more precise analysis of the closed-loop system since in contrast to the second condition in Proposition 5.1 the effect of every communication link is analyzed individually.

Disturbance propagation. Consider that subsystem Σ_{17} is disturbed, i.e., $\mathcal{D} = \{17\}$, then the propagation of $\boldsymbol{d}_{17}(t)$ is indicated by

$$\bar{\boldsymbol{v}}_{\mathrm{D}17}(\mathcal{D}) = 7.4\,\mathrm{cm} \tag{5.68a}$$

$$\bar{\boldsymbol{v}}_{\mathrm{D}k}(\mathcal{D}) = 1.6\,\mathrm{cm}, \quad \forall k \in \{18, 21\} \tag{5.68b}$$

$$\bar{\boldsymbol{v}}_{\mathrm{D}11}(\mathcal{D}) = 1.2\,\mathrm{cm} \tag{5.68c}$$

$$\bar{v}_{\mathrm{D}j}(\mathcal{D}) = 1.1\,\mathrm{cm}, \quad \forall j \in \{12, 19, 22\} \tag{5.68d}$$

$$\bar{v}_{\mathrm{D}i}(\mathcal{D}) = 1.0\,\mathrm{cm}, \quad \forall i \in \{1 \ldots 10, 13 \ldots 16, 20, 23\} \tag{5.68e}$$

resulting from the analysis in Proposition 5.2. This result shows that there is only a small propagation of the disturbance within the overall system. Mainly the neighboring subsystems of Σ_{17} are influenced by the disturbance $\boldsymbol{d}_{17}(t)$. For the other subsystems the bound $\bar{v}_{\mathrm{D}i}(\mathcal{D})$ on the performance output $\boldsymbol{v}_i(t)$ is similar to the desired maximal difference performance output $\bar{\boldsymbol{v}}^*_{\Delta i} = 1\,\mathrm{cm}$ claimed in (5.65) since $\bar{v}_{\mathrm{D}i}(\mathcal{D})$ and $\bar{\boldsymbol{v}}^*_{\Delta i}$ are identical if no subsystem is disturbed, i.e., $\mathcal{D} = \emptyset$.

The maximal communication graph $\bar{\mathcal{G}}_{\mathrm{CD}}(\mathcal{D} = \{17\})$ resulting from the analysis of the communication activation with Proposition 5.3 for $\mathcal{D} = \{17\}$ is depicted in Fig. 5.17 (a). A part of that analysis is shown in Table 5.4. The analysis shows that in the worst case C_{17} transmits information to C_{11}, C_{12}, C_{14}, C_{16}, C_{18}, C_{19}, C_{21} and C_{22}. Most of the other local control units do not send any information.

Table 5.4: Analysis of the communication activation of the self-organizing controlled tanks for $\mathcal{D} = \{17\}$ by Proposition 5.3.

i	$\bar{\mathcal{F}}_{\mathrm{C}i}$	$\bar{\mathcal{F}}_{\mathrm{CD}i}$	h	$\boldsymbol{\epsilon}_{hi}^{\mathrm{T}}$	$\bar{\boldsymbol{u}}_{\mathrm{D}hi}^{\mathrm{T}}(\mathcal{D})$	$\boldsymbol{\epsilon}_{hi} \leq \bar{\boldsymbol{u}}_{\mathrm{D}hi}(\mathcal{D})$
17	$\mathcal{N}\backslash\{17\}$	$\{11, 12, 14, 16, 18, 19, 21, 22\}$	1	(0.000049 0.000835)	(0.000000 0.000000)	False
			2	(0.000029 0.000758)	(0.000000 0.000000)	False
			3	(0.000011 0.000500)	(0.000000 0.000003)	False
			4	(0.000053 0.000918)	(0.000000 0.000000)	False
			5	(0.000028 0.000710)	(0.000000 0.000000)	False
			6	(0.000011 0.000500)	(0.000000 0.000003)	False
			7	(0.000015 0.000704)	(0.000000 0.000013)	False
			8	(0.000015 0.000704)	(0.000000 0.000013)	False
			9	(0.000003 0.000355)	(0.000002 0.000083)	False
			10	(0.000006 0.000860)	(0.000002 0.000418)	False
			11	(0.000037 0.005150)	(0.000015 0.018594)	True
			12	(0.000087 0.003913)	(0.000021 0.016526)	True
			13	(0.000046 0.001003)	(0.000015 0.000776)	False
			14	(0.000548 0.025184)	(0.000428 0.158668)	True
			15	(0.000461 0.010015)	(0.000003 0.000163)	False
			16	(0.000775 0.035109)	(0.000428 0.158171)	True
			18	(0.001219 0.031652)	(0.001043 0.073877)	True
			19	(0.000254 0.005173)	(0.000210 0.011395)	True
			20	(0.000050 0.000941)	(0.000016 0.000354)	False
			21	(0.001177 0.030568)	(0.001043 0.073877)	True
			22	(0.000254 0.005173)	(0.000210 0.011395)	True
			23	(0.000047 0.000886)	(0.000016 0.000354)	False

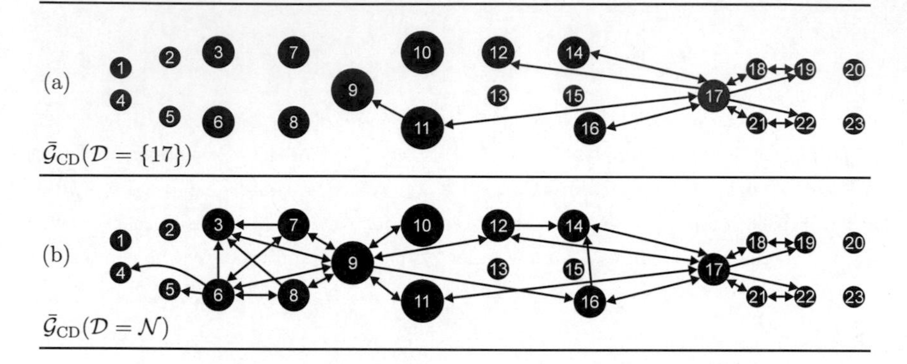

Figure 5.17: Maximal communication graph $\bar{\mathcal{G}}_{\mathrm{CD}}(\mathcal{D})$ for $\mathcal{D} = \{17\}$ and $\mathcal{D} = \mathcal{N}$ of the self-organizing controlled tanks Σ.

Reduction of the communication structure. If all subsystems are considered to be disturbed $\mathcal{D}=\mathcal{N}$ the determination of $\bar{\mathcal{G}}_{\mathrm{CD}}(\mathcal{D}=\mathcal{N})$ with Proposition 5.3 leads to the graph in Fig. 5.17 (b). According to Theorem 5.6, the self-organizing controlled tanks Σ with this reduced maximal communication graph $\bar{\mathcal{G}}_{\mathrm{C}}^{*} = \bar{\mathcal{G}}_{\mathrm{CD}}(\mathcal{D}=\mathcal{N})$ has the same performance as the closed-loop system with the original maximal communication graph $\bar{\mathcal{G}}_{\mathrm{C}}$ for $\boldsymbol{x}_0 = \mathbf{0}$. Figure 5.17 (b) shows that only the depicted communication links $\bar{\mathcal{G}}_{\mathrm{CD}}(\mathcal{D}=\mathcal{N})$ are necessary to guarantee the required control aim in (5.65) for $\boldsymbol{x}_0 = \mathbf{0}$. The graph $\bar{\mathcal{G}}_{\mathrm{CD}}(\mathcal{D}=\mathcal{N})$ has 43 communication links which is an enormous reduction compared to the $23 \cdot 22 = 506$ communication links in the original maximal communication graph $\bar{\mathcal{G}}_{\mathrm{C}}$. Hence, only a few of the communication links of centralized controller C_{C} are necessary to mimic the behavior of the centralized control system Σ_{C} with a permanent communication among all local control units. Note that in the following section the original maximal communication graph $\bar{\mathcal{G}}_{\mathrm{C}}$ and not the reduced one is used.

5.5.3 Behavior of the self-organizing controlled tanks

This section verifies the analysis results in the previous section by evaluating a simulation of the self-organizing controlled tanks. At first, the general behavior of the closed-loop system is presented and discussed.

Behavior of the self-organizing controlled tanks. In Fig. 5.18 the behavior of the water supply system for $\boldsymbol{x}_{0i} = \mathbf{0}$, $(\forall i \in \mathcal{N})$, using the designed local control units C_i in Section 5.5.1 is depicted. The topmost plot shows the disturbances $\boldsymbol{d}_i(t) = d_{\mathrm{TB}i}(t)$ affecting the subsystems Σ_i, where the disturbances affecting Σ_6, Σ_9 and Σ_{17} are colored. The small disturbances affecting

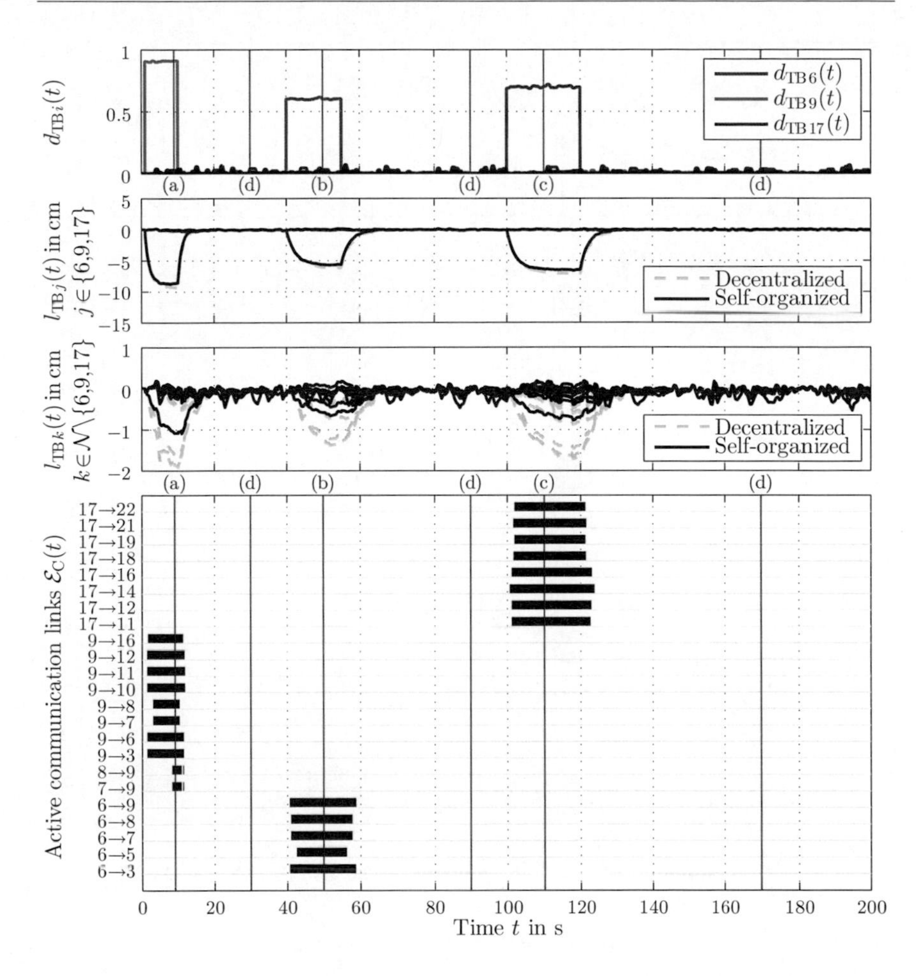

Figure 5.18: Behavior of the water supply system controlled by the self-organizing controller C defined in (5.17) with parameters given in Section 5.5.1.

the other subsystems are black. The second and third plot display the level $l_{\mathrm{TB}i}(t)$ of tank TB while using the self-organizing controller C (solid line) and the decentralized controller C_{D} (dashed line). The transmission of information among the local control units is shown in the lower plot. The black bars indicate at which time the communication links $(i \rightarrow j)$ are in the set $\mathcal{E}_{\mathrm{C}}(t)$ of communicational edges which means that C_i transmits information to C_j. Since there are 506 possible communication links, the lower plot only shows the links that are

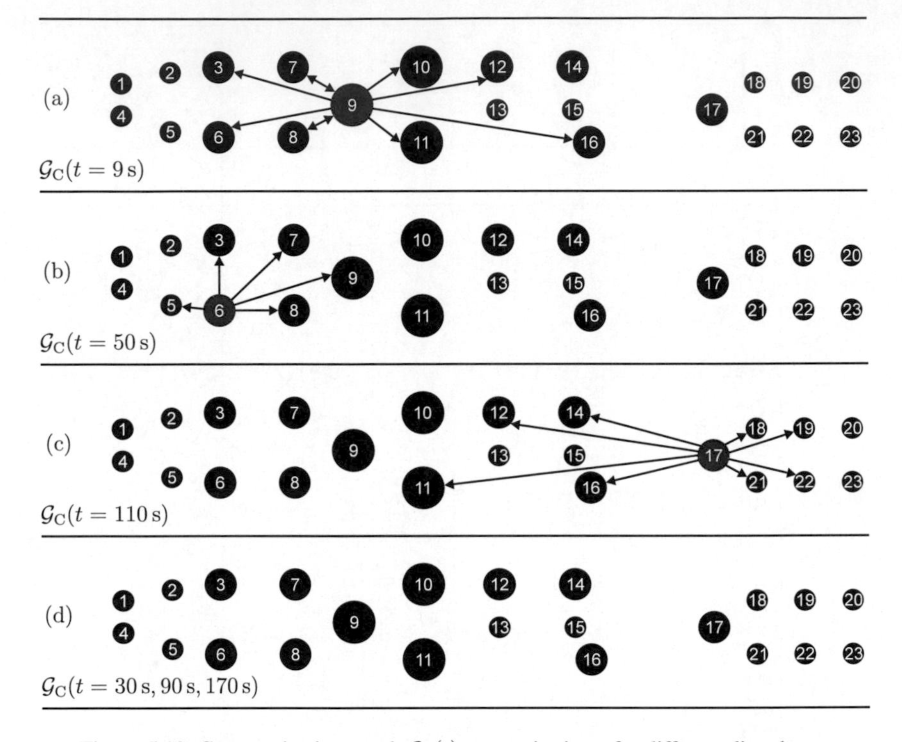

Figure 5.19: Communication graph $\mathcal{G}_C(t)$ at certain times for different disturbance situations in Fig. 5.18 (disturbed subsystems are red).

activated during runtime. An overview about all communication links is given in Fig. B.3 in the appendix. This shows that only a few of the possible communication links are used. The thin red vertical lines refer to the corresponding communication graphs $\mathcal{G}_C(t)$ for different times shown in Fig. 5.19.

Figure 5.18 shows that the self-organizing controller C leads to a better performance compared to the usage of a decentralized controller C_D. In particular, the deviation of the maximal level $l_{TBk}(t)$, $(\forall k \in \mathcal{N}\backslash\{6, 9, 17\})$, of the subsystems with a small perturbation is much smaller while using C (third plot). The performance of the disturbed subsystems Σ_6, Σ_9 and Σ_{17} is similar for both control concepts since the effect of a disturbance at Σ_i can not be reduced by sending information to C_i or sending information from C_i to other local control units.

At most times there is no communication among the local control units C_i (lower plot of Fig. 5.18 and Fig. 5.19 (d)). The communication is only activated if a subsystem is disturbed significantly. For example, if subsystem Σ_6 is disturbed, the corresponding local control unit

C_6 sends information to C_3, C_5, C_7, C_8 and C_9 (Fig. 5.19 (b)). The communication structures in Fig. 5.19 show that it is sufficient that the local control units of the disturbed subsystems send information. Only if Σ_9 is disturbed, C_7 and C_8 send also information to C_9.

In summary, the self-organizing controlled tanks show characteristic properties of self-organization since the communication graph $\mathcal{G}_C(t)$ adapts to the currently acting disturbances to guarantee a similar performance while using the centralized controller C_C.

Comparison to decentralized control and centralized control. Figure 5.20 depicts the physical coupling graph $\mathcal{G}_P$ and the maximal values $\max_{t\geq 0} |l_{TBi}(t)|$ of the tanks levels in TBi for all $t \geq 0$ while using the self-organizing controller C (black bar), the decentralized controller C_D (gray bar) and the centralized controller C_C (blue bar) without giving the actual amount of the values. This depiction indicates the benefit of the self-organizing control concept. The performance with the self-organizing controller C is similar to the performance with the centralized controller C_C. The performance is improved by the self-organizing controller C compared to the decentralized controller C_D, where the communication is only activated if a subsystem is disturbed (cf. Fig. 5.19). Hence, the self-organizing controller C combines the benefit of the decentralized controller – a low communication effort – and the benefit of the centralized controller – a good disturbance attenuation.

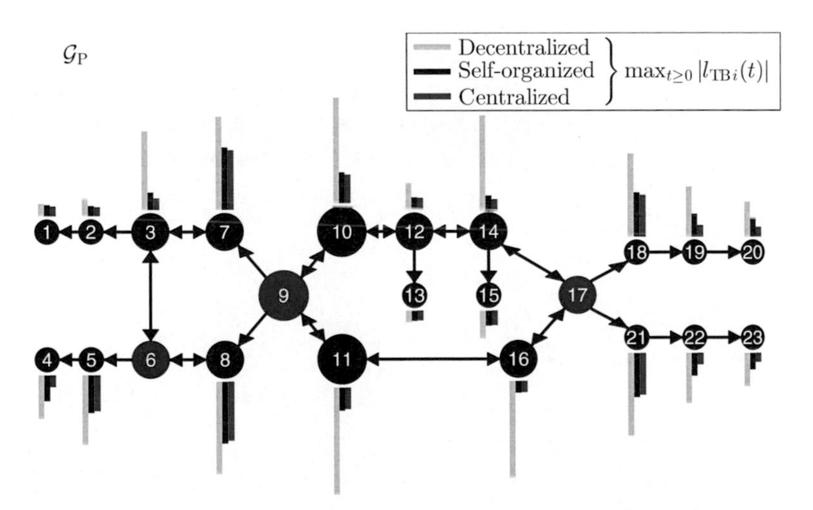

Figure 5.20: Physical coupling graph $\mathcal{G}_P$ with the maximal values of the tank levels $\max_{t\geq 100} |l_{TBi}|$, $(i = 1, \ldots, 23)$, for decentralized control, self-organized control and centralized control w.r.t the behavior depicted in Fig. 5.18.

Control aim: Desired difference behavior. The control aim in (5.65) is guaranteed for all levels $l_{\mathrm{TB}i}(t)$, $(\forall i \in \mathcal{N})$, (cf. Fig. 5.21). The difference $\boldsymbol{v}_{\Delta i}(t) = l_{\mathrm{TB}i}(t) - l_{\mathrm{TBC}i}(t)$ between the level $l_{\mathrm{TB}i}(t) = \boldsymbol{v}_i(t)$ while using the self-organizing controller C and the level $l_{\mathrm{TBC}i}(t) = \boldsymbol{v}_{\mathrm{C}i}(t)$ while using the centralized controller C_{C} is smaller than the desired maximal difference performance output of $\bar{\boldsymbol{v}}^*_{\Delta i} = 1\,\mathrm{cm}$ for all $t \geq 0$.

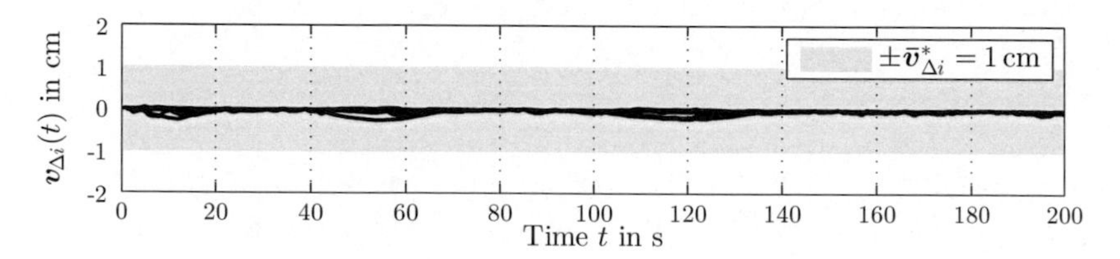

Figure 5.21: Verification of the control aim in (5.65); Difference behavior of the self-organizing controlled tanks Σ.

Verification of the analysis results. In the following the analysis results for the behavior of the self-organizing controlled tanks Σ in Section 5.5.2 is compared with the actual behavior of Σ:

- **Practical stability**: Since the initial states are zero, the overall state $\boldsymbol{x}(t)$ should stay in the set $\mathcal{A}$ defined in (5.67). Figure 5.22 shows the verification of the practical stability. The gray area is the compact set $\mathcal{A}$ defined in (5.67). The solid lines are the phase portrait of $x_{i,1}(t)$, $(\forall i \in \mathcal{N})$, and $x_{i,2}(t)$ for each subsystem. The subsystem states do not leave the set $\mathcal{A}$ for all $t \geq 0$. Hence, the analysis results are confirmed.

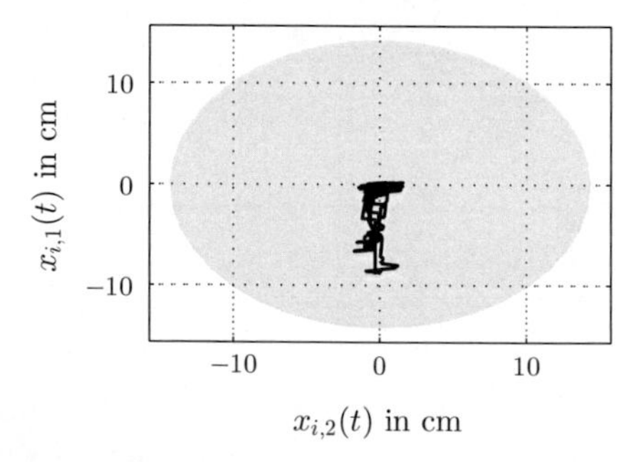

Figure 5.22: Verification of the practical stability of the self-organizing controlled tanks Σ according to Theorem 5.2.

- **Asymptotic stability**: For nearly vanishing disturbances the levels $l_{\mathrm{TB}i}$ converge to zero (cf. Fig. 5.18). The same holds for the other states of the subsystems which are not plotted in Fig. 5.18. Hence, Σ is asymptotically stable for $\boldsymbol{d}_i(t) = \mathbf{0}$, $(\forall i \in \mathcal{N})$. This approves the result in the Tables B.7 - B.14.

- **Disturbance propagation**: For the evaluation of the disturbance propagation of $\boldsymbol{d}_{17}(t)$ the behavior of the tanks after 100 s is investigated in more detail since in that time interval subsystem Σ_{17} is disturbed only. The analysis result in (5.68) is confirmed, e.g.,

$$\max_{t \geq 100} |t_{\mathrm{TB}17}(t)| = 6.6\,\mathrm{cm} \leq \bar{\boldsymbol{v}}_{\mathrm{D}17}(\mathcal{D}) = 7.4\,\mathrm{cm}$$

$$\max_{t \geq 100} |t_{\mathrm{TB}i}(t)| = 0.5\,\mathrm{cm} \leq \bar{\boldsymbol{v}}_{\mathrm{D}i}(\mathcal{D}) = 1\,\mathrm{cm}, \quad \forall i \in \{1 \ldots 10, 13 \ldots 16, 20, 23\}.$$

The disturbance propagation of $\boldsymbol{d}_{17}(t)$ is illustrated in Fig. 5.23 which depicts the physical coupling graph $\mathcal{G}_{\mathrm{P}}$ and the maximal values $\max_{t \geq 100} |t_{\mathrm{TB}i}(t)|$ of the tank levels for all $t \geq 100$ s from the behavior of the self-organizing controlled tanks Σ shown in Fig. 5.18. With a decentralized controller C_{D} the disturbance $\boldsymbol{d}_{17}(t)$ has a great influence on subsystems that are close to Σ_{17}. With the self-organizing controller C this influence is reduced.

The analysis result of the communication activation depicted in Fig. 5.17 (a) matches with the behavior of the self-organizing controlled tanks Σ. There only is communication from C_{17} to C_{11}, C_{12}, C_{14}, C_{16}, C_{18}, C_{19}, C_{21} and C_{22} if Σ_{17} is disturbed (cf. Fig. 5.19 (c)).

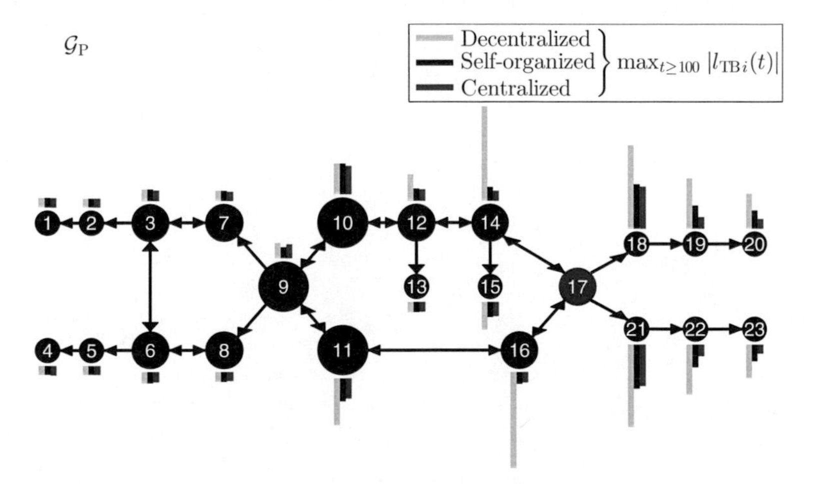

Figure 5.23: Physical coupling graph $\mathcal{G}_{\mathrm{P}}$ with the maximal values of the tank levels $\max_{t \geq 100} |l_{\mathrm{TB}i}|$, $(i = 1, \ldots, 23)$, for $t \geq 100$ s.

- **Reduction of the communication structure**: Figure 5.19 approves the analysis results illustrated in Fig. 5.17 (b) since only the determined communication links within $\bar{\mathcal{G}}_{\mathrm{CD}}(\mathcal{D}=\mathcal{N})$ are used by the local control units. Hence, the other communication links in $\bar{\mathcal{G}}_{\mathrm{C}}$ are not necessary and can be canceled out while guaranteeing the same performance.

5.5.4 Trade-off between performance and communication effort

Consider different self-organizing controllers C designed for certain desired maximal difference performance outputs $\bar{\boldsymbol{v}}^*_{\Delta i}$, $(\forall i \in \mathcal{N})$, equal for every subsystem that lead to different self-organizing control systems Σ. Figure 5.24 compares the performance of the different Σ indicated by $\bar{\boldsymbol{v}}^*_{\Delta i}$, $(\forall i \in \mathcal{N})$, and the communication effort indicated by the overall communication time t_Σ for the same disturbance situation as depicted in Fig. 5.18. The time t_Σ is the sum of all lengths of the communication intervals for each communication link $(i \rightarrow j) \in \mathcal{E}_{\mathrm{C}}$. Roughly speaking, t_Σ is the length of all black bars in the lower plot of Fig. 5.18 in a row. For example, for $\bar{\boldsymbol{v}}^*_{\Delta i} = 1\,\mathrm{cm}$ the overall communication time is $t_\Sigma \approx 6$ min (cf. Fig. 5.18). Figure 5.24 shows that even for small values of $\bar{\boldsymbol{v}}^*_{\Delta i}$ there is a huge reduction of the communication effort compared to a permanent communication within the centralized controller C_{C}. Hence, the desired maximal difference performance outputs $\bar{\boldsymbol{v}}^*_{\Delta i} = 1\,\mathrm{cm}$, $(\forall i \in \mathcal{N})$, claimed in (5.65) ensures a good compromise between a low communication effort and a reasonable deviation between the performance of Σ and the performance of the centralized control system Σ_{C}.

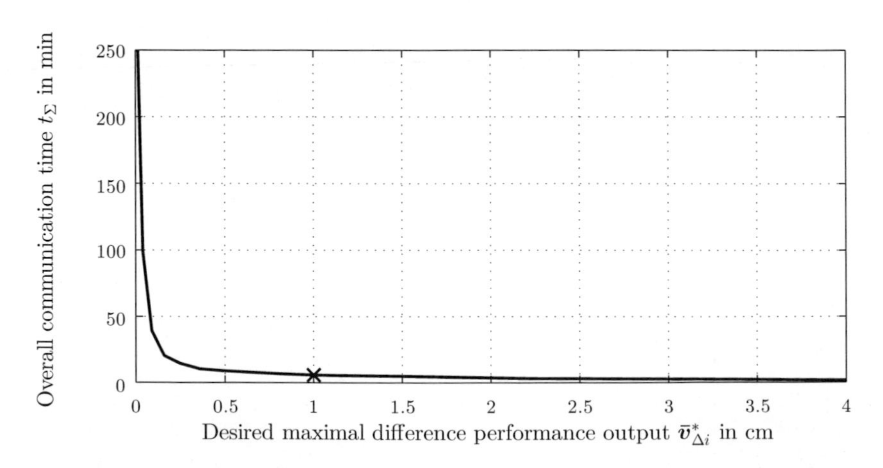

Figure 5.24: Communication time for certain self-organizing systems Σ of the form (5.19) resulting from different desired maximal difference performance outputs $\bar{\boldsymbol{v}}^*_{\Delta i}$.

5.6 Summary and evaluation of the proposed control concept

The main properties of the self-organizing control concept with situation-dependent transmission of control signals for the control of physically interconnected subsystems can be summarized as follows:

- Every centralized controller C_{C} with a static state feedback according to (5.1) can be realized as a self-organizing controller C with a situation-dependent communication graph $\mathcal{G}_{\mathrm{C}}(t)$ formalized in (5.17), i.e.,

$$C_{\mathrm{C}} : \boldsymbol{u}_{\mathrm{C}}(t) = -\boldsymbol{K}_{\mathrm{C}}\boldsymbol{x}_{\mathrm{C}}(t) \quad \Rightarrow \quad C \text{ with } \mathcal{G}_{\mathrm{C}}(t).$$

- The difference between the behavior of a given centralized control system Σ_{C} and the behavior of the resulting self-organizing control system Σ is bounded and can be made arbitrarily small by the choice of the switching thresholds (cf. Theorem 5.1 and Theorem 5.5).

- If the centralized controller C_{C} asymptotically stabilizes the undisturbed physically interconnected subsystems, then the deduced self-organizing controller C practically stabilizes the interconnected subsystems (cf. Theorem 5.2 and Theorem 5.3), i.e.,

$$\Sigma_{\mathrm{C}} \text{ is asymtotically stable.} \quad \Rightarrow \quad \Sigma \text{ is practically stable.}$$

- If the subsystems are not disturbed and the self-organizing controller C can asymptotically stabilize the overall plant P without communication among the local control units (decentralized control system Σ_{D}), then the self-organizing control system Σ is asymptotically stable for small feedback gains generated by the situation-dependent communication (cf. Theorem 5.4 and Proposition 5.1).

- Unnecessary communication links resulting from the structure of the centralized controller C_{C}, can be canceled out of the self-organizing controller C (cf. Theorem 5.6).

In summary, the proposed self-organizing control concept is applicable for various types of physically interconnected systems for which a trade-off between a good control performance and a low communication effort is desired.

Properties of self-organization. The claimed properties of a self-organizing control system in Section 1.2 for the proposed control concept are verified as follows:

- **Flexibility.** Consider an asymptotically stable self-organizing control system Σ. Then the communication within the self-organizing controller C is only activated if the subsystems are disturbed. Furthermore, it is shown that the activation of the communication links depends on the disturbance situation. If, for example, only one subsystem is disturbed, the communication links that might be activated in the worst case can be predicted by an algebraic relation (Proposition 5.3). In summary, this situation-dependent communication is sufficient, to mimic the behavior of the centralized control system Σ_{C}.

- **Scalability.** The local control unit C_i does not include any model information from subsystem Σ_i or from any other subsystems. The parameters of C_i just follow from the structure and the values of the central feedback gain matrix $\boldsymbol{K}_{\mathrm{C}}$ of C, i.e.,

$$\boldsymbol{K}_{\mathrm{C}} \text{ within } C_{\mathrm{C}} \quad \Rightarrow \quad \text{Parameters of } C_i,\ \forall i \in \mathcal{N}.$$

  Specifically, the structure of $\boldsymbol{K}_{\mathrm{C}}$ specifies the possible communication among the local control units C_i (cf. Fig. 5.6). Hence, the complexity of the local control units scales linearly with the number of subsystems. For the design of the local control units and the analysis of the behavior of the resulting closed-loop system global model information is needed. Due to this property, the application example in Section 5.5 shows that an analysis of the behavior of every controlled subsystem might be very complex, but, nevertheless, such a detailed analysis is possible for this control concept.

- **Fault tolerance.** The proposed control concept is not analyzed w.r.t. faults within the subsystems. Since the self-organizing control system Σ mimics the behavior of the centralized control system Σ_{C}, faults have a similar impact on the behavior of Σ as faults have on the behavior of Σ_{C}. If a fault leads to instability of Σ_{C}, than generally also Σ would be unstable. A solution for both control concepts could be a redesign of the central feedback gain matrix $\boldsymbol{K}_{\mathrm{C}}$.

In summary, the proposed self-organizing control concept shows the first two desired self-organizing properties. The property of fault tolerance is similar to a classical central feedback approach. This is not surprising since the self-organizing controller tries to mimic the behavior of such a controller.

6 Comparison of the control concepts for physically interconnected systems

In this section the main properties of the two self-organizing control concepts for physically interconnected systems presented in Chapter 4 and Chapter 5 are compared. Note that the comparison refers to the state feedback approaches of both concepts. Therefore, Table 6.1 presents an overview about the most important aspects.

Main aim and basic idea. The concepts have different aims. The concept with the situation-dependent compensation of physical coupling inputs aims to bound the disturbance propagation within the closed-loop system (cf. Chapter 4). In the following this concept is denoted as *first concept*. The aim of the concept presented in Chapter 5 is to mimic the behavior of a centralized control system Σ_{C}. This concept is denoted as *second concept*. The basic ideas of both concepts are different. The general representation of the local control input $\boldsymbol{u}_i(t)$ in Table 6.1 shows that both concepts use a permanent local state feedback and a situation-dependent part which covers different tasks. Within the first concept the local control unit C_j transmits the locally determined coupling input $\hat{\boldsymbol{s}}_{ij}(t) = \boldsymbol{s}_{ij}(t) = \boldsymbol{L}_{ij}\boldsymbol{C}_{zj}\boldsymbol{x}_j(t)$ to C_i whenever the coupling input $\boldsymbol{s}_{ij}(t)$ exceeds the switching threshold ϵ_{ij} ($\alpha_{ij}(t) = 1$) in order to compensate this impact by adapting the local control input $\boldsymbol{u}_i(t)$ using the decoupling matrix $\boldsymbol{K}_{\mathrm{D}i}$. The second concept is a distributed implementation of centralized controller C_{C} defined in (5.1) including a central feedback gain matrix $\boldsymbol{K}_{\mathrm{C}} = (\boldsymbol{K}_{\mathrm{C}ij})$. The determined control signal $\hat{\boldsymbol{u}}_{ij}(t) = \boldsymbol{K}_{\mathrm{C}ij}\boldsymbol{x}_j(t)$ is sent from C_j to C_i whenever some element of $\hat{\boldsymbol{u}}_{ij}(t)$ exceeds the switching threshold $\boldsymbol{\epsilon}_{ij}$ to mimic the control input in the centralized control case. Hence, the basis of the first concept is the physical coupling structure among the subsystems, whereas the basis of the second concept is a given centralized controller. With this, the components of the local control units have different tasks which are compared in the following.

Tasks of the local control units. The feedback units F_i have a similar structure (cf. (4.11) and (5.13)). Both include a static local state feedback and in addition, the communicated control input $\tilde{\boldsymbol{u}}_i(t)$ determined by the decision unit D_i. However, the tasks are different. For the first concept the local state feedback has to stabilize the isolated subsystems (cf. Assumption 4.1) which is not necessary for the second concept. For this concept the centralized controller C_{C} with a static feedback of the overall plant state has to stabilize the plant P (cf. Assumption 5.1).

Table 6.1: Comparison of control concepts for physically interconnected systems.

Concept	1. Disturbance attenuation by compensating physical couplings	2. Mimicry of a centralized control system Σ_{C}
Idea	Situation-dependent exchange of determined coupling inputs to compensate the physical coupling.	Distributed implementation of C_{C} with situation-dependent transmission of the control signals.
General $\boldsymbol{u}_i(t) =$	$\underbrace{-\boldsymbol{K}_{\mathrm{C}i}\boldsymbol{x}_i(t)}_{\text{permanent}} \underbrace{-\boldsymbol{K}_{\mathrm{D}i} \sum_{j \in \mathcal{N} \setminus \{i\}} \alpha_{ij}(t)\hat{\boldsymbol{s}}_{ij}(t)}_{\text{situation-dependent}}$	$\underbrace{-\boldsymbol{K}_{\mathrm{C}ii}\boldsymbol{x}_i(t)}_{\text{permanent}} \underbrace{- \sum_{j \in \mathcal{N} \setminus \{i\}} \alpha_{ij}(t)\hat{\boldsymbol{u}}_{ij}(t)}_{\text{situation-dependent}}$
Switching law	$\alpha_{ij}(t) = \begin{cases} 1 & \text{if } \Vert\hat{\boldsymbol{s}}_{ij}(t)\Vert \geq \epsilon_{ij} \\ 0 & \text{else} \end{cases}$	$\alpha_{ij}(t) = \begin{cases} 1 & \text{if } \vert\hat{\boldsymbol{u}}_{ij}(t)\vert \overset{\exists}{\geq} \boldsymbol{\epsilon}_{ij} \\ 0 & \text{else} \end{cases}$
Feedback unit F_i	Decentralized static state feedback with additionally communicated control input $\tilde{\boldsymbol{u}}_i(t)$: $\boldsymbol{u}_i(t) = -\boldsymbol{K}_{\mathrm{C}i}\boldsymbol{x}_i(t) - \tilde{\boldsymbol{u}}_i(t)$	
Observation unit O_i	Determine coupling input $\hat{\boldsymbol{s}}_{ji}(t) = \boldsymbol{L}_{ji}\boldsymbol{C}_{\mathrm{z}i}\boldsymbol{x}_i(t)$	Determine local control signal $\hat{\boldsymbol{u}}_{ji}(t) = \boldsymbol{K}_{\mathrm{C}ji}\boldsymbol{x}_i(t)$
Decision unit D_i	Send $\hat{\boldsymbol{s}}_{ji}(t)$ to C_j if $\Vert\hat{\boldsymbol{s}}_{ji}(t)\Vert \geq \epsilon_{ji}$	Send $\hat{\boldsymbol{u}}_{ji}(t)$ to C_j if $\vert\hat{\boldsymbol{u}}_{ji}(t)\vert \overset{\exists}{\geq} \boldsymbol{\epsilon}_{ji}$
	Calculate communicated control input $\tilde{\boldsymbol{u}}_i(t)$: $\tilde{\boldsymbol{u}}_i(t) = \boldsymbol{K}_{\mathrm{D}i} \sum_{j \in \mathcal{P}_{\mathrm{C}i}(t)} \hat{\boldsymbol{s}}_{ij}(t)$	$\tilde{\boldsymbol{u}}_i(t) = \sum_{j \in \mathcal{P}_{\mathrm{C}i}(t)} \hat{\boldsymbol{u}}_{ij}(t)$
Communication structure	$\boldsymbol{L} \Rightarrow \bar{\mathcal{G}}_{\mathrm{C}} = (\mathcal{V}_{\mathrm{C}}, \bar{\mathcal{E}}_{\mathrm{C}})$	$\boldsymbol{K}_{\mathrm{C}} \Rightarrow \bar{\mathcal{G}}_{\mathrm{C}} = (\mathcal{V}_{\mathrm{C}}, \bar{\mathcal{E}}_{\mathrm{C}})$
Main requirements	-Subsystems Σ_i are stabilizable -Σ_i can be physically decoupled	-Plant P is stabilizable
Asymptotic stability	Sufficient condition (Theorem 4.3)	Sufficient condition (Theorem 5.4)
Practical stability	All $\bar{\Sigma}_i$ are stable $\Rightarrow \Sigma$ is stable (Theorem 4.1 and Theorem 4.2)	Σ_{C} is stable $\Rightarrow \Sigma$ is stable (Theorem 5.2 and Theorem 5.3)
Difference behavior to Σ_{C}	Bounded and arbitrarily adjustable by thresholds (Theorem 4.4)	Bounded and arbitrary adjustable by thresholds (Theorem 5.1)
Disturbance behavior	Disturbance propagation arbitrarily adjustable (Theorem 4.5)	Disturbance propagation is similar as in Σ_{C} (Proposition 5.2)
Analysis and design with ...	**Local** model information	**Global** model information
Properties of self-organization	-Flexibility ✓ -Scalability✓ -Fault-tolerance✓	-Flexibility ✓ -Scalability✓ -Fault-tolerance

For the first concept, the observation unit O_i determines the coupling inputs $\boldsymbol{s}_{ji} = \boldsymbol{L}_{ji}\boldsymbol{C}_{zi}\boldsymbol{x}_i(t)$ to the subsystems Σ_j on which its corresponding subsystem Σ_i has a physical impact. For the second concept, O_i determines the control signals $\hat{\boldsymbol{u}}_{ji}(t) = \boldsymbol{K}_{\mathrm{C}ji}\boldsymbol{x}_i(t)$ using the matrices $\boldsymbol{K}_{\mathrm{C}ji}$ which build the central feedback-gain matrix $\boldsymbol{K}_{\mathrm{C}}$. Hence, O_i either detects the current physical coupling situation of the subsystem (first concept) or the current control situation compared to a centralized control system Σ_{C} (second concept).

The first task ot the decision unit D_i is similar for both concepts. D_i decides whether the coupling input $\boldsymbol{s}_{ji}(t)$ or the control input $\hat{\boldsymbol{u}}_{ji}(t)$ has such a significant impact that it has to be known by the corresponding local control unit C_j. In particular, $\boldsymbol{s}_{ji}(t)$ is sent if it exceeds the switching threshold ϵ_{ji} and $\hat{\boldsymbol{u}}_{ji}(t)$ is sent if it exceeds the switching threshold $\boldsymbol{\epsilon}_{ji}$. The usage of the received information to determine the communicated control input $\tilde{\boldsymbol{u}}_i(t)$ is different (second task). For the first concept the received coupling input $\hat{\boldsymbol{s}}_{ij}(t)$ is multiplied with the decoupling matrix $\boldsymbol{K}_{\mathrm{D}i}$ such that it compensates the actual coupling input $\boldsymbol{s}_{ij}(t)$. The received control signal $\hat{\boldsymbol{u}}_{ij}$ are just added up to mimic the overall local control input $\boldsymbol{u}_{\mathrm{C}i}(t)$ in the centralized case (second concept).

Communication structure. Recall that the set $\mathcal{E}_{\mathrm{C}}(t)$ of communication links within the communication graph $\mathcal{G}_{\mathrm{C}}(t)$ has an upper limit and a lower limit for all presented control concepts (cf. (2.25) in Section 2.3). The lower limit, the set $\underline{\mathcal{E}}_{\mathrm{C}}$ of minimal communication links within the minimal communication graph $\underline{\mathcal{G}}_{\mathrm{C}} = (\mathcal{V}_{\mathrm{C}}, \underline{\mathcal{E}}_{\mathrm{C}})$, is identical for both concepts, namely $\underline{\mathcal{E}}_{\mathrm{C}} = \emptyset$, which implies a decentralized control structure (cf. (3.19) and (3.24)). For the first concept, the set $\bar{\mathcal{E}}_{\mathrm{C}}$ of maximal communication links results from the physical coupling structure among the subsystems (cf. (4.10)), where for the second concept $\bar{\mathcal{E}}_{\mathrm{C}}$ follows from the structure of the central feedback-gain matrix $\boldsymbol{K}_{\mathrm{C}}$ (cf. (5.12)), i.e.,

$$\underline{\mathcal{E}}_{\mathrm{C}} = \emptyset \subseteq \mathcal{E}_{\mathrm{C}}(t) \subseteq \bar{\mathcal{E}}_{\mathrm{C}} = \{(j \to i) \mid \|\boldsymbol{L}_{ij}\| \neq 0,\ i,j \in \mathcal{V}_{\mathrm{C}},\ i \neq j\} \qquad \text{(first concept)}$$
$$\underline{\mathcal{E}}_{\mathrm{C}} = \emptyset \subseteq \mathcal{E}_{\mathrm{C}}(t) \subseteq \bar{\mathcal{E}}_{\mathrm{C}} = \{(j \to i) \mid \|\boldsymbol{K}_{\mathrm{C}ij}\| \neq 0,\ i,j \in \mathcal{V}_{\mathrm{C}},\ i \neq j\} \qquad \text{(second concept)}$$

hold true for all $t \geq 0$. Recall the physical coupling among the subsystems is represented by the physical interconnection matrix $\boldsymbol{L} = (\boldsymbol{L}_{ij})$ which is build of the matrices $\boldsymbol{L}_{ij}$ (cf. (3.3)).

Main requirements. The self-organizing control concepts are applicable for interconnected subsystem with different properties. The first concept can be used for subsystems Σ_i which are stabilizable (cf. Assumption 3.2) which is a usual requirement and, furthermore, there has to be an opportunity to completely compensate the physical coupling input $\boldsymbol{s}_i(t)$ into the subsystems Σ_i (cf. Assumption 4.2) which is an unusual requirement and more restrictive than the first one. To apply the second control concept the overall plant P has to be stabilizable (cf.

Assumption 3.5). In contrast to the first control concept where the requirements correspond to properties of the subsystems Σ_i, the requirement for the second control concept corresponds to the overall system to be controlled. In summary, the requirements for the first control concept are more restrictive than for the second one, but they can be checked individually for every subsystem Σ_i without building the overall model of the plant P.

Comparison of the main results. The following properties of the self-organizing control system Σ resulting from the two control concepts are compared:

- **Asymptotic stability (Control aim 3.1)**: Both concepts use the logical relation

$$(\Sigma_{\mathrm{D}} \text{ asympt. stable}) \wedge (\Sigma \text{ with } \mathcal{E}_{\mathrm{C}}(t) = \emptyset, \forall t \geq t_{\mathrm{D}}) \Rightarrow (\Sigma \text{ asympt. stable})$$

  that if the decentralized control system Σ_{D} is asymptotically stable and the communication among the local control units will be deactivated permanently after a finite time t_{D} to derive a sufficient condition for asymptotic stability of the corresponding self-organizing control system Σ (cf. Theorem 4.3 for the first concept and Theorem 5.4 for the second concept). The conditions for both concepts mainly check if the feedback gain of each communication link individually is sufficiently small which is a kind of small-gain condition.

- **Practical stability (Control aim 3.2)**: Using the first control concept, the decentralized controlled subsystems $\bar{\Sigma}_i$ have to be asymptotically stable to practically stabilize the corresponding self-organizing control system Σ (cf. Theorem 4.1 and Theorem 4.2). In contrast, with the second concept the corresponding undisturbed centralized control system Σ_{C} has to be asymptotically stable to obtain practical stability of Σ (cf. Theorem 5.2 and Theorem 5.3). Hence, for the first concept stability of Σ can be achieved by stabilizing the subsystems Σ_i from a local perspective, whereas for the second concept Σ is stabilized by designing an overall centralized controller C_{C} based on a global perspective on the overall plant P.

- **Disturbance attenuation (Control aim 3.3)**: With the first control concept the disturbance propagation within the closed-loop system Σ can be arbitrarily adjusted by the switching thresholds (cf. Theorem 4.5 and Theorem 4.6). By using the second concept the disturbance propagation can not be bounded (cf. Proposition 5.2 and Theorem 5.5). In fact the disturbance behavior is similar as in the corresponding centralized control system Σ_{C}.

- **Difference behavior (Control aim 3.4)**: The application of both control concepts leads to an upper bounded difference behavior between Σ and its corresponding centralized control system Σ_{C} which depends on the switching thresholds. Furthermore, the difference between Σ and Σ_{C} can be made arbitrarily small by a reduction of the switching thresholds (cf. Theorem 4.4 for the first concept and Theorem 5.1 for the second concept). A method for choosing the thresholds to guarantee a desired difference behavior for the second concepts is presented in Theorem 5.5. The main difference between both concept is that for the first concept the desired difference behavior can be adjusted for a certain subsystem Σ_i individually by designing the corresponding local control unit C_i with local model information about Σ_i only. For the second concept the local control units C_i are designed from global perspective by using model information about the overall plant P.

Properties of self-organization.

- **Flexibility**: Both concepts are adapting the communication among the local control units to the current disturbance situation to guarantee a desired performance. Similar methods for analyzing which communication links are activated for specific disturbance situations are derived (cf. Proposition 4.2 for first concept and Proposition 5.3 for second concept). The application of this analysis method to the water supply system example has shown that a disturbance affecting a single subsystem leads to a communication among neighboring local control units only.

- **Scalability**: The first concept is applicable for systems with a large number of subsystem (complex systems) since the analysis and design methods can be performed in a distributed way while using local model information of the subsystems only. Therefore, also new subsystems can be easily integrated into a present closed-loop system Σ without an overall redesign of the local control units (cf. Section 4.7). These properties are not given for using the second control concept. Nevertheless, the structure of the local control units is quite simple since they do not need any model information about the other subsystems.

- **Fault-tolerance**: With the first control concept the propagation of a fault is bounded which is some kind of passive fault-tolerance of the self-organizing control system Σ (cf. Section 4.8). Faults in a single subsystem do not affect the desired performance of all other controlled subsystem. The second concept has similar passive fault-tolerance capabilities as the corresponding centralized control system Σ_{C} which might be bad.

In summary, both concepts show characteristic properties of self-organizing systems where for the first concept these properties are more pronounced than for the second concept. Hence, if the actual subsystem can be physical decoupled according to (4.3), then the first concept would be the best choice for application. Nevertheless, the second concept is a good alternative if

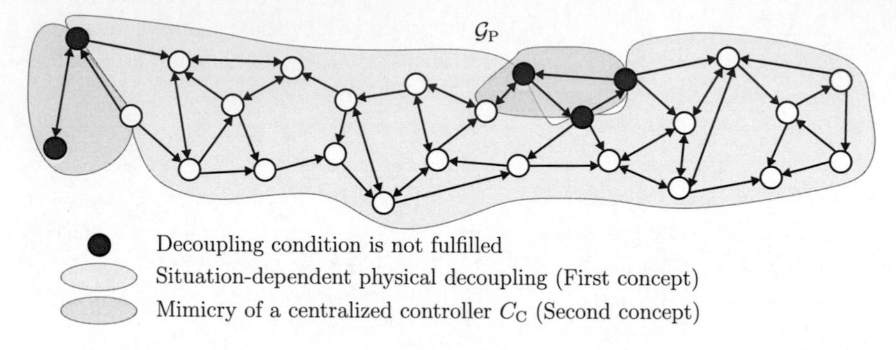

Figure 6.1: Joint usage of the two control concepts.

the decoupling condition can not be fulfilled. Furthermore, the concepts can be used together, which is discussed in the following.

Joint usage of the two control concepts. If the decoupling condition is not fulfilled for all subsystems Σ_i, then the first control concept can still be used. The subsystems which fulfill the decoupling condition can be controlled by the local control units C_i defined in (4.14) belonging to the first control concept, where the rest of the subsystems are controlled by the local control units C_i defined in (5.16) belonging to the second control concept (cf. Fig. 6.1). In particular, the local control units C_i whose subsystems have a physical impact on subsystem for which the decoupling condition is not fulfilled have to include also parts of the local control units of the second concept and vice versa. Hence, such local control units are kind of hybrid since they include components from both concepts. With this, the advantages of both control concepts can be used together.

Part II

Self-organizing control of multi-agent systems

7 Modeling and control of multi-agent systems

This chapter introduces the self-organizing control structure for multi-agent systems within a leader-follower structure used in this part of the thesis. The model of the leader agent and the follower agent are presented. A general model for the self-organizing control system is derived which includes the specific constrained communication structure for leader-follower synchronization. A formal definition of the control aims concerning the asymptotic synchronization and the synchronization performance of the agents are given. The general self-organizing control system is analyzed with respect to these control aims. Finally, a robot formation problem is introduced as a running example.

Chapter contents

7.1 Synchronization of agents within a leader-follower structure

This part of the thesis concerns the control of multi-agent systems, where the agents are interconnected within a so-called leader-follower structure (Fig. 7.1). The multi-agent system (plant P) consists of the leader Σ_0 and N agents Σ_i, ($\forall i \in \mathcal{N} = \{1, 2, \ldots, N\}$) which have individual dynamics and are the followers of the leader Σ_0. The task is that the output $y_i(t)$ of the agents Σ_i tracks the reference signal $y_{\rm s}(t)$ generated by the leader Σ_0 which is well-known as asymptotic synchronization.

Definition 7.1 ***(Asymptotic synchronization)*** *The agents Σ_i, ($\forall i \in \mathcal{N}$), defined in* (7.2) *are said to be asymptotically synchronized to the leader Σ_0 defined in* (7.3), *if the relation*

$$\lim_{t\to\infty} |y_i(t) - y_{\rm s}(t)| = 0, \quad \forall i \in \mathcal{N} = \{1, \ldots, N\} \tag{7.1}$$

holds.

This aim shall be achieved by using a self-organizing controller C. Every agent Σ_i, ($\forall i \in \mathcal{N}$), is controlled by a local control unit C_i which exchange their outputs $y_i(t)$ over the communication network. The task of the local control unit C_0 of the leader Σ_0 is to send the reference signal $y_{\rm s}(t)$ to a selection of C_i, ($i \in \mathcal{N}$). Hence, the local control units C_i, ($\forall i \in \mathcal{N}$), which are desired to have a situation-dependent communication, build the self-organizing controller C. In summary, the multi-agent system (plant P) and the communicating local control units C_i, ($i = 0, 1, 2 \ldots N$), (self-organizing controller C) build the self-organizing control system Σ.

Figures 7.2-7.3 illustrate the asymptotic synchronization by means of a robot formation problem without presenting the applied controller concept. Note that the robot formation is intro-

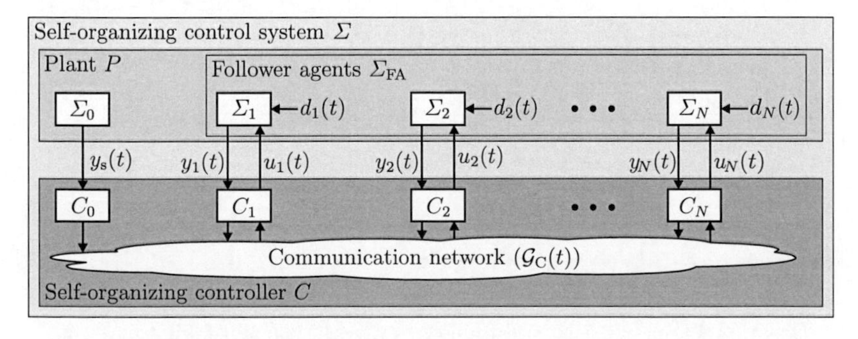

Figure 7.1: Self-organizing control of multi-agent systems within a leader-follower structure.

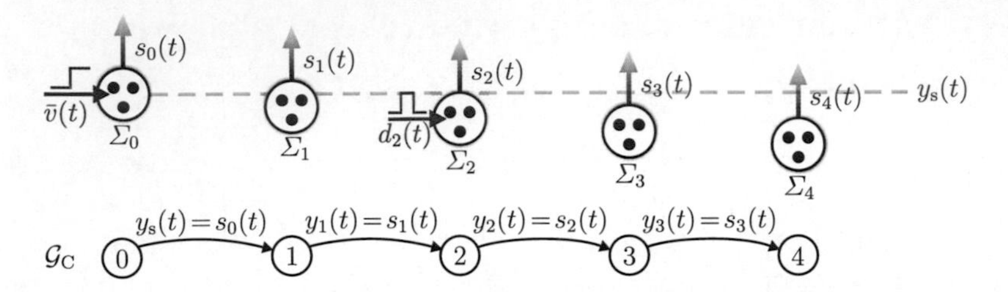

Figure 7.2: Robot formation with serial communication graph $\mathcal{G}_C$.

duced in more detail in Section 7.7. The formation consists of a leading robot Σ_0 with the position $s_0(t) = y_s(t)$ and four follower robots Σ_i, $(i = 1, \ldots, 4)$, whose positions $s_i(t) = y_i(t)$ shall be synchronized to the position $s_0(t)$ of the leader Σ_0. The local control units C_i of the robots communicate within a fixed serial communication graph $\mathcal{G}_C$ (Fig. 7.2).

Figure 7.3 shows the robot formation in two different situations, where in both situations the robots initially track the position of the leading robot Σ_0. In the left part Σ_0 gets a changed reference velocity $\bar{v}(t)$, which causes a control error $y_{\Delta i}(t) = y_i(t) - y_s(t)$ describing the difference between the reference position $s_0(t)$ and the position $s_i(t)$ of the robots. The control error

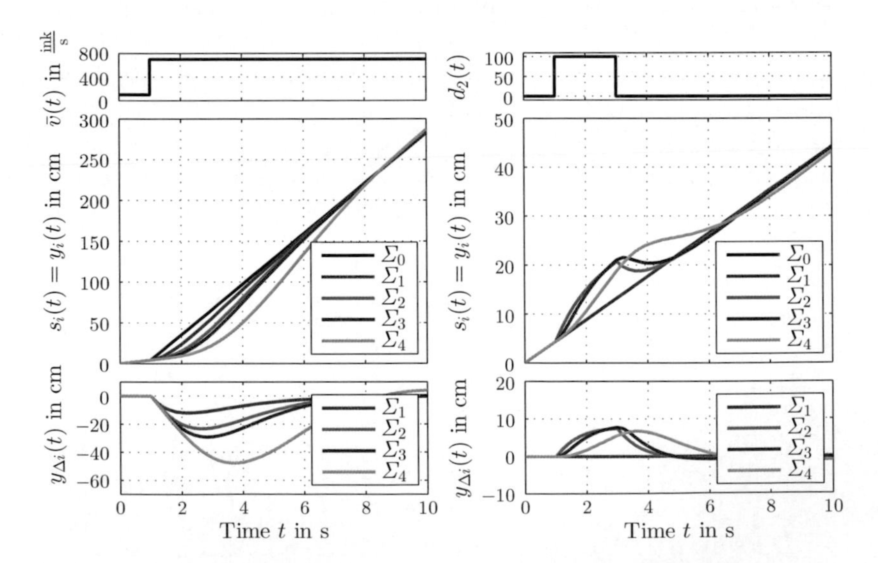

Figure 7.3: Example for the leader-follower synchronization of a formation with five robots with a change of the reference velocity $\bar{v}(t)$ (left) and a disturbance affecting robot Σ_2 (right).

$y_{\Delta i}(t)$ vanishes asymptotically. Hence, the robots asymptotically synchronize with the leading robot Σ_0 (Definition 7.1). In the right part of Fig. 7.3 the robot Σ_2 is affected by a disturbance $d_2(t)$ which leads to a control error $y_{\Delta i}(t)$ of robot Σ_2. But also the robots Σ_3 and Σ_4 are influenced by $d_2(t)$ since Σ_3 receives the disturbed position $s_2(t)$ of Σ_2. Since the disturbance $d_2(t)$ vanishes, the control errors $y_{\Delta i}(t)$ also vanish.

Both situations show that the control error $y_{\Delta i}(t)$ of the robots could be reduced by an adaption of the communication graph. If, e.g., all robots receive direct information from the leader Σ_0, the performance would be improved for a changing reference velocity $\bar{v}(t)$ compared to a serial communication. Furthermore, if, e.g., robot Σ_3 receives the position $s_1(t)$ of Σ_1 instead of $s_2(t)$ of Σ_2, sa disturbed robot Σ_2 would not affect the performance of the robots Σ_3 and Σ_4.

With all these considerations in mind, the main task for self-organizing control of multi-agent systems can be summarized as follows:

> Develop local control units C_i that can recognize impacts on the behavior of the agents that should track the reference signal $y_s(t)$ to react to these impacts by adapting the communication structure to guarantee a desired performance of the overall closed-loop system Σ.

Since the agents have a common control aim (cf. Definition 7.1), generally the local control units C_i can not be designed individually to guarantee a desired performance of the overall closed-loop system Σ.

Before explaining the actual control concepts in Chapters 8-10, at first the following general aspects are introduced:

- The model of the leader Σ_0 and the model of the agents Σ_i building the overall plant P are introduced in Section 7.2, where the agents Σ_i are assumed to include the dynamics of the leader Σ_0.
- Section 7.3 introduces a general model of the self-organizing controller C, where basic components of the local control units C_i are presented which are identical for all particular control concepts presented in Chapters 8-10.
- Section 7.4 presents a general model of the self-organizing control system Σ.
- Different control aims claimed for the resulting self-organizing control system Σ are introduced in Section 7.5.
- The behavior of the general self-organizing control system Σ is analyzed in Section 7.6.
- The robot formation problem is introduced as a running example in Section 7.7.

7.2 Modeling of the agents

In this section the model of the multi-agent system (plant P) is derived (Fig. 7.1). Section 7.2.1 presents the model of the individual follower agents Σ_i and the model of the leader Σ_0. The inclusion of the dynamics of the leader Σ_0 within the dynamics of the individual agents Σ_i for guaranteeing synchronization is concerned in Section 7.2.2 which is commonly known under the keyword *internal-model principle*. The resulting model of the overall plant P is given in Section 7.2.3.

7.2.1 Model of the leader and the follower agents

Figure 7.1 shows that the agents are divided into the leader (agent) Σ_0 and the follower agents Σ_i, $(\forall i \in \mathcal{N})$. The models of both types of agents are defined in this section.

Model of the follower agents. The individual agents Σ_i, $(i \in \mathcal{N})$ are represented by the linear state-space model

$$\Sigma_i : \begin{cases} \dot{\boldsymbol{x}}_i(t) = \boldsymbol{A}_i \boldsymbol{x}_i(t) + \boldsymbol{b}_i u_i(t) + \boldsymbol{g}_i d_i(t), & \boldsymbol{x}_i(0) = \boldsymbol{x}_{0i} \\ y_i(t) = \boldsymbol{c}_i^{\mathrm{T}} \boldsymbol{x}_i(t) \end{cases} \tag{7.2}$$

with the state $\boldsymbol{x}_i(t) \in \mathbb{R}^{n_i}$, the initial state $\boldsymbol{x}_{0i} \in \mathbb{R}^{n_i}$, the scalar input $u_i(t) \in \mathbb{R}$, the scalar disturbance $d_i(t) \in \mathbb{R}$, and the scalar output $y_i(t) \in \mathbb{R}$. It is assumed that there is no direct throughput of $u_i(t)$ and $d_i(t)$ towards $y_i(t)$. Furthermore, the following assumptions are made.

Assumption 7.1 *All pairs* $(\boldsymbol{A}_i, \boldsymbol{b}_i)$, $(\forall i \in \mathcal{N})$, *are controllable and all pairs* $(\boldsymbol{A}_i, \boldsymbol{c}_i^{\mathrm{T}})$, $(\forall i \in \mathcal{N})$, *are observable.*

Assumption 7.2 *The disturbances* $d_i(t)$, $(\forall i \in \mathcal{N})$, *are bounded, i.e.,* $|d_i(t)| < \infty$ *for all* $t \geq 0$.

Model of the leader. The leader Σ_0 is described by

$$\Sigma_0 : \begin{cases} \dot{\boldsymbol{x}}_{\mathrm{s}}(t) = \boldsymbol{A}_{\mathrm{s}} \boldsymbol{x}_{\mathrm{s}}(t), & \boldsymbol{x}_{\mathrm{s}}(0) = \boldsymbol{x}_{\mathrm{s}0} \\ y_{\mathrm{s}}(t) = \boldsymbol{c}_{\mathrm{s}}^{\mathrm{T}} \boldsymbol{x}_{\mathrm{s}}(t) \end{cases} \tag{7.3}$$

with the state $\boldsymbol{x}_{\mathrm{s}}(t) \in \mathbb{R}^{n_{\mathrm{s}}}$, the initial state $\boldsymbol{x}_{\mathrm{s}0} \in \mathbb{R}^{n_{\mathrm{s}}}$ and the reference signal $y_{\mathrm{s}}(t) \in \mathbb{R}$ as output. The leader Σ_0 is assumed to be undisturbed. Note that reference signal $y_{\mathrm{s}}(t)$ is sometimes denoted as $y_0(t)$, i.e., $y_0(t) \equiv y_{\mathrm{s}}(t)$. To avoid trivial solutions for the synchronization problem the following assumption is made.

Assumption 7.3 *The system matrix $\boldsymbol{A}_\mathrm{s} \in \mathbb{R}^{n_\mathrm{s}\times n_\mathrm{s}}$ of the leader Σ_0 defined in (7.3) is assumed to have at least one eigenvalue with nonnegative real part and the pair $(\boldsymbol{A}_\mathrm{s}, \boldsymbol{c}_\mathrm{s}^\mathrm{T})$ is completely observable.*

Due to Assumption 7.3, the reference signal $y_\mathrm{s}(t)$ does not converge to zero for almost all initial states $\boldsymbol{x}_{\mathrm{s}0} \neq \mathbf{0}$.

7.2.2 System inclusion for synchronization of the agents

In [89] it is proven that for a leader-follower synchronization all agents Σ_i, $(\forall i \in \mathcal{N})$ are required to include the model of the leader Σ_0. Hence, the agents Σ_i have to satisfy the internal-model principle [83, 134]. The definition of *system inclusion* in [89] can be adapted to the presented setup as follows.

Definition 7.2 *([89]* ***System inclusion****) The agent Σ_i, $(i \in \mathcal{N})$, defined in (7.2) is said to include the leader Σ_0 defined in (7.3) (in symbols: $\Sigma_0 \subseteq \Sigma_i$), if for every initial state $\boldsymbol{x}_{\mathrm{s}0} \in \mathbb{R}^{n_\mathrm{s}}$ there exists an initial state $\boldsymbol{x}_{0i} \in \mathbb{R}^n$ such that the relation*

$$y_i(t) = y_\mathrm{s}(t), \quad i \in \mathcal{N}, \quad \forall t \geq 0 \tag{7.4}$$

holds true.

A sufficient condition for system inclusion in [89] is adapted to the introduced setup in the following theorem.

Theorem 7.1 *([89]* ***System inclusion****) The agent Σ_i, $(i \in \mathcal{N})$, defined in (7.2) includes the leader Σ_0 defined in (7.3), if there exists a regular matrix $\boldsymbol{T}_i \in \mathbb{R}^{n_i}$ such that the following relations hold*

$$\boldsymbol{T}_i^{-1}\boldsymbol{A}_i\boldsymbol{T}_i = \begin{pmatrix} \boldsymbol{A}_\mathrm{s} & \boldsymbol{O} \\ \boldsymbol{O} & \boldsymbol{A}_{\mathrm{p}i} \end{pmatrix} =: \tilde{\boldsymbol{A}}_i, \quad i \in \mathcal{N}$$

$$\boldsymbol{c}_i^\mathrm{T}\boldsymbol{T}_i = \begin{pmatrix} \boldsymbol{c}_\mathrm{s}^\mathrm{T} & \boldsymbol{c}_{\mathrm{p}i}^\mathrm{T} \end{pmatrix} =: \tilde{\boldsymbol{c}}_i^\mathrm{T}, \quad i \in \mathcal{N}$$

with $\boldsymbol{A}_{\mathrm{p}i} \in \mathbb{R}^{(n_i-n_\mathrm{s})\times(n_i-n_\mathrm{s})}$ and $\boldsymbol{c}_{\mathrm{p}i}^\mathrm{T} \in \mathbb{R}^{(n_i-n_\mathrm{s})}$.

Consequently, the following assumption is weak since otherwise the agents Σ_i can never be synchronized with the leader Σ_0.

Assumption 7.4 *All agents Σ_i, $(\forall i \in \mathcal{N})$, defined in (7.2) are assumed to include the dynamics of the leader Σ_0 defined in (7.3) to satisfy the conditions in Theorem 7.1, i.e.,*

$$\Sigma_0 \subseteq \Sigma_i, \quad \forall i \in \mathcal{N}. \tag{7.5}$$

Theorem 7.1 and Assumption 7.4 yield the transformed model of the agents

$$\tilde{\Sigma}_i : \begin{cases} \frac{\mathrm{d}}{\mathrm{d}t}\tilde{\boldsymbol{x}}_i(t) = \tilde{\boldsymbol{A}}_i\tilde{\boldsymbol{x}}_i(t) + \tilde{\boldsymbol{b}}_i\boldsymbol{u}_i(t) + \tilde{\boldsymbol{g}}_i d_i(t), & \tilde{\boldsymbol{x}}_i(0) = \tilde{\boldsymbol{x}}_{0i} \\ \boldsymbol{y}_i(t) = \tilde{\boldsymbol{c}}_i^{\mathrm{T}}\tilde{\boldsymbol{x}}_i(t), \end{cases} \tag{7.6}$$

with

$$\begin{aligned} \tilde{\boldsymbol{b}}_i &:= \boldsymbol{T}_i^{-1}\boldsymbol{b}_i = \begin{pmatrix} \boldsymbol{b}_{\mathrm{q}i} \\ \boldsymbol{b}_{\mathrm{p}i} \end{pmatrix} \in \mathbb{R}^{n_i} \\ \tilde{\boldsymbol{g}}_i &:= \boldsymbol{T}_i^{-1}\boldsymbol{g}_i \in \mathbb{R}^{n_i} \\ \tilde{\boldsymbol{x}}_i(t) &:= \boldsymbol{T}_i^{-1}\boldsymbol{x}_i(t) = \begin{pmatrix} \boldsymbol{x}_{\mathrm{q}i} \\ \boldsymbol{x}_{\mathrm{p}i} \end{pmatrix} \in \mathbb{R}^{n_i} \\ \tilde{\boldsymbol{x}}_{0i} &:= \boldsymbol{T}_i^{-1}\boldsymbol{x}_i(0) = \begin{pmatrix} \boldsymbol{x}_{\mathrm{q}0i} \\ \boldsymbol{x}_{\mathrm{p}0i} \end{pmatrix} \in \mathbb{R}^{n_i}, \end{aligned}$$

where $\tilde{\boldsymbol{x}}_i(t)$ is the transformed state and $\tilde{\boldsymbol{x}}_{0i}$ is the transformed initial state.

7.2.3 Overall model of the plant

Considering (7.2), the model of the *follower agents* Σ_{FA} results to

$$\Sigma_{\mathrm{FA}} : \begin{cases} \dot{\boldsymbol{x}}(t) = \boldsymbol{A}\boldsymbol{x}(t) + \boldsymbol{B}\boldsymbol{u}(t) + \boldsymbol{G}\boldsymbol{d}(t), & \boldsymbol{x}(0) = \boldsymbol{x}_0 \\ \boldsymbol{y}(t) = \boldsymbol{C}\boldsymbol{x}(t) \end{cases} \tag{7.7}$$

with

$$\begin{aligned} \boldsymbol{A} &= \mathrm{diag}(\boldsymbol{A}_1 \ldots \boldsymbol{A}_N) \in \mathbb{R}^{n\times n}, & \boldsymbol{B} &= \mathrm{diag}(\boldsymbol{b}_1 \ldots \boldsymbol{b}_N) \in \mathbb{R}^{n\times m} \\ \boldsymbol{G} &= \mathrm{diag}(\boldsymbol{g}_1 \ldots \boldsymbol{g}_N) \in \mathbb{R}^{n\times m_{\mathrm{d}}}, & \boldsymbol{C} &= \mathrm{diag}(\boldsymbol{c}_1^{\mathrm{T}} \ldots \boldsymbol{c}_N^{\mathrm{T}}) \in \mathbb{R}^{r\times n} \end{aligned}$$

$$\boldsymbol{x}(t)=\begin{pmatrix} \boldsymbol{x}_1(t) \\ \vdots \\ \boldsymbol{x}_N(t) \end{pmatrix}, \quad \boldsymbol{x}_0=\begin{pmatrix} \boldsymbol{x}_{01} \\ \vdots \\ \boldsymbol{x}_{0N} \end{pmatrix}, \quad \boldsymbol{y}(t)=\begin{pmatrix} y_1(t) \\ \vdots \\ y_N(t) \end{pmatrix}, \quad \boldsymbol{u}(t)=\begin{pmatrix} u_1(t) \\ \vdots \\ u_N(t) \end{pmatrix}, \quad \boldsymbol{d}(t)=\begin{pmatrix} d_1(t) \\ \vdots \\ d_N(t) \end{pmatrix},$$

where $\boldsymbol{x}(t) \in \mathbb{R}^n$, $\boldsymbol{x}_0 \in \mathbb{R}^n$, $\boldsymbol{y}(t) \in \mathbb{R}^r$, $\boldsymbol{u}(t) \in \mathbb{R}^m$ and $\boldsymbol{d}(t) \in \mathbb{R}^{n_{\mathrm{d}}}$ denote the overall state, the overall initial state, the overall measured output, the overall input and the overall disturbance, respectively.

The resulting plant P includes the leader Σ_0 and the follower agents Σ_{FA}, i.e.,

$$P: \begin{cases} \dot{\boldsymbol{x}}_{\mathrm{os}}(t) = \boldsymbol{A}_{\mathrm{os}}\boldsymbol{x}_{\mathrm{os}}(t) + \boldsymbol{G}_{\mathrm{os}}\boldsymbol{d}(t), & \boldsymbol{x}_{\mathrm{os}}(0) = \boldsymbol{x}_{\mathrm{os0}} \\ \boldsymbol{y}(t) = \boldsymbol{C}_{\mathrm{os}}\boldsymbol{x}_{\mathrm{os}}(t) \end{cases} \tag{7.8}$$

with

$$\boldsymbol{A}_{\mathrm{os}} = \begin{pmatrix} \boldsymbol{A}_{\mathrm{s}} & \boldsymbol{O}_{n_{\mathrm{s}}\times n} \\ \boldsymbol{O}_{n\times n_{\mathrm{s}}} & \boldsymbol{A} \end{pmatrix}, \quad \boldsymbol{C}_{\mathrm{os}} = \begin{pmatrix} \boldsymbol{O}_{r\times n_{\mathrm{s}}} & \boldsymbol{C} \end{pmatrix}, \quad \boldsymbol{G}_{\mathrm{os}} = \begin{pmatrix} \boldsymbol{O}_{n_{\mathrm{s}}\times n_{\mathrm{d}}} \\ \boldsymbol{G} \end{pmatrix}$$

$$\boldsymbol{x}_{\mathrm{os}} = \begin{pmatrix} \boldsymbol{x}_{\mathrm{s}}(t) \\ \boldsymbol{x}(t) \end{pmatrix}, \quad \boldsymbol{x}_{\mathrm{os0}} = \begin{pmatrix} \boldsymbol{x}_{\mathrm{s0}} \\ \boldsymbol{x}_0 \end{pmatrix}.$$

7.3 Model of the general self-organizing controller

This section introduces the general communication among the local control units C_i and describes how the transmitted outputs are used to synchronize the agents. The general parts of the local control units C_i presented in Section 7.3.2 are identical in all control concepts introduced in Chapters 8-10. Finally, Section 7.3.3 presents the general structure of the self-organizing controller C.

7.3.1 Cycle-free communication among the local control units

The communication of the outputs $y_i(t)$ among the local control units C_i of the agents is represented by the communication graph

$$\mathcal{G}_{\mathrm{C}}(t) = (\mathcal{V}_{\mathrm{C}}, \mathcal{E}_{\mathrm{C}}(t), \boldsymbol{K}) \tag{7.9}$$

which is a directed time-varying labeled graph. Compared to the general definition of the communication graph $\mathcal{G}_{\mathrm{C}}(t)$ in (2.16), $\mathcal{G}_{\mathrm{C}}(t)$ is labeled by the weighting matrix $\boldsymbol{K}$ composed of the elements $k_{ij} \in \mathbb{R}^+$, which weight the edges $(j \to i) \in \mathcal{E}_{\mathrm{C}}(t)$. In particular, element $k_{ij} \in \mathbb{R}^+$ weights the communicated outputs $y_j(t)$ from C_j to C_i. The set of communicational vertices $\mathcal{V}_{\mathrm{C}}$ is defined by $\mathcal{V}_{\mathrm{C}} = \{0\} \cup \mathcal{N}$, where the vertex $0 \in \mathcal{V}_{\mathrm{C}}$ represents the leader Σ_0. The number of agents including the leader is denoted by $V = |\mathcal{V}_{\mathrm{C}}| = N+1$. Clearly, for Σ_0 the relation $\mathcal{P}_{\mathrm{C}0}(t) = \emptyset$ holds.

Limitation of the communication links. Similar to the general definition of the communication graph $\mathcal{G}_{\mathrm{C}}(t)$ in Section 2.3 a maximal communication graph $\bar{\mathcal{G}}_{\mathrm{C}} = (\mathcal{V}_{\mathrm{C}}, \bar{\mathcal{E}}_{\mathrm{C}}, \boldsymbol{K})$ and a minimal communication graph $\underline{\mathcal{G}}_\mathrm{C} = (\mathcal{V}_\mathrm{C}, \underline{\mathcal{E}}_\mathrm{C}, \boldsymbol{K})$ are defined. Both are labeled graphs. The weighting matrix $\boldsymbol{K}$ is the same as for $\mathcal{G}_\mathrm{C}(t)$ defined in (7.9). According (2.25), the set $\mathcal{E}_\mathrm{C}(t)$ of communicational edges is lower bounded by the set $\underline{\mathcal{E}}_\mathrm{C}$ of minimal communication links defined in (2.23) and upper bounded by the set $\bar{\mathcal{E}}_\mathrm{C}$ of maximal communication links defined in (2.21). Recall that $\underline{\mathcal{E}}_\mathrm{C}$ results from the set $\underline{\mathcal{P}}_{\mathrm{C}i}$ of minimal communicational predecessors and that $\bar{\mathcal{E}}_\mathrm{C}$ results from the set $\bar{\mathcal{P}}_{\mathrm{C}i}$ of maximal communicational predecessors. Because of the desired cycle free communication the sets $\underline{\mathcal{P}}_{\mathrm{C}i}$ and $\bar{\mathcal{P}}_{\mathrm{C}i}$ are restricted by

$$\bar{\mathcal{P}}_{\mathrm{C}i} \subseteq \{j \in \mathcal{V}_\mathrm{C} \mid j < i\}, \quad \forall i \in \mathcal{N} \tag{7.10}$$

$$\underline{\mathcal{P}}_{\mathrm{C}i} \subseteq \bar{\mathcal{P}}_{\mathrm{C}i}, \quad \forall i \in \mathcal{N}. \tag{7.11}$$

With this, according to (2.17) and (2.25) the set $\mathcal{P}_{\mathrm{C}i}(t)$ of communicational predecessors is also restricted

$$\mathcal{P}_{\mathrm{C}i}(t) \subseteq \{j \in \mathcal{V}_\mathrm{C} \mid j < i\}, \quad \forall i \in \mathcal{N}, \quad \forall t \geq 0. \tag{7.12}$$

For set $\mathcal{E}_\mathrm{C}(t)$ of communication links the relation

$$\mathcal{E}_\mathrm{C}(t) \subseteq \{(j \to i) \mid i, j \in \mathcal{V}_\mathrm{C},\ j < i\}, \quad \forall t \geq 0 \tag{7.13}$$

holds true. This boundedness of $\mathcal{G}_\mathrm{C}(t)$ is illustrated in Fig. 7.4.

Due to the boundedness of the communication links, for the elements k_{ji} of $\boldsymbol{K}$ the relation

$$k_{ij} = 0, \quad \forall i, j \in \mathcal{V}_\mathrm{C}, \quad j \notin \bar{\mathcal{P}}_{\mathrm{C}i} \tag{7.14}$$

holds true.

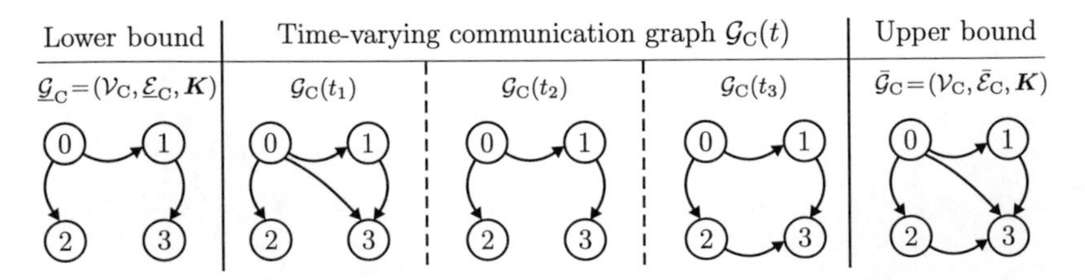

Figure 7.4: Limitation of the communication graph $\mathcal{G}_\mathrm{C}(t)$ for the self-organizing control of multi-agent systems.

Connectedness of the agents. It is well-known that the agents have to be connected to the leader which means that there has to exist a spanning tree with the leader as root node (cf. [89]). In particular, within $\mathcal{G}_{\rm C}(t)$ there has to exist a path from node 0 to every other node i, ($\forall i \in \mathcal{N}$):

$$\exists\, \mathrm{Path}(0 \rightarrow i) \in \mathcal{G}_{\rm C}(t), \quad \forall i \in \mathcal{N}, \quad \forall t \geq 0. \tag{7.15}$$

The following lemma states that for the concerned communication structure all agents have to have a predecessor to be connected with the leader.

Lemma 7.1 ***(Spanning tree within the communication graph*** $\mathcal{G}_{\rm C}(t)$***)*** *Consider the communication graph* $\mathcal{G}_{\rm C}(t)$ *defined in* (7.9) *with the communication links in* $\mathcal{E}_{\rm C}(t)$ *restricted according to* (7.13). *Relation* (7.15) *is fulfilled which means that there exists a spanning tree with the leader as root node in* $\mathcal{G}_{\rm C}(t)$, *if and only if all agents* Σ_i *have a predecessor, i.e.,*

$$\mathcal{P}_{{\rm C}i}(t) \neq \emptyset, \quad \forall i \in \mathcal{N}, \quad \forall t \geq 0.$$

Proof. The proof is given in Appendix A.10. □

Lemma 7.1 presents a necessary and sufficient condition for the existence of a spanning tree in $\mathcal{G}_{\rm C}(t)$ and, therewith, also a condition for the ability to synchronize the agents Σ_i with the leader Σ_0.

Communication cost. Each agent Σ_i is assumed to be located in a three-dimensional space relatively to leader Σ_0 indicated by the coordinates X_i, Y_i and Z_i, which is illustrated in Fig. 7.5 for a formation control of quadcopters. Hence, the distance among Σ_i and Σ_j is defined by

$$\delta_{ij} = \sqrt{(X_i - X_j)^2 + (Y_i - Y_j)^2 + (Z_i - Z_j)^2)}, \quad \forall i, j \in \mathcal{V}_{\rm C}.$$

The *communication cost* c_{ij} from C_j and C_i is defined to be the square of their distance

$$c_{ij} = \delta_{ij}^2, \quad \forall i, j \in \mathcal{V}_{\rm C}. \tag{7.16}$$

Since possible motions of the agents are assumed to be small compared to their actual distance to each other, the communication costs c_{ij} are set to be constant.

The possible predecessors of Σ_i are presorted in the function $c_{{\rm MIN}i}(j)$, $(j \in \mathcal{V}, j < i)$ by increasing order of their communication cost c_{ij}. The function $z = c_{{\rm MIN}i}(j)$ returns agent Σ_z to which agent Σ_i has the j-th smallest communication cost. Obviously for $z = c_{{\rm MIN}i}(1)$, Σ_z is the agent with the smallest communication cost and for $l = c_{{\rm MIN}i}(i)$, Σ_l is the agent with the highest communication cost.

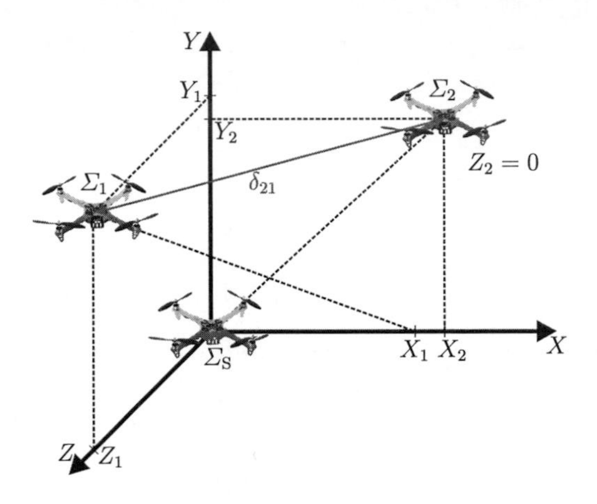

Figure 7.5: Orientation of the agents in a three-dimensional space.

7.3.2 General model of the local control units

Recall that the local control units C_i include a feedback unit F_i, an observation unit O_i and a decision unit D_i (Fig. 7.6). The feedback unit F_i is used to track the local reference signal $y_{si}(t)$ and is identical for all control concepts presented in Chapters 8–10. The general task of the decision unit D_i is to handle the incoming and outgoing information. The following paragraphs describe the structure of the feedback unit F_i and the realization of the general tasks of the decision unit D_i. The tasks and the structure of the observation unit O_i is different for every

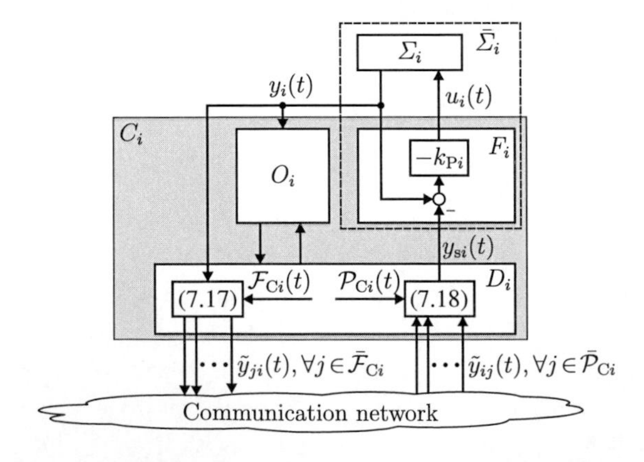

Figure 7.6: Local control unit C_i for situation-dependent communication within the control of multi-agent system.

control concept. Finally, the components are combined to define the general local control units C_i, ($\forall i \in \mathcal{N}$), of the agents Σ_i and the general local control unit C_0 of the leader Σ_0.

Decision unit D_i. The decision unit D_i has two general tasks. The first task is to send the output $y_i(t)$ to its current followers C_j, ($\forall j \in \mathcal{F}_{\mathrm{C}i}(t)$), i.e.,

$$\tilde{y}_{ji}(t) = \begin{cases} y_i(t) & \text{if } j \in \mathcal{F}_{\mathrm{C}i}(t) \\ \eta & \text{else,} \end{cases} \qquad \forall j \in \bar{\mathcal{F}}_{\mathrm{C}i}, \quad \forall i \in \mathcal{V}_{\mathrm{C}}. \tag{7.17}$$

The signal $\tilde{y}_{ji}(t)$ is the *transmitted output* which only includes $y_i(t)$, if C_j is a follower of C_i, i.e., $j \in \mathcal{F}_{\mathrm{C}i}(t)$. The symbol η is a placeholder for the signal $\tilde{y}_{ji}(t)$ and means that no signal is sent from C_i to C_j. The second task is to determine the *local reference signal*

$$y_{\mathrm{s}i}(t) = \begin{cases} \sum\limits_{j \in \mathcal{P}_{\mathrm{C}i}(t)} \tilde{k}_{ij}(t) y_j(t) & \text{if } \mathcal{P}_{\mathrm{C}i}(t) \neq \emptyset, \\ y_{\mathrm{s}i}(t^-) & \text{else,} \end{cases} \qquad \forall i \in \mathcal{N} \tag{7.18}$$

while using the incoming outputs $y_j(t)$, ($\forall j \in \mathcal{P}_{\mathrm{C}i}(t)$), from the current predecessors of C_i included in the transmitted outputs $\tilde{y}_{ij}(t)$. The elements $\tilde{k}_{ij}(t)$ are defined by

$$\tilde{k}_{ij}(t) = \begin{cases} \dfrac{k_{ij}}{\sum_{l \in \mathcal{P}_{\mathrm{C}i}(t)} k_{il}} & \text{if } j \in \mathcal{P}_{\mathrm{C}i}(t) \\ 0 & \text{else,} \end{cases} \qquad \forall i, j \in \mathcal{V}_{\mathrm{C}}. \tag{7.19}$$

If there are no incoming outputs, i.e., $\mathcal{P}_{\mathrm{C}i}(t) = \emptyset$, the local reference signal $y_{\mathrm{s}i}(t)$ is kept constant at its previous value $y_{\mathrm{s}i}(t^-)$, where $y_{\mathrm{s}i}(t^-) = 0$ for $t = 0$. The elements $\tilde{k}_{ij}(t)$ build the *situation-dependent communication matrix*

$$\tilde{\boldsymbol{K}}(t) = (\tilde{k}_{ij}(t)), \quad \forall (j \to i) \in \mathcal{E}_{\mathrm{C}}(t).$$

Equation (7.14) implies that $\tilde{k}_{ij}(t) = 0$, ($\forall i, j \in \mathcal{V}_{\mathrm{C}}, j \notin \bar{\mathcal{P}}_{\mathrm{C}i}, \forall t \geq 0$). Due to (7.19), the weightings $\tilde{k}_{ij}(t)$ satisfy the relation

$$\sum_{j \in \mathcal{P}_{\mathrm{C}i}(t)} \tilde{k}_{ij}(t) = 1 \quad \text{for} \quad \mathcal{P}_{\mathrm{C}i}(t) \neq \emptyset. \tag{7.20}$$

With this, the general model of the decision units D_i results to

$$D_i \begin{cases} \text{(7.18) generating } y_{si}(t) \text{ using } \mathcal{P}_{Ci}(t) \text{ and } \tilde{y}_{ij}(t), (\forall j \in \bar{\mathcal{P}}_{Ci}) \\ \text{(7.17) generating } \tilde{y}_{ji}(t), (\forall j \in \bar{\mathcal{F}}_{Ci}), \text{ using } \mathcal{F}_{Ci}(t) \text{ and } y_i(t) \\ \text{The generation of } \mathcal{P}_{Ci}(t) \text{ and } \mathcal{F}_{Ci}(t) \text{ is undefined yet,} \end{cases} \quad \forall i \in \mathcal{N}. \tag{7.21}$$

The generation of the set $\mathcal{F}_{Ci}(t)$ of followers and the set $\mathcal{P}_{Ci}(t)$ of predecessors differs among the specific control concepts presented in Chapters 8–10 and is therefore delayed to these chapters.

Feedback unit F_i. The feedback unit

$$F_i : u_i(t) = -k_{Pi}(y_i(t) - y_{si}(t)), \quad \forall i \in \mathcal{N}, \tag{7.22}$$

generates the input $u_i(t)$ for the agents Σ_i using the feedback gain k_{Pi} to track the local reference signal $y_{si}(t)$ determined by the decision units D_i according to (7.18) (cf. Fig. 7.6). Recall that the structure of the feedback unit F_i is identical for all concepts presented in Chapters 8–10.

Local control units C_i of the follower agents. Summarizing the descriptions of the general tasks of the decision unit D_i and the feedback unit F_i leads to the following three general tasks of the local control units C_i:

1. The decision unit D_i sends the output $y_i(t)$ to its current followers C_j, $(\forall j \in \mathcal{F}_{Ci}(t))$, by means of the transmitted output $\tilde{y}_{ji}(t)$ defined in (7.17).

2. The decision unit D_i uses the incoming outputs to determine the local reference signal $y_{si}(t)$ defined in (7.18).

3. The feedback unit F_i defined in (7.22) generates the input $u_i(t)$ for agent Σ_i to track the local reference signal $y_{si}(t)$.

With this, the general model of the local control units C_i results to

$$C_i : \begin{cases} F_i \text{ defined in (7.22)} \\ O_i \text{ to be defined in Chapters 8–10} \\ D_i \text{ partly defined in (7.21) (remaining part in Chapters 8–10),} \end{cases} \quad \forall i \in \mathcal{N}. \tag{7.23}$$

Local control unit C_0 of the leader Σ_0. There is no feedback unit and observation unit within the local control unit C_0 of the leader Σ_0. That is, the local control unit

$$C_0 = D_0 : \begin{cases} \text{(7.17) generating } \tilde{y}_{j0}(t), (\forall j \in \bar{\mathcal{F}}_{\mathrm{C}0}), \text{ using } \mathcal{F}_{\mathrm{C}0}(t) \text{ and } y_0(t) = y_{\mathrm{s}}(t) \\ \text{Generation of } \mathcal{F}_{\mathrm{C}0}(t) \text{ is undefined yet} \end{cases} \tag{7.24}$$

includes the decision unit D_0 only. C_0 sends the reference signal $y_{\mathrm{s}}(t)$ to the followers Σ_j, $(\forall j \in \bar{\mathcal{F}}_{\mathrm{C}0})$, of Σ_0 within the transmitted outputs $\tilde{y}_{j0}(t)$. The generation of the set $\mathcal{F}_{\mathrm{C}0}(t)$ of followers of Σ_0 differs for the specific control concepts. Hence, C_0 includes no feedback controller.

7.3.3 Self-organizing controller with undefined change of the communication

The general description of the communication structure in Section 7.3.1 and the definition of the general structure of the local control units C_i in Section 7.3.2 leads the general self-organizing controller

$$C: \; C_i, (\forall i \in \mathcal{V}_{\mathrm{C}}), \text{ partly defined in (7.23)-(7.24) with } \mathcal{E}_{\mathrm{C}}(t) \text{ bounded according to (7.13),} \tag{7.25}$$

where the actual situation-dependent change of the communication graph $\mathcal{G}_{\mathrm{C}}(t) = (\mathcal{V}_{\mathrm{C}}, \mathcal{E}_{\mathrm{C}}(t), \boldsymbol{K})$ is not yet defined. Hence, the general difference between the developed control concepts is the adjustment of the communication links $\mathcal{E}_{\mathrm{C}}(t)$ such that the overall closed-loop system Σ adapts to its current situation.

7.4 Model of the general self-organizing control system

This section presents a general model of the self-organizing control system Σ which is identical for all control concepts presented in Chapters 8-10. First, the model of the follower agents controlled by the feedback unit F_i is derived (Section 7.4.1). Second, the controlled agents are combined with the communication generated by the decision units D_i to obtain the general model of the self-organizing control system Σ (Section 7.4.2). Section 7.4.3 summarizes the main properties and assumption for Σ.

7.4.1 Controlled agents

The model of controlled agent $\bar{\Sigma}_i$, $(\forall i \in \mathcal{N})$, which consists of the agents Σ_i defined in (7.2) and the feedback unit F_i defined in (7.22), is given by

$$\bar{\Sigma}_i : \begin{cases} \dot{\boldsymbol{x}}_i(t) = \bar{\boldsymbol{A}}_i \boldsymbol{x}_i(t) + \boldsymbol{b}_i k_{\mathrm{P}i} y_{\mathrm{s}i}(t) + \boldsymbol{g}_i d_i(t), & \boldsymbol{x}_i(0) = \boldsymbol{x}_{0i} \\ y_i(t) = \boldsymbol{c}_i^{\mathrm{T}} \boldsymbol{x}_i(t) \end{cases} \tag{7.26}$$

with $\bar{\boldsymbol{A}}_i = \boldsymbol{A}_i - \boldsymbol{b}_i k_{\mathrm{P}i} \boldsymbol{c}_i^{\mathrm{T}}$. To stabilize $\bar{\Sigma}_i$, the feedback gain $k_{\mathrm{P}i}$ must satisfy the following assumption.

Assumption 7.5 *The feedback gain $k_{\mathrm{P}i}$, $(i \in \mathcal{N})$, is chosen such that $\bar{\boldsymbol{A}}_i = \boldsymbol{A}_i - \boldsymbol{b}_i k_{\mathrm{P}i} \boldsymbol{c}_i^{\mathrm{T}}$ is a Hurwitz matrix.*

The model of the *transformed controlled agent* results from Theorem 7.1, Assumption 7.4 and (7.26) to

$$\tilde{\bar{\Sigma}}_i : \begin{cases} \frac{\mathrm{d}}{\mathrm{d}t} \tilde{\boldsymbol{x}}_i(t) = \tilde{\bar{\boldsymbol{A}}}_i \tilde{\boldsymbol{x}}_i(t) + k_{\mathrm{P}i} \tilde{\boldsymbol{b}}_i y_{\mathrm{s}i}(t) + \tilde{\boldsymbol{g}}_i d_i(t), & \tilde{\boldsymbol{x}}_i(0) = \tilde{\boldsymbol{x}}_{0i} \\ y_i(t) = \tilde{\boldsymbol{c}}_i^{\mathrm{T}} \tilde{\boldsymbol{x}}_i(t) \end{cases} \tag{7.27}$$

with the transformed system matrix

$$\tilde{\bar{\boldsymbol{A}}}_i = \tilde{\boldsymbol{A}}_i - k_{\mathrm{P}i} \tilde{\boldsymbol{b}}_i \tilde{\boldsymbol{c}}_i^{\mathrm{T}} = \begin{pmatrix} \boldsymbol{A}_{\mathrm{s}} - k_{\mathrm{P}i} \boldsymbol{b}_{\mathrm{q}i} \boldsymbol{c}_{\mathrm{s}}^{\mathrm{T}} & -k_{\mathrm{P}i} \boldsymbol{b}_{\mathrm{q}i} \boldsymbol{c}_{\mathrm{p}i}^{\mathrm{T}} \\ -k_{\mathrm{P}i} \boldsymbol{b}_{\mathrm{p}i} \boldsymbol{c}_{\mathrm{s}}^{\mathrm{T}} & \boldsymbol{A}_{\mathrm{p}i} - k_{\mathrm{P}i} \boldsymbol{b}_{\mathrm{p}i} \boldsymbol{c}_{\mathrm{p}i}^{\mathrm{T}} \end{pmatrix}.$$

Controlled follower agents. Equation (7.18) and (7.26) yield the relation

$$\dot{\boldsymbol{x}}_i(t) = \bar{\boldsymbol{A}}_i \boldsymbol{x}_i(t) + \boldsymbol{b}_i k_{\mathrm{P}i} \sum_{j \in \mathcal{P}_{\mathrm{C}i}(t)} \tilde{k}_{ij}(t) y_j(t) + \boldsymbol{g}_i d_i(t), \quad \forall i \in \mathcal{N}.$$

With this, the overall model of the *controlled follower agents* $\bar{\Sigma}_{\mathrm{FA}}$ is given by

$$\bar{\Sigma}_{\mathrm{FA}} : \begin{cases} \dot{\boldsymbol{x}}(t) = \bar{\boldsymbol{A}}(t) \boldsymbol{x}(t) + \boldsymbol{B} \boldsymbol{K}_{\mathrm{P}} \tilde{\boldsymbol{k}}_{\mathrm{s}}(t) y_{\mathrm{s}}(t) + \boldsymbol{G} \boldsymbol{d}(t), & \boldsymbol{x}(0) = \boldsymbol{x}_0 \\ \boldsymbol{y}(t) = \boldsymbol{C} \boldsymbol{x}(t) \end{cases} \tag{7.28}$$

with

$$\begin{aligned} \bar{\boldsymbol{A}}(t) &= \boldsymbol{A} + \boldsymbol{B} \boldsymbol{K}_{\mathrm{P}} \tilde{\boldsymbol{K}}_{\mathrm{F}}(t) \boldsymbol{C} \\ \boldsymbol{K}_{\mathrm{P}} &= \mathrm{diag}(k_{\mathrm{P}1} \dots k_{\mathrm{P}N}) \\ \tilde{\boldsymbol{K}}_{\mathrm{F}}(t) &= \left(\tilde{\boldsymbol{K}}^*(t) - \boldsymbol{I}_N \right) \end{aligned} \tag{7.29}$$

$$\tilde{\boldsymbol{K}}^*(t) = \begin{pmatrix} \mathbf{0}_N^{\mathrm{T}} \\ \boldsymbol{I}_N \end{pmatrix} \tilde{\boldsymbol{K}}(t) \begin{pmatrix} \mathbf{0}_N & \boldsymbol{I}_N \end{pmatrix}$$

$$\tilde{\boldsymbol{k}}_{\mathrm{s}}(t) = \begin{pmatrix} \tilde{k}_{01}(t) \\ \vdots \\ \tilde{k}_{0N}(t) \end{pmatrix}.$$

The system matrix $\bar{\boldsymbol{A}}(t) = (\bar{\boldsymbol{A}}_{ij}(t))$ of $\bar{\Sigma}_{\mathrm{FA}}$ has the blocks

$$\bar{\boldsymbol{A}}_{ii}(t) = (\boldsymbol{A}_i - \boldsymbol{b}_i k_{\mathrm{P}i} \boldsymbol{c}_i^{\mathrm{T}}), \quad \forall i \in \mathcal{N}$$
$$\bar{\boldsymbol{A}}_{ij}(t) = \tilde{k}_{ij}(t) \boldsymbol{b}_i k_{\mathrm{P}i} \boldsymbol{c}_j^{\mathrm{T}}, \quad i \neq j, \quad \forall i, j \in \mathcal{N}.$$

Note that only the non-diagonal blocks depend upon the switching of the communication within $\tilde{\boldsymbol{K}}(t)$. Due to (7.14), the relation

$$\bar{\boldsymbol{A}}_{ij}(t) = \boldsymbol{O}_{n_i \times n_j}, \quad \forall i, j \in \mathcal{N}, \quad i < j$$

holds, which means that the matrix $\bar{\boldsymbol{A}}(t)$ is a block lower triangular matrix for all $t \geq 0$.

The model $\bar{\Sigma}_{\mathrm{FA}}$ defined in (7.28) describes the behavior of the controlled agents $\bar{\Sigma}_i$, ($\forall i \in \mathcal{N}$), connected by a situation-dependent communication represented by $\tilde{\boldsymbol{K}}(t)$. The controlled follower agents $\bar{\Sigma}_{\mathrm{FA}}$ are connected with the leader Σ_0 over the reference signal $y_{\mathrm{s}}(t)$.

7.4.2 Overall closed-loop system

Combining the model (7.28) of controlled follower agents $\bar{\Sigma}_{\mathrm{FA}}$ with the model (7.3) of the leader Σ_0 leads to the general model of the self-organizing control system

$$\Sigma : \begin{cases} \dot{\boldsymbol{x}}_{\mathrm{os}}(t) = \bar{\boldsymbol{A}}_{\mathrm{os}}(t) \boldsymbol{x}_{\mathrm{os}}(t) + \boldsymbol{G}_{\mathrm{os}} \boldsymbol{d}(t), \quad \boldsymbol{x}_{\mathrm{os}}(0) = \boldsymbol{x}_{\mathrm{os}0} \\ \boldsymbol{y}(t) = \boldsymbol{C}_{\mathrm{os}} \boldsymbol{x}_{\mathrm{os}}(t) \\ \left. \begin{array}{l} O_i, \ \forall i \in \mathcal{N} \\ D_i, \ \forall i \in \mathcal{N} \end{array} \right\} \text{generating } \bar{\boldsymbol{A}}_{\mathrm{os}}(t) \text{ with } \mathcal{G}_{\mathrm{C}}(t) \text{ defined in (7.9)} \end{cases} \tag{7.30}$$

with the system matrix

$$\bar{\boldsymbol{A}}_{\mathrm{os}}(t) = \begin{pmatrix} \boldsymbol{A}_{\mathrm{s}} & \boldsymbol{O}_{n_{\mathrm{s}} \times n} \\ \boldsymbol{B}_{\mathrm{F}} \boldsymbol{K}_{\mathrm{P}} \tilde{\boldsymbol{k}}_{\mathrm{s}}(t) \boldsymbol{c}_{\mathrm{s}}^{\mathrm{T}} & \bar{\boldsymbol{A}}(t) \end{pmatrix}.$$

Due to the situation-dependent communication represented by the communication graph $\mathcal{G}_{\mathrm{C}}(t)$ defined in (7.9), the overall system matrix $\bar{\boldsymbol{A}}_{\mathrm{os}}(t)$ is time-variant. Hence, the change of $\mathcal{G}_{\mathrm{C}}(t)$ changes the behavior of Σ. The decision units D_i use information from the observation units O_i to decide about the change of $\mathcal{G}_{\mathrm{C}}(t)$. In Chapters 8-10 different structures for D_i and O_i are proposed, to guarantee the control aims claimed in Section 7.5.

7.4.3 Summary of the basic properties and assumptions

The following list summarizes the main properties and assumption for the self-organizing control system Σ defined in (7.30):

- The local control units C_i, $(\forall i \in \mathcal{N})$, exchange the output $y_i(t)$ of their corresponding agent Σ_i over the communication network.
- The agents Σ_i are completely controllable and completely observable (Assumption 7.1).
- The disturbances $d_i(t)$, $(\forall i \in \mathcal{N})$, affecting the agents Σ_i are bounded (Assumption 7.2).
- The system matrix $\boldsymbol{A}_{\mathrm{s}}$ of the leader Σ_0 defined in (7.3) has at least one eigenvalue with nonnegative real part (Assumption 7.3).
- All agents Σ_i, $(\forall i \in \mathcal{N})$, include the dynamics of the leader Σ_0 (Assumption 7.4).
- The communication graph $\mathcal{G}_{\mathrm{C}}(t)$ defined in (7.9) is restricted to be cycle free (cf. (7.12)).
- The communication graph $\mathcal{G}_{\mathrm{C}}(t)$ defined in (7.9) has a spanning tree with the leader as root node (cf. (7.15) and Lemma 7.1).
- The decision unit D_i handles the communication of the outputs among the local control units C_i and determines the local reference signal $y_{\mathrm{s}i}(t)$ for the feedback unit F_i (cf. Section 7.3.2).
- The feedback unit F_i defined in (7.22) is identical for all control concepts within Part II.
- The controlled agents $\bar{\Sigma}_i$, $(\forall i \in \mathcal{N})$, are assumed to be stable (Assumption 7.5).

In summary, Fig. 7.7 shows that the overall interaction among the agents Σ_i within the self-organizing control system Σ is described by the communication graph $\mathcal{G}_{\mathrm{C}}(t)$ since, in contrast to Part I, there is no physical interconnection among the agents Σ_i.

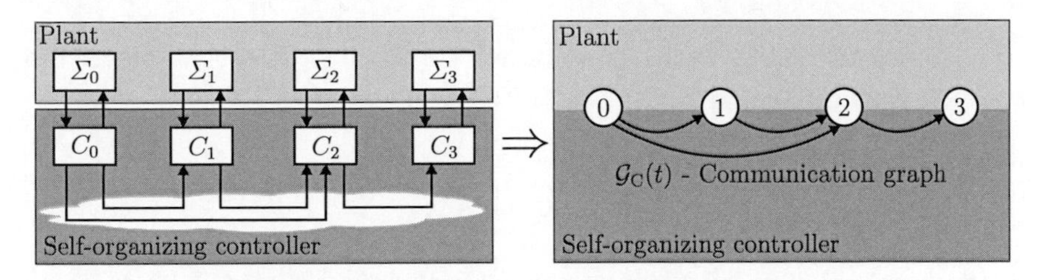

Figure 7.7: Structural representation of the self-organizing control system Σ defined in (7.30) for an example with four agents at time t.

7.5 Control aims: Asymptotic synchronization and synchronization performance

The local control units C_i shall be designed such that the self-organizing control systems Σ satisfy the following control aims.

The most essential aim, the asymptotic synchronization of the agents which is claimed for all control concepts, is presented in Chapters 8-10.

Control aim 7.1 (Asymptotic synchronization) *The agents Σ_i, ($\forall i \in \mathcal{N}$), defined in (7.2) within the self-organizing control system Σ defined in (7.30) have to be asymptotically synchronized to the leader Σ_0 defined in (7.3) according to Definition 7.1.*

General conditions for the asymptotic synchronization of the agents are presented in Section 7.6.2, where in the chapters presenting the respective control concepts it is shown that the resulting self-organizing control system Σ satisfies these conditions.

Synchronization performance. The following three control aims concern the adherence to a desired maximal control error for different situations.

In Chapter 8 a desired maximal control error is claimed for the case that the leader Σ_0 and the followers Σ_i have different initial states. Consider a robot formation as presented in Section 7.7.3, then, e.g., a stepwise change of the reference velocity of the leader robot could cause such difference in the initial state.

Control aim 7.2 (Desired maximal control error) *Consider the self-organizing control system Σ defined in* (7.30) *for which the initial state $\boldsymbol{x}_{s0}$ of the leader Σ_0 is different from the initial state of the agents Σ_i*

$$\boldsymbol{x}_{s0} \neq \boldsymbol{x}_{0i}, \quad \forall i \in \mathcal{N}. \tag{7.31}$$

If the outputs $y_i(t)$, $(\forall i \in \mathcal{N})$, of all agents Σ_i initially track the reference signal $y_s(t)$

$$y_i(0) = y_s(0), \quad \forall i \in \mathcal{N}, \tag{7.32}$$

the control errors $y_{\Delta i}(t)$, $(\forall i \in \mathcal{N})$, have to be smaller than the desired maximal control error *$\bar{y}^*_{\Delta i}$, i.e.,*

$$|y_{\Delta i}(t)| = |y_i(t) - y_s(t)| \leq \bar{y}^*_{\Delta i}, \quad \forall i \in \mathcal{N}, \quad \forall t \geq 0. \tag{7.33}$$

Chapter 9 concerns the situation that the agents Σ_i are affected by a disturbance. Therefore, the disturbance propagation should be bounded such that the undisturbed agents have to guarantee a desired maximal control error.

Control aim 7.3 (Disturbance attenuation) *Consider the situation, where due to appropriate initial states $\boldsymbol{x}_{i0}$, $(\forall i \in \mathcal{N})$, undisturbed agents Σ_i within the self-organizing control system Σ defined in* (7.30) *would follow the reference signal $y_s(t)$ of the leader Σ_0, i.e.,*

$$y_i(t) = y_s(t), \quad \forall i \in \mathcal{N}, \quad \forall t \geq 0. \tag{7.34}$$

Then the synchronization error $y_{\Delta i}(t)$ of all agents Σ_i that are not directly disturbed have to be smaller than the desired maximal control error *$\bar{y}^*_{\Delta}$, i.e.,*

$$|y_{\Delta i}(t)| = |y_i(t) - y_s(t)| \leq \bar{y}^*_{\Delta}, \quad \forall i \in \mathcal{U}, \quad \forall t \geq 0. \tag{7.35}$$

where $\mathcal{U}$ is the set of undisturbed agents *with*

$$\mathcal{U} = \{i \in \mathcal{N} \mid d_i(t) = 0, \forall t \geq 0\}. \tag{7.36}$$

The occurrence of faulty communication links is investigated in Chapter 10. For such situations the aim is to preserve the maximal control error which occurs in the fault-free case for different initial states of Σ_0 and Σ_i.

Control aim 7.4 (Preserve performance in case of faulty communication links) *Consider the self-organizing control system Σ defined in (7.30), where the agents Σ_i, $(\forall i \in \mathcal{N}_{\rm F})$ are faulty in the way that the corresponding local control units C_i can not transmit any information. Then the faulty communication links have to be replaced by admissible new communication links such that the following two requirements are guaranteed:*

1. *The control errors $y_{\Delta i}(t)$, $(\forall i \in \mathcal{N})$, have to be smaller than or equal to the maximal control error $\bar{y}^*_{\Delta \mathrm{B}i}$ while using the basic communication graph $\mathcal{G}_{\rm B}$, i.e.,*

$$|y_{\Delta i}(t)| = |y_i(t) - y_{\rm s}(t)| \leq \bar{y}^*_{\Delta \mathrm{B}i}, \quad \forall i \in \mathcal{N}, \quad \forall t \geq 0 \tag{7.37}$$

   if the initial state $\boldsymbol{x}_{\rm s0}$ of the leader Σ_0 is different from the initial state of the agents Σ_i according to (7.31) and the outputs $y_i(t)$, $(\forall i \in \mathcal{N})$, of all agents Σ_i initially track the reference signal $y_{\rm s}(t)$ according to (7.32).

2. *A new communication link $(j \rightarrow i)$ from C_j to C_i has a minimal communication cost, i.e.,*

$$j = \arg \min_{j \in \mathcal{V}_{\rm C} \backslash \mathcal{N}, j < i} c_{ij}, \quad \forall i \in \mathcal{N}. \tag{7.38}$$

All three aims require different control strategies presented in the following chapters.

7.6 Analysis of the general self-organizing control system

This section investigates the asymptotic synchronization of the general self-organizing control system Σ defined in (7.30) as desired in Control aim 7.1 and the synchronization performance of Σ for a fixed communication structure as desired in Control aim 7.2. Therefore, the section is divided into three parts. First, important difference signals used to derive an overall difference system $\tilde{\Sigma}_\Delta$ are presented which describes the difference between the behavior of the leader Σ_0 and the behavior of the controlled agents $\bar{\Sigma}_i$, $(\forall i \in \mathcal{N})$, (Section 7.6.1). Second, conditions for the asymptotic synchronization of the controlled agents $\bar{\Sigma}_i$ (Control aim 7.1) interconnected by a situation-dependent communication structure are deduced while investigating the asymptotic stability of the overall difference system $\tilde{\Sigma}_\Delta$ (Section 7.6.2). Third, an upper bound on the control error $y_{\Delta i}(t)$ of the agents is derived, where the local control units of the agents communicate within a fixed communication structure (Section 7.6.3).

7.6.1 Difference behavior between leader and followers

This section derives an overall difference system whose output is the *overall control error*

$$\boldsymbol{y}_\Delta(t) = \begin{pmatrix} y_{\Delta 1}(t) \\ \vdots \\ y_{\Delta N}(t) \end{pmatrix} \in \mathbb{R}^r.$$

Initially, important difference signals are introduced.

Important difference signals. Figure 7.8 shows the relation between three difference signals

$$y_{\Delta i}(t) = y_i(t) - y_{\mathrm{s}}(t) \tag{7.39}$$

$$y_{\Delta \mathrm{s}i}(t) = y_{\mathrm{s}i}(t) - y_{\mathrm{s}}(t) \tag{7.40}$$

$$e_i(t) = y_{\mathrm{s}i}(t) - y_i(t) \tag{7.41}$$

for a controlled agent $\bar{\Sigma}_i$. The course of the local reference signal $y_{\mathrm{s}i}(t)$ is continuous since no switching in the communication structure is assumed. Recall that the control error $y_{\Delta i}(t)$ indicates the current performance of the agents. The *local reference error* $y_{\Delta \mathrm{s}i}(t) \in \mathbb{R}$ is the difference between the actual reference signal $y_{\mathrm{s}}(t)$ and the local reference signal $y_{\mathrm{s}i}(t)$ of the local control unit C_i. The *local control error* $e_i(t)$ is the current deviation between the local reference signal $y_{\mathrm{s}i}(t)$ and the output $y_i(t)$ of Σ_i. Hence, $e_i(t)$ indicates the local performance of the controlled agents $\bar{\Sigma}_i$.

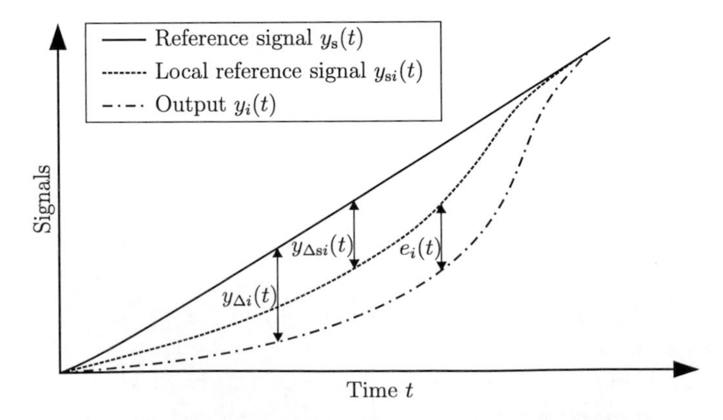

Figure 7.8: Important difference signals for a single controlled agent $\bar{\Sigma}_i$.

Transformed difference system. To analyze the synchronization behavior of the controlled agents $\bar{\Sigma}_i$, a system is derived which has the overall control error $\boldsymbol{y}_{\Delta}(t)$ as an output. Therefore, the transformed model $\tilde{\bar{\Sigma}}_i$ of controlled agents defined in (7.27) is used. Recall that $\tilde{\bar{\Sigma}}_i$ is divided into a part which is identical to the dynamics of the leader Σ_0 and a remaining part which describes the difference to the dynamics of the leader. Due to this representation of the agents, a difference system can be derived which describes the difference between the behavior of the controlled agents $\bar{\Sigma}_i$ and the behavior of the leader Σ_0.

Consider the *transformed difference state*

$$\tilde{\boldsymbol{x}}_{\Delta i}(t) = \begin{pmatrix} \boldsymbol{x}_{\mathrm{q}i}(t) - \boldsymbol{x}_{\mathrm{s}}(t) \\ \boldsymbol{x}_{\mathrm{p}i}(t) \end{pmatrix} \in \mathbb{R}^{n_i} \tag{7.42}$$

describing the difference between the transformed controlled agents $\tilde{\bar{\Sigma}}_i$ defined in (7.27) and the leader Σ_0 defined in (7.3). Then the control error $y_{\Delta i}$ defined in (7.39) is the output of the *transformed controlled difference subsystem*

$$\tilde{\bar{\Sigma}}_{\Delta i}: \begin{cases} \frac{\mathrm{d}}{\mathrm{d}t}\tilde{\boldsymbol{x}}_{\Delta i}(t) = \tilde{\bar{\boldsymbol{A}}}_i \tilde{\boldsymbol{x}}_{\Delta i}(t) + k_{\mathrm{P}i}\tilde{\boldsymbol{b}}_i y_{\Delta \mathrm{s}i}(t) + \tilde{\boldsymbol{g}}_i d_i(t), \quad \tilde{\boldsymbol{x}}_{\Delta i}(0) = \tilde{\boldsymbol{x}}_{\Delta 0i} \\ y_{\Delta i}(t) = \tilde{\boldsymbol{c}}_i^{\mathrm{T}} \tilde{\boldsymbol{x}}_{\Delta i}(t), \end{cases} \tag{7.43}$$

for all $i \in \mathcal{N}$, where $\frac{\mathrm{d}}{\mathrm{d}t}\Delta\tilde{\boldsymbol{x}}_i(t)$ results from

$$\begin{aligned} \frac{\mathrm{d}}{\mathrm{d}t}\Delta\tilde{\boldsymbol{x}}_i(t) &= \frac{\mathrm{d}}{\mathrm{d}t}\tilde{\boldsymbol{x}}_i(t) - \dot{\boldsymbol{x}}_{\mathrm{s}}(t) \\ &= \begin{pmatrix} \boldsymbol{A}_{\mathrm{s}} - k_{\mathrm{P}i}\boldsymbol{b}_{\mathrm{q}i}\boldsymbol{c}_{\mathrm{s}}^{\mathrm{T}} & -k_{\mathrm{P}i}\boldsymbol{b}_{\mathrm{q}i}\boldsymbol{c}_{\mathrm{p}i}^{\mathrm{T}} \\ -k_{\mathrm{P}i}\boldsymbol{b}_{\mathrm{p}i}\boldsymbol{c}_{\mathrm{s}}^{\mathrm{T}} & \boldsymbol{A}_{\mathrm{p}i} - k_{\mathrm{P}i}\boldsymbol{b}_{\mathrm{p}i}\boldsymbol{c}_{\mathrm{p}i}^{\mathrm{T}} \end{pmatrix} \begin{pmatrix} \boldsymbol{x}_{\mathrm{q}i}(t) \\ \boldsymbol{x}_{\mathrm{p}i}(t) \end{pmatrix} - \begin{pmatrix} \boldsymbol{A}_{\mathrm{s}} & \boldsymbol{O} \\ \boldsymbol{O} & \boldsymbol{O} \end{pmatrix} \begin{pmatrix} \boldsymbol{x}_{\mathrm{s}}(t) \\ \boldsymbol{0} \end{pmatrix} \\ &\quad + \begin{pmatrix} k_{\mathrm{P}i}\boldsymbol{b}_{\mathrm{q}i} \\ k_{\mathrm{P}i}\boldsymbol{b}_{\mathrm{q}i} \end{pmatrix} \underbrace{y_{\mathrm{s}i}(t)}_{y_{\mathrm{s}}(t) + y_{\Delta \mathrm{s}i}(t)} + \tilde{\boldsymbol{g}}_i d_i(t) \\ &= \begin{pmatrix} \boldsymbol{A}_{\mathrm{s}} - k_{\mathrm{P}i}\boldsymbol{b}_{\mathrm{q}i}\boldsymbol{c}_{\mathrm{s}}^{\mathrm{T}} & -k_{\mathrm{P}i}\boldsymbol{b}_{\mathrm{q}i}\boldsymbol{c}_{\mathrm{p}i}^{\mathrm{T}} \\ -k_{\mathrm{P}i}\boldsymbol{b}_{\mathrm{p}i}\boldsymbol{c}_{\mathrm{s}}^{\mathrm{T}} & \boldsymbol{A}_{\mathrm{p}i} - k_{\mathrm{P}i}\boldsymbol{b}_{\mathrm{p}i}\boldsymbol{c}_{\mathrm{p}i}^{\mathrm{T}} \end{pmatrix} \begin{pmatrix} \boldsymbol{x}_{\mathrm{q}i}(t) \\ \boldsymbol{x}_{\mathrm{p}i}(t) \end{pmatrix} - \begin{pmatrix} \boldsymbol{A}_{\mathrm{s}} & \boldsymbol{O} \\ \boldsymbol{O} & \boldsymbol{O} \end{pmatrix} \begin{pmatrix} \boldsymbol{x}_{\mathrm{s}}(t) \\ \boldsymbol{0} \end{pmatrix} \\ &\quad + \begin{pmatrix} k_{\mathrm{P}i}\boldsymbol{b}_{\mathrm{q}i}\boldsymbol{c}_{\mathrm{s}}^{\mathrm{T}} & \boldsymbol{O} \\ k_{\mathrm{P}i}\boldsymbol{b}_{\mathrm{p}i}\boldsymbol{c}_{\mathrm{s}}^{\mathrm{T}} & \boldsymbol{O} \end{pmatrix} \begin{pmatrix} \boldsymbol{x}_{\mathrm{s}}(t) \\ \boldsymbol{0} \end{pmatrix} + \begin{pmatrix} k_{\mathrm{P}i}\boldsymbol{b}_{\mathrm{q}i} \\ k_{\mathrm{P}i}\boldsymbol{b}_{\mathrm{p}i} \end{pmatrix} y_{\Delta \mathrm{s}i}(t) + \tilde{\boldsymbol{g}}_i d_i(t) \\ &= \underbrace{\begin{pmatrix} \boldsymbol{A}_{\mathrm{s}} - k_{\mathrm{P}i}\boldsymbol{b}_{\mathrm{q}i}\boldsymbol{c}_{\mathrm{s}}^{\mathrm{T}} & -k_{\mathrm{P}i}\boldsymbol{b}_{\mathrm{q}i}\boldsymbol{c}_{\mathrm{p}i}^{\mathrm{T}} \\ -k_{\mathrm{P}i}\boldsymbol{b}_{\mathrm{p}i}\boldsymbol{c}_{\mathrm{s}}^{\mathrm{T}} & \boldsymbol{A}_{\mathrm{p}i} - k_{\mathrm{P}i}\boldsymbol{b}_{\mathrm{p}i}\boldsymbol{c}_{\mathrm{p}i}^{\mathrm{T}} \end{pmatrix}}_{\tilde{\bar{\boldsymbol{A}}}_i} \underbrace{\begin{pmatrix} \boldsymbol{x}_{\mathrm{q}i}(t) - \boldsymbol{x}_{\mathrm{s}}(t) \\ \boldsymbol{x}_{\mathrm{p}i}(t) \end{pmatrix}}_{\tilde{\boldsymbol{x}}_{\Delta i}} + k_{\mathrm{P}i} \underbrace{\begin{pmatrix} \boldsymbol{b}_{\mathrm{q}i} \\ \boldsymbol{b}_{\mathrm{p}i} \end{pmatrix}}_{\tilde{\boldsymbol{b}}_i} y_{\Delta \mathrm{s}i}(t) + \tilde{\boldsymbol{g}}_i d_i(t). \end{aligned}$$

The input to $\tilde{\bar{\Sigma}}_{\Delta i}$ is the local reference error $y_{\Delta si}(t)$. Due to (7.18) and (7.20), the relation

$$\begin{aligned} y_{\Delta si}(t) &= y_{si}(t) - y_s(t) \\ &= \sum_{j \in \mathcal{P}_{Ci}(t)} \tilde{k}_{ij}(t) y_j(t) - \sum_{j \in \mathcal{P}_{Ci}(t)} \tilde{k}_{ij}(t) y_s(t) \\ &= \sum_{j \in \mathcal{P}_{Ci}(t)} \tilde{k}_{ij}(t)(y_j(t) - y_s(t)) \\ &= \sum_{j \in \mathcal{P}_{Ci}(t)} \tilde{k}_{ij}(t) y_{\Delta j}(t) \end{aligned} \tag{7.44}$$

holds true for $\mathcal{P}_{Ci}(t) \neq \emptyset$. Note that for $\mathcal{P}_{Ci}(t) = \emptyset$, $y_{\Delta si}(t)$ only depends on $y_s(t)$. For this, the local reference error $y_{\Delta si}(t)$ can not be expressed by the control errors $y_{\Delta j}(t)$.

The overall transformed difference system results from (7.43) and (7.44) to

$$\tilde{\Sigma}_\Delta : \begin{cases} \dot{\boldsymbol{x}}_\Delta(t) = \tilde{\bar{\boldsymbol{A}}}(t)\tilde{\boldsymbol{x}}_\Delta(t) + \tilde{\boldsymbol{G}}\boldsymbol{d}(t), & \tilde{\boldsymbol{x}}_\Delta(0) = \tilde{\boldsymbol{x}}_{\Delta 0} \\ \boldsymbol{y}_\Delta(t) = \tilde{\boldsymbol{C}}\tilde{\boldsymbol{x}}_\Delta(t) \end{cases} \tag{7.45}$$

with

$$\begin{aligned} \tilde{\bar{\boldsymbol{A}}}(t) &= \tilde{\boldsymbol{A}} + \tilde{\boldsymbol{B}}\boldsymbol{K}_P\tilde{\boldsymbol{K}}_F(t)\tilde{\boldsymbol{C}} \\ \tilde{\boldsymbol{A}} &= \text{diag}(\tilde{\boldsymbol{A}}_1 \dots \tilde{\boldsymbol{A}}_N) \\ \tilde{\boldsymbol{B}} &= \text{diag}(\tilde{\boldsymbol{b}}_1 \dots \tilde{\boldsymbol{b}}_N) \\ \tilde{\boldsymbol{C}} &= \text{diag}(\tilde{\boldsymbol{c}}_1^T \dots \tilde{\boldsymbol{c}}_N^T) \\ \tilde{\boldsymbol{G}} &= \text{diag}(\tilde{\boldsymbol{g}}_1 \dots \tilde{\boldsymbol{g}}_N) \\ \tilde{\boldsymbol{x}}_\Delta(t) &= \begin{pmatrix} \tilde{\boldsymbol{x}}_{\Delta 1}(t) \\ \vdots \\ \tilde{\boldsymbol{x}}_{\Delta N}(t) \end{pmatrix}, \quad \tilde{\boldsymbol{x}}_0 = \begin{pmatrix} \tilde{\boldsymbol{x}}_{\Delta 01} \\ \vdots \\ \tilde{\boldsymbol{x}}_{\Delta 0N} \end{pmatrix}, \end{aligned} \tag{7.46}$$

where $\tilde{\boldsymbol{x}}_\Delta(t) \in \mathbb{R}^n$ is the overall transformed difference state and $\tilde{\boldsymbol{x}}_{\Delta 0} \in \mathbb{R}^n$ is the overall transformed initial difference state of $\tilde{\Sigma}_\Delta$.

Similar to the system matrix $\bar{\boldsymbol{A}}(t)$ of the controlled follower agents $\bar{\Sigma}_{FA}$ the system matrix $\tilde{\bar{\boldsymbol{A}}}(t) = (\bar{\boldsymbol{A}}_{ij}(t))$ of the overall transformed difference system $\tilde{\Sigma}_\Delta$ has the blocks

$$\begin{aligned} \tilde{\bar{\boldsymbol{A}}}_{ii}(t) &= (\tilde{\boldsymbol{A}}_i - \tilde{\boldsymbol{b}}_i k_{Pi}\tilde{\boldsymbol{c}}_i^T), \quad \forall i \in \mathcal{N}, \quad \forall t \geq 0 \\ \tilde{\bar{\boldsymbol{A}}}_{ij}(t) &= \tilde{k}_{ij}(t)\tilde{\boldsymbol{b}}_i k_{Pi}\tilde{\boldsymbol{c}}_j^T, \quad i \neq j, \quad \forall i,j \in \mathcal{N}, \quad \forall t \geq 0 \end{aligned}$$

Hence, only the non-diagonal blocks depend upon the switching of the communication within

$\tilde{\boldsymbol{K}}(t)$. Furthermore, also the relation

$$\tilde{\tilde{\boldsymbol{A}}}_{ij}(t) = \boldsymbol{O}_{n_i \times n_j}, \quad \forall i,j \in \mathcal{N}, \quad i < j, \quad \forall t \geq 0$$

holds according to (7.14). Hence, $\tilde{\tilde{\boldsymbol{A}}}(t)$ is a block lower triangular matrix for all $t \geq 0$.

In the following, this overall transformed difference system $\tilde{\Sigma}_\Delta$ is used to derive conditions on the asymptotic synchronization of the controlled agents $\bar{\Sigma}_i$.

7.6.2 Asymptotic synchronization with situation-dependent communication

This section investigates the asymptotic synchronization of the general self-organizing control system Σ defined in (7.30) as desired in Control aim 7.1. The analysis is based on the results in [89] which are adapted for the used notation and the occurring situation-dependent communication. To fulfill Control aim 7.1, the overall control error $\boldsymbol{y}_\Delta(t)$ has to vanish for $t \to \infty$, i.e.,

$$\lim_{t\to\infty} |\boldsymbol{y}_\Delta(t)| = \lim_{t\to\infty} \begin{pmatrix} |y_1(t) - y_s(t)| \\ \vdots \\ |y_N(t) - y_s(t)| \end{pmatrix} = \lim_{t\to\infty} \begin{pmatrix} |y_{\Delta 1}(t)| \\ \vdots \\ |y_{\Delta N}(t)| \end{pmatrix} = \boldsymbol{0}. \tag{7.47}$$

Consider the overall transformed difference system $\tilde{\Sigma}_\Delta$ defined in (7.45). Then the overall difference output $\boldsymbol{y}_\Delta(t)$ vanishes according to (7.47), if $\tilde{\Sigma}_\Delta$ is asymptotically stable. To obtain asymptotic stability of $\tilde{\Sigma}_\Delta$, it is not enough to check if $\tilde{\tilde{\boldsymbol{A}}}(t)$ is a Hurwitz matrix since $\tilde{\Sigma}_\Delta$ is a switched linear system. Furthermore, generally the overall disturbance $\boldsymbol{d}(t)$ has to vanish to guarantee asymptotic stability of $\tilde{\Sigma}_\Delta$. With this, the following lemma presents general conditions for the asymptotic synchronization of the agents Σ_i, $(\forall i \in \mathcal{N})$.

Lemma 7.2 (Asymptotic synchronization with situation-dependent communication) *Consider the general self-organizing control system Σ defined in (7.30). The agents Σ_i defined in (7.2) are asymptotically synchronized to the leader Σ_0 defined in (7.3) as desired in Control aim 7.1, if the following conditions hold true after a finite time $\bar{t} \in \mathbb{R}_+$:*

1. *Every agent Σ_i, $(\forall i \in \mathcal{N})$, has a predecessor, i.e.,*

$$\mathcal{P}_{\mathrm{C}i}(t) \neq \emptyset, \quad \forall i \in \mathcal{N}, \quad \forall t \geq \bar{t} \geq 0. \tag{7.48}$$

2. *The disturbances $d_i(t)$, $(\forall i \in \mathcal{N})$, affecting the agents Σ_i vanish, i.e.,*

$$d_i(t) = 0, \quad \forall i \in \mathcal{N}, \quad \forall t \geq \bar{t} \geq 0. \tag{7.49}$$

Proof. The proof is given in Appendix A.11. □

Lemma 7.2 shows that the agents Σ_i can synchronize to the leader Σ_0 due to the following three properties of Σ:

1. The agents Σ_i include the dynamics of the leader Σ_0 (Assumption 7.4) such that they can generate the same motion as Σ_0.
2. Due to the cycle free communication, only the controlled agents $\bar{\Sigma}_i$ have to be asymptotically stable (Assumption 7.5) to track their local reference signal $y_{si}(t)$ and, hence, also track the reference signal $y_s(t)$ generated by the leader Σ_0.
3. The well-known requirement of a spanning tree with the leader as root node is required by condition (7.48) (cf. Lemma 7.1).

In summary, the following property of the general self-organizing control system Σ defined in (7.30) resulting from Lemma 7.2 is highlighted

> If all agents have a predecessor and no disturbance occurs, then the agents Σ_i, $(\forall i \in \mathcal{N})$, asymptotically synchronize with the leader Σ_0.

Hence, the control concepts presented in the following chapters have to guarantee that despite the situation-dependent communication every agent still has a predecessor. In particular, the condition (7.48) has to be fulfilled by all concepts.

Example 7.1 *Robot formation with arbitrary switching of the communication*

Figure 7.9 shows the behavior of the robot formation with four robots (upper part of Fig 7.2) controlled by the general self-organizing controller C defined in (7.25) for an arbitrary switching of the communication links $\mathcal{E}_C$. Note that the model of the robots and corresponding feedback unit F_i are presented in Section 7.7. It can be seen that despite the arbitrary switching of the set $\mathcal{P}_{Ci}$ of predecessors the robots (second plot) still asymptotically synchronize to the reference position s_0 of the leading robot Σ_0 since every robot always has a predecessor, i.e., $\mathcal{P}_{Ci}(t) \neq \emptyset$ for all $t \geq 0$ (cf. (7.48)). □

7.6.3 Synchronization performance within a fixed communication structure

In this section the performance of the controlled agents interconnected via a fixed communication graph $\mathcal{G}_B$ is analyzed w.r.t the maximal control error $\max_{t\geq 0} |y_{\Delta i}(t)|$, $(\forall i \in \mathcal{N})$. It is shown that there exists an upper bound on $\max_{t\geq 0} |y_{\Delta i}(t)|$ which depends on the choice of the basic communication graphs $\mathcal{G}_B$ (Lemma 7.3). Initially, the basic communication graph $\mathcal{G}_B$ and the resulting overall closed-loop system Σ_B using $\mathcal{G}_B$ are introduced. Furthermore, the situation to be investigated is presented which is equivalent to the one in Control aim 7.2.

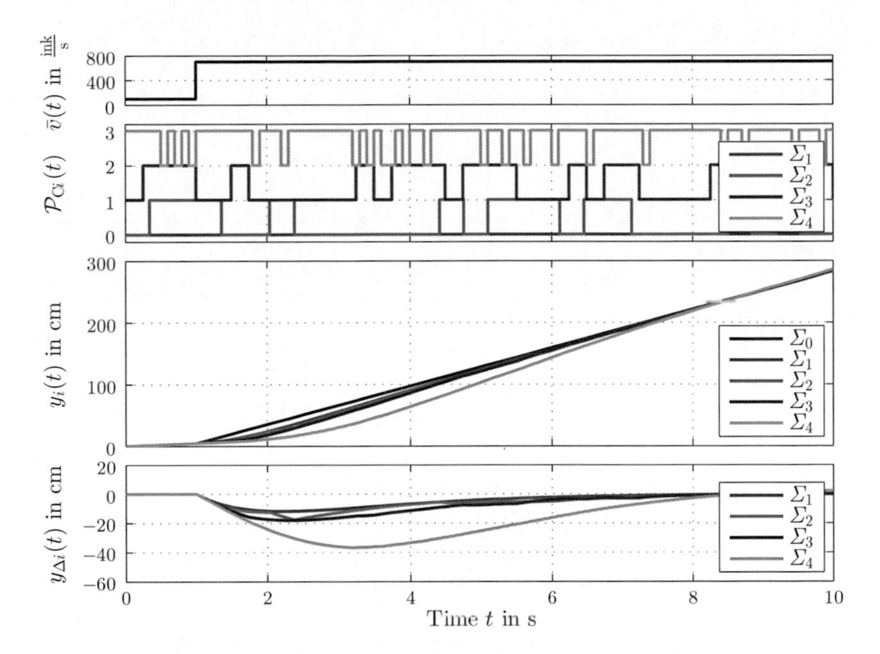

Figure 7.9: Asymptotic synchronization with arbitrary switching.

Basic communication graph $\mathcal{G}_{\mathrm{B}}$. The *basic communication graph* $\mathcal{G}_{\mathrm{B}}$ is a time-invariant labeled graph defined by

$$\mathcal{G}_{\mathrm{B}} = (\mathcal{V}_{\mathrm{C}}, \mathcal{E}_{\mathrm{B}}, \boldsymbol{K}), \tag{7.50}$$

where the weighting matrix $\boldsymbol{K}$ is the same as for the communication graph $\mathcal{G}_{\mathrm{C}}(t)$ defined in (7.9). The corresponding *set $\mathcal{E}_{\mathrm{B}}$ of basic edges* is defined by

$$\mathcal{E}_{\mathrm{B}} = \{(j \to i) | j \in \mathcal{P}_{\mathrm{B}i}, i \in \mathcal{V}_{\mathrm{C}}\}$$

where $\mathcal{P}_{\mathrm{B}i}$ is the *set of basic predecessors* with

$$\mathcal{P}_{\mathrm{B}i} \subseteq \{j \in \mathcal{V}_{\mathrm{C}} \mid j < i\}, \quad \forall i \in \mathcal{N}.$$

The *basic communication graph* $\mathcal{G}_{\mathrm{B}}$ is required to have a spanning tree with the leader Σ_0 as root node which is guaranteed by the following assumption (cf. Lemma 7.1).

Assumption 7.6 *Within the basic communication graph* $\mathcal{G}_{\mathrm{B}}$ *defined in* (7.50) *every agent has* a predecessors, i.e., $\mathcal{P}_{\mathrm{B}i} \neq \emptyset$ for all $i \in \mathcal{N}$.

With this, the *set $\mathcal{F}_{\mathrm{B}i}$ of basic followers* is defined by

$$\mathcal{F}_{\mathrm{B}i} = \{j \in \mathcal{V}_{\mathrm{C}} \mid i \in \mathcal{P}_{\mathrm{B}j}\}, \quad i \in \mathcal{V}_{\mathrm{C}}$$

similar to (2.18). According to (7.19), the *basic communication matrix* $\boldsymbol{K}_{\mathrm{B}} = (k_{\mathrm{B}ij}) \in \mathbb{R}^{V \times V}$ is build by the elements

$$k_{\mathrm{B}ij} = \begin{cases} \dfrac{k_{ij}}{\sum_{l \in \mathcal{P}_{\mathrm{B}i}} k_{il}} & \text{if } j \in \mathcal{P}_{\mathrm{B}i} \\ 0 & \text{else,} \end{cases} \qquad i, j \in \mathcal{V}_{\mathrm{C}}.$$

Closed-loop system with basic communication graph $\mathcal{G}_{\mathrm{B}}$. The overall closed-loop system Σ_{B} using the basic communication graph $\mathcal{G}_{\mathrm{B}}$ is defined by

$$\Sigma_{\mathrm{B}}: \quad \Sigma \text{ defined in (7.30) with } \mathcal{G}_{\mathrm{C}}(t) = \mathcal{G}_{\mathrm{B}} \text{ for all } t \geq 0. \tag{7.51}$$

Hence, Σ_{B} has a fixed communication graph which means that the communication links do not change, i.e., $\mathcal{E}(t) = \mathcal{E}_{\mathrm{B}}$ for all $t \geq 0$.

Situation to be investigated. Consider Control aim 7.2 with the situation that the transformed initial states of the agents Σ_i, $(\forall i \in \mathcal{N})$, are identical

$$\tilde{\boldsymbol{x}}_{0i} = \boldsymbol{T}_i^{-1} \boldsymbol{x}_{0i} \boldsymbol{T}_i = \begin{pmatrix} \boldsymbol{x}_{\mathrm{q}0i} \\ \mathbf{0} \end{pmatrix} =: \begin{pmatrix} \boldsymbol{x}_{\mathrm{q}0} \\ \mathbf{0} \end{pmatrix}, \quad \forall i \in \mathcal{N}, \tag{7.52}$$

indicating that initially the outputs $y_i(t)$, $(\forall i \in \mathcal{N})$ of the agents are identical (7.32). Furthermore, the initial state $\boldsymbol{x}_{\mathrm{s}0}$ of the leader of Σ_0 is assumed to be different from the initial state $\boldsymbol{x}_{\mathrm{q}0}$, i.e.,

$$\boldsymbol{x}_{\mathrm{s}0} \neq \boldsymbol{x}_{\mathrm{q}0}. \tag{7.53}$$

With this, the initial state

$$\tilde{\boldsymbol{x}}^*_{\Delta 0} := \tilde{\boldsymbol{x}}_{\Delta 0i} = \begin{pmatrix} \boldsymbol{x}_{\mathrm{q}0} - \boldsymbol{x}_{\mathrm{s}0} \\ \mathbf{0} \end{pmatrix} = \begin{pmatrix} \boldsymbol{x}_{\Delta \mathrm{q}0} \\ \mathbf{0} \end{pmatrix} \tag{7.54}$$

represents the difference between the initial state $\boldsymbol{x}_{s0}$ of the leader Σ_0 in (7.3) and the transformed initial state $\tilde{\boldsymbol{x}}_{0i}$ of the agents in (7.52).

Maximal control error using the basic communication graph $\mathcal{G}_B$. According to (7.27), (7.40) and (7.51), for Σ_B the relation

$$y_{\Delta i}(t) = \tilde{\boldsymbol{c}}_i^T e^{\tilde{\bar{\boldsymbol{A}}}_i t} \tilde{\boldsymbol{x}}_{\Delta 0i} + \int_0^t \tilde{\boldsymbol{c}}_i^T e^{\tilde{\bar{\boldsymbol{A}}}_i (t-\tau)} k_{Pi} \tilde{\boldsymbol{b}}_i y_{\Delta si}(\tau) d\tau, \quad \forall i \in \mathcal{N} \tag{7.55}$$

$$= \tilde{\boldsymbol{c}}_i^T e^{\tilde{\bar{\boldsymbol{A}}}_i t} \tilde{\boldsymbol{x}}_{\Delta 0i} + \int_0^t \tilde{\boldsymbol{c}}_i^T e^{\tilde{\bar{\boldsymbol{A}}}_i (t-\tau)} k_{Pi} \tilde{\boldsymbol{b}}_i \sum_{j \in \mathcal{P}_{Bi}} k_{Bij} y_{\Delta j}(\tau) d\tau, \quad \forall i \in \mathcal{N} \tag{7.56}$$

holds true for all $t \geq 0$. With this, the following lemma derives an upper bound on the maximal control errors $y_{\Delta i}(t)$, $(\forall i \in \mathcal{N})$ of the agents within Σ_B.

Lemma 7.3 *Consider the overall closed-loop system Σ_B with a fixed basic communication graph defined in* (7.51) *and the initial situation in* (7.52) *and* (7.53)*. The control error $y_{\Delta i}(t)$ of the agents Σ_i defined in* (7.2) *is bounded by the* maximal basic control error $\bar{y}_{\Delta Bi}$, *i.e.,*

$$\max_{t \geq 0} |y_{\Delta i}(t)| \leq \bar{y}_{\Delta Bi} = m_{y0Bi} + m_{yyi} \sum_{j \in \mathcal{P}_{Bi}} k_{Bij} \bar{y}_{\Delta Bj}, \quad \forall i \in \mathcal{N} \tag{7.57}$$

with $\bar{y}_{\Delta B0} = 0$ and

$$m_{yyi} = \int_0^\infty \left| \tilde{\boldsymbol{c}}_i^T e^{\tilde{\bar{\boldsymbol{A}}}_i \tau} k_{Pi} \tilde{\boldsymbol{b}}_i \right| d\tau, \quad \forall i \in \mathcal{N} \tag{7.58}$$

$$m_{y0Bi} = \boldsymbol{m}_{y0i}^T \bar{\boldsymbol{x}}_{\Delta 0}, \quad \forall i \in \mathcal{N}$$

$$\boldsymbol{m}_{y0i}^T = \max_{t \geq 0} \left| \tilde{\boldsymbol{c}}_i^T e^{\tilde{\bar{\boldsymbol{A}}}_i t} \right|, \quad \forall i \in \mathcal{N}$$

and $\bar{\boldsymbol{x}}_{\Delta 0} = |\tilde{\boldsymbol{x}}_{\Delta 0}^*|$. Hence, the overall control error $\boldsymbol{y}_\Delta(t)$ is bounded by the overall maximal basic control error $\bar{\boldsymbol{y}}_{\Delta B} \in \mathbb{R}^N$:

$$\max_{t \geq 0} |\boldsymbol{y}_\Delta(t)| \leq \bar{\boldsymbol{y}}_{\Delta B} = (\boldsymbol{I} - \boldsymbol{M}_{yy} \boldsymbol{K}_B^*)^{-1} \boldsymbol{m}_{y0B} \tag{7.59}$$

with

$$\bar{\boldsymbol{y}}_{\Delta B} = \begin{pmatrix} \bar{y}_{\Delta B1} \\ \vdots \\ \bar{y}_{\Delta BN} \end{pmatrix}, \quad \boldsymbol{m}_{y0B} = \begin{pmatrix} m_{y0B1} \\ \vdots \\ m_{y0BN} \end{pmatrix} \tag{7.60}$$

$$\boldsymbol{M}_{\text{yy}} = \text{diag}(m_{\text{yy}1} \dots m_{\text{yy}N}) \tag{7.61}$$

$$\boldsymbol{K}_{\text{B}}^* = \begin{pmatrix} \boldsymbol{0}_N^{\text{T}} \\ \boldsymbol{I}_N \end{pmatrix} \boldsymbol{K}_{\text{B}} \begin{pmatrix} \boldsymbol{0}_N & \boldsymbol{I}_N \end{pmatrix}.$$

Proof. According to (7.56), the relation

$$\max_{t \geq 0} |y_{\Delta i}(t)| \leq \left| \tilde{\boldsymbol{c}}_i^{\text{T}} \text{e}^{\tilde{\tilde{\boldsymbol{A}}}_i t} \right| |\tilde{\boldsymbol{x}}_{\Delta 0i}| + \int_0^{\infty} \left| \tilde{\boldsymbol{c}}_i^{\text{T}} \text{e}^{\tilde{\tilde{\boldsymbol{A}}}_i \tau} k_{\text{P}i} \tilde{\boldsymbol{b}}_i \right| \text{d}\tau \sum_{j \in \mathcal{P}_{\text{B}i}} k_{\text{B}ij} \max_{t \geq 0} |y_{\Delta j}(t)|, \quad \forall i \in \mathcal{N}$$

holds true. The scalars $m_{\text{yy}i}$, $(\forall i \in \mathcal{N})$, and the vectors $\boldsymbol{m}_{\text{y}0i}^{\text{T}}$ are finite since $\tilde{\tilde{\boldsymbol{A}}}_i$ are Hurwitz matrices according to Assumption 7.5. The relation $\tilde{\boldsymbol{x}}_{\Delta 0}^* = \tilde{\boldsymbol{x}}_{\Delta 0i}$, $(\forall i \in \mathcal{N})$, holds true according to (7.54) which proves the relations in (7.57). The composition of relations in (7.57) leads to

$$\max_{t \geq 0} |\boldsymbol{y}_{\Delta}(t)| \leq \boldsymbol{m}_{\text{y0B}} + \boldsymbol{M}_{\text{yy}} \boldsymbol{K}_{\text{B}}^* \max_{t \geq 0} |\boldsymbol{y}_{\Delta}(t)|. \tag{7.62}$$

Solving (7.62) for $\max_{t \geq 0} |\boldsymbol{y}_{\Delta}(t)|$ yields the bound in (7.59). The inverse matrix $(\boldsymbol{I} - \boldsymbol{M}_{\text{yy}} \boldsymbol{K}_{\text{B}}^*)^{-1}$ exists, because $\boldsymbol{K}_{\text{B}}^*$ is a lower triangular matrix with vanishing diagonal elements. Hence, the matrix $\boldsymbol{I} - \boldsymbol{M}_{\text{yy}} \boldsymbol{K}_{\text{B}}^*$ has N identical eigenvalues with the value 1, i.e., $\lambda_i(\boldsymbol{I} - \boldsymbol{M}_{\text{yy}} \boldsymbol{K}_{\text{B}}^*) = 1$ for all $i \in \mathcal{N}$. Furthermore, the matrix $\boldsymbol{I} - \boldsymbol{M}_{\text{yy}} \boldsymbol{K}_{\text{B}}^*$ is an M-Matrix and, hence, its inverse a non-negative matrix. Consequently, the inequality sign remains when multiplying with the inverse matrix. □

Lemma 7.3 gives an upper bound $\bar{y}_{\Delta \text{B}i}$ for the maximum control error $\max_{t \geq 0} |y_{\Delta i}(t)|$ which depends on the dynamics of the agents Σ_i, $(\forall i \in \mathcal{N})$, the initial state $\tilde{\boldsymbol{x}}_{\Delta 0}^*$ as well as on the choice of the basic communication graph $\mathcal{G}_{\text{B}}$. This shows that the synchronization performance can be adapted by the choice of the basic communication graph $\mathcal{G}_{\text{B}}$. Figure 7.10 illustrates this property for a system with two agents. Since C_2 receives the output $y_1(t)$ of Σ_1 within

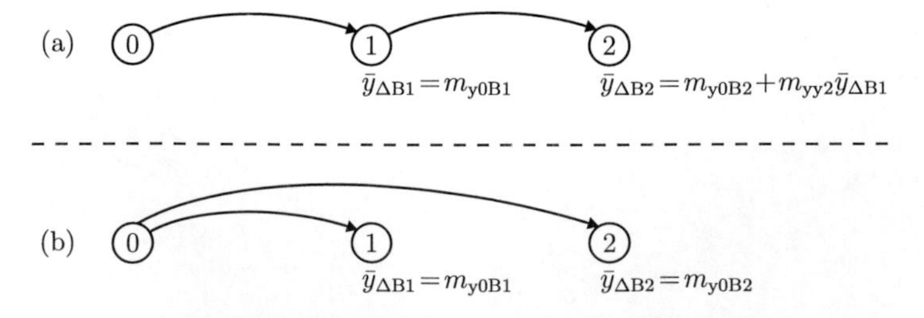

Figure 7.10: Maximal basic control error $\bar{y}_{\Delta \text{B}i}$ for two different basic communication graphs $\mathcal{G}_{\text{B}}$.

the basic communication graph $\mathcal{G}_{\rm B}$ in Fig. 7.10 (a), the maximal basic control error $\bar{y}_{\Delta \rm B2}$ of Σ_2 depends on the maximal basic control error $\bar{y}_{\Delta \rm B1}$ of Σ_1. Depending on the dynamics of Σ_1, this might lead to a bound $\bar{y}_{\Delta \rm B2}$ which is much bigger than using $\mathcal{G}_{\rm B}$ depicted in Fig. 7.10 (b). With direct information from Σ_0 the bound $\bar{y}_{\Delta \rm B2}$ only depends on the dynamics of Σ_2 represented by $m_{\rm y0B2}$.

7.7 Running example: Robot formation

This section introduces a robot formation as a running example, where the number of robots varies for the specific examples. The formation problem and basic structure of the concerned lab plant is presented (Section 7.7.1). According to the general model of the agents presented in Section 7.2.1, the models of the robots are given in Section 7.7.2. Section 7.7.3 introduces a specific robot formation with 21 robots.

7.7.1 Formation problem and structure of the lab plant

Figure 7.11 shows the lab plant SAMS (abbreviation for "**S**ynchronisation of **A**utonomous **M**obile **S**ystems) at the Institute of Automation and Computer Control at the Ruhr-University Bochum, Germany. The plant consists of mobile robots moving on a table. The robots are filmed by a camera whose pictures are evaluated at a computer to track the position of the robots. To detect the orientation of the robots, every robot is equipped with a white disk with three black dots on top. The applied controllers are implemented in MATLAB®/Simulink. The

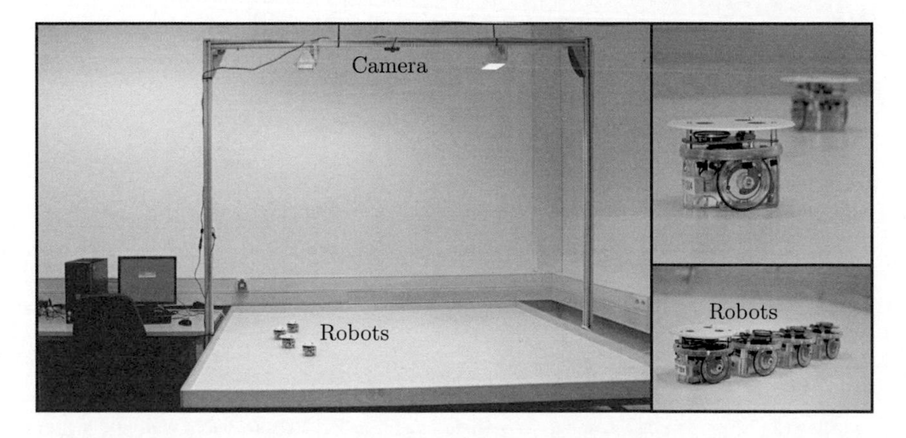

Figure 7.11: Structure of the lab plant SAMS.

(a)

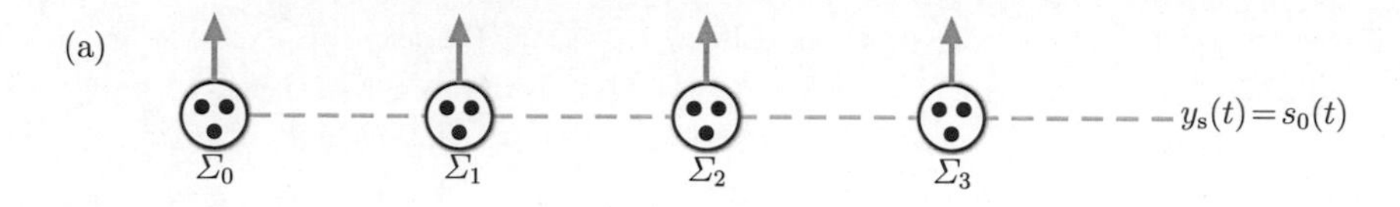

(b)

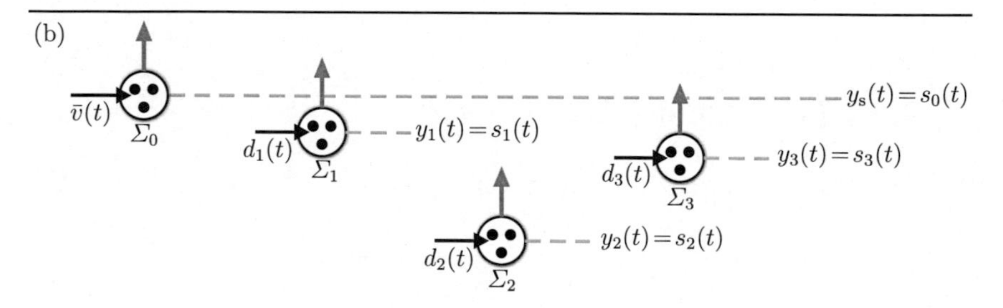

Figure 7.12: Robot formation problem.

information exchange between the computer and the robots is wireless.

The aim is to control the robots such that they move in a desired formation, e.g., depicted in Fig. 7.12 (a). Therefore, the robots have to track the position $s_0(t) = y_s(t)$ of the leading robot Σ_0. Due to a change of the reference velocity $\bar{v}(t)$ at the leading robot Σ_0 or a disturbance $d_i(t)$ affecting the follower agent Σ_i, there occurs a deviation between $s_0(t) = y_s(t)$ and the position $s_i(t) = y_i(t)$ of Σ_i (Fig. 7.12 (b)), which is described by the control error $y_{\Delta i}(t) = y_i(t) - y_s(t) = s_i(t) - s_0(t)$.

7.7.2 Model of the leading robot and the follower robots

The models of the *follower robots* Σ_i, $(\forall i \in \mathcal{N})$, are given by

$$\Sigma_i : \begin{cases} \dot{\boldsymbol{x}}_i(t) = \underbrace{\begin{pmatrix} 0 & 2\alpha_{\mathrm{A}i} k_{\mathrm{I}i} \\ 0 & 0 \end{pmatrix}}_{\boldsymbol{A}_i} \underbrace{\begin{pmatrix} v_i(t) \\ s_i(t) \end{pmatrix}}_{\boldsymbol{x}_i(t)} + \underbrace{\begin{pmatrix} 2\alpha_{\mathrm{A}i} k_{\mathrm{m}i} \\ k_{\mathrm{m}i} \end{pmatrix}}_{\boldsymbol{b}_i} u_i(t) + \underbrace{\begin{pmatrix} -2\alpha_{\mathrm{A}i} k_{\mathrm{m}i} \\ -k_{\mathrm{m}i} \end{pmatrix}}_{\boldsymbol{g}_i} d_i(t) \\ \boldsymbol{x}_i(0) = \boldsymbol{x}_{0i} \\ y_i(t) = \underbrace{\begin{pmatrix} 2.5 & 0 \end{pmatrix}}_{\boldsymbol{c}_i^{\mathrm{T}}} \boldsymbol{x}_i(t) \end{cases} \tag{7.63}$$

according to (7.2), where the parameters are listed in Table 7.1. Robots with different dynamics are investigated:

- An "average robot" Σ_i, $(\forall i \in \mathcal{N}_{\text{average}})$, has the motor constant $k_{\text{m}i} = 1.0$, where the indices of such robots are included in the *set* $\mathcal{N}_{\text{average}}$ *of average robots.*

- A "fast robot" Σ_i, $(\forall i \in \mathcal{N}_{\text{fast}})$, has the motor constant $k_{\text{m}i} = 1.5$, where the indices of such robots are included in the *set* $\mathcal{N}_{\text{fast}}$ *of fast robots.*

- A "slow robot" Σ_i, $(\forall i \in \mathcal{N}_{\text{slow}})$, has the motor constant $k_{\text{m}i} = 0.5$, where the indices of such robots are included in the *set* $\mathcal{N}_{\text{slow}}$ *of slow robots.*

The *leading robot* Σ_0 has no input $u_i(t)$, i.e.,

$$\Sigma_0 : \begin{cases} \dot{\boldsymbol{x}}_{\text{s}}(t) = \underbrace{\begin{pmatrix} 0 & 2\alpha_{\text{A}0}k_{\text{I}0} \\ 0 & 0 \end{pmatrix}}_{\boldsymbol{A}_{\text{s}}} \boldsymbol{x}_{\text{s}}(t), \quad \boldsymbol{x}_{\text{s}}(0) = \underbrace{\begin{pmatrix} 0 \\ \bar{v} \end{pmatrix}}_{\boldsymbol{x}_{\text{s}0}} \\ y_{\text{s}}(t) = \underbrace{\begin{pmatrix} 2.5 & 0 \end{pmatrix}}_{\boldsymbol{c}_{\text{s}}^{\text{T}}} \boldsymbol{x}_{\text{s}}(t) \end{cases} \tag{7.64}$$

where its parameters are listed in Table 7.1. The leading robot Σ_0 is assumed to get a step-wise constant reference velocity $\bar{v}(t)$ which is modeled in the initial state $\boldsymbol{x}_{\text{s}0}$. Note that the measuring unit for $\bar{v}(t)$ is increments per second ($\frac{\text{Ink}}{\text{s}}$), where one increment is $0.132\,\text{cm}$.

Model of the overall plant. The resulting multi-agent system (plant P) defined by

$$P : \begin{cases} \Sigma_0 \text{ defined in (7.64)} \\ \Sigma_i, (\forall i \subset \mathcal{N}), \text{ defined in (7.63)} \end{cases} \tag{7.65}$$

consists of the leading robot Σ_0 and the follower agents Σ_i, $(\forall i \in \mathcal{N})$.

System inclusion. Since $\boldsymbol{A}_i = \boldsymbol{A}_{\text{s}}$ and $\boldsymbol{c}_i^{\text{T}} = \boldsymbol{c}_{\text{s}}^{\text{T}}$ holds true for all $i \in \mathcal{N}$, Assumption 7.4 is fulfilled for all robots Σ_i. Hence, all robots Σ_i include the dynamics of the leading robot

Table 7.1: Parameters of the agents.

$i \in$	$\alpha_{\text{A}i}$	$k_{\text{I}i}$	$k_{\text{m}i}$
$\mathcal{N}_{\text{average}}$	0.0221	0.4	1
$\mathcal{N}_{\text{fast}}$	0.0221	0.4	1.5
$\mathcal{N}_{\text{slow}}$	0.0221	0.4	0.5
$\{0\}$	0.0221	0.4	1

Σ_0. Furthermore, the transformed model of the robots $\tilde{\Sigma}_i$ is identical to the model of the robots presented in (7.63), i.e.,

$$\tilde{\boldsymbol{A}}_i = \boldsymbol{A}_i, \quad \forall i \in \mathcal{N} \tag{7.66a}$$
$$\tilde{\boldsymbol{b}}_i = \boldsymbol{b}_i, \quad \forall i \in \mathcal{N} \tag{7.66b}$$
$$\tilde{\boldsymbol{g}}_i = \boldsymbol{g}_i, \quad \forall i \in \mathcal{N} \tag{7.66c}$$
$$\tilde{\boldsymbol{c}}_i^{\mathrm{T}} = \boldsymbol{c}_i^{\mathrm{T}}, \quad \forall i \in \mathcal{N}. \tag{7.66d}$$

Controlled agents. The feedback unit F_i defined in (7.22) is identical for all robots Σ_i, ($\forall i \in \mathcal{N}$), i.e.,

$$F_i : u_i(t) = -\underbrace{15}_{k_{\mathrm{P}i}}(y_i(t) - y_{\mathrm{s}i}(t)), \quad \forall i \in \mathcal{N}. \tag{7.67}$$

Hence, the model $\bar{\Sigma}_i$ of the controlled robots (agents) resulting from (7.63) and (7.67) is given by

$$\bar{\Sigma}_i : \begin{cases} \dot{\boldsymbol{x}}_i(t) = \underbrace{\begin{pmatrix} 2.5k_{\mathrm{P}i}2\alpha_{\mathrm{A}i}k_{\mathrm{m}i} & 2\alpha_{\mathrm{A}i}k_{\mathrm{I}i} \\ 2.5k_{\mathrm{P}i}k_{\mathrm{m}i} & 0 \end{pmatrix}}_{\bar{\boldsymbol{A}}_i} \boldsymbol{x}_i(t) + \boldsymbol{g}_i d_i(t) + k_{\mathrm{P}i}\boldsymbol{b}_i y_{\mathrm{s}i}(t), \quad \boldsymbol{x}_i(0) = \boldsymbol{x}_{0i} \\ y_i(t) = \underbrace{\begin{pmatrix} 2.5 & 0 \end{pmatrix}}_{\boldsymbol{c}_i^{\mathrm{T}}} \boldsymbol{x}_i(t). \end{cases} \tag{7.68}$$

According to (7.66), the relation

$$\tilde{\bar{\boldsymbol{A}}}_i = \bar{\boldsymbol{A}}_i, \quad \forall i \in \mathcal{N}$$

holds true for the transformed controlled model $\tilde{\bar{\Sigma}}_i$ of the robots defined in (7.27).

All elements k_{ij}, ($\forall i, j \in \mathcal{V}_{\mathrm{C}}, j < i$), are set to be one, i.e., $k_{ij} = 1$ for all $i, j \in \mathcal{V}_{\mathrm{C}}$ and $j < i$.

Stepwise change of the reference velocity. Consider the initial situation stated in Control aim 7.2 and Control aim 7.4. The relation (7.32) matches with the situation that the initial positions s_{0i} of the robots Σ_i are equal to the initial position $s_{0\mathrm{s}}$ of the leading robot Σ_0, i.e.,

$$s_{0\mathrm{s}} = s_{0i}, \quad \forall i \in \mathcal{N}$$

Relation (7.31) results from a difference between the identical initial velocity v_{0i} of the robots Σ_i and the reference velocity $\bar{v}$ of the leading robot Σ_0, i.e.,

$$\bar{v} \neq v_{0i} = v^*, \quad \forall i \in \mathcal{N}.$$

In the following this entire initial situation is denoted by a stepwise change of the reference velocity $\bar{v}$. In summary, Control aim 7.2 and Control aim 7.4 claim a desired maximal control error $y_{\Delta i}(t) = y_i(t) - y_s(t) = s_i(t) - s_0(t)$ for a stepwise change of the reference velocity $\bar{v}$ for the leading robot Σ_0.

7.7.3 Large robot formation

At the end of Chapters 8–10 the presented control concepts are demonstrated by their application to the robot formation illustrated in Fig. 7.13. The dynamics and position of the robots are

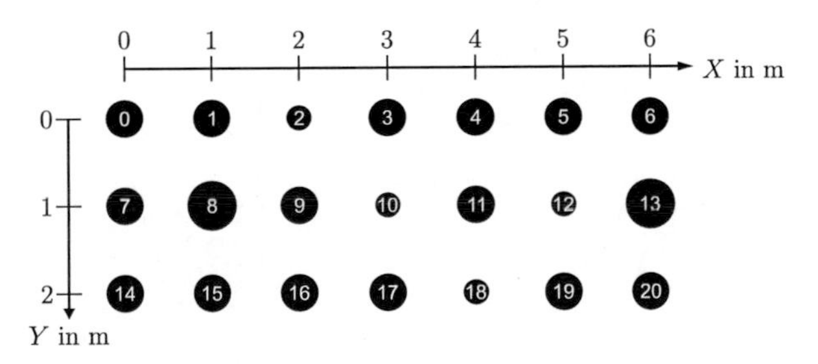

Figure 7.13: Robot formation with 20 follower robots.

listed in Table B.15 which leads to the sets

$$\begin{aligned} \mathcal{N}_{\text{average}} &= \{1, 3, 4, 5, 6, 7, 9, 11, 15, 16, 17, 19, 20\} \\ \mathcal{N}_{\text{fast}} &= \{2, 10, 12, 18\} \\ \mathcal{N}_{\text{slow}} &= \{8, 13, 14\}. \end{aligned}$$

Note that the different dynamics of the robots are indicated by the size of their corresponding circles in Fig. 7.13. Agents with slow dynamics are represented by large circles and agents with fast dynamics have small circles. According to (7.16), the communication costs c_{ij} among the the local control units C_i of the robots are listed in Table B.16

8 Performance improvement by requesting additional information

This chapter presents a control concept which tries to improve the synchronization performance compared to a fixed basic communication in case of set-point changes at the leader agent. The local control units request additional information, whenever the local control error exceeds a given threshold. Methods for guaranteeing a desired maximal value for the control error of the agents by choosing the thresholds and the additional communication links are derived. The control concept is applied to a robot formation.

Chapter contents

8.1 Situation-dependent information request

Consider the two different fixed communication graphs $\mathcal{G}_\mathrm{B}$ and $\mathcal{G}_\mathrm{A}$ depicted in the left part of Fig. 8.1 for synchronizing the agents. On the one hand , within the basic communication graph $\mathcal{G}_\mathrm{B}$ the information about $y_\mathrm{s}(t)$ has to pass all agents Σ_j, $(j=1\dots i-1)$, to get to Σ_i and the corresponding delay obviously induces a bad synchronization performance. On the other hand, within $\mathcal{G}_\mathrm{B}$ there is communication only to directly neighboring agents which leads to a low communication effort. Using the *adjusted communication graph* $\mathcal{G}_\mathrm{A}$, every agent directly receives information from the leader Σ_0 and, hence, reaches quickly the reference signal $y_\mathrm{s}(t)$. This direct communication in $\mathcal{G}_\mathrm{A}$ induces a high communication effort. Based on these considerations, the main **aim** of the control concept to be developed in this chapter is summarized as follows:

> Develop a self-organizing controller C that leads to a self-organizing control systems Σ that combines the benefits of communication graphs like $\mathcal{G}_\mathrm{B}$ and $\mathcal{G}_\mathrm{A}$ depicted in Fig. 8.1 which are a low communication effort ($\mathcal{G}_\mathrm{B}$) and a good synchronization performance ($\mathcal{G}_\mathrm{A}$).

It seems sufficient that the control units C_i, $(\forall i \in \mathcal{N})$ communicate in a row if the output $y_i(t)$ of the agents is identical with the reference signal $y_\mathrm{s}(t)$ (cf. $\mathcal{G}_\mathrm{B}$ in Fig. 8.1). However, if the agents do not track $y_\mathrm{s}(t)$, it is reasonable that the control units C_i, $(\forall i \in \mathcal{N})$ request additional information from Σ_0 or agents that are closer to Σ_0. Therefore, switching among communication links of the following communication structures defines the situation-dependent

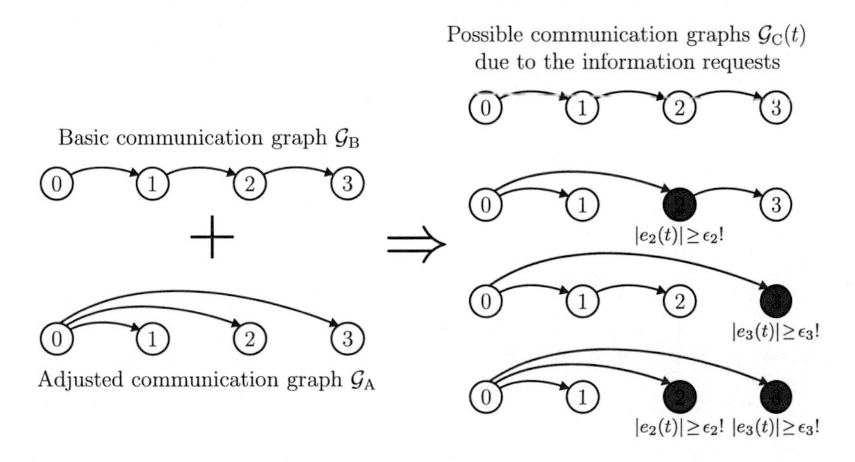

Figure 8.1: Idea of the situation-dependent information request.

communication graph structure $\mathcal{G}(t)$ used for this control concept (e.g. Fig. 8.1 right):

- The links from the basic communication graph $\mathcal{G}_{\rm B}$ are used to recognize the behavior of Σ_0 and has low communication effort.
- The links from the *adjusted communication graph* $\mathcal{G}_{\rm A}$ are used to reduce the control errors with higher communication effort.

Recalling the local reference signal

$$y_{{\rm s}i}(t) = \sum_{j\in\mathcal{P}_{{\rm C}i}(t)} \tilde{k}_{ij}(t)y_j(t), \qquad \forall i\in\mathcal{N} \tag{8.1}$$

defined in (7.18), the corresponding set of predecessors

$$\mathcal{P}_{{\rm C}i}(t) = \begin{cases} \mathcal{P}_{{\rm A}i} & \text{if } |e_i(t)| \geq \epsilon_i \\ \mathcal{P}_{{\rm B}i} & \text{else,} \end{cases} \qquad \forall i\in\mathcal{N}. \tag{8.2}$$

changes from the set $\mathcal{P}_{{\rm B}i}$ of basic predecessors to the *set* $\mathcal{P}_{{\rm A}i}$ *of adjusted predecessor*, whenever the local control error $e_i(t) = y_i(t) - y_{{\rm s}i}(t)$ exceeds the switching threshold ϵ_i. Note that $\mathcal{P}_{{\rm A}i}$, $(\forall i \in \mathcal{N})$, corresponds to the adjusted communication graph $\mathcal{G}_{\rm A}$ which will be defined in Section 8.2.1.

Based on these considerations, the **idea** of the self-organizing control concept proposed in this chapter can be summarized as follows:

> The incoming communication links to Σ_i are changed from links within the basic communication graph $\mathcal{G}_{\rm B}$ to links within the adjusted communication graph $\mathcal{G}_{\rm A}$ whenever the local control error $e_i(t)$ exceeds the switching threshold ϵ_i by requesting information through the local control unit C_i to guarantee a desired synchronization performance.

Figure 8.1 illustrates the idea of the situation-dependent information request for a multi-agent system with three follower agents. If the local control error $e_2(t)$ exceeds the switching threshold ϵ_2, then C_2 deactivates the transmission of $y_1(t)$ from C_1 to C_2 and requests a transmission of $y_{\rm s}(t)$ from C_0 to C_2 instead. Due to this request, C_2 receives direct information from C_0 and, hence, reaches the reference signal $y_{\rm s}(t)$ quicker than with its basic predecessor C_1.

With this, the **main tasks** of the local control units C_i can be summarized as follows:

- As explained in Section 7.3, the feedback unit F_i tracks the local reference signal $y_{{\rm s}i}(t)$.
- The observation unit O_i determines the local control error $e_i(t)$ to indicate the current situation of the agent Σ_i.

- The decision unit D_i decides about requesting information based on the current value of the local control error $e_i(t)$.
- The decision unit D_i uses the additional information to adapt its local reference signal $y_{si}(t)$ and thereby reduces the control error $y_{\Delta i}(t)$.

In the following the structure of the local control units that fulfills these four tasks is illustrated for a multi-agent system with two followers.

Structure of the self-organizing controller for two followers. The basic communication graph $\mathcal{G}_B$ and the adjusted communication graph $\mathcal{G}_A$ for controlling the multi-agent system P defined in (7.8) consisting of the leader Σ_0 and two follower agents Σ_1 and Σ_2 is given in Fig. 8.2. Furthermore, Fig. 8.2 depicts the two possible communication graphs $\mathcal{G}_C(t)$ resulting from $\mathcal{G}_B$ and $\mathcal{G}_A$.

Figure 8.3 shows the resulting structure of the self-organizing control concept with a situation-dependent request of information. Consider that the agents Σ_1 and Σ_2 initially track the reference signal $y_s(t)$, i.e., $y_s(0) = y_1(0) = y_2(0)$ (cf. (7.32)). In case of a difference between the initial state $\boldsymbol{x}_{s0}$ of the leader Σ_0 and the initial states $\boldsymbol{x}_{01} = \boldsymbol{x}_{02}$ of the agents Σ_1 and Σ_2, i.e, $\boldsymbol{x}_{s0} \neq \boldsymbol{x}_{01} = \boldsymbol{x}_{02}$ (cf. (7.31)), the components of the local control units C_0, C_1 and C_2 work as follows:

- The **feedback units** F_i, $(i = 1, 2)$ compare the corresponding output $y_i(t)$ with the local reference signal $y_{si}(t)$ to generate the input $u_i(t)$ to the agents Σ_i to track the local reference signal $y_{si}(t)$ (1. Task). Recall that F_i is already defined in (7.22).
- The **observation unit** O_2 calculates the local control error $e_2(t) = y_2(t) - y_{s2}(t)$ for D_2 by using the output $y_2(t)$ and the local reference signal $y_{s2}(t)$ (2. Task).
- The **decision unit** D_2 sends a request $r_{02}(t_{k,2})$ at the time instance $t_{k,2}$, $(k \in \mathbb{N})$, to the local control unit C_0 if the local control error $e_2(t)$ exceeds the switching threshold ϵ_2,

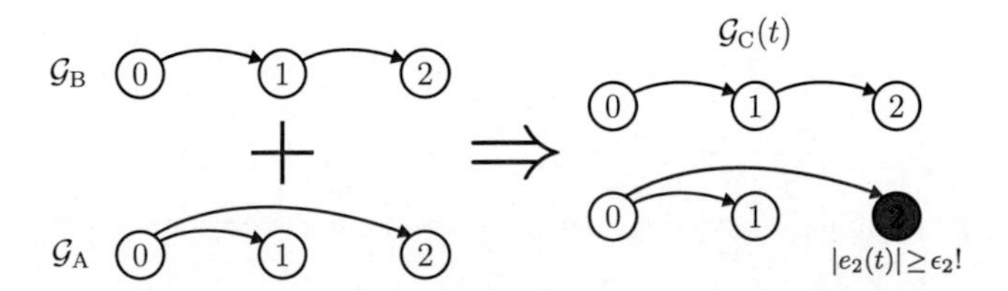

Figure 8.2: Basic communication graph $\mathcal{G}_B$ and the adjusted communication graph $\mathcal{G}_A$ for the self-organizing control of a multi-agent system with two followers and resulting possible communication graphs $\mathcal{G}_C(t)$.

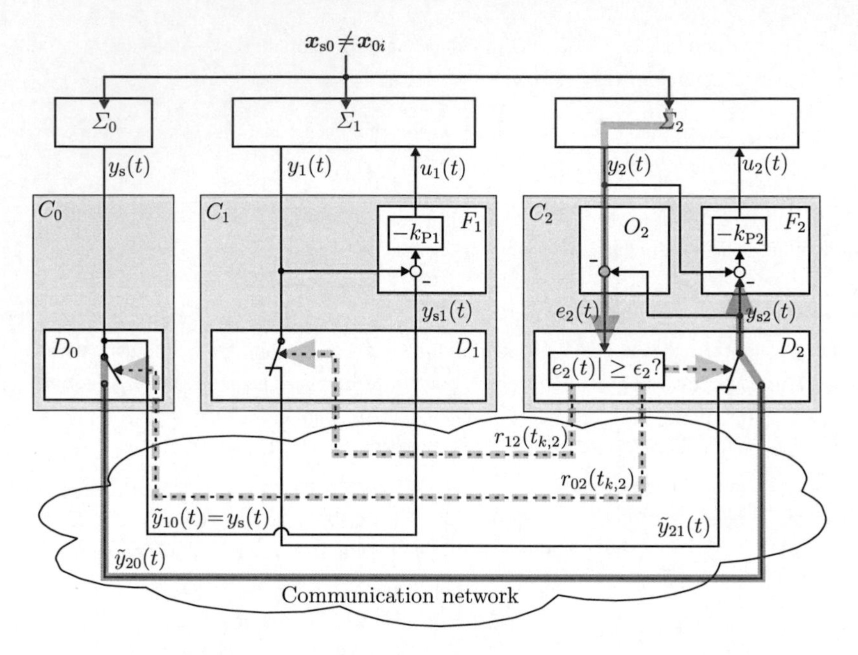

Figure 8.3: Self-organizing control concept with situation-dependent information request for two follower agents using graphs $\mathcal{G}_B$ and $\mathcal{G}_A$ from Fig. 8.2.

whereupon C_0 sends the reference signal $y_s(t)$ to C_2 (3. Task). Furthermore, D_2 sends a request $r_{12}(t_{k,2})$ to the local control unit C_1 to interrupt the transmission of $y_1(t)$ from C_1 to C_2.

- The **decision unit** D_2 in the local control unit C_2 uses the reference signal $y_s(t)$ to adapt its local reference signal to

$$y_{s2}(t) = \begin{cases} y_s(t) & \text{if } |e_2(t)| \geq \epsilon_2 \\ y_1(t) & \text{else.} \end{cases} \tag{8.3}$$

The effect of this control strategy is illustrated by the behavior of the controlled agent $\bar{\Sigma}_2$ defined in (7.26) with the input $y_{s2}(t)$:

$$\bar{\Sigma}_2 : \begin{cases} \dot{\boldsymbol{x}}_2(t) = \bar{\boldsymbol{A}}_i \boldsymbol{x}_2(t) + k_{P2} \boldsymbol{b}_2 y_{s2}(t), & \boldsymbol{x}_2(0) = \boldsymbol{x}_{02} \\ y_2(t) = \boldsymbol{c}_2^{\mathrm{T}} \boldsymbol{x}_2(t). \end{cases} \tag{8.4}$$

Due to the information request, the corresponding local reference signal $y_{s2}(t)$ is the reference signal $y_s(t)$ to be tracked (cf. (8.3)). Hence, the deviation between the local reference signal $y_{s2}(t)$ using the situation-dependent communication graph $\mathcal{G}_C(t)$ and the actual reference signal $y_s(t)$ can be adjusted by the choice of the switching threshold ϵ_2. The smaller the threshold ϵ_2, the smaller is the difference between both signals and, hence, the smaller is the difference between the performance with the self-organizing controller C and the performance with a controller which permanently uses the links of $\mathcal{G}_A$. The following example illustrates this property.

Example 8.1 *Situation-dependent information request within a formation of three robots*

In this example the self-organizing control concept with situation-dependent information request illustrated in Fig. 8.3 is used to build a formation of three robots as defined in Section 7.7. The applied basic communication graph $\mathcal{G}_B$ and the applied adjusted communication graph $\mathcal{G}_A$ are shown in the left part of Fig. 8.2. The follower robots Σ_1 and Σ_2 are both average robots, i.e., $\mathcal{N}_{\text{average}} = \{1, 2\}$. The example is subdivided into two parts: First, the construction of the local control units C_0, C_1 and C_2 is presented. Second, the behavior of the resulting self-organizing control system Σ is investigated.

Local control units. Recall that the feedback units F_1 and F_2 use the feedback gain $k_{P1} = k_{P2} = 15$ (cf. Section 7.7). Since C_1 does not request any information, there is not switching threshold ϵ_1. For C_2 the switching threshold is set to be $\epsilon_2 = 0.5$.

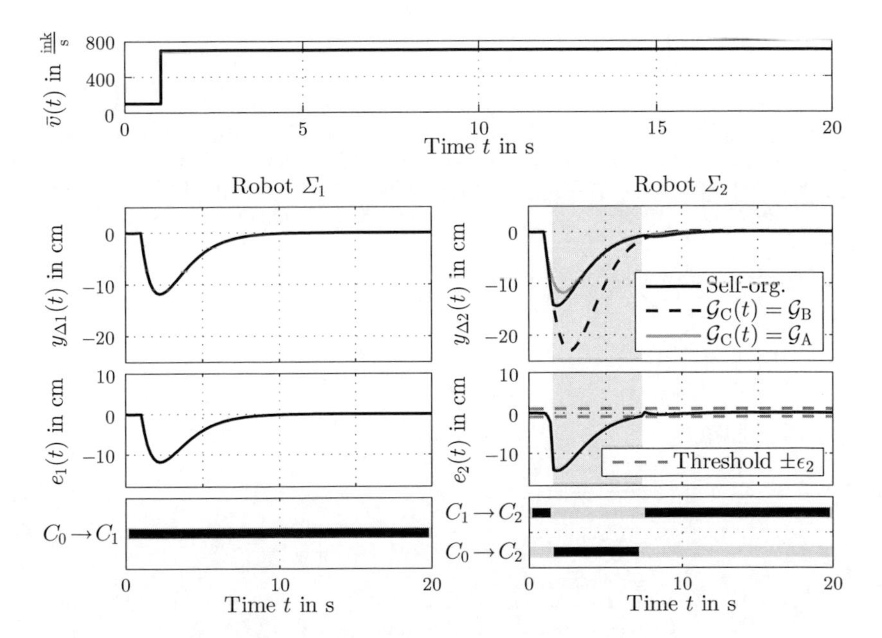

Figure 8.4: Behavior of the robot formation defined in (7.65) with $N = 2$ controlled be the local control units C_i specified in Fig. 8.3.

Behavior of the self-organizing control system Σ. The behavior of Σ with

$$\boldsymbol{x}_{s0} = \boldsymbol{x}_{01} = \boldsymbol{x}_{02} = \begin{pmatrix} 0 \\ 100 \end{pmatrix}$$

and a stepwise change of the reference velocity $\bar{v}(t)$ from $100\,\frac{\text{Ink}}{\text{s}}$ to $700\,\frac{\text{Ink}}{\text{s}}$ at $t = 1\,\text{s}$ is depicted in the upper plot of Fig. 8.4 with the black solid line. The change of the reference velocity $\bar{v}(t)$ leads to a deviation between the reference position $y_s(t)$ and the position $s_i(t)$ of the robots which is indicated by the control errors $y_{\Delta 1}(t)$ and $y_{\Delta 2}(t)$ in the second row of plots. Therefore, the local control errors $e_1(t)$ and $e_2(t)$ increase (third row of plots). When the local control error $e_2(t)$ exceeds the switching threshold ϵ_2, the local control unit C_2 sends an information request to C_0 which is indicated by the black bars in the lower row of plots. In this time interval C_2 receives the position $s_0(t)$ of the leader Σ_0 as local reference signal $y_{s2}(t)$. If the synchronization is completed, in particular, if the local control error $e_2(t)$ falls below the threshold ϵ_2, the communication graph $\mathcal{G}_C(t)$ becomes identical to $\mathcal{G}_B$.

To illustrate the performance improvement, the control error $y_{\Delta 2}(t)$ is depicted for robot Σ_2 while using the basic communication graph $\mathcal{G}_C(t) = \mathcal{G}_B$ only (dashed black line) and while using the adjusted communication graph $\mathcal{G}_C(t) = \mathcal{G}_A$ only (gray solid line). With the information request to the leader Σ_0 for this small time interval the performance of the self-organizing control system Σ is similar to the performance of the formation while using $\mathcal{G}_A$ at all time. Hence, the control error $y_{\Delta 2}(t)$ can be noticeably reduced.

Figure 8.5 shows interesting properties of the self-organizing control system Σ. The left plot depicts the maximal value $\max_{t\geq 0} |y_{\Delta\Delta 2}(t)|$ of the difference control error $y_{\Delta\Delta 2}(t) = y_{\Delta 2}(t) - y_{\Delta A2}(t)$ for different stepwise changes of the reference velocity $v_{\text{Step}} = \bar{v}(5\,\text{s}) - \bar{v}(0\,\text{s})$ as in Fig. 8.4 with the threshold $\epsilon_2 = 0.5$ as stated above. $y_{\Delta A2}$ is the control error while using the adjusted communication graph $\mathcal{G}_C(t) = \mathcal{G}_A$ only (gray solid line in Fig. 8.4). Hence, $y_{\Delta\Delta 2}(t)$ describes the performance difference between the self-organizing controlled robots Σ and robots which communicate within the adjusted communication graph $\mathcal{G}_A$ all of the time. The maximal difference control error $\max_{t\geq 0} |y_{\Delta\Delta 2}(t)|$ with the self-organizing controller C is bounded (solid line), whereas with a fixed communication graph $\mathcal{G}_C(t) = \mathcal{G}_B$ $\max_{t\geq 0} |y_{\Delta\Delta 2}(t)|$ (dashed line) it increases linearly with v_{Step}. To achieve this boundedness, the request time increases with v_{Step} (dot-dashed line). The *request time*, depicted in Fig. 8.5, is the length of the time interval in which C_0 transmits $y_s(t)$ to C_2 which is the length of the lower black

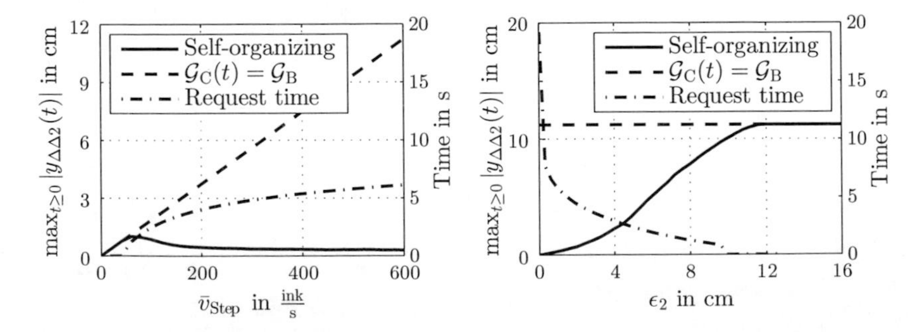

Figure 8.5: Relation between the reference velocity v_{Step} and the maximal value of the difference control error $y_{\Delta\Delta 2}(t) = y_{\Delta 2}(t) - y_{\Delta A2}(t)$ of Σ_2 (left); Relation between the switching threshold ϵ_2 in C_2 and the maximal value of the difference control error $y_{\Delta\Delta 2}(t)$ of Σ_2 (right).

bar in the lower right plot of Fig. 8.4.

The relation between $\max_{t\geq 0}|y_{\Delta\Delta 2}(t)|$ and the switching threshold ϵ_2 is depicted in right part of Fig. 8.5. The value for $\max_{t\geq 0}|y_{\Delta\Delta 2}(t)|$ increases with switching threshold ϵ_2 until $\epsilon_2 > 11.5$. Thereafter, $\max_{t\geq 0}|y_{\Delta\Delta 2}(t)|$ is equal to the value for a fixed communication graph $\mathcal{G}_{\rm C}(t) = \mathcal{G}_{\rm B}$ (dashed line) since for large thresholds, ϵ_2 is not exceeded by the local control error $e_2(t)$ at all. If $\epsilon_2 = 0$ holds, then C_0 transmits information to C_2 at all times. For small values of the threshold ϵ_2 there only is a small time interval in which information is requested but a huge performance improvement at Σ_2 compared to a fixed basic communication. Hence, the communication structure is adapted only for a small time interval to improve the synchronization performance. This shows that there is a trade-off between the maximal value $\max_{t\geq 0}|y_{\Delta\Delta 2}(t)|$ of the difference control error and the time of requesting information.

In summary, this example shows that the proposed self-organizing control concept with the situation-dependent request of information has the following interesting properties:

- The local control units request information only if there is a significant local control error which is a situation-dependent communication (cf. Fig. 8.4).

- The deviation between the performance of the self-organizing control system Σ and a multi-agent system which uses the adjusted communication graph $\mathcal{G}_{\rm A}$ at all times can be adjusted by the choice of the switching thresholds ϵ_i (cf. Fig. 8.5).

- There is a trade-off between the request time and the performance deviation (cf. Fig. 8.5 right).

□

Overview of the chapter. In the following sections it is shown that the mentioned properties of the proposed control concept also exist if it is applied to an arbitrary number of agents. Therefore, the content of the next sections is as follows:

- The structure of the self-organizing controller C is extended to an arbitrary number of agents (Section 8.2).

- An analysis of the self-organizing control system Σ yields a condition on the asymptotic synchronization of the agents, and an upper bound on the control error of the self-organizing controlled agents. (Section 8.3).

- Two methods for designing the adjusted communication graph $\mathcal{G}_{\rm A}$ guaranteeing a desired maximal control error as claimed in Control aim 7.2 are given (Section 8.4).

- The self-organizing control concept is demonstrated by its application to a robot formation (Section 8.5).

- The self-organizing properties of the closed-loop system with the situation-dependent information request are evaluated (Section 8.6).

8.2 Structure of the self-organizing controller and model of the closed-loop system

This section presents a formal definition of the self-organizing controller C (Section 8.2.3). Before, the structure of the local control units C_i, $(\forall i \in \mathcal{V}_{\mathrm{C}})$ is presented (Section 8.2.2) which perform a situation-dependent communication defined in Section 8.2.1. Finally, the multi-agent system P and the self-organizing controller C are assembled to derive the model of the overall closed-loop system Σ (Section 8.2.4).

8.2.1 Communication structure with information requests

This section analyzes the boundedness of the communication among the local control units C_i indicated by the set $\mathcal{E}_{\mathrm{C}}(t)$ of communication links, where at first the adjusted communication graph $\mathcal{G}_{\mathrm{A}}$ is defined.

Adjusted communication graph. The *adjusted communication graph* $\mathcal{G}_{\mathrm{A}}$ is a time-invariant labeled graph defined by

$$\mathcal{G}_{\mathrm{A}} = (\mathcal{V}_{\mathrm{C}}, \mathcal{E}_{\mathrm{A}}, \boldsymbol{K}), \tag{8.5}$$

where the weighting matrix $\boldsymbol{K}$ is the same as for the communication graph $\mathcal{G}_{\mathrm{C}}(t)$ defined in (7.9). The corresponding *set* $\mathcal{E}_{\mathrm{A}}$ *of adjusted edges* is defined by

$$\mathcal{E}_{\mathrm{A}} = \{(j \to i) | j \in \mathcal{P}_{\mathrm{A}i}, i \in \mathcal{V}_{\mathrm{C}}\},$$

where $\mathcal{P}_{\mathrm{A}i}$ is the *set of adjusted predecessors* with

$$\mathcal{P}_{\mathrm{A}i} \subseteq \{j \in \mathcal{V}_{\mathrm{C}} \mid j < i\}, \quad \forall i \in \mathcal{N}.$$

The *adjusted communication graph* $\mathcal{G}_{\mathrm{A}}$ is required to have a spanning tree with the leader Σ_0 as root node which is guaranteed by the following assumption (cf. Lemma 7.1).

Assumption 8.1 *Within the adjusted communication graph $\mathcal{G}_{\mathrm{A}}$ defined in* (8.5) *every agent has a predecessors, i.e., $\mathcal{P}_{\mathrm{A}i} \neq \emptyset$ for all $i \in \mathcal{N}$.*

With this, the *set* $\mathcal{F}_{\mathrm{A}i}$ *of adjusted followers* is defined by

$$\mathcal{F}_{\mathrm{A}i} = \{j \in \mathcal{V}_{\mathrm{C}} \mid i \in \mathcal{P}_{\mathrm{A}j}\}, \quad i \in \mathcal{V}_{\mathrm{C}}$$

similar to (2.18). According to (7.19), the *adjusted communication matrix* $\boldsymbol{K}_{\mathrm{A}} = (k_{\mathrm{A}ij}) \in$

$\mathbb{R}^{V \times V}$ is built by the elements $k_{\mathrm{A}ij}$ with

$$k_{\mathrm{A}ij} = \begin{cases} \dfrac{k_{ij}}{\sum_{l \in \mathcal{P}_{\mathrm{A}i}} k_{il}} & \text{if } j \in \mathcal{P}_{\mathrm{A}i} \\ 0 & \text{else,} \end{cases} \qquad i, j \in \mathcal{V}_{\mathrm{C}}.$$

Situation-dependent communication graph $\mathcal{G}_{\mathrm{C}}(t)$. The change of the communication graph $\mathcal{G}_{\mathrm{C}}(t) = (\mathcal{V}_{\mathrm{C}}, \mathcal{E}_{\mathrm{C}}(t), \boldsymbol{K})$ depends on the change of set $\mathcal{E}_{\mathrm{C}}(t)$ communication links defined in (2.17), where $\mathcal{E}_{\mathrm{C}}(t)$ changes due to the situation-dependent information request indicated by the sets $\mathcal{P}_{\mathrm{C}i}(t)$, ($\forall i \in \mathcal{V}_{\mathrm{C}}$), of predecessors defined in (8.2).

According to (8.2), the set $\underline{\mathcal{P}}_{\mathrm{C}i}$ of minimal communicational predecessors and the set $\bar{\mathcal{P}}_{\mathrm{C}i}$ of maximal communicational predecessors are given by

$$\underline{\mathcal{P}}_{\mathrm{C}i} = \mathcal{P}_{\mathrm{B}i} \cap \mathcal{P}_{\mathrm{A}i}, \quad \forall i \in \mathcal{N} \tag{8.6}$$

$$\bar{\mathcal{P}}_{\mathrm{C}i} = \mathcal{P}_{\mathrm{B}i} \cup \mathcal{P}_{\mathrm{A}i}, \quad \forall i \in \mathcal{N}. \tag{8.7}$$

Therefore, the set $\mathcal{P}_{\mathrm{C}i}(t)$ of predecessors is bounded by

$$\underbrace{\mathcal{P}_{\mathrm{B}i} \cap \mathcal{P}_{\mathrm{A}i}}_{\underline{\mathcal{P}}_{\mathrm{C}i}} \subseteq \mathcal{P}_{\mathrm{C}i}(t) \subseteq \underbrace{\mathcal{P}_{\mathrm{B}i} \cup \mathcal{P}_{\mathrm{A}i}}_{\bar{\mathcal{P}}_{\mathrm{C}i}}, \qquad \forall i \in \mathcal{N}, \quad \forall t \geq 0.$$

The set $\bar{\mathcal{E}}_{\mathrm{C}}$ of maximal communication links and set $\underline{\mathcal{E}}_{\mathrm{C}}$ of minimal communication links follow from (2.21), (2.23), (8.6) and (8.7) to

$$\underline{\mathcal{E}}_{\mathrm{C}} = \mathcal{E}_{\mathrm{B}} \cap \mathcal{E}_{\mathrm{A}}$$

$$\bar{\mathcal{E}}_{\mathrm{C}} = \mathcal{E}_{\mathrm{B}} \cup \mathcal{E}_{\mathrm{A}}.$$

With this, the set $\mathcal{E}_{\mathrm{C}}(t)$ of communication links within the communication graph $\mathcal{G}_{\mathrm{C}}(t) = (\mathcal{V}_{\mathrm{C}}, \mathcal{E}_{\mathrm{C}}(t), \boldsymbol{K})$ is bounded, i.e.,

$$\underbrace{\mathcal{E}_{\mathrm{B}} \cap \mathcal{E}_{\mathrm{A}}}_{\underline{\mathcal{E}}_{\mathrm{C}}} \subseteq \mathcal{E}(t) \subseteq \underbrace{\mathcal{E}_{\mathrm{B}} \cup \mathcal{E}_{\mathrm{A}}}_{\bar{\mathcal{E}}_{\mathrm{C}}}, \quad t \geq 0. \tag{8.8}$$

8.2.2 Structure of the local control units

Recall that the local control units C_i include a feedback unit F_i, an observation unit O_i and a decision unit D_i (Fig. 8.6). The feedback unit F_i is used to track the local reference signal $y_{\mathrm{s}i}(t)$. The observation unit O_i detects the current situation of the agent Σ_i indicated by the

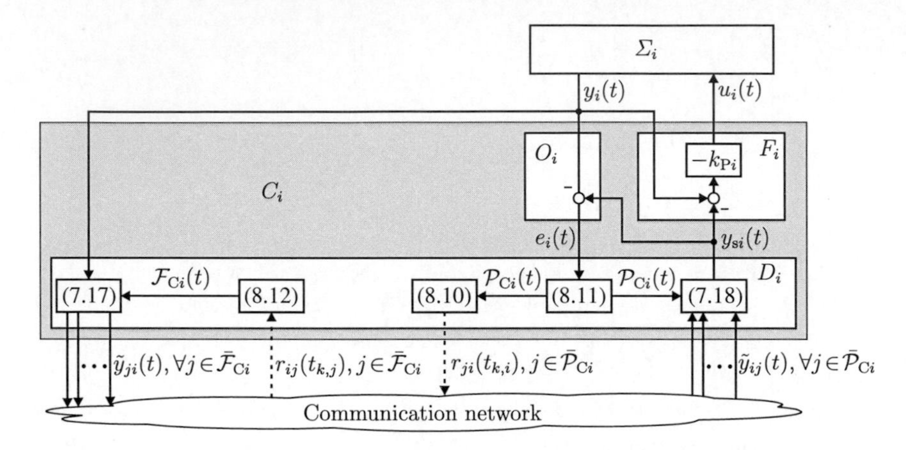

Figure 8.6: Local control unit C_i, $(\forall i \in \mathcal{N})$, for requesting additional information.

local control error $e_i(t)$. The decision unit D_i handles the incoming and outgoing information and requests additional information depending on the local control error $e_i(t)$. The following paragraphs describes the observation unit O_i and the decision unit D_i. The feedback unit F_i is already defined in (7.22) in Section 7.3.2. Finally, the components are combined to define the local control units C_i, $(\forall i \in \mathcal{N})$, of the agents Σ_i and the local control unit C_0 of the leader Σ_0.

Observation unit O_i. The task of the observation unit O_i is to provide the local control error $e_i(t)$ to the decision unit D_i:

$$O_i \ : \ e_i(t) = y_{si}(t) - y_i(t), \quad i \in \mathcal{N}, \tag{8.9}$$

where $e_i(t)$ is used as a measure to evaluate the necessity for requesting additional information (cf. (8.11)).

Decision unit D_i. The decision unit performs three tasks.

First, it determines the local reference signal $y_{si}(t)$ for the feedback unit F_i according to (7.18) by using the transmitted outputs $\tilde{y}_{ij}$, $(\forall j \in \bar{\mathcal{P}}_{Ci})$, and the set $\mathcal{P}_{Ci}(t)$ of predecessors. This task is explained in more detail in Section 7.3.2.

Second, D_i sends an *information request* $r_{ji}(t_{k,i})$, $(k \in \mathbb{N}_0)$, at the *request time instance* $t_{k,i}$ to the local control unit C_j, $(j \in \bar{\mathcal{P}}_{Ci})$, if its set $\mathcal{P}_{Ci}(t)$ of predecessors changes. An information request is defined by

$$r_{ji}(t_{k,i}) = \begin{cases} 1 & \text{if } j \in \mathcal{P}_{\mathrm{C}i}(t_{k,i}) \backslash \mathcal{P}_{\mathrm{C}i}(t_{k-1,i}) \\ 0 & \text{if } j \in \mathcal{P}_{\mathrm{C}i}(t_{k-1,i}) \backslash \mathcal{P}_{\mathrm{C}i}(t_{k,i}) \\ \eta & \text{else,} \end{cases} \qquad j \in \bar{\mathcal{P}}_{\mathrm{C}i}, \quad k > 0, \quad i \in \mathcal{N} \tag{8.10a}$$

with the request time instance

$$t_{0,i} = 0 \tag{8.10b}$$

$$t_{k+1,i} = \inf\{t > t_{k,i} \mid \mathcal{P}_{\mathrm{C}i}(t) \neq \mathcal{P}_{\mathrm{C}i}(t_{k,i})\}. \tag{8.10c}$$

A change in the set of predecessors

$$\mathcal{P}_{\mathrm{C}i}(t) = \begin{cases} \mathcal{P}_{\mathrm{A}i} & \text{if } |e_i(t)| \geq \epsilon_i \\ \mathcal{P}_{\mathrm{B}i} & \text{if } |e_i(t)| \leq \bar{\epsilon}_i \\ \mathcal{P}_{\mathrm{C}i}(t^-) & \text{else,} \end{cases} \qquad \forall i \in \mathcal{N} \tag{8.11}$$

occurs if the local control error $e_i(t)$ exceeds the *switching threshold* ϵ_i. The symbol t^- denotes the time instance directly before the time t with $\mathcal{P}_{\mathrm{C}i}(t^-) = \mathcal{P}_{\mathrm{B}i}$ for $t = 0$. Recall that the switching condition $|e_i(t)| \geq \epsilon_i$ in (8.11) is the main reason for requesting information. To prevent Zeno behavior, $\mathcal{P}_{\mathrm{C}i}(t)$ changes to $\mathcal{P}_{\mathrm{B}i}$ only if $e_i(t)$ falls below the threshold $\bar{\epsilon}_i$ with $\bar{\epsilon}_i < \epsilon_i$.

Third, D_i sends the output $y_i(t)$ to its current followers C_j, $(\forall j \in \mathcal{F}_{\mathrm{C}i}(t))$, with the transmitted output $\tilde{y}_{ji}(t)$ according to (7.17) explained in Section 7.3.2. The set of followers

$$\mathcal{F}_i(t) = \begin{cases} \mathcal{F}_i(t^-) \cup \{j\} & \text{if } (\exists k \in \mathbb{N}, j \in \bar{\mathcal{F}}_{\mathrm{C}i}) \ t = t_{k,j} \wedge r_{ij}(t_{k,j}) = 1 \\ \mathcal{F}_i(t^-) \backslash \{j\} & \text{if } (\exists k \in \mathbb{N}, j \in \bar{\mathcal{F}}_{\mathrm{C}i}) \ t = t_{k,j} \wedge r_{ij}(t_{k,j}) = 0 \\ \mathcal{F}_i(t^-) & \text{else,} \end{cases} \qquad \forall i \in \mathcal{V}_{\mathrm{C}} \tag{8.12}$$

changes if there is an information request $r_{ij}(t_{k,j})$, $(j \in \bar{\mathcal{F}}_{\mathrm{C}i})$, from the local control unit C_j at the request time instance $t_{k,j}$. Note that $\mathcal{F}_{\mathrm{C}i}(t^-) = \mathcal{F}_{\mathrm{B}i}$ for $t = 0$. For example, agent Σ_4 is added to the set $\mathcal{F}_2(t^-) = \{3\}$ of followers of agent Σ_2 for the information request $r_{2,4}(t = t_{1,4}) = 1$ at the first request time instance $t_{1,4}$ of C_4, such that its new set of followers is $\mathcal{F}_2(t) = \{3, 4\}$.

In summary, the decision unit D_i is defined by

$$D_i : \begin{cases} \text{(8.11) generating } \mathcal{P}_{\mathrm{C}i}(t) \text{ using } e_i(t) \\ \text{(7.18) generating } y_{\mathrm{s}i}(t) \text{ using } \mathcal{P}_{\mathrm{C}i}(t) \text{ and } \tilde{y}_{ij}(t), (\forall j \in \bar{\mathcal{P}}_{\mathrm{C}i}) \\ \text{(8.10) generating } r_{ji}(t_{k,i}), (j \in \bar{\mathcal{P}}_{\mathrm{C}i}), \text{ using } \mathcal{P}_{\mathrm{C}i}(t) \\ \text{(8.12) generating } \mathcal{F}_{\mathrm{C}i}(t) \text{ using } r_{ij}(t_{k,j}), (j \in \bar{\mathcal{F}}_{\mathrm{C}i}) \\ \text{(7.17) generating } \tilde{y}_{ji}(t), (\forall j \in \bar{\mathcal{F}}_{\mathrm{C}i}), \text{ using } \mathcal{F}_{\mathrm{C}i}(t) \text{ and } y_i(t), \end{cases} \quad \forall i \in \mathcal{N}. \tag{8.13}$$

Local control units C_i of the agents Σ_i. The feedback unit F_i, the observation unit O_i and the decision unit D_i build the local control units

$$C_i : \begin{cases} F_i \text{ defined in (7.22)} \\ O_i \text{ defined in (8.9)} \\ D_i \text{ defined in (8.13)}, \end{cases} \quad \forall i \in \mathcal{N} \tag{8.14}$$

of the agents Σ_i.

Local control unit C_0 of the leader Σ_0. There is no feedback unit and observation unit within the local control unit C_0 of the leader Σ_0. The local control unit

$$C_0 = D_0 : \begin{cases} \text{(8.12) generating } \mathcal{F}_{\mathrm{C}0}(t) \text{ using } r_{0j}(t_{k,j}), (j \in \bar{\mathcal{F}}_{\mathrm{C}0}) \\ \text{(7.17) generating } \tilde{y}_{j0}(t), (\forall j \in \bar{\mathcal{F}}_{\mathrm{C}0}), \text{ using } \mathcal{F}_{\mathrm{C}0}(t) \text{ and } y_0(t) = y_{\mathrm{s}}(t) \end{cases} \tag{8.15}$$

only includes the decision unit D_0 which handles the information requests $r_{0i}(t_{k,i})$, $(i \in \bar{\mathcal{F}}_{\mathrm{C}0})$, from the local control units C_i of the agents Σ_i.

8.2.3 Self-organizing controller with situation-dependent information requests

The self-organizing controller C in which the local control units C_i request additional information to improve the synchronization performance results to

$$C : \; C_i, (\forall i \in \mathcal{V}_{\mathrm{C}}), \text{ defined in (8.14)-(8.15) with } \mathcal{E}_{\mathrm{C}}(t) \text{ bounded according to (8.8).} \tag{8.16}$$

In particular $\mathcal{E}_{\mathrm{C}}(t)$ changes due to the situation-dependent information request performed by the decision units D_i defined in (8.13) using the local control error $e_i(t)$ determined by the observation units O_i defined in (8.9).

8.2.4 Overall self-organizing control system

The control of the multi-agent system P defined in (7.8) by the self-organizing controller C defined in (8.16) leads to the overall closed-loop system

$$\Sigma : \begin{cases} \dot{\boldsymbol{x}}_{\text{os}}(t) = \bar{\boldsymbol{A}}_{\text{os}}(t)\boldsymbol{x}_{\text{os}}(t), \quad \boldsymbol{x}_{\text{os}}(0) = \boldsymbol{x}_{\text{os}0} \\ \boldsymbol{y}(t) = \boldsymbol{C}_{\text{os}}\boldsymbol{x}_{\text{os}}(t) \\ \left.\begin{array}{l} O_i, (\forall i \in \mathcal{N}) \text{ defined in (8.9)} \\ D_i, (\forall i \in \mathcal{N}) \text{ defined in (8.13)} \\ C_0, \text{ defined in (8.15)} \end{array}\right\} \text{generating } \bar{\boldsymbol{A}}_{\text{os}}(t) \text{ with } \mathcal{P}_{\text{C}i}(t) \text{ defined in (8.11).} \end{cases} \tag{8.17}$$

Note that the general model of the self-organizing control system Σ defined in (7.30) and Σ with the information request only differ in the structure of the observation unit O_i and the decision unit D_i since in the general model O_i and D_i are not defined. Furthermore, the agents Σ_i are assumed to be undisturbed, i.e., $\boldsymbol{d}(t) = \boldsymbol{0}$ for all $t \geq 0$.

8.3 Analysis of the overall self-organizing control system

In this section the asymptotic synchronization (Control aim 7.1) and the synchronization performance (Control aim 7.2) of the self-organizing control system Σ with situation-dependent request of information is investigated. It is shown that asymptotic synchronization of the agents with the proposed controller is always guaranteed (Theorem 8.1 in Section 8.3.1). An upper bound on the maximal control error $\max_{t\geq 0} |y_{\Delta i}(t)|$ is derived which depends on the choice of the communication graphs $\mathcal{G}_{\text{A}}$ and $\mathcal{G}_{\text{B}}$ and the switching thresholds ϵ_i (Theorem 8.2 in Section 8.3.2). This bound is used in Section 8.4 to design the self-organizing controller C such that the resulting self-organizing control system Σ satisfies Control aim 7.2. Furthermore, specific communication structures with one predecessor for each agent are investigated in Section 8.3.3, where it is shown that the maximal control error of an adjusted communication graph $\mathcal{G}_{\text{A}}$ with one predecessor for each agent can always be as small as with an adjusted communication graph $\mathcal{G}_{\text{A}}$ with multiple predecessor for each agent (Proposition 8.1). In addition, a condition under which the situation-dependent information request leads to smaller control errors compared to a permanent information exchange within the basic communication graph $\mathcal{G}_{\text{B}}$ is derived (Proposition 8.2).

8.3.1 Asymptotic synchronization with information requests

This section investigates Control aim 7.1 for the self-organizing control system Σ defined in (8.17). It has to be shown that despite the situation-dependent information requests among the local control units C_i the agents Σ_i still asymptotically synchronize to the leader Σ. Therefore, the conditions in Lemma 7.2 for the general self-organizing control system Σ defined in (7.30) are verified for the specific self-organizing control system Σ with situation-dependent information request.

Theorem 8.1 (Asymptotic synchronization with information requests) *Consider the self-organizing control system Σ defined in (8.17). Then the agents Σ_i, $(\forall i \in \mathcal{N})$, defined in (7.2) are asymptotically synchronized to the leader Σ_0 defined in (7.3) as desired in Control aim 7.1 for all possible graphs $\mathcal{G}_{\mathrm{A}}$ and $\mathcal{G}_{\mathrm{B}}$.*

Proof. The general self-organizing control system Σ defined in (7.30) and the self-organizing control system Σ defined in (8.17) with information requests are identical besides the change of the communication graph $\mathcal{G}_{\mathrm{C}}(t)$. Hence, the two conditions Lemma 7.2 have to be fulfilled to prove the asymptotic synchronization of the agents:

1. Since set $\mathcal{P}_{\mathrm{C}i}(t)$ of predecessor is either $\mathcal{P}_{\mathrm{B}i}$ or $\mathcal{P}_{\mathrm{A}i}$ according to (8.11) and $\mathcal{P}_{\mathrm{B}i}$ and $\mathcal{P}_{\mathrm{A}i}$ are not empty according to Assumptions 7.6 and 8.1, the first condition in Lemma 7.2 is always fulfilled.

2. The second condition in Lemma 7.2 is fulfilled since $\boldsymbol{d}(t) = \mathbf{0}$ holds true for all $t \geq 0$.

Since Assumptions 7.1–7.5 also hold for Σ defined in (8.17), the proof is completed. □

Theorem 8.1 shows that the choice of the current set $\mathcal{P}_{\mathrm{C}i}(t)$ of predecessor within the decision unit D_i guarantees that the sets $\mathcal{P}_{\mathrm{C}i}(t)$, $(\forall i \in \mathcal{N})$, of predecessors are always not empty, i.e., $\mathcal{P}_{\mathrm{C}i}(t) \neq \emptyset$ for all $i \in \mathcal{N}$ and all $t \geq 0$. Therefore, the communication graph $\mathcal{G}_{\mathrm{C}}(t)$ always has a spanning tree with the leader as root node (cf. Lemma 7.1) which leads to the following fact:

> The self-organizing controlled agents performing information request always asymptotically synchronize with the leader.

8.3.2 Synchronization performance with information requests

Since the switching of the communication structure occurs due to the current value of the local control error $e_i(t)$ and not due to the current value of the control error $y_{\Delta i}(t)$, deriving an upper bound on $\max_{t \geq 0} |y_{\Delta i}(t)|$ which depends on the switching threshold ϵ_i is a difficult task.

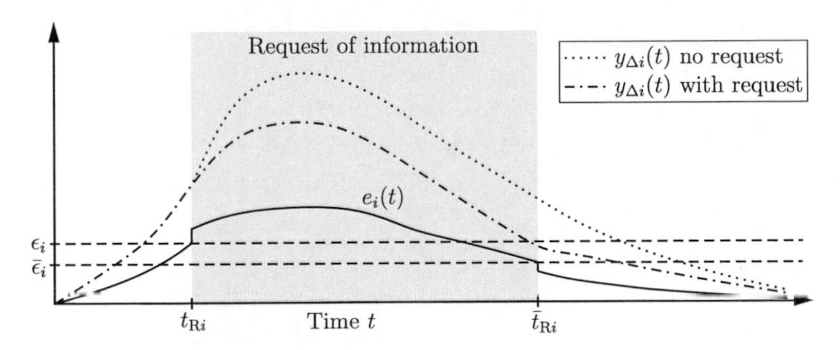

Figure 8.7: Typical course of the control error $y_{\Delta i}(t)$ and the local control error $e_i(t)$ due to an information request.

To illustrate the idea, a typical course of $e_i(t)$ and $y_{\Delta i}(t)$ in the case of the initial situation in (7.31) and (7.52) is depicted in Fig. 8.7. At time t_{Ri} the local control error $e_i(t)$ (solid line) jumps to a greater value since the corresponding local control unit C_i requests new information. The new information generally leads to a greater local reference signal $y_{si}(t)$ which is closer to the reference signal $y_s(t)$. Hence, the control error $y_{\Delta i}(t)$ (dotted dashed line) with information request generally is smaller than the control error $y_{\Delta i}(t)$ while using the basic communication links only (dotted line).

Due to the switching of the communication structure, the actual times t_{Ri} at which the local control error $e_i(t)$ exceeds the switching threshold ϵ_i are unknown. Therefore, the times t_{Bi} at which the local control errors $e_i(t)$ would exceed the switching thresholds ϵ_i while using the fixed basic communication, i.e.,

$$t_{Bi} = \min_{t \geq 0} t : |e_i(t)| \geq \epsilon_i, \quad i \in \mathcal{N}, \tag{8.18}$$

are used as an upper bound on the time t_{Ri}. With this, the following assumption is made.

Assumption 8.2 *Consider the overall closed-loop system Σ_B with a fixed basic communication defined in* (7.51) *and the self-organizing control system Σ with the situation-dependent communication graph $\mathcal{G}_C(t)$ defined in* (8.17), *where both have the initial situation in* (7.52) *and* (7.53). *Assume the behavior of the local control error $e_i(t)$ and control error $y_{\Delta i}(t)$ for Σ illustrated Fig. 8.7. The switching times t_{Ri} at which the local control errors $e_i(t)$, $(\forall i \in \mathcal{N})$, of Σ exceed the thresholds ϵ_i for the first time are assumed to be lower than the given times t_{Bi} at which the local control errors $e_i(t)$ of Σ_B exceed the thresholds ϵ_i for the first time, i.e.,*

$$t_{Ri} \leq t_{Bi}, \quad \forall i \in \mathcal{N}.$$

The following lemma derives a method for determining the times instances $t_{\mathrm{B}i}$.

Lemma 8.1 (Exceedance of switching thresholds with basic communication only) *Consider the overall closed-loop system Σ_{B} with a fixed basic communication defined in (7.51) with the initial situation in (7.52) and (7.53). According to (8.18), the times $t_{\mathrm{B}i}$, $(\forall i \in \mathcal{N})$, at which the local control errors $e_i(t)$ of the agents Σ_i exceed the corresponding threshold ϵ_i are given by*

$$t_{\mathrm{B}i} = \arg\min_{t \geq 0} \left\{ \left| \boldsymbol{\alpha}_i^{\mathrm{T}} (\boldsymbol{K}_{\mathrm{B}}^* - \boldsymbol{I}_N) \tilde{\boldsymbol{C}} \mathrm{e}^{\tilde{\bar{\boldsymbol{A}}}_{\mathrm{B}} t} \tilde{\boldsymbol{x}}_{\Delta 0} \right| = \epsilon_i \right\}, \quad i \in \mathcal{N} \tag{8.19}$$

with $\boldsymbol{\alpha}_i^{\mathrm{T}} = \begin{pmatrix} \alpha_0 & \alpha_1 & \dots & \alpha_N \end{pmatrix}$, $\alpha_i = 1$ *and* $\alpha_j = 0$, $(j \neq i, \forall j \in \mathcal{V})$.

Proof. The local control error $e_i(t)$ defined in (7.41) can be written as

$$e_i(t) = \underbrace{y_{\mathrm{s}i}(t) - y_{\mathrm{s}}(t)}_{y_{\Delta \mathrm{s}i}(t)} - \underbrace{(y_i(t) - y_{\mathrm{s}}(t))}_{y_{\Delta i}(t)}$$

Hence, the overall local control error results to

$$\boldsymbol{e}(t) = \boldsymbol{y}_{\Delta \mathrm{s}}(t) - \boldsymbol{y}_{\Delta}(t)$$

with

$$\boldsymbol{e}(t) = \begin{pmatrix} e_1(t) \\ \vdots \\ e_N(t) \end{pmatrix}, \quad \boldsymbol{y}_{\Delta \mathrm{s}}(t) = \begin{pmatrix} y_{\Delta \mathrm{s}1}(t) \\ \vdots \\ y_{\Delta \mathrm{s}N}(t) \end{pmatrix}.$$

Due to the basic communication graph $\mathcal{G}_{\mathrm{B}}$ defined in (7.50), the relation

$$\boldsymbol{y}_{\Delta \mathrm{s}}(t) = \boldsymbol{K}_{\mathrm{B}}^* \boldsymbol{y}_{\Delta}(t)$$

holds true which leads to the overall local control error

$$\boldsymbol{e}(t) = (\boldsymbol{K}_{\mathrm{B}}^* - \boldsymbol{I}_N) \boldsymbol{y}_{\Delta}(t). \tag{8.20}$$

The combination of (7.45) and (8.20) leads to the calculation of the times $t_{\mathrm{B}i}$ in (8.19). Note that the transposed vectors $\boldsymbol{\alpha}_i^{\mathrm{T}}$ select the values of the local control errors $e_i(t)$ from an overall control error $\boldsymbol{e}(t)$. □

With Assumption 8.2 and Lemma 8.1 an upper bound on the control error $y_{\Delta i}(t)$ can be calculated as presented in the following theorem.

Theorem 8.2 (Maximal control error with information requests) *Consider the self-organizing control system Σ defined in (8.17) with the initial situation in (7.52) and (7.53). Then the control errors $y_{\Delta i}(t)$ of the agents Σ_i are bounded by the maximal adjusted control error $\bar{y}_{\Delta \mathrm{A}i}$:*

$$\max_{t\geq 0} |y_{\Delta i}(t)| \leq \bar{y}_{\Delta \mathrm{A}i} = m_{\mathrm{y0A}i} + m_{\mathrm{yy}i} \sum_{j \in \mathcal{P}_{\mathrm{A}i}} k_{\mathrm{A}ij} \bar{y}_{\Delta \mathrm{A}j}, \quad \forall i \in \mathcal{N} \tag{8.21}$$

with $\bar{y}_{\Delta \mathrm{A}0} = 0$ and

$$\begin{aligned} m_{\mathrm{y0A}i} &= \boldsymbol{m}_{\mathrm{y0}i}^{\mathrm{T}} \left(\boldsymbol{m}_{\mathrm{x0}i} \cdot \bar{\boldsymbol{x}}_{\Delta 0} + \boldsymbol{m}_{\mathrm{xe}i} \cdot \epsilon_i \right), && \forall i \in \mathcal{N} \\ \boldsymbol{m}_{\mathrm{xe}i} &= \int_0^{t_{\mathrm{B}i}} \left| \mathrm{e}^{\tilde{\boldsymbol{A}}_i \tau} k_{\mathrm{P}i} \tilde{\boldsymbol{b}}_i \right| \mathrm{d}\tau, && \forall i \in \mathcal{N} \\ \boldsymbol{m}_{\mathrm{x0}i} &= \max_{t \in [0, t_{\mathrm{B}i}]} \left| \mathrm{e}^{\tilde{\boldsymbol{A}}_i t} \right|, && \forall i \in \mathcal{N}. \end{aligned} \tag{8.22}$$

Hence, the overall control error $\boldsymbol{y}_\Delta(t)$ is bounded by the overall maximal control error $\bar{\boldsymbol{y}}_{\Delta \mathrm{A}} \in \mathbb{R}^N$, i.e.,

$$\max_{t \geq 0} |\boldsymbol{y}_\Delta(t)| \leq \bar{\boldsymbol{y}}_{\Delta \mathrm{A}} = (\boldsymbol{I} - \boldsymbol{M}_{\mathrm{yy}} \boldsymbol{K}_{\mathrm{A}}^*)^{-1} \boldsymbol{m}_{\mathrm{y0A}} \tag{8.23}$$

with

$$\begin{aligned} \bar{\boldsymbol{y}}_{\Delta \mathrm{A}} &= \begin{pmatrix} \bar{y}_{\Delta \mathrm{A}1} \\ \vdots \\ \bar{y}_{\Delta \mathrm{A}N} \end{pmatrix}, \quad \boldsymbol{m}_{\mathrm{y0A}} = \begin{pmatrix} m_{\mathrm{y0A}1} \\ \vdots \\ m_{\mathrm{y0A}N} \end{pmatrix} \\ \boldsymbol{K}_{\mathrm{A}}^* &= \begin{pmatrix} \boldsymbol{0}_N^{\mathrm{T}} \\ \boldsymbol{I}_N \end{pmatrix} \boldsymbol{K}_{\mathrm{A}} \begin{pmatrix} \boldsymbol{0}_N & \boldsymbol{I}_N \end{pmatrix}. \end{aligned} \tag{8.24}$$

Proof. The proof is given in Appendix A.12. □

Theorem 8.2 gives an upper bound $\bar{\boldsymbol{y}}_{\Delta \mathrm{A}}$ for the maximal control error $\max_{t \geq 0} |\boldsymbol{y}_\Delta(t)|$ while using a situation-dependent information request which depends on the switching thresholds ϵ_i, $(\forall i \in \mathcal{N})$, the initial state $\bar{\boldsymbol{x}}_{\Delta 0}$, the basic communication graph $\mathcal{G}_{\mathrm{B}}$ and the adjusted communication graph $\mathcal{G}_{\mathrm{A}}$. Hence, the following is highlighted:

> The maximal control error $\bar{\boldsymbol{y}}_{\Delta \mathrm{A}}$ can be adjusted by the choice of the adjusted communication graph $\mathcal{G}_{\mathrm{A}}$ and the switching thresholds ϵ_i, $(\forall i \in \mathcal{N})$.

Note that the illustration for the maximal basic control error $\bar{y}_{\Delta\mathrm{B}i}$ in Fig. 7.10 would be similar for the maximal control error $\bar{y}_{\Delta\mathrm{A}i}$ in Theorem 8.2.

Equation (8.21) shows that $\bar{y}_{\Delta\mathrm{A}i}$ is composed of two parts. The first addend $m_{\mathrm{y0A}i}$ in (8.21) depends on the dynamic of the controlled agents $\bar{\Sigma}_i$, the initial state $\tilde{\boldsymbol{x}}_{\Delta 0}$, the basic communication graph $\mathcal{G}_\mathrm{B}$ and the switching threshold ϵ_i. It can be seen that for $\epsilon_i \to 0$, $m_{\mathrm{y0A}i}$ only depends on the initial state $\bar{\boldsymbol{x}}_{\Delta 0}$, i.e.,

$$\lim_{\epsilon_i \to 0} m_{\mathrm{y0A}i} = \boldsymbol{m}_{\mathrm{y0}i}^{\mathrm{T}} \bar{\boldsymbol{x}}_{\Delta 0}, \tag{8.25}$$

because the second addend of $m_{\mathrm{y0A}i}$ goes to zero and the first addend goes to $\bar{\boldsymbol{x}}_{\Delta 0}$ since the switching time $t_{\mathrm{B}i}$ is zero for $\epsilon_i = 0$. Hence, for $\epsilon_i = 0$, $(\forall i \in \mathcal{N})$, the maximal control error $\bar{y}_{\Delta i}$ only depends on the dynamics of the controlled agents $\bar{\Sigma}_i$, the initial state $\bar{\boldsymbol{x}}_{\Delta 0}$, and the adjusted communication graph $\mathcal{G}_\mathrm{A}$, which is reasonable since the adjusted communication $\mathcal{G}_\mathrm{A}$ is active at all times. This would lead to an identical upper bound as for the fixed basic communication presented in Lemma 7.3.

All these explanations do not give a precise answer to the question: *When does the self-organizing controller with a situation-dependent communication structure lead to a better performance than a control within a fixed basic communication structure?* This question will be answered in the following section while considering that every agent has exactly one predecessor.

8.3.3 Behavior while using one predecessor for each agent

To simplify the analysis of the self-organizing control system Σ defined in (8.17) and the design of the local control units it is concerned that every agent has exactly one predecessor.

Assumption 8.3 *Every agent Σ_i has exactly one predecessor*

$$|\mathcal{P}_{\mathrm{A}i}| = |\mathcal{P}_{\mathrm{B}i}| = 1, \quad \forall i \in \mathcal{N}. \tag{8.26}$$

According to the switching condition in (8.11), Assumption 8.3 implies

$$|\mathcal{P}_{\mathrm{C}i}(t)| = 1, \quad \forall i \in \mathcal{N}, \quad \forall t \geq 0.$$

Hence, the agents Σ_i are always synchronized to the leader Σ_0 (cf. Theorem 8.1).

With this assumption the remaining section is divided into three parts:

1. The results on the upper bound on the control error in Section 8.3.2 are adapted for agents with one predecessor.

2. It is shown that Assumption 8.3 leads to no restriction in the synchronization performance of the self-organizing control system Σ.
3. A condition is derived under which the situation-dependent information request guarantees a better performance than with a fixed basic communication.

Maximal control error. With Assumption 8.3 the relation in (8.21) for the boundedness of the control error $y_{\Delta i}(t)$ in Theorem 8.2 can be written as follows.

Corollary 8.1 (Maximal control error using one predecessor) *Consider the self-organizing control system Σ defined in* (8.17) *with the initial situation in* (7.52) *and* (7.53). *The control error $y_{\Delta i}(t)$ of agent Σ_i that only has one predecessor (cf. Assumption 8.3) is bounded by the maximal adjusted control error $\bar{y}_{\Delta \mathrm{A}i}$:*

$$\max_{t\geq 0} |y_{\Delta i}(t)| \leq \bar{y}_{\Delta \mathrm{A}i} = m_{\mathrm{y0A}i} + m_{\mathrm{yy}i}\bar{y}_{\Delta \mathrm{A}j}, \quad j \in \mathcal{P}_{\mathrm{A}i}, \quad \forall i \in \mathcal{N}. \tag{8.27}$$

Note that an extension of Corollary 8.1 for considering the overall closed-loop system Σ_{B} with a fixed basic communication graph $\mathcal{G}_{\mathrm{B}}$ defined in (7.51) is straight-forward.

No restriction in the synchronization performance. Corollary 8.1 can be used to show that Assumption 8.3 is no restriction for guaranteeing the same upper bounds on the control error compared to a communication structure with multiple predecessor for each agent Σ_i.

Proposition 8.1 (Maintenance of synchronization performance by using one predecessor) *Consider the self-organizing control system Σ defined in* (8.17) *with the initial situation in* (7.52) *and* (7.53). *There exists an adjusted communication graph $\tilde{\mathcal{G}}_{\mathrm{A}} = (\mathcal{V}_{\mathrm{C}}, \tilde{\mathcal{E}}_{\mathrm{A}}, \boldsymbol{K})$ where each agent has only one predecessor (Assumption 8.3). Its set $\tilde{\mathcal{E}}_{\mathrm{A}}$ of adjusted communication links is a subset of the set $\mathcal{E}_{\mathrm{A}}$ of communication links within an adjusted communication graph $\mathcal{G}_{\mathrm{a}} = (\mathcal{V}_{\mathrm{C}}, \mathcal{E}_{\mathrm{A}}, \boldsymbol{K})$ with multiple predecessors for each agent, i.e., $\tilde{\mathcal{E}}_{\mathrm{A}} \subseteq \mathcal{E}_{\mathrm{A}}$. With this, the relation*

$$\bar{y}_{\Delta \mathrm{A}i} \geq \tilde{\bar{y}}_{\Delta \mathrm{A}i}, \quad \forall i \in \mathcal{N} \tag{8.28}$$

holds true, where $\bar{y}_{\Delta i}$ is the maximal control error resulting from $\mathcal{G}_{\mathrm{A}}$ and $\tilde{\bar{y}}_{\Delta i}$ is the maximal control error resulting from $\tilde{\mathcal{G}}_{\mathrm{A}}$.

Proof. The proof is given in Appendix A.13. □

Proposition 8.1 shows that for a given adjusted communication structure $\mathcal{G}_{\mathrm{A}}$ the elements $k_{\mathrm{A}ij}$ can be chosen such that the maximal control errors $\bar{y}_{\Delta \mathrm{A}i}$ are minimal, where the resulting

$k_{\mathrm{A}ij}$ imitates a communication structure with one predecessor according to Assumption 8.3. Furthermore, it is sufficient to have one predecessor for guaranteeing a desired maximal control error $\bar{y}_{\Delta \mathrm{A}i}$.

The extension of Proposition 8.1 for considering the overall closed-loop system Σ_{B} with a fixed basic communication graph $\mathcal{G}_{\mathrm{B}}$ defined in (7.51) is straight-forward.

Performance improvement with situation-dependent information request. The following proposition presents a condition under which the self-organizing controller with a situation-dependent communication structure leads to a lower maximal control error than a controller with a fixed basic communication structure.

Proposition 8.2 (Performance improvement with situation-dependent information request) *Consider the overall closed-loop system Σ_{B} with a fixed basic communication defined in* (7.51) *and the self-organizing control system Σ with the situation-dependent communication graph $\mathcal{G}_{\mathrm{C}}(t)$ defined in* (8.17), *where both are in the initial situation defined in* (7.52) *and* (7.53). *Then the inequality*

$$\bar{y}_{\Delta \mathrm{A}i} \leq \bar{y}_{\Delta \mathrm{B}i}, \quad \forall i \in \mathcal{N} \tag{8.29}$$

holds true if the relation

$$\bar{y}_{\Delta \mathrm{A}r} \leq \frac{1}{m_{\mathrm{yy}i}} (m_{\mathrm{y0B}i} - m_{\mathrm{y0A}i}) + \bar{y}_{\Delta \mathrm{B}j}, \quad j \in \mathcal{P}_{\mathrm{B}i}, \quad r \in \mathcal{P}_{\mathrm{A}i}, \quad \forall i \in \mathcal{N} \tag{8.30}$$

is fulfilled.

Proof. According to (7.57) and (8.21), relation (8.29) is fulfilled for

$$m_{\mathrm{y}\epsilon \mathrm{B}i} + m_{\mathrm{yy}i}\bar{y}_{\Delta \mathrm{B}j} \leq m_{\mathrm{y}\epsilon i} + m_{\mathrm{yy}i}\bar{y}_{\Delta \mathrm{A}r}, \quad j \in \mathcal{P}_{\mathrm{B}i}, \quad r \in \mathcal{P}_{\mathrm{A}i}, \quad \forall i \in \mathcal{N}. \tag{8.31}$$

Solving (8.31) for $\bar{y}_{\Delta \mathrm{A}r}$ yields condition (8.30). □

Proposition 8.2 – in particular (8.30) – shows that $\mathcal{P}_{\mathrm{A}i}$ has to be chosen such that the relation

$$\bar{y}_{\Delta \mathrm{A}r} \leq \bar{y}_{\Delta \mathrm{B}j}, \quad j \in \mathcal{P}_{\mathrm{B}i}, \quad r \in \mathcal{P}_{\mathrm{A}i}, \quad \forall i \in \mathcal{N}$$

holds true since generally $(m_{\mathrm{y0B}i} - m_{\mathrm{y0A}i})$ is smaller than zero according to Lemma 7.3 and Theorem 8.2. For $\epsilon_i = 0$ the relation $(m_{\mathrm{y0B}i} - m_{\mathrm{y0A}i}) = 0$ holds true. However, due to the communication effort and due to noise effects, it is not appropriate to set the value of ϵ_i to zero. If ϵ_i is enlarged, the first addend of (8.30) increases. Hence, $\bar{y}_{\Delta \mathrm{A}r}$, $(r \in \mathcal{P}_{\mathrm{A}i})$, has to decrease. In summary, (8.30) gives a condition for the choice of $\mathcal{P}_{\mathrm{A}i}$ for which the self-organizing controller guarantees a smaller maximal control error than when using a controller with the fixed basic communication. However, Proposition 8.2 does not state a method for designing the communication graphs. Such methods are presented in the following section.

8.4 Design of the local control units

This section presents methods for designing the parameters of the local control units C_i such that the resulting self-organizing control system Σ satisfies Control aim 7.1 and Control aim 7.2. Two algorithms for designing the adjusted communication graph $\mathcal{G}_{\mathrm{A}}$ are presented while considering that the basic communication graph $\mathcal{G}_{\mathrm{B}}$ and the switching thresholds ϵ_i are given. Both are search algorithms which are aimed to satisfy Control aim 7.2 by guaranteeing a minimal communication effort. For the design of the communication graphs it is assumed that every agent has only one predecessor (cf. Assumption 8.3). The first algorithm presented in Section 8.4.1 is performed from a global perspective on the overall plant P, uses global model information and guarantees a globally minimal communication effort. The second algorithm in Section 8.4.2 is performed from a local perspective on the controlled agents $\bar{\Sigma}_i$, uses local model information only and, hence, guarantees a locally minimal communication effort. Section 8.4.3 presents a method for designing the switching thresholds to guarantee that for a certain initial situation no information is requested by the local control units while considering that the basic communication graph $\mathcal{G}_{\mathrm{B}}$ is given. The results are summarized in an overall design algorithm presented in Section 8.4.4.

8.4.1 Design of the communication structure from a global perspective

This section presents a method for designing the adjusted communication graph $\mathcal{G}_{\mathrm{A}}$ by performing a search which uses global model information to guarantee the following three requirements:

1. The asymptotic synchronization of the agents Σ_i according to Control aim 7.1.

2. The adherence to a desired maximal control error $\bar{\boldsymbol{y}}^{*}_{\Delta\mathrm{A}}$ according to Control aim 7.2.

3. A minimal communication cost

$$c(\mathcal{G}_{\mathrm{A}}) := \sum_{i\in\mathcal{N}} \sum_{j\in\mathcal{P}_{\mathrm{A}i}} c_{ij} \tag{8.32}$$

within the adjusted communication graph $\mathcal{G}_{\mathrm{A}} = (\mathcal{V}_{\mathrm{C}}, \mathcal{E}_{\mathrm{A}}, \boldsymbol{K})$.

The basic communication graph $\mathcal{G}_{\mathrm{B}}$ and the switching thresholds ϵ_i, $(\forall i \in \mathcal{N})$, are assumed to be given. These requirements can be summarized to the following problem:

Problem 8.1 (Adjusted communication graph $\mathcal{G}_{\mathrm{A}}$ which guarantees desired synchronization performance and a globally minimal communication cost) *Consider the self-organizing control system Σ defined in (8.17) with the initial situation in (7.52) and (7.53), where $\tilde{\boldsymbol{x}}^*_{\Delta 0} = \tilde{\boldsymbol{x}}_{\Delta 0\mathrm{A}}$. Given is the basic communication graph $\mathcal{G}_{\mathrm{B}}$ and the switching thresholds ϵ_i, $(\forall i \in \mathcal{N})$, where the sets $\mathcal{P}_{\mathrm{B}i}$ of basic predecessors within $\mathcal{G}_{\mathrm{B}}$ are assumed to fulfill Assumption 7.6. Find the adjusted communication graph $\mathcal{G}_{\mathrm{A}}$ to fulfill Control aim 7.2 that solves the optimization problem*

$$\min_{\mathcal{E}_{\mathrm{A}} \subseteq \{(j \to i) | i,j \in \mathcal{V}_{\mathrm{C}},\ j<i\}} c(\mathcal{G}_{\mathrm{A}}) = \sum_{i \in \mathcal{N}} \sum_{j \in \mathcal{P}_{\mathrm{A}i}} c_{ij} \tag{8.33a}$$

$$\text{s.t.} \quad \bar{\boldsymbol{y}}_{\Delta\mathrm{A}}(\mathcal{G}_{\mathrm{A}}) \leq \bar{\boldsymbol{y}}^*_{\Delta\mathrm{A}} \tag{8.33b}$$

$$|\mathcal{P}_{\mathrm{A}i}| = 1, \quad \forall i \in \mathcal{N}. \tag{8.33c}$$

According to Theorem 8.1, constraint (8.33c) follows from the first requirement which claims Control aim 7.1 which also guarantees that the sets $\mathcal{P}_{\mathrm{A}i}$, $(\forall i \in \mathcal{N})$, of adjusted predecessors fulfill Assumption 8.1. Constraint (8.33b) follows from the desired maximal control error $\bar{\boldsymbol{y}}^*_{\Delta\mathrm{A}}$ claimed in Control aim 7.2 and Theorem 8.2. Furthermore, the minimization of $c(\mathcal{G}_{\mathrm{A}})$ follows from the third requirement. The notation $\bar{\boldsymbol{y}}_{\Delta\mathrm{A}}(\mathcal{G}_{\mathrm{A}})$ indicates that the maximal control error depends on the choice of the adjusted communication graph $\mathcal{G}_{\mathrm{A}}$, where $\bar{\boldsymbol{y}}_{\Delta\mathrm{A}}(\mathcal{G}_{\mathrm{A}})$ is determined with (8.23) in Theorem 8.2.

With this, the remaining part of the section is divided into three parts:

1. The general formulation of a search problem is recalled.
2. The structure of a search algorithm is presented which designs an adjusted communication graph $\mathcal{G}_{\mathrm{A}}$ that solves Problem 8.1. Due to the requirement on a global minimal communication cost in (8.33a), the algorithm uses global model information. Furthermore, the algorithm derives a graph $\mathcal{G}_{\mathrm{A}}$ such that every agent has only one predecessor (cf. Assumption 8.3).
3. The search algorithm is analyzed with respect to the solution of Problem 8.1.

Solving problems by search. For solving problems by search, the problem has to be formulated such that a search algorithm can solve it. In [115] it is stated that a problem consists of four parts:

- The search starts from the *initial state* A.
- The *set of actions* defines what set of *states* X are reachable from the current state S by executing these actions.
- The *goal test function* is applied to all found states X to determine if they are goal states.
- The *path cost function* $c(X)$ assigns the cost of the path.

The initial state A and the set of actions define the *state space* $\mathcal{X}$ that is the set of all states X reachable from A by any sequence of actions. The problem can be interpreted as the input to the search algorithm. The output is a solution of the problem that is a path to a *goal state* B.

Design of the adjusted communication graph $\mathcal{G}_\mathrm{A}$ by search. In the following the problem of finding $\mathcal{G}_\mathrm{A}$ is formulated as a problem that can be solved by a uniform-cost search, which is similar to the breath-first-search, but always expands to a new state X with the lowest communication cost. The notation is taken from [115].

A state $X = (\mathcal{P}_{\mathrm{x}1}, \ldots, \mathcal{P}_{\mathrm{x}N})$ represents a specific adjusted communication graph $\mathcal{G}_\mathrm{A}$, where the sets $\mathcal{P}_{\mathrm{x}i}$, $(\forall i \in \mathcal{N})$, are related to the certain sets $\mathcal{P}_{\mathrm{A}i}$ of predecessors describing $\mathcal{G}_\mathrm{A}$. With this, the four parts which define the search problem for designing $\mathcal{G}_\mathrm{A}$ are given by:

- The *initial state* A is described by the predecessors $\mathcal{P}_{0i}$, i.e.,

  $$A = (\mathcal{P}_{01}, \ldots, \mathcal{P}_{0N}), \text{ with } \mathcal{P}_{0i} = \{c_{\mathrm{MIN}i}(1)\}, \ \forall i \in \mathcal{N}$$

  and is a communication graph with the lowest communication cost: $c(A) = \min_{\mathcal{E}_\mathrm{A}} c(\mathcal{G}_\mathrm{A})$. Recall that the function $c_{\mathrm{MIN}i}(j)$, $(j \in \mathcal{V}, j < i)$, presorts possible predecessors of Σ_i by increasing order of their communication cost c_{ij}.

- The *set of actions* consists of a single rule:

  Rule 8.1 *Consider the current state $S = (\mathcal{P}_{\mathrm{s}1}, \ldots, \mathcal{P}_{\mathrm{s}N})$, where $\mathcal{P}_{\mathrm{s}i} = \{p_{\mathrm{s}i}\}$ for all $i \in \mathcal{N}$. If there exists*

  $$\mathcal{P}_{\mathrm{s}i} \neq \{0\}, \quad i \in \mathcal{N},$$

  generate the new state $X = (\mathcal{P}_{\mathrm{x}1}, \ldots, \mathcal{P}_{\mathrm{x}N})$ with

  $$\mathcal{P}_{\mathrm{x}i} = \left\{ c_{\mathrm{MIN}i}\left(\arg\min_{l \in \{1,\ldots,i\}} \left(c_{i,p_{\mathrm{s}i}} \leq c_{i,c_{\mathrm{MIN}i}(l)} \right) \right) \right\},$$
  $$\mathcal{P}_{\mathrm{x}j} = \{p_{\mathrm{s}j}\}, \quad \forall j \in \mathcal{N} \backslash \{i\}$$

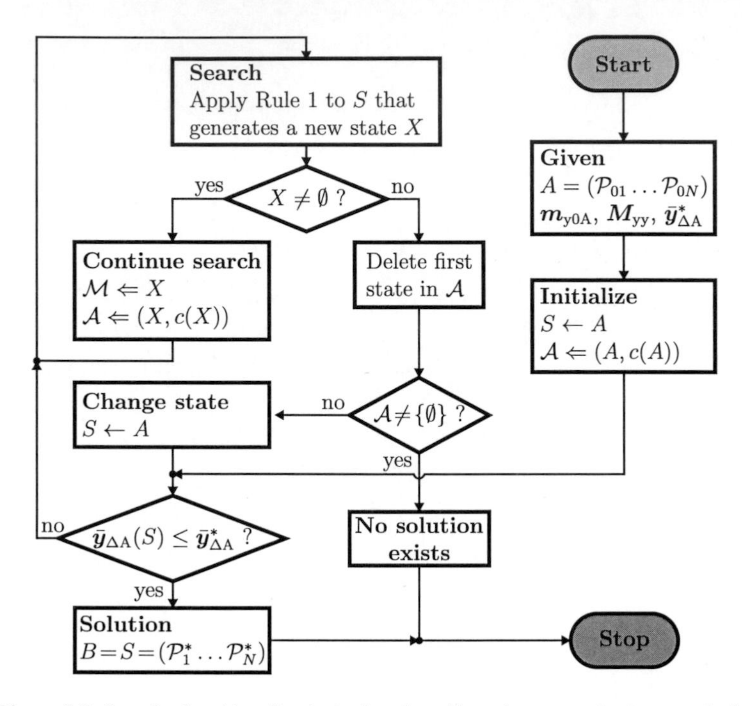

Figure 8.8: Search algorithm for designing the adjusted communication graph $\mathcal{G}_{\mathrm{A}}$ using global model information.

which does not belong to the list $\mathcal{M}$ of marked states: $X\notin\mathcal{M}$. If no such state X exists, then X is defined by $X=\emptyset$.

- The *goal test function* is (8.33b), where $\boldsymbol{y}_{\Delta\mathrm{A}}(\mathcal{G}_{\mathrm{A}})$ results from Theorem 8.2.
- The *path cost function* is the communication cost $c(X)$ defined in (8.32).

The algorithm for designing $\mathcal{G}_{\mathrm{A}}$ is summarized in Fig. 8.8. The input to the algorithm is the initial state $A=(\mathcal{P}_{01}\dots\mathcal{P}_{0N})$, the desired maximal control error $\bar{\boldsymbol{y}}^*_{\Delta\mathrm{A}}$ and the bounds $\boldsymbol{m}_{\mathrm{y0A}}$ from (8.24) as well as $\boldsymbol{M}_{\mathrm{yy}}$ from (7.61) on the dynamics of the overall system.

The output of the algorithm is a goal state $B=(\mathcal{P}^*_1,\dots,\mathcal{P}^*_N)$ containing the goal predecessors $\mathcal{P}^*_i$, $(\forall i\in\mathcal{N})$, that describe the desired communication structures $\mathcal{G}^*_{\mathrm{A}}$. Initially the current state S is set to $A=(\mathcal{P}_{01},\dots,\mathcal{P}_{0N})$ (indicated by $S\leftarrow A$).

Furthermore, A with its cost $c(A)$ is written at the first position of the *list $\mathcal{A}$ of active* states. The states in $\mathcal{A}$ are sorted by increasing order of the cost c in (8.32). The first step is to check the goal condition (8.33b) for A. If the condition is not fulfilled, the search starts.

In the search step, Rule 8.1 is applied to the current state S to find a new state X that is not in the *list* $\mathcal{M}$ *of marked states*. If such a state X exists, i.e., $X \neq \emptyset$, X is added to $\mathcal{M}$ (indicated by $\mathcal{M} \Leftarrow X$) and sorted in the list $\mathcal{A}$ of active states (indicated by $\mathcal{A} \Leftarrow (A, c(A))$). This step is labeled *Continue search*. Afterwards, the next search step is performed.

If $X = \emptyset$ holds, the first state of $\mathcal{A}$ is deleted. If $\mathcal{A}$ is empty, i.e., $\mathcal{A} = \{\emptyset\}$, there is no solution ($B = \emptyset$) for Control aim 7.2. Otherwise, S is changed to the first state in $\mathcal{A}$ (indicated by $S \leftarrow \mathcal{A}$).

If the current graph $\mathcal{G}_{\mathrm{A}}$ represented by S fulfills the goal condition in (8.33b), then S is the solution $B = S = (\mathcal{P}_1^*, \ldots, \mathcal{P}_N^*)$. Otherwise, the next search step is performed.

Analysis of the search algorithm. The following analysis concerns the correctness of the search algorithm presented in Fig. 8.8.

Theorem 8.3 (Correctness of the search algorithm for designing $\mathcal{G}_{\mathrm{A}}$) *The adjusted communication graph $\mathcal{G}_{\mathrm{A}}$ resulting from the search algorithm presented in Fig. 8.8 is a solution for Problem 8.1 if the desired maximal control errors $\bar{y}^*_{\Delta \mathrm{A}i}$, $(\forall i \in \mathcal{N})$, are chosen such that the relations*

$$\bar{y}^*_{\Delta \mathrm{A}i} \geq m_{\mathrm{y0A}i}, \quad \forall i \in \mathcal{N}, \tag{8.34}$$

are satisfied.

Proof. The proof is divided into three parts:

1. To fulfill the asymptotic synchronization of the agents (Control aim 7.1) it is shown that within the designed adjusted communication graph $\mathcal{G}_{\mathrm{A}}$ every agent has a predecessor which is claimed by the constraint in (8.33c).

2. It is shown that there exists an adjusted communication graph $\mathcal{G}_{\mathrm{A}}$ defined in (8.5) which satisfies the goal condition (8.33b) if condition (8.34) is satisfied.

3. It is proven that the search algorithm in Fig. 8.8 finds a solution. Therefore, it has to be shown that Rule 8.1 can generate any possible communication structure while starting from A since the uniform-cost search guarantees that if B is connected to A, it finds the shortest path to B (cf. [115]). Furthermore, in [115] it is claimed that the application of Rule 8.1 to S always causes a nonnegative cost from S to X, i.e.,

$$c(S) \leq c(X). \tag{8.35}$$

Asymptotic synchronization. In the initial state A all sets $\mathcal{P}_{0i}$ are non-empty, i.e., $\mathcal{P}_{0i} \neq \emptyset$ for all $i \in \mathcal{N}$ since the function $c_{\mathrm{MIN}i}(1)$ is defined for all $i \in \mathcal{N}$. With this, and since Rule 8.1 guarantees that for possible new predecessors $\mathcal{P}_{\mathrm{x}i}$ the relation $\mathcal{P}_{\mathrm{x}i} \neq \emptyset$ holds true for all $i \in \mathcal{N}$, it is guaranteed that for every graph $\mathcal{G}_{\mathrm{A}}$, which is a solution of search algorithm, relation (8.33c) is satisfied.

Existence of a solution. According to Theorem 8.2, the goal condition (8.33b) is satisfied if the relation

$$\max_{t \geq 0} |y_{\Delta i}(t)| \leq \bar{y}_{\Delta \mathrm{A}i} = m_{\mathrm{y0A}i} + m_{\mathrm{yy}i} \sum_{j \in \mathcal{P}_{\mathrm{A}i}} k_{\mathrm{A}ij} \bar{y}_{\Delta \mathrm{A}j} \leq \bar{y}^{*}_{\Delta \mathrm{A}i}$$

holds true for all $i \in \mathcal{N}$. The smallest possible value of $\bar{y}_{\Delta \mathrm{A}i}$ is $m_{\mathrm{y0A}i}$ for the worst case that all agents receive information from the leader Σ_0, i.e., $\mathcal{P}_{\mathrm{A}i} = \{0\}$ for all $i \in \mathcal{N}$, since $\bar{y}_{\Delta \mathrm{A0}} = 0$ holds true. In summary, relation (8.34) guarantees the existence of graph $\mathcal{G}_{\mathrm{A}}$ that satisfies the goal condition (8.33b).

Find a solution. Consider the initial state A with $\mathcal{P}_{0i} = \{p_{0i}\}$, the current state S with $\mathcal{P}_{\mathrm{s}i} = \{p_{\mathrm{s}i}\}$ and the desired state B with $\mathcal{P}_{\mathrm{b}i} = \{p_{\mathrm{b}i}\}$ for all $i \in \mathcal{N}$. Rule 8.1 has to be applied at most $(p_{0i} - p_{\mathrm{b}i})$-times to $\mathcal{P}_{\mathrm{s}i}$ starting with $\mathcal{P}_{0i}$ to get $\mathcal{P}_{\mathrm{b}i}$ since with Rule 8.1 all possible predecessor $\mathcal{P}_{\mathrm{x}i} = \{p_{\mathrm{x}i}\}$, $(0 \leq p_{\mathrm{x}i} < i)$, are checked by increasing order of their communication cost to Σ_i. Hence, any possible communication structure can be generated by multiple application of Rule 8.1 which leads to the state space

$$\mathcal{X} = \{X = (\mathcal{P}_{\mathrm{x}1}, \ldots, \mathcal{P}_{\mathrm{x}N}) \mid \mathcal{P}_{\mathrm{x}i} = \{j\},\ 0 \leq j < i,\ i \in \mathcal{N}\}.$$

The application of Rule 8.1 to the sets $\mathcal{P}_{\mathrm{s}r} = \{p_{\mathrm{s}r}\}$, $(r \in \mathcal{N})$, of S, the consideration of the relation $c(S) \leq c(X)$ and the communication cost in (8.32) yields

$$\sum_{i \in \mathcal{N} \setminus \{r\}} \sum_{j \in \mathcal{P}_{\mathrm{s}i}} c_{ij} + c_{r p_{\mathrm{s}r}} \leq \sum_{i \in \mathcal{N} \setminus \{r\}} \sum_{j \in \mathcal{P}_{\mathrm{s}i}} c_{ij} + c_{r c_{\mathrm{MIN}r}(l)}.$$

Since the generation of $\mathcal{P}_{\mathrm{x}r}$ with Rule 8.1 requires $c_{r p_{\mathrm{s}r}} \leq c_{r c_{\mathrm{MIN}r}(l)}$, the proof is completed. □

Every agent Σ_i, $\forall i \in \mathcal{N}$, has i possible predecessors. Hence, there exist $N!$ possible communication structures which leads to a complexity of $O(N!)$ for the search algorithm in Fig. 8.8. The benefit of the search algorithm is that not necessarily all possible communication structures have to be built and to be checked for fulfilling the three requirements claimed in Problem 8.1 at the beginning of this section. Note that $c(X)$ in (8.32) could be replaced by any other cost function that fulfills the relation $c(S) \leq c(X)$. Hence, there is no need to know the exact communication cost among the agents. It is sufficient to know an abstract value that provides a measure of the communication effort among the local control units.

Design of the basic communication graph. The search algorithm in Fig 8.8 for the determination of $\mathcal{G}_{\mathrm{A}}$ can be easily adapted to determine the basic communication graph $\mathcal{G}_{\mathrm{B}}$ as well. For this, $\bar{\boldsymbol{y}}_{\Delta\mathrm{A}}(S)$, $\bar{\boldsymbol{y}}^*_{\Delta\mathrm{A}}$, $\tilde{\boldsymbol{x}}_{\Delta0\mathrm{A}}$ and $\boldsymbol{m}_{\mathrm{y}0\mathrm{A}}$ have to be replaced by $\bar{\boldsymbol{y}}_{\Delta\mathrm{B}}(S)$, $\bar{\boldsymbol{y}}^*_{\Delta\mathrm{B}}$, $\tilde{\boldsymbol{x}}_{\Delta0\mathrm{B}}$ and $\boldsymbol{m}_{\mathrm{y}0\mathrm{B}}$, respectively. Then, $\mathcal{G}_{\mathrm{B}}$ instead of $\mathcal{G}_{\mathrm{A}}$ is the result which guarantees that Control aim 7.1 and Control aim 7.2 are fulfilled for the overall closed-loop system Σ_{B} using the basic communication graph $\mathcal{G}_{\mathrm{B}}$ instead of the self-organizing control system Σ. Furthermore, the communication cost $c(\mathcal{G}_{\mathrm{B}})$ is similar to (8.33a). Note that $\bar{\boldsymbol{y}}^*_{\Delta\mathrm{B}}$ is the *desired overall maximal basic control error.*

Example 8.2 *Situation-dependent information request within a formation of six robots*

Consider a robot formation with six robots with different dynamics. The robots Σ_i, $(\forall i \in \mathcal{N}_{\mathrm{avarage}} = \{1,3,4\})$ are average robots, where Σ_2 is a fast robot ($\mathcal{N}_{\mathrm{fast}} = \{2\}$ indicated by index "f") and Σ_5 is a slow robot ($\mathcal{N}_{\mathrm{slow}} = \{5\}$ indicated by index "s"). Furthermore, the basic communication graph $\mathcal{G}_{\mathrm{B}}$ depicted at the top of Fig. 8.9 and the switching thresholds $\epsilon_i = 0.1\,\mathrm{cm}$, $(\forall i \in \{2,\dots,5\})$, are given. The aim is to design the adjusted communication graph $\mathcal{G}_{\mathrm{A}}$ that solves Problem 8.1 with the desired maximal control errors

$$\begin{aligned} \bar{y}^*_{\Delta\mathrm{A}1} &= \bar{y}_{\Delta\mathrm{B}1,\mathrm{serial}} \\ \bar{y}^*_{\Delta\mathrm{A}j} &= 0.75 \cdot \bar{y}_{\Delta\mathrm{B}j,\mathrm{serial}}, \quad j = 2,\dots 5, \end{aligned}$$

for a stepwise change of the reference velocity $\bar{v}(t)$ by $100\,\frac{\mathrm{Ink}}{\mathrm{s}}$, i.e., $\tilde{\boldsymbol{x}}_{\Delta0} = \begin{pmatrix} 0 & 100 \end{pmatrix}^{\mathrm{T}}$. Hence, the maximal control error of the robots Σ_i, $(\forall i \in \{2,\dots,5\})$, has to be reduced by $25\,\%$ compared to a serial communication between the robots. The maximal basic control errors $\bar{y}_{\Delta\mathrm{B}i,\mathrm{serial}}$ result from Lemma 7.3 and bound the control errors $y_{\Delta i}(t)$ of the overall closed-loop system Σ_{B} with basic communication which uses $\mathcal{G}_{\mathrm{B}}$ in Fig. 8.9. The overall maximal basic control error is given by

$$\boldsymbol{y}_{\Delta\mathrm{B},\mathrm{serial}} = (1.97,\ \ 3.38,\ \ 6.23,\ \ 9.84,\ \ 17.34\,)^{\mathrm{T}}\mathrm{cm}\,.$$

According to Lemma 7.3 and Theorem 8.2, relation (8.34) is fulfilled for all desired maximal control errors $\bar{y}^*_{\Delta i}$. Hence, the search algorithm presented in Fig. 8.8 can solve Problem 8.1 according to Theorem 8.3, where the resulting adjusted communication graph $\mathcal{G}_{\mathrm{A}}$ is shown in the upper part of Fig. 8.9. The remaining part of Fig. 8.9 shows the behavior of the self-organizing controlled robots using the depicted graphs $\mathcal{G}_{\mathrm{B}}$ and $\mathcal{G}_{\mathrm{A}}$. Note that the lower plot shows the currently activated communication of each local control unit C_i, where the gray bar indicates that C_i gets the output $y_j(t)$ from its basic predecessor C_j, $(j \in \mathcal{P}_{\mathrm{B}i})$, and the blue bar indicates that C_i gets the output $y_k(t)$ from its adjusted predecessor C_k, $(k \in \mathcal{P}_{\mathrm{A}i})$. For example, the local control units C_i, $(i > 2)$, receive the output from Σ_j, $(j \in \mathcal{P}_{\mathrm{A}i})$, at the time t =5 s.

Figure 8.10 shows the desired maximal control errors $\bar{y}^*_{\Delta\mathrm{A}i}$ for the robots Σ_i, $(i = 2,\dots,5)$, (red dashed lines) and the actual control errors $y_{\Delta i}(t)$ as a phase portrait (solid blue line). The blue dashed lines are the designed maximal control errors $\bar{y}_{\Delta\mathrm{A}i}$, which lie beneath the desired maximal control error $\bar{y}^*_{\Delta\mathrm{A}i}$. Hence, the designed $\mathcal{G}_{\mathrm{A}}$ solves Problem 8.1 and, therefore, guarantees a desired performance of the robot formation. □

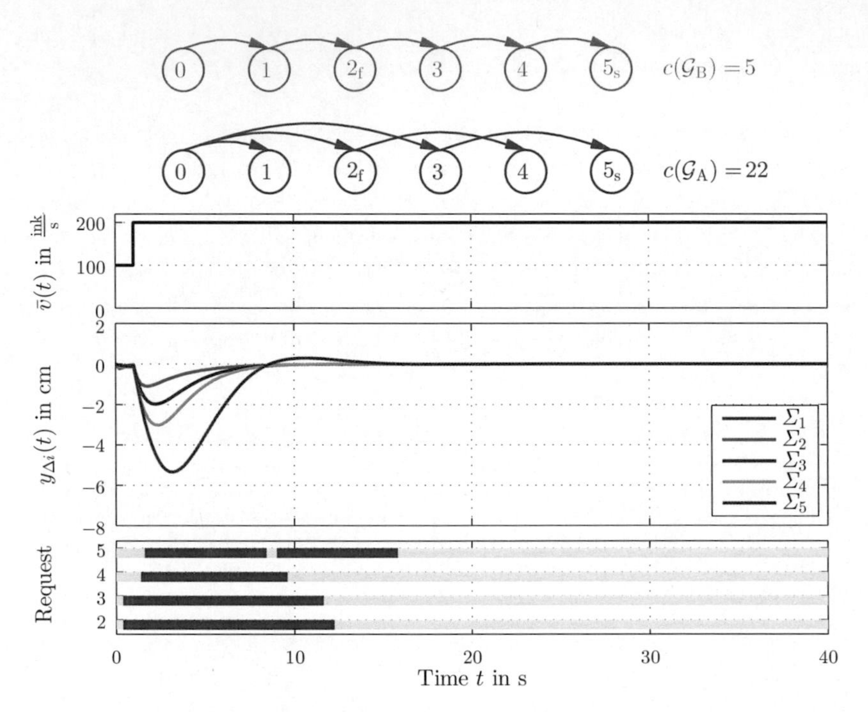

Figure 8.9: Designed adjusted communication graph $\mathcal{G}_A$ from a global perspective and behavior of the resulting self-organizing controlled robots.

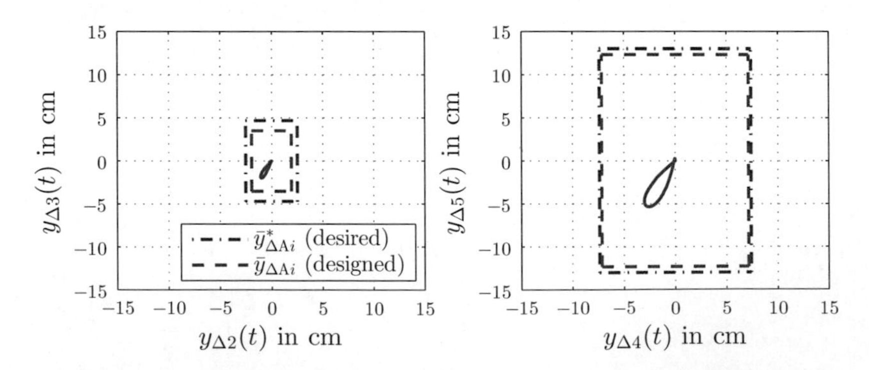

Figure 8.10: Verification of the compliance of the desired maximal control errors $\bar{y}^*_{\Delta j}$ defined in (8.36) according to the behavior of the robot formation depicted in Fig. 8.9.

8.4.2 Design of the communication structure from a local perspective

This section presents a method for choosing the set $\mathcal{P}_{\mathrm{A}i}$ of adjusted predecessors for each agent Σ_i individually. The design is performed from a local perspective of each agent by using local model information only. Therefore, the following three requirements have to be satisfied:

1. The asymptotic synchronization of the agents according to Control aim 7.1.
2. The adherence to a desired maximal control error $\bar{y}^*_{\Delta \mathrm{A}i}$ for each agent Σ_i according to Control aim 7.2.
3. A minimal *local communication cost*

$$c_i(\mathcal{P}_{\mathrm{A}i}) := \sum_{j \in \mathcal{P}_{\mathrm{A}i}} c_{ij}, \quad \forall i \in \mathcal{N} \tag{8.37}$$

   for every agent Σ_i, $(\forall i \in \mathcal{N})$.

The sets $\mathcal{P}_{\mathrm{B}i}$, $(\forall i \in \mathcal{N})$, of basic predecessors and the switching thresholds ϵ_i, $(\forall i \in \mathcal{N})$, are assumed to be given. These requirements can be summarized to the following problem:

Problem 8.2 (Adjusted communication graph $\mathcal{G}_{\mathrm{A}}$ which guarantees desired synchronization performance and a locally minimal communication cost) *Consider the self-organizing control system Σ defined in* (8.17) *with the initial situation in* (7.52) *and* (7.53), *where $\tilde{x}^*_{\Delta 0} = \tilde{x}_{\Delta 0 \mathrm{A}}$. Given are the sets $\mathcal{P}_{\mathrm{B}i}$, $(\forall i \in \mathcal{N})$, of basic predecessors and the switching thresholds ϵ_i, $(\forall i \in \mathcal{N})$, where the sets $\mathcal{P}_{\mathrm{B}i}$ are assumed to fulfill Assumption 7.6. Find the sets $\mathcal{P}_{\mathrm{A}i}$, $(\forall i \in \mathcal{N})$, of adjusted predecessors to fulfill Control aim 7.2 that solve the optimization problem*

$$\min_{\mathcal{P}_{\mathrm{A}i} \subseteq \{j \in \mathcal{V}_{\mathrm{C}} | j < i\}} \quad c_i(\mathcal{P}_{\mathrm{A}i}) = \sum_{j \in \mathcal{P}_{\mathrm{A}i}} c_{ij} \tag{8.38a}$$

$$\text{s.t.} \quad \bar{y}_{\Delta \mathrm{A}i}(\mathcal{P}_{\mathrm{A}i}) \leq \bar{y}^*_{\Delta \mathrm{A}i} \tag{8.38b}$$

$$|\mathcal{P}_{\mathrm{A}i}| = 1 \tag{8.38c}$$

for all $i \in \mathcal{N}$.

According to Theorem 8.1, constraint (8.38c) follows from the first requirement which claims Control aim 7.1 which also guarantees that the sets $\mathcal{P}_{\mathrm{A}i}$, $(\forall i \in \mathcal{N})$, of adjusted predecessors fulfill Assumption 8.1. Constraint (8.38b) follows from the desired maximal control errors $\bar{y}^*_{\Delta \mathrm{A}i}$ claimed in Control aim 7.2 and Theorem 8.2. Furthermore, the minimization of $c_i(\mathcal{P}_{\mathrm{A}i})$ follows from the third requirement. The notation $\bar{y}_{\Delta\mathrm{A}}(\mathcal{P}_{\mathrm{A}i})$ indicates that the maximal control error depends on the choice of the set $\mathcal{P}_{\mathrm{A}i}$ of adjusted predecessors, where $\bar{y}_{\Delta\mathrm{A}}(\mathcal{P}_{\mathrm{A}i})$ is determined with (8.21) in Theorem 8.2.

With this, the remaining part of the section is divided into two parts:

1. The structure of a design algorithm which chooses the sets $\mathcal{P}_{\mathrm{A}i}$, $(\forall i \in \mathcal{N})$, of adjusted predecessors from a local perspective of each agent is presented.

2. The algorithm is analyzed with respect to the fulfillment of Problem 8.2.

Algorithm for local design of the sets $\mathcal{P}_{\mathrm{A}i}$ of adjusted predecessors. A method for determining the sets $\mathcal{P}_{\mathrm{A}i}$, $(\forall i \in \mathcal{N})$, of adjusted predecessors is given by Algorithm 8.1. This algorithm consists of an outer for-loop and an inner for-loop. The outer for-loop guarantees that every agent Σ_i is picked a single time. The inner for-loop choses the set $\mathcal{P}_{\mathrm{A}i}$ of agent Σ_i by finding a predecessor Σ_j for Σ_i with the lowest communication cost c_{ij} which guarantees the adherence to the desired maximal control error $\bar{y}^*_{\Delta\mathrm{A}i}$ of Σ_i. Note that all inner for-loops could also be processed in parallel since the results are independent. The inner for-loop uses the following information:

- The desired maximal control error $\bar{y}^*_{\Delta\mathrm{A}i}$ for Σ_i.

- The scalars $m_{\mathrm{y0A}i}$ from (8.22) and $m_{\mathrm{yy}i}$ from (7.58) represent an upper bound on the behavior of $\bar{\Sigma}_i$ which result from local model information of $\bar{\Sigma}_i$.

- The function $c_{\mathrm{MIN}i}$ for Σ_i that presorts possible predecessors of Σ_i by increasing order of their communication cost.

Algorithm 8.1: Design of the sets $\mathcal{P}_{\mathrm{A}i}$, $(\forall i \in \mathcal{N})$, from a local perspective

Given: $\bar{y}^*_{\Delta\mathrm{A}i}$, $c_{\mathrm{MIN}i}$, $m_{\mathrm{y0A}i}$, $m_{\mathrm{yy}i}$, $\forall i \in \mathcal{N}$

```
1  for i to N do
       // Local design based on ȳ*_ΔAi, m_yyi, m_y0Ai, c_MINi and ȳ*_ΔAj, ∀j∈{0,1...i−1}
2      for l = 1 to i do
3          Determine current maximal control error: ȳ_Δ = m_y0Ai + m_yyi ȳ*_ΔAj for j = c_MINi(l)
4          if ȳ_Δ ≤ ȳ*_ΔAi then
5              break
6          end
7      end
8      Define the predecessor of Σ_i by setting P_Ai = {c_MINi(l)}.
9  end
```

Result: $\mathcal{P}_{\mathrm{A}i}$ for all $i \in \mathcal{N}$ defining the adjusted communication graph $\mathcal{G}_{\mathrm{A}}$.

- The desired maximal control errors $\bar{y}^*_{\Delta Aj}$, $(j \in \{0, 1 \ldots i-1\})$, of all possible predecessors of Σ_i.

The function $c_{\mathrm{MIN}i}$ for Σ_i needs global information about the communication cost c_{ij} of agent Σ_i to possible predecessors Σ_j, $(j \in \{0, 1 \ldots i-1\})$. Note that this costs c_{ij} do not have to be an exact value of the actual communication cost. It could also be an abstract value that represents the communication cost.

The desired maximal control errors $\bar{y}^*_{\Delta Aj}$, $(j \in \{0, 1 \ldots i-1\})$, of possible predecessors of Σ_i are no global information about the dynamics of the agents or the communication among the local control units, but rather information about control aims of each agent. Hence, the assumption of knowing the aims $\bar{y}^*_{\Delta Aj}$, $(j \in \{0, 1 \ldots i-1\})$, for a distributed realization of Algorithm 8.1 is weak.

Analysis of the design algorithm. The following theorem concerns the correctness of Algorithm 8.1 to fulfill the three requirements claimed at the beginning of this section.

Theorem 8.4 (Correctness of Algorithm 8.1 designing the sets $\mathcal{P}_{\mathrm{A}i}$ of adjusted predecessors by using local information) *The sets $\mathcal{P}_{\mathrm{A}i}$, $(\forall i \in \mathcal{N})$, of adjusted predecessors resulting from Algorithm 8.1 are a solution for Problem 8.2 if the desired maximal control errors $\bar{y}^*_{\Delta \mathrm{A}i}$, $(\forall i \in \mathcal{N})$, are chosen such that relation (8.34) is satisfied.*

Proof. Since the function $c_{\mathrm{MIN}i}(l)$ always derives a single predecessor $\mathcal{P}_{\mathrm{A}i} = \{j\}$ with $j \in \{0, 1 \ldots i-1\}$ (cf. Step 3 and Step 8), the constraint in (8.38c) is fulfilled. The break condition in Steps 4-6 satisfies the constraint in (8.38b). Furthermore, the iterative checking of the predecessors in Step 3 by increasing order of their communication cost guarantees that the local communication cost $c_i(\mathcal{P}_{\mathrm{A}i})$ is minimal (cf. (8.38)). According to Theorem 8.2, the smallest possible value of $\bar{y}_{\Delta \mathrm{A}i}$ is $m_{\mathrm{y0A}i}$ for $\mathcal{P}_{\mathrm{A}i} = \{0\}$ since $\bar{y}_{\Delta \mathrm{A}0} = 0$ holds true. Hence, the relation (8.34) guarantees the existence of a set $\mathcal{P}_{\mathrm{A}i}$ that solves the optimization problem in (8.38). With this, the proof is completed. □

Theorem 8.4 shows that the self-organizing controller C which uses the sets $\mathcal{P}_{\mathrm{A}i}$, $(\forall i \in \mathcal{N})$, of adjusted predecessors resulting from Algorithm 8.1 leads to a self-organizing control system Σ defined in (8.17) that satisfies Control aim 7.1, Control aim 7.2 and a minimal local communication cost $c_i(\mathcal{P}_{\mathrm{A}i})$ for every agent Σ_i.

Since every agent checks i possible predecessors, Algorithm 8.1 performs $\sum_{i \in \mathcal{N}} i = \frac{N(N-1)}{2}$ iterations. Hence, Algorithm 8.1 has a quadratic complexity, i.e., $O(N^2)$.

In summary, the following property of Algorithm 8.1 is highlighted:

Algorithm 8.1 can be implemented in a decentralized framework, where the sets $\mathcal{P}_{\mathrm{A}i}$, $(\forall i \in \mathcal{N})$, of adjusted predecessors can be determined locally at each agent to build an overall adjusted communication graph $\mathcal{G}_{\mathrm{A}}$ with which a desired maximal control error $\bar{y}^*_{\Delta\mathrm{A}i}$ is guaranteed for each agent Σ_i.

Design of the sets $\mathcal{P}_{\mathrm{B}i}$ of basic predecessors. Algorithm 8.1 which choses the sets $\mathcal{P}_{\mathrm{A}i}$ of adjusted predecessors can be easily adapted to determine the sets $\mathcal{P}_{\mathrm{B}i}$ of basic predecessors as well. For this, $\bar{y}^*_{\Delta\mathrm{A}i}$, $\tilde{\boldsymbol{x}}_{\Delta 0\mathrm{A}}$ and $m_{\mathrm{y}0\mathrm{A}i}$ have to be replaced by $\bar{y}^*_{\Delta\mathrm{B}i}$, $\tilde{\boldsymbol{x}}_{\Delta 0\mathrm{B}}$ and $m_{\mathrm{y}0\mathrm{B}i}$, respectively. Then, the sets $\mathcal{P}_{\mathrm{B}i}$ of basic predecessors instead of the sets $\mathcal{P}_{\mathrm{A}i}$ adjusted predecessors result which guarantees that Control aim 7.1 and Control aim 7.2 are fulfilled for the overall closed-loop system Σ_{B} with basic communication instead for the self-organizing control system Σ. Furthermore, the local communication costs $c_i(\mathcal{G}_{\mathrm{B}})$, $(\forall i \in \mathcal{N})$, are similar to (8.38a). Note that $\bar{y}^*_{\Delta\mathrm{B}i}$, $(\forall i \in \mathcal{N})$, are the desired maximal basic control errors.

Example 8.2 (cont.) *Situation-dependent information request within a formation of six robots*

Consider the same requirements as for the design of the adjusted communication graph $\mathcal{G}_{\mathrm{A}}$ from a global perspective in Example 8.2 on page 277, In the following $\mathcal{G}_{\mathrm{A}}$ is designed by Algorithm 8.1 to solve Problem 8.2. According to Lemma 7.3 and Theorem 8.2, relation (8.34) is fulfilled for all desired maximal control errors $\bar{y}^*_{\Delta\mathrm{A}i}$. Hence, Algorithm 8.1 can solve Problem 8.1 according to Theorem 8.4. The resulting adjusted communication graph $\mathcal{G}_{\mathrm{A}}$ is shown in the upper part of Fig. 8.12. Compared to $\mathcal{G}_{\mathrm{A}}$ in Fig. 8.9 the graph $\mathcal{G}_{\mathrm{A}}$ in Fig. 8.12 has a higher communication cost $c(\mathcal{G}_{\mathrm{A}})$ since Algorithm 8.1 guarantees a locally minimal cost $c_i(\mathcal{P}_{\mathrm{A}i})$ only. Nevertheless, the desired maximal control errors $\bar{y}^*_{\Delta j}$ defined in (8.36) are not exceeded by the actual control errors $y_{\Delta i}(t)$ of the robots (cf. Fig. 8.11) without knowing the overall model of the plant P. □

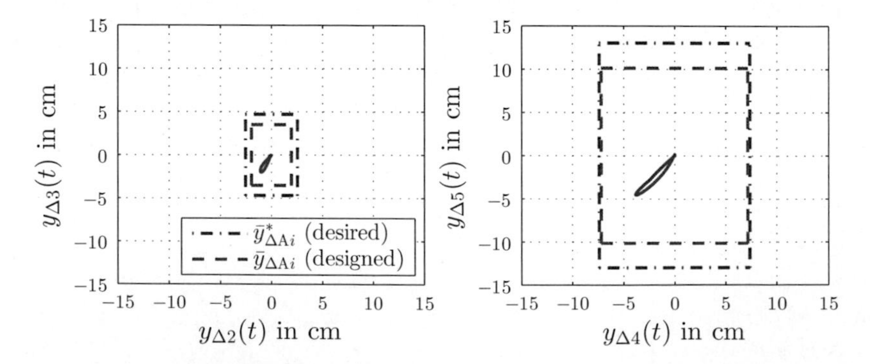

Figure 8.11: Verification of the compliance of the desired maximal control errors $\bar{y}^*_{\Delta j}$ defined in (8.36) according to the behavior of the robot formation depicted in Fig. 8.12.

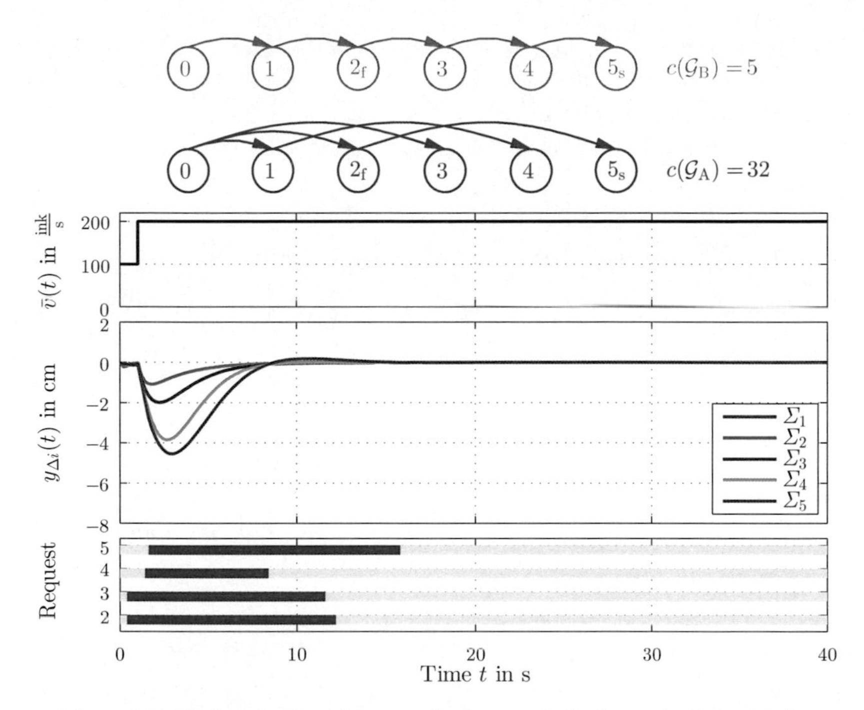

Figure 8.12: Designed adjusted communication graph $\mathcal{G}_A$ from a local perspective and behavior of the resulting self-organizing controlled robots.

8.4.3 Choice of the switching thresholds

This section concerns the design of the switching thresholds ϵ_i which should be designed such that no information is requested for a small difference between the initial state of the leader Σ_0 and the initial state of the agents Σ_i.

Problem 8.3 (Switching thresholds which avoid requests) *Consider the self-organizing control system Σ defined in* (8.17) *with the initial situation in* (7.52) *and* (7.53). *Given is the basic communication graph $\mathcal{G}_B$, where the sets $\mathcal{P}_{Bi}$ within $\mathcal{G}_B$ are assumed to fulfill Assumption 7.6. Find the switching thresholds ϵ_i, ($\forall i \in \mathcal{N}$), such that no information is requested for the initial state $\tilde{\boldsymbol{x}}^*_{\Delta 0} = \tilde{\boldsymbol{x}}_{\Delta 0\text{min}}$, which means that $e_i(t)$ does not exceeded the switching threshold ϵ_i, i.e.,*

$$|e_i(t)| < \epsilon_i, \quad \forall i \in \mathcal{N}, \quad t \geq 0. \tag{8.39}$$

Considering the robot formation problem, a solution to Problem 8.3 would guarantee that there is no information request for small changes of the reference velocity $\bar{v}(t)$. This requirement is appropriate since information should only be requested to obtain a significant improvement on the overall system performance. For small changes of the reference velocity the improvement is marginal.

According to (7.45) and (8.20), the overall local control $\boldsymbol{e}(t)$ for the overall closed-loop system Σ_{B} defined in (7.51) with basic communication is bounded by $\bar{\boldsymbol{e}}_{\mathrm{maxB}}$:

$$\max_{t\geq 0} |\boldsymbol{e}(t)| \leq \bar{\boldsymbol{e}}_{\mathrm{maxB}} = \max_{t\geq 0} \left| (\boldsymbol{K}_{\mathrm{B}}^{*} - \boldsymbol{I}_{N}) \tilde{\boldsymbol{C}} \mathrm{e}^{\tilde{\bar{\boldsymbol{A}}}_{\mathrm{B}} t} \tilde{\boldsymbol{x}}_{\Delta 0} \right| . \tag{8.40}$$

Based on this relation, the following proposition derives a condition for designing the switching thresholds ϵ_i.

Proposition 8.3 (Design of switching thresholds to guarantee no requests) *Consider the self-organizing control system Σ defined in* (8.17) *with the initial situation in* (7.52) *and* (7.53), *where $\tilde{\boldsymbol{x}}_{\Delta 0}^{*} = \tilde{\boldsymbol{x}}_{\Delta 0\mathrm{min}}$. If the switching thresholds ϵ_i are designed such that the relation*

$$\boldsymbol{\epsilon} = \begin{pmatrix} \epsilon_1 \\ \vdots \\ \epsilon_N \end{pmatrix} < \boldsymbol{e}_{\mathrm{maxB}}, \tag{8.41}$$

holds true, then the switching thresholds ϵ_i, $(\forall i \in \mathcal{N})$, are a solution for Problem 8.3.

Proof. Since $\boldsymbol{e}_{\mathrm{maxB}}$ is an upper bound for the overall local control error $\boldsymbol{e}(t)$, the relation $|\boldsymbol{e}(t)| < \boldsymbol{\epsilon}$ holds true for all $t \geq 0$. □

Compared to other methods for designing the switching thresholds, Proposition 8.3 derives no condition for which situations the information should be exchanged, but a condition in which situations an information exchange is unnecessary.

8.4.4 Overall design algorithm

Combining the design of the communication graphs $\mathcal{G}_{\mathrm{A}}$ and $\mathcal{G}_{\mathrm{B}}$ in Sections 8.4.1 and 8.4.2 and the design of the switching thresholds ϵ_i, $(\forall i \in \mathcal{N})$, in Section 8.4.3 leads to Algorithm 8.2 that determines the parameters of the self-organizing controller C defined in (8.16).

Corollary 8.2 (Design parameters of local control units C_i performing information requests) *If the parameters $\mathcal{G}_{\mathrm{A}}$, $\mathcal{G}_{\mathrm{B}}$ and $\boldsymbol{\epsilon}$ of the self-organizing controller C defined in* (8.16) *are designed by Algorithm 8.2 and relation* (8.34) *is satisfied, then the self-organizing control system Σ defined in* (8.17) *fulfills Control aim 7.1 and Control aim 7.2.*

Algorithm 8.2: Design local control units C_i defined in (8.14) performing information requests to guarantee a desired maximal control error $\bar{y}^*_{\Delta i} = \bar{y}^*_{\Delta \mathrm{A}i}$ according to Control aim 7.2

Given: $\tilde{\tilde{\Sigma}}_i$, $(i \in \mathcal{N})$, $\bar{\boldsymbol{y}}^*_{\Delta\mathrm{B}}$, $\bar{\boldsymbol{y}}^*_{\Delta\mathrm{A}}$, $\bar{\boldsymbol{x}}_{\Delta 0\mathrm{B}}$, $\bar{\boldsymbol{x}}_{\Delta 0\mathrm{A}}$, $\bar{\boldsymbol{x}}_{\Delta 0\mathrm{min}}$

1. Design $\mathcal{G}_\mathrm{B}$ using the search algorithm in Fig. 8.8 or Algorithm 8.1 with $\boldsymbol{M}_\mathrm{yy}$ defined (7.61) and $\boldsymbol{m}_\mathrm{y0B}$ defined in (7.60), where $\tilde{\boldsymbol{x}}_{\Delta 0} = \bar{\boldsymbol{x}}_{\Delta\mathrm{B}0}$.
2. Chose ϵ_i, $(\forall i \in \mathcal{N})$, such that relation (8.41) in Proposition 8.3 is fulfilled with $\boldsymbol{e}_\mathrm{maxB}$ defined in (8.40), where $\tilde{\boldsymbol{x}}_{\Delta 0} = \bar{\boldsymbol{x}}_{\Delta 0\mathrm{min}}$.
3. Design $\mathcal{G}_\mathrm{A}$ using the search algorithm in Fig. 8.8 or Algorithm 8.1 with $\boldsymbol{M}_\mathrm{yy}$ defined (7.61) and $\boldsymbol{m}_\mathrm{y0A}$ defined (8.24), where $\tilde{\boldsymbol{x}}_{\Delta 0} = \bar{\boldsymbol{x}}_{\Delta\mathrm{A}0}$.
4. Construct F_i, $(\forall i \in \mathcal{N})$, defined in (7.22) by using the feedback gains $k_{\mathrm{P}i}$ from Σ_i.
5. Construct O_i, $(\forall i \in \mathcal{N})$, defined in (8.9) without any parametrization.
6. Construct D_i, $(\forall i \in \mathcal{N})$, defined in (8.13) by using ϵ_i as well as $\mathcal{P}_{\mathrm{B}i}$, $\mathcal{F}_{\mathrm{B}i}$, $\mathcal{P}_{\mathrm{A}i}$ and $\mathcal{F}_{\mathrm{A}i}$ extracted from $\mathcal{G}_\mathrm{B}$ and $\mathcal{G}_\mathrm{A}$.

Result: C_i, $(\forall i \in \mathcal{N})$, consisting of F_i, O_i and D_i which guarantee that the control error $y_{\Delta i}(t)$ of the agents is smaller than the desired maximal control error $\bar{y}^*_{\Delta i} = \bar{y}^*_{\Delta\mathrm{A}i}$.

Proof. Consider Theorem 8.3 and Theorem 8.4. The search algorithm in Fig. 8.8 and Algorithm 8.1 guarantee that Assumptions 7.6 and 8.1 are fulfilled. Hence, the self-organizing control system Σ fulfills Control aim 7.1 according to Theorem 8.1. Furthermore, both algorithms guarantee that with the adjusted communication graph $\mathcal{G}_\mathrm{A}$ the self-organizing control system Σ fulfills Control aim 7.2, because $\bar{y}^*_{\Delta i} = \bar{y}^*_{\Delta\mathrm{A}i} \leq m_{\mathrm{y0A}i}$, $(\forall i \in \mathcal{N})$. □

Note that if for example $\mathcal{G}_\mathrm{B}$ or/and ϵ are given, the corresponding design steps in Algorithm 8.2 can be skipped.

Example 8.2 (cont.) *Situation-dependent information request within a formation of six robots*

In the following all parameters $\mathcal{G}_\mathrm{B}$, $\mathcal{G}_\mathrm{A}$ and ϵ for the self-organizing controller C defined in (8.16) are designed for the formation of six robots. The overall desired maximal basic control error $\bar{\boldsymbol{y}}^*_{\Delta\mathrm{B}}$ is equal to the maximal control error $\bar{\boldsymbol{y}}_{\Delta\mathrm{B,IS,10}}$ of a fixed serial basic communication graph $\mathcal{G}_\mathrm{B}$ with identical average robots, i.e.,

$$\bar{\boldsymbol{y}}^*_{\Delta\mathrm{B}} := \bar{\boldsymbol{y}}_{\Delta\mathrm{B,IS,10}} = (0.20,\ \ 0.45,\ \ 0.77,\ \ 1.17,\ \ 1.67\,)^\mathrm{T}\mathrm{cm} \tag{8.42}$$

which results from a stepwise change of the reference velocity of $\bar{v} = 10\ \mathrm{Ink/s}$, i.e., $\bar{\boldsymbol{x}}_{\Delta 0\mathrm{B}} = \begin{pmatrix} 0 & 10 \end{pmatrix}^\mathrm{T}$. The desired maximal control errors are given by

$$\begin{aligned} \bar{y}^*_{\Delta 1} &= \bar{y}_{\Delta\mathrm{B1,IS,100}} \\ \bar{y}^*_{\Delta j} &= 0.75 \cdot \bar{y}_{\Delta\mathrm{B}j\mathrm{,IS,100}}, \quad j = 2, \dots 5, \end{aligned}$$

for a stepwise change by the reference velocity $\bar{v}(t)$ by $100\,\frac{\mathrm{Ink}}{\mathrm{s}}$, i.e., $\bar{\boldsymbol{x}}_{\Delta 0\mathrm{A}} = \begin{pmatrix} 0 & 100 \end{pmatrix}^\mathrm{T}$. The overall maximal basic control error for a fixed serial basic communication graph $\mathcal{G}_\mathrm{B}$ is given by

$$\bar{\boldsymbol{y}}_{\Delta\mathrm{B,IS,100}} = (1.97,\ \ 4.46,\ \ 7.61,\ \ 11.57,\ \ 16.58)^\mathrm{T}\mathrm{cm}\,.$$

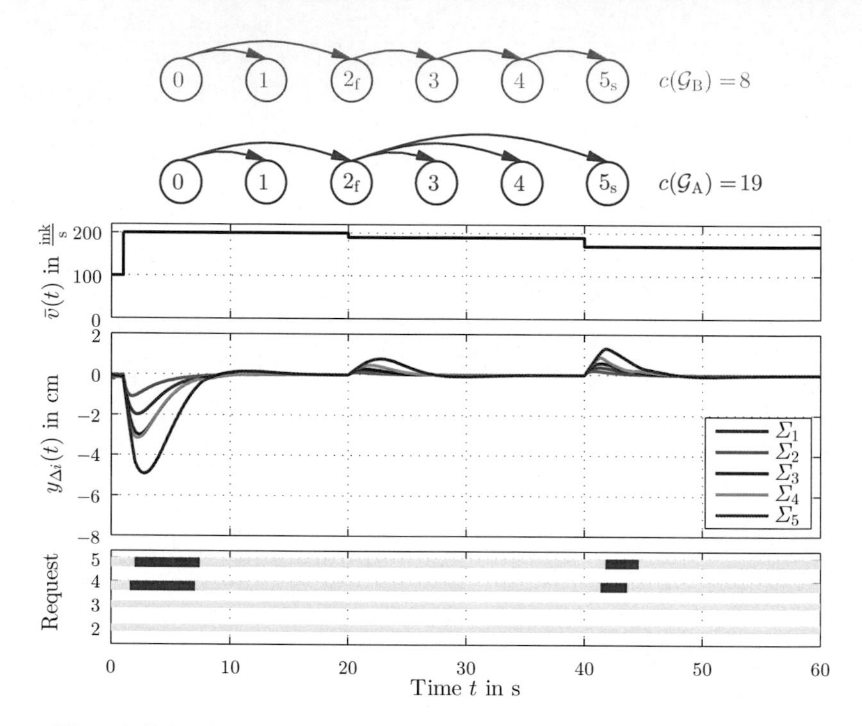

Figure 8.13: Designed graph $\mathcal{G}_{\mathrm{A}}$ and $\mathcal{G}_{\mathrm{B}}$ by Algorithm 8.2 and behavior of the resulting self-organizing controlled robots.

Furthermore, no information should be requested if the reference velocity $\bar{v}$ changes by less than $10\,\frac{\mathrm{Ink}}{\mathrm{s}}$, i.e., $\bar{\boldsymbol{x}}_{\Delta 0\min} = \begin{pmatrix} 0 & 10 \end{pmatrix}^{\mathrm{T}}$.

Design of the parameters of C. For the design of the graphs $\mathcal{G}_{\mathrm{A}}$ and $\mathcal{G}_{\mathrm{B}}$ with Algorithm 8.2 the search algorithm presented in Fig. 8.8 is used, where the resulting graphs $\mathcal{G}_{\mathrm{B}}$ and $\mathcal{G}_{\mathrm{A}}$ are presented in the upper part of Fig. 8.13. It can be seen that in $\mathcal{G}_{\mathrm{B}}$, compared to a serial communication, robot Σ_2 gets information from Σ_0 and not from Σ_1. At the first view, this is surprising since Σ_2 is a "fast robot" and should not need information from Σ_0. But $\mathcal{G}_{\mathrm{B}}$ in Fig. 8.13 can guarantee the desired maximal basic control error $\bar{\boldsymbol{y}}^*_{\Delta\mathrm{B}}$ in the time interval $t \in [20\,\mathrm{s}, 40\,\mathrm{s}]$ since with the improvement of $\bar{y}_{\Delta\mathrm{B}2}$ also the following maximal control errors decrease compared to a serial communication (cf. Fig. 8.15). For $\mathcal{G}_{\mathrm{A}}$ it can be seen that all local control units C_i, $(i > 2)$, are requesting information from C_2 since Σ_2 is a fast robot. In Fig. 8.14 it is shown that the control errors $y_{\Delta i}(t)$ stay below $\bar{y}^*_{\Delta\mathrm{A}i}$ over the complete time interval.

With the designed switching thresholds

$$\boldsymbol{\epsilon} = (0,\ 0,\ 0,\ 0.22,\ 0.43)^{\mathrm{T}}\,\mathrm{cm}$$

no information is requested after the stepwise change of $\bar{v}(t)$ of $-10\,\frac{\mathrm{Ink}}{\mathrm{s}}$ at the time $t = 20\,\mathrm{s}$ (cf. Fig. 8.13). □

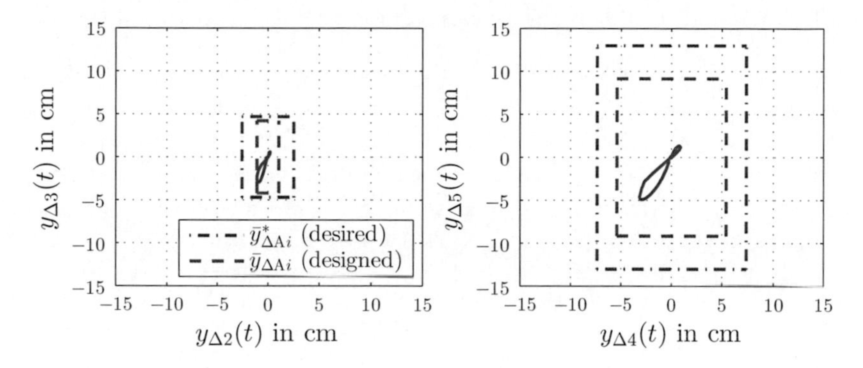

Figure 8.14: Verification of the compliance of the desired maximal control errors $\bar{y}^*_{\Delta Ai}$ from (8.43) according to the behavior of the robots in Fig. 8.13.

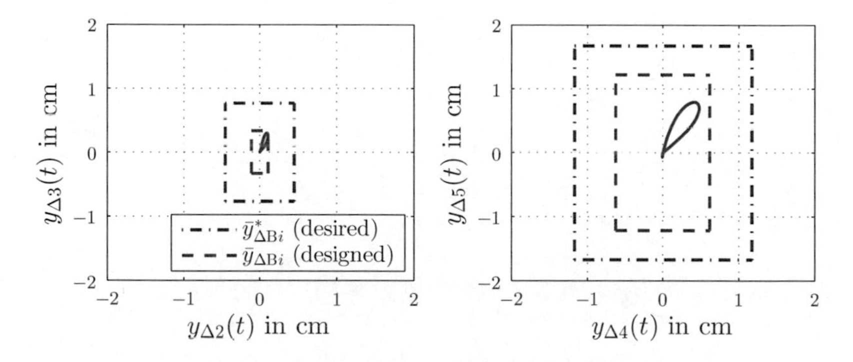

Figure 8.15: Verification of the compliance of the desired basic maximal control errors $\bar{y}^*_{\Delta Bi}$ defined in (8.42) in the time interval $t \in [20\,\text{s}, 40\,\text{s}]$ according to the behavior of the robots in Fig. 8.13.

8.5 Application example: Large robot formation

In this section the proposed self-organizing control concept with a situation-dependent information request is applied to the robot formation consisting of 21 robots presented in Section 7.7.3. Section 8.5.1 presents the basic communication graph $\mathcal{G}_B$, the control aim and the resulting parameters of the local control units C_i. The behavior of the closed-loop system is presented in Section 8.5.2. Section 8.5.3 illustrates the relation between the performance and the communication effort of the self-organizing control system Σ.

8.5.1 Formation problem and parameters of the local control units

Consider the basic communication graph $\mathcal{G}_{\rm B}$ depicted in the upper part of Fig. 8.16. The maximal basic control errors $y_{\Delta {\rm B}i}$ are listed in Table 8.1 for a stepwise change of the reference velocity $\bar{v}(t)$ by $200\frac{\rm ink}{\rm s}$ (cf. Lemma 7.3). According to Control aim 7.2, the local control errors $y_{\Delta i}(t)$, $(\forall i \in \mathcal{N})$, of the robots Σ_i will not exceed the desired maximal control errors $\bar{y}^*_{\Delta i} = \bar{y}^*_{\Delta {\rm A}i}$ listed in Table 8.1 for a stepwise change of the reference velocity $\bar{v}(t)$ by $200\frac{\rm inc}{\rm s}$, i.e., $\bar{\boldsymbol{x}}_{\Delta 0{\rm A}} = \left(0 \;\; 200\frac{\rm ink}{\rm s}\right)^{\rm T}$. The values for $\bar{y}^*_{\Delta {\rm A}i}$ follow from

$$\bar{y}^*_{\Delta {\rm A}i} = \mu_i \bar{y}_{\Delta {\rm B}i}, \quad \forall i \in \mathcal{N},$$

where the scalars μ_i are listed in Table 8.1. For example, the scalar $\mu_{20} = 0.6$ means that the maximal control error $\bar{y}_{\Delta {\rm A}20}$ of robot Σ_{20} while using the self-organizing controller C should be at least 40% smaller than the maximal basic control error $\bar{y}_{\Delta {\rm B}20}$.

Furthermore, no information shall be requested if the reference velocity $\bar{v}$ changes by $20\,\frac{\rm Ink}{\rm s}$, i.e., $\bar{\boldsymbol{x}}_{\Delta 0{\rm min}} = \left(0 \;\; 20\frac{\rm ink}{\rm s}\right)^{\rm T}$.

Recall that the local control units C_i defined in (8.14) for the robots Σ_i can be constructed by using Algorithm 8.2 for a given basic communication graph $\mathcal{G}_{\rm B}$. Hence, the following six steps are performed to define the feedback unit F_i, the observation unit O_i and the decision unit D_i within C_i:

1. The first step can be skipped since $\mathcal{G}_{\rm B}$ is given.

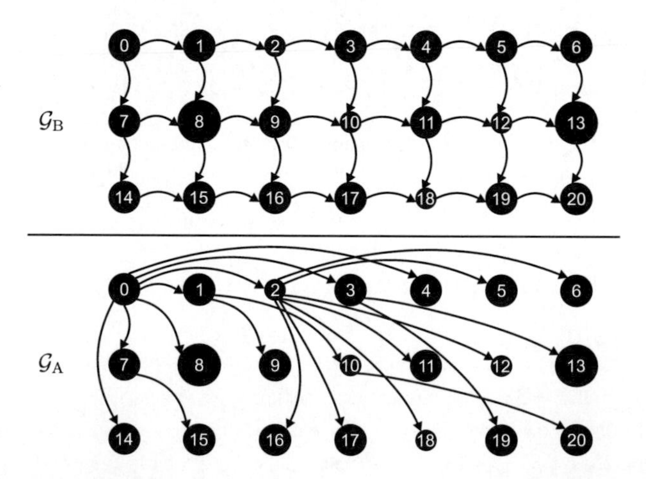

Figure 8.16: Basic communication graph $\mathcal{G}_{\rm B}$ for a robot formation of 21 robots with individual dynamics.

2. The switching thresholds ϵ_i resulting from (8.41) in Proposition 8.3 are listed in Table 8.1.

3. The adjusted communication graph $\mathcal{G}_{\mathrm{A}}$ designed by Algorithm 8.1 from a local perspective of the robots is depicted in the lower part of Fig. 8.16.

4. Recall that all **feedback units** F_i, $(\forall i \in \mathcal{N})$, defined in (7.67) have the same feedback gains $k_{\mathrm{P}i} = 15$.

5. The **observation units** O_i, $(\forall i \in \mathcal{N})$, defined in (8.9) can be constructed without any parametrization.

6. The **decision units** D_i, $(\forall i \in \mathcal{N})$, defined in (8.13) are constructed by using the sets $\mathcal{P}_{\mathrm{B}i}$ of basic predecessors, the sets $\mathcal{F}_{\mathrm{B}i}$ of basic followers, the sets $\mathcal{P}_{\mathrm{A}i}$ of adjusted predecessors and the sets $\mathcal{F}_{\mathrm{A}i}$ of adjusted followers, listed in Table 8.1 which are extracted from the

Table 8.1: Parameters of the local control units C_i defined in (8.14) for controlling the robot formation using the given basic communication graph $\mathcal{G}_{\mathrm{B}}$ depicted in Fig. 8.16 by situation-dependent information request.

i	$\mathcal{P}_{\mathrm{B}i}$	$\mathcal{F}_{\mathrm{B}i}$	$\mathcal{P}_{\mathrm{A}i}$	$\mathcal{F}_{\mathrm{A}i}$	$y_{\Delta\mathrm{B}i}$ in cm	μ_i	$y^*_{\Delta\mathrm{A}i}$ in cm	$y_{\Delta\mathrm{A}i}$ in cm	ϵ in cm
0	$\emptyset$	$\{1,7\}$	$\emptyset$	$\{1,2,3,4,7,8,14\}$	-	-	-	-	-
1	$\{0\}$	$\{2,8\}$	$\{0\}$	$\{9,10\}$	3.95	1.0	3.95	3.95	-
2	$\{1\}$	$\{3,9\}$	$\{0\}$	$\{5,6,11,12,16,17,18\}$	6.75	0.9	6.07	4.05	0.24
3	$\{2\}$	$\{4,10\}$	$\{0\}$	$\{13,19\}$	12.47	0.9	11.22	8.32	0.51
4	$\{3\}$	$\{5,11\}$	$\{0\}$	$\emptyset$	19.68	0.8	15.74	11.34	0.54
5	$\{4\}$	$\{6,12\}$	$\{2\}$	$\emptyset$	28.79	0.8	23.02	19.73	0.58
6	$\{5\}$	$\{13\}$	$\{2\}$	$\emptyset$	40.28	0.7	28.19	23.10	0.62
7	$\{0\}$	$\{8,14\}$	$\{0\}$	$\{15\}$	3.95	1.0	3.95	3.95	-
8	$\{1,7\}$	$\{9,15\}$	$\{0\}$	$\emptyset$	12.49	0.9	11.24	10.43	0.90
9	$\{2,8\}$	$\{10,16\}$	$\{1\}$	$\emptyset$	16.09	0.9	14.48	13.99	0.47
10	$\{3,9\}$	$\{11,17\}$	$\{1\}$	$\{20\}$	18.73	0.8	14.98	13.37	0.26
11	$\{4,10\}$	$\{12,18\}$	$\{2\}$	$\emptyset$	28.19	0.8	22.54	19.22	0.56
12	$\{5,11\}$	$\{13,19\}$	$\{2\}$	$\emptyset$	35.20	0.7	24.63	19.37	0.30
13	$\{6,12\}$	$\{20\}$	$\{3\}$	$\emptyset$	60.12	0.7	42.08	37.11	1.27
14	$\{7\}$	$\{15\}$	$\{0\}$	$\emptyset$	12.49	0.9	11.24	10.43	0.90
15	$\{8,14\}$	$\{16\}$	$\{7\}$	$\emptyset$	19.71	0.9	17.74	15.62	0.48
16	$\{9,15\}$	$\{17\}$	$\{2\}$	$\emptyset$	26.54	0.8	21.23	18.09	0.51
17	$\{10,16\}$	$\{18\}$	$\{2\}$	$\emptyset$	32.51	0.8	26.00	19.83	0.54
18	$\{1117\}$	$\{19\}$	$\{2\}$	$\emptyset$	37.36	0.7	26.15	19.38	0.29
19	$\{12,18\}$	$\{20\}$	$\{3\}$	$\emptyset$	49.73	0.7	34.81	30.40	0.63
20	$\{13,19\}$	$\emptyset$	$\{10\}$	$\emptyset$	73.26	0.6	43.95	41.36	0.67

given basic communication graph $\mathcal{G}_{\mathrm{B}}$ and the designed adjusted communication graph $\mathcal{G}_{\mathrm{A}}$. Furthermore, D_i includes the switching thresholds ϵ_i listed in Table 8.1.

With this, the local control units C_i, $(\forall i \in \mathcal{N})$, for controlling the robots within the formation are completely defined. In the following the robots controlled by these local control units C_i are denoted as self-organizing controlled robots Σ.

8.5.2 Behavior of the self-organizing controlled robot formation

This section analyzes a simulation of the self-organizing controlled robots for a specific course of the reference velocity $\bar{v}(t)$ of the leading robot Σ_0.

In Fig. 8.17 the behavior of the robot formation using the local control units C_i, $(\forall i \in \mathcal{N})$, specified in Section 8.5.1 is depicted. The topmost plot shows the reference velocity $\bar{v}(t)$ of the leading robot Σ_0. The second plot shows the control error $y_{\Delta i}(t)$ of the robots using the self-organizing controller C (black solid line) and a controller with a fixed basic communication $\mathcal{G}_{\mathrm{C}}(t) = \mathcal{G}_{\mathrm{B}}$ (gray dashed line). The transmission of information among the local control units is shown in the lower plot. The bars indicate at which time the communication links $(i \rightarrow j)$ are in the set $\mathcal{E}_{\mathrm{C}}(t)$ of communicational edges which means that C_i transmits information to C_j. Note that the communication links within the basic communication graph $\mathcal{G}_{\mathrm{B}}$ are marked by gray bars. Furthermore, communication links within the adjusted communication graph $\mathcal{G}_{\mathrm{A}}$ are black. Since there are 210 possible communication links, the lower plot only shows the links that are activated during runtime. The thin red vertical lines refer to the corresponding communication graphs $\mathcal{G}_{\mathrm{B}}$ and $\mathcal{G}_{\mathrm{A}}$ depicted in Fig. 8.16.

Figure 8.17 shows that the self-organizing controller C leads to a better performance compared to the usage of a controller with a fixed basic communication graph $\mathcal{G}_{\mathrm{B}}$. In particular, the control error $y_{\Delta i}(i)$, $(\forall i \in \mathcal{U})$, of the disturbed robots which request information is much smaller while using C (second plot).

At most times all local control units C_i transmit information to their basic followers (lowest plot of Fig. 8.17). Information is only requested if there is a significant change of the reference velocity $\bar{v}(t)$. For the first change of $\bar{v}(t)$ all local control units that can request information are requesting information. Hence, for a short time interval the communication graph $\mathcal{G}_{\mathrm{C}}(t)$ is equal to $\mathcal{G}_{\mathrm{A}}$. After approximately 10 seconds the adjusted communication is deactivated since the robots have tracked the position of the leading robot Σ_0 again. As required in Section 8.5.1 for the second small stepwise change of the reference velocity $\bar{v}$ by 20 $\frac{\mathrm{Ink}}{\mathrm{s}}$ no information is requested.

In summary, the self-organizing controlled robots show characteristic properties of self-organization since the communication graph $\mathcal{G}_{\mathrm{C}}(t)$ adapts to the current reference velocity to reduce the control error compared to the fixed usage of $\mathcal{G}_{\mathrm{B}}$.

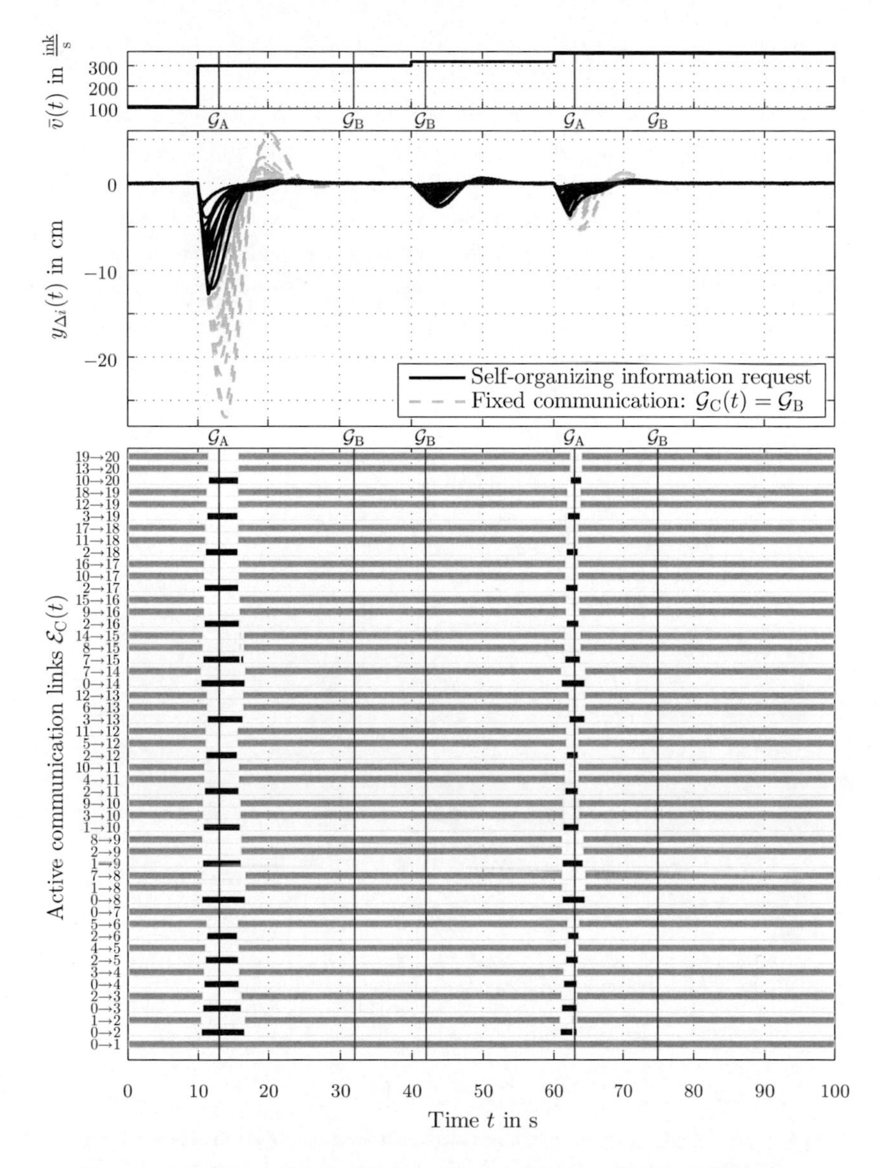

Figure 8.17: Behavior of the robot formation controlled by the self-organizing controller C defined in (8.16) with the parameters presented in Section 8.5.1.

Comparison to fixed basic communication and fixed adjusted communication. Figure 8.18 depicts the maximal values of the control errors $\max_{t\geq 0} |y_{\Delta i}(t)|$, $(\forall i \in \mathcal{N})$, while using the self-organizing controller C (black bar), a controller with a fixed basic communication graph $\mathcal{G}_{\mathrm{C}}(t) = \mathcal{G}_{\mathrm{B}}$ (gray bar) and a controller with a fixed adjusted communication graph $\mathcal{G}_{\mathrm{C}}(t) = \mathcal{G}_{\mathrm{A}}$ (blue bar) without giving the actual amount of the values. This depiction indicates the benefit of the self-organizing control concept. The performance with the self-organizing controller C is similar to the performance of the closed-loop system using the adjusted communication graph $\mathcal{G}_{\mathrm{A}}$ at all time. The performance is improved by the self-organizing controller C compared to the control by the basic communication graph $\mathcal{G}_{\mathrm{B}}$, where with C the communication is only activated if there is a significant change of the reference velocity $\bar{v}(t)$ (cf. Fig. 8.17). Hence, the self-organizing controller C combines the benefit of the basic communication graph $\mathcal{G}_{\mathrm{B}}$ – a low communication effort – and the benefit of the adjusted communication graph $\mathcal{G}_{\mathrm{A}}$ – a good synchronization performance. Recall that this property was the main aim claimed at the beginning of this chapter.

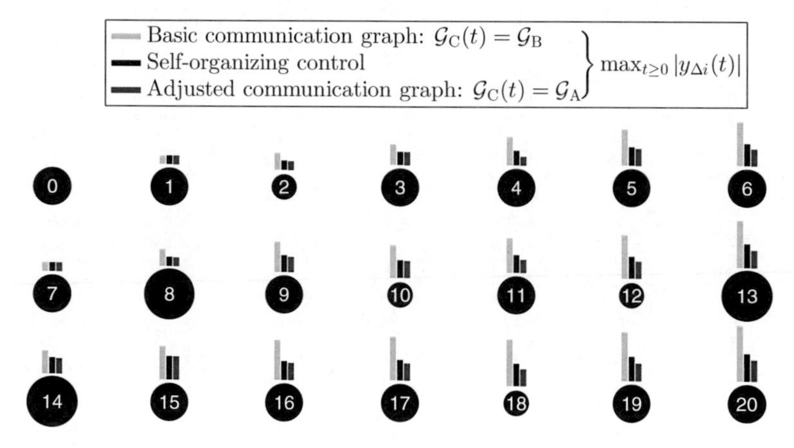

Figure 8.18: Maximal values of the control errors $\max_{t\geq 0} |y_{\Delta i}(t)|$, $(\forall i \in \mathcal{N})$, of the following robots w.r.t the behavior depicted in Fig. 8.17.

Control aims: Asymptotic synchronization and desired maximal control error. Figure 8.17 shows that all robots Σ_i, $(\forall i \in \mathcal{N})$, asymptotically track the position of the leading robot Σ_0.

In Fig. 8.19 it is shown that the control errors $y_{\Delta i}(t)$, $(\forall i \in \{2, 3, 15, 16\})$, stay below the desired maximal control error $\bar{y}^*_{\Delta \mathrm{A} i}$ over the complete time interval. The same holds for the other robots. Hence, Control aim 7.2 is fulfilled.

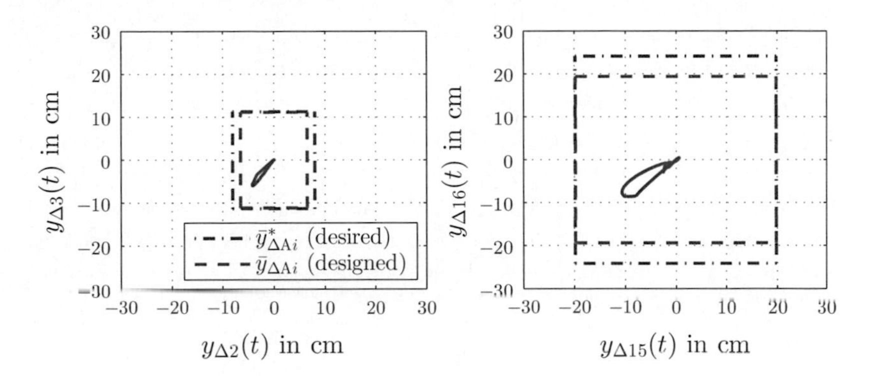

Figure 8.19: Verification of the compliance of the desired maximal control errors $\bar{y}^*_{\Delta Aj}$ listed in Table 8.1 according to the behavior of the robot formation depicted in Fig. 8.17.

8.5.3 Trade-off between performance and communication effort

Consider different self-organizing controllers C with different switching thresholds ϵ_i equal for every local control unit that lead to different self-organizing control systems Σ, where C uses the graphs $\mathcal{G}_B$ and $\mathcal{G}_A$ depicted in Fig. 8.16. Figure 8.20 shows the relation between the performance of Σ with the different thresholds indicated by the sum of the maximum values of the control error, i.e., $\sum_{i\in\mathcal{N}} \max_{t\geq 0} |y_{\Delta i}(t)|$, and the communication effort among the local control units C_i indicated by the request time $t_{\Sigma R}$ for the same change of the reference velocity $\bar{v}(t)$ depicted in Fig. 8.17. The request time t_Σ is the sum of all lengths of the communication

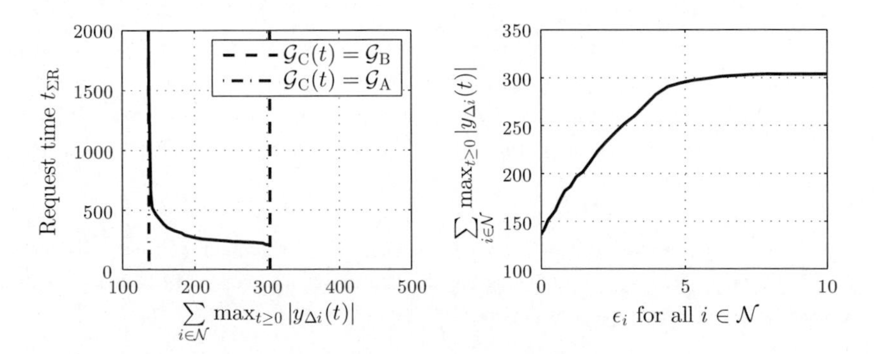

Figure 8.20: Request time for certain self-organizing systems Σ of the form (8.17) resulting from different switching thresholds ϵ_i.

intervals for each communication link $(i \to j) \in \mathcal{E}_{\mathrm{A}}$, which is the length of all black bars in the lower plot of Fig. 8.17 in a row. The dashed line is the value for $\sum_{i \in \mathcal{N}} \max_{t \geq 0} |y_{\Delta i}(t)|$ while using the basic communication graph $\mathcal{G}_{\mathrm{B}}$ all over the time, whereas the dotted dashed line shows the value for $\sum_{i \in \mathcal{N}} \max_{t \geq 0} |y_{\Delta i}(t)|$ using only $\mathcal{G}_{\mathrm{A}}$.

Figure 8.20 shows that even for a small increase of $\sum_{i \in \mathcal{N}} \max_{t \geq 0} |y_{\Delta i}(t)|$ compared to the value for using $\mathcal{G}_{\mathrm{A}}$ there is a huge reduction of the communication effort compared to a permanent communication within $\mathcal{G}_{\mathrm{A}}$. Hence, the presented control concept ensures a good compromise between a low communication effort and a reasonable deviation to the synchronization performance while using $\mathcal{G}_{\mathrm{A}}$ at all time.

8.6 Summary and evaluation of the proposed control concept

The main properties of the self-organizing control concept with situation-dependent information request for the control of a multi-agent system within a leader-follower structure can be summarized as follows:

- The concept combines the benefit a low communication effort provided by the basic communication graph $\mathcal{G}_{\mathrm{B}}$ and the benefit of a good synchronization performance given by the adjusted communication graph $\mathcal{G}_{\mathrm{A}}$.
- The maximal control error of the agents can be adjusted by the choice of the adjusted communication graph $\mathcal{G}_{\mathrm{A}}$ and the switching thresholds ϵ_i, $(\forall i \in \mathcal{N})$ (Theorem 8.2).
- The self-organizing controlled agents Σ_i always asymptotically synchronize with the leader Σ_0 (Theorem 8.1).
- The adjusted communication graph $\mathcal{G}_{\mathrm{A}}$ can be designed to guarantee a desired maximal control error for the agents. Two design methods are presented: The first one uses global model information and guarantees a global minimal communication cost (Fig. 8.8). The second one uses local model information and guarantees a local minimal communication cost (Algorithm 8.1).

In summary, the proposed self-organizing control concept is applicable for multi-agent systems for which the communication effort among the agents is of major interest but which also might have a desired synchronization performance.

Properties of self-organization. The claimed properties of a self-organizing control system in Section 1.2 for the proposed control concept are verified as follows:

- **Flexibility.** The communication structure is adapting to the current difference between the situation of the leader and the agents. In particular, information is requested in case of large local control errors of the agents. Considering the presented robot formation information is only requested for significant changes of the reference velocity of the leading robot.

- **Scalability.** The local control units C_i do not include any model information from agent Σ_i or from any other agents. The parameters of C_i follow from the basic communication graph $\mathcal{G}_\mathrm{B}$ and the adjusted communication graph $\mathcal{G}_\mathrm{A}$, i.e.,

$$\left.\begin{aligned} \mathcal{G}_\mathrm{B} &\Rightarrow \mathcal{P}_{\mathrm{B}i} \text{ and } \mathcal{F}_{\mathrm{B}i} \\ \mathcal{G}_\mathrm{A} &\Rightarrow \mathcal{P}_{\mathrm{A}i} \text{ and } \mathcal{F}_{\mathrm{A}i} \\ &\epsilon_i \end{aligned}\right\} \text{ Parameters of } C_i,\ \forall i \in \mathcal{N}.$$

  Hence, the complexity of the local control units scales linearly with the number of subsystems. For the design of the local control units and the analysis of the behavior of the resulting closed-loop system global model information is needed. Design methods for $\mathcal{G}_\mathrm{A}$ using global model information or local model information are presented.

- **Fault tolerance.** The proposed control concept is not analyzed w.r.t. faults within the agents or local control units. Since the basic communication graph $\mathcal{G}_\mathrm{B}$ is used most of the time, faulty links within $\mathcal{G}_\mathrm{B}$ could lead to an enormous deterioration of the synchronization. Note that Chapter 10 provides a control concept which solves such problems, where the combination of both concepts is discussed in more detail in Chapter 11.

In summary, the proposed self-organizing control concept shows flexibility and a good scalability.

9 Disturbance attenuation by communication interruption

This chapter presents a control concept for disturbance attenuation. The local control units interrupt the communication to their followers, whenever the disturbance impact on the corresponding agent exceeds a given threshold. Therefore, the maximal value for the control error of the undisturbed agents can be arbitrarily adjusted by the choice of the switching threshold. The effectiveness of the concept is demonstrated by its application to a robot formation.

Chapter contents

9.1 Situation-dependent interruption of communication links

Consider the situation that the local control units C_i communicate within the basic communication graph $\mathcal{G}_{\rm B}$ i.e., $\mathcal{G}_{\rm C}(t) = \mathcal{G}_{\rm B}$, and a disturbance $d_i(t)$ affects an agent Σ_i (Fig. 9.1). Due to the transmission of the output $y_i(t)$ to the followers of Σ_i, the effect of the disturbance $d_i(t)$ spreads on the other agents. Depending on the amount of the disturbance there might occur large control errors for the affected agents. Since the disturbance $d_i(t)$ may be very large, a reasonable control error can only be satisfied for the undisturbed agents. Based in this problem formulation, the main **aim** of the control concept that will be presented in this chapter can be summarized as follows:

> Develop a self-organizing controller C with a situation-dependent communication which bounds the disturbance propagation on the undisturbed self-organizing controlled agents (cf. Control aim 7.3).

To quantify the impact of the disturbance on the nominal behavior of agent Σ_i, the output $y_i(t)$ can be divided into two parts, i.e.,

$$y_i(t) = y_{{\rm U}i}(t) + y_{{\rm d}i}(t), \quad \forall i \in \mathcal{N}. \tag{9.1}$$

The signal $y_{{\rm U}i}(t)$ represents the undisturbed behavior of the agent Σ_i, whereas $y_{{\rm d}i}(t)$ is the impact of the disturbance $d_i(t)$ on $y_i(t)$. Using this quantification, the local control units C_i, $(\forall i \in \mathcal{I}(t))$, are assumed to interrupt the communication to its followers C_j, $(\forall j \in \mathcal{F}_{{\rm B}i})$, whenever the disturbance impact $y_{{\rm d}i}(t)$ exceeds the switching threshold ϵ, where the set $\mathcal{I}(t)$ of

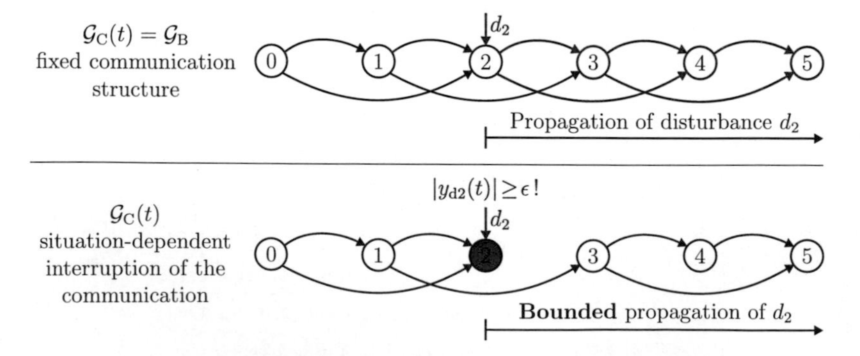

Figure 9.1: Idea of the situation-dependent interruption of the communication.

interrupting local control units is specified as

$$\mathcal{I}(t) = \{i \in \mathcal{N} \mid |y_{\mathrm{d}i}(t)| > \epsilon\}. \tag{9.2}$$

The switching threshold ϵ defines a maximal allowable disturbance impact $y_{\mathrm{d}i}(t)$ which leads to tolerable influences of the disturbance $d_i(t)$ on the followers of the disturbed agent Σ_i.

Due to this interruption, the determination of the local reference signal $y_{\mathrm{s}i}(t)$ for the agents Σ_i results to

$$y_{\mathrm{s}i}(t) = \sum_{j \in \mathcal{P}_{\mathrm{B}i} \setminus \mathcal{I}(t)} \tilde{k}_{ij}(t) y_j(t), \qquad \forall i \in \mathcal{N}. \tag{9.3}$$

With this, no agent receives the output $y_i(t)$ of an agent with a large disturbance impact since all indices $i \in \mathcal{I}(t)$ are excluded from the set $\mathcal{P}_{\mathrm{B}i}$ of basic predecessors. Hence, the propagation of the disturbance is bounded.

Based on these considerations, the **idea** of the self-organizing control concept presented in this chapter can be summarized as follows:

> Whenever the disturbance impact $y_{\mathrm{d}i}(t)$ on the output $y_i(t)$ of the disturbed agent Σ_i exceeds the switching threshold ϵ, the corresponding local control unit C_i interrupts the transmission of the output $y_i(t)$ to its basic followers C_j, $(\forall j \in \mathcal{P}_{\mathrm{B}i})$.

Figure 9.1 illustrates the idea of this situation-dependent interruption for a disturbance affecting agent Σ_2. If the disturbance impact $y_{\mathrm{d}2}(t)$ exceeds the switching threshold ϵ, then C_2 interrupts is communication to C_3 and C_4. Due to the interruption of the communication, the propagation of the disturbance $d_2(t)$ on Σ_3, Σ_4 and Σ_5 can be bounded.

Equations (9.2) and (9.3) characterize the fundamental operating principle of the local control units C_i, $(\forall i \in \mathcal{N})$, see Fig. 9.1. The actual realization of C_i is different since the decision of interrupting the communication is made individually at the local control units C_i. Furthermore, the disturbance impact $y_{\mathrm{d}i}(t)$ has to be reconstructed by using locally available information only. In summary, the basic **tasks** of the local control units C_i, $(\forall i \in \mathcal{N})$, can be defined as follows:

1. As explained in Section 7.3, the feedback unit F_i tracks the local reference signal $y_{\mathrm{s}i}(t)$.
2. The observation unit O_i reconstructs the disturbance impact $y_{\mathrm{d}i}(t)$ to indicate the current disturbance situation of agent Σ_i.
3. The decision unit D_i interrupts the communication to the followers C_j, $(\forall j \in \mathcal{F}_{\mathrm{B}i})$, if the disturbance impact $y_{\mathrm{d}i}(t)$ exceeds the switching threshold ϵ (cf. (9.2)).

4. The decision unit D_i adapts the local reference signal $y_{si}(t)$ depending on the received outputs $y_j(t)$, $(\forall j \in \mathcal{P}_{Ci}(t))$, see (9.3).

In the following the structure of the local control units that fulfill these four tasks is illustrated for a multi-agent system with two followers.

Structure of the self-organizing controller for two followers. The concerned basic communication graph $\mathcal{G}_B$ for controlling the multi-agent system P defined in (7.8) consisting of the leader Σ_0 and two follower agents Σ_1 and Σ_2 is depicted in Fig. 9.2. Furthermore, Fig. 9.2 shows the possible communication graphs $\mathcal{G}_C(t)$ for a disturbance $d_1(t)$ affecting agent Σ_1.

Figure 9.3 depicts the resulting structure of the self-organizing control system Σ with a situation-dependent interruption of the communication. Consider a situation, where due to appropriate initial states $\boldsymbol{x}_{10}$ and $\boldsymbol{x}_{20}$ the undisturbed agents Σ_1 and Σ_2 track the reference signal $y_s(t)$, i.e., $y_s(t) = y_1(t) = y_2(t)$ for all $t \geq 0$ (cf. (7.34)). In case of a disturbance $d_1(t)$ affecting Σ_1 the components of the local control units C_0, C_1 and C_2 work as follows:

- The **feedback units** F_i, $(i = 1, 2)$, generate the inputs $u_i(t)$ to the agents Σ_i by comparing the corresponding output $y_i(t)$ with the local reference signal $y_{si}(t)$ to track the local reference signal $y_{si}(t)$ (1. Task). Recall that F_i is defined in (7.22).

- The **observation unit** O_1 is an observer which reconstructs the disturbance impact $y_{d1}(t)$ on Σ_1 by evaluating the output $y_1(t)$ and the input $u_1(t)$ (2. Task). The output of O_1 is the *reconstructed disturbance impact* $\hat{y}_{d1}(t)$ determined for the decision unit D_1. Note that the observation unit O_i will be defined in Section 9.2.2.

- The **decision unit** D_1 sends the output $y_1(t)$ to C_2 as long as the reconstructed disturbance impact $\hat{y}_{d1}(t)$ is below the switching threshold ϵ. If $\hat{y}_{d1}(t)$ exceeds ϵ, the transmission of

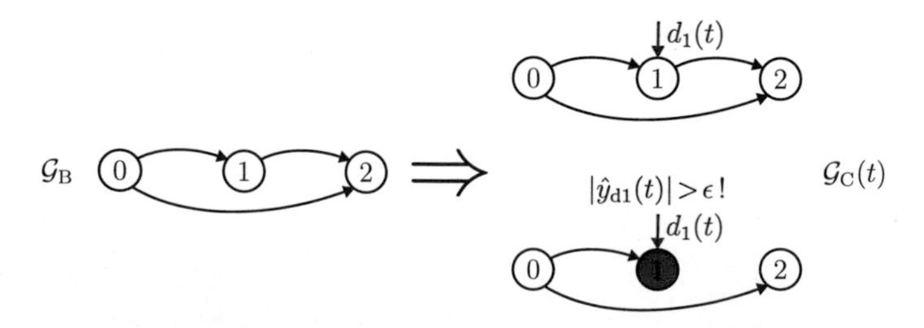

Figure 9.2: Basic communication graph $\mathcal{G}_B$ for the self-organizing control of a multi-agent system with two followers and possible communication graphs $\mathcal{G}_C(t)$ resulting from the interruption of the communication.

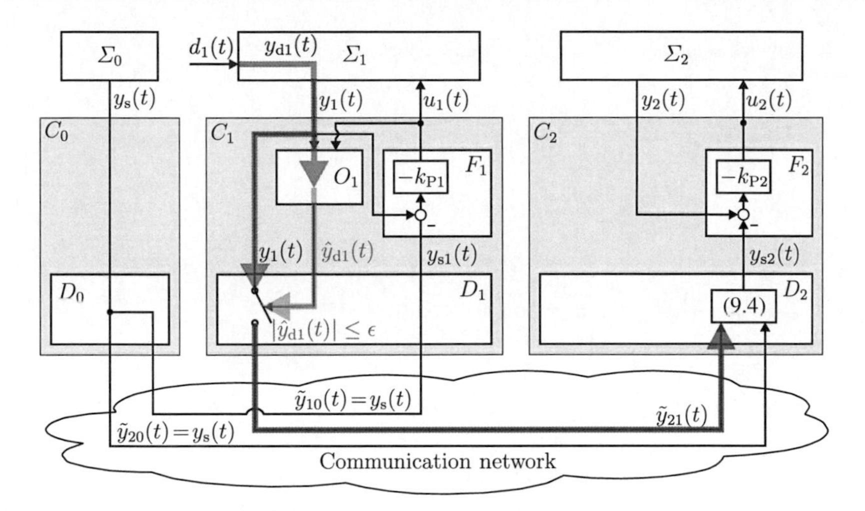

Figure 9.3: Self-organizing control concept with situation-dependent interruption of the communication for two followers.

$y_1(t)$ to C_2 is interrupted (3. Task), i.e.,

$$\tilde{y}_{21}(t) = \begin{cases} y_1(t) & \text{if } |\hat{y}_{d1}(t)| \leq \epsilon \\ \eta & \text{else.} \end{cases}$$

Recall that $\tilde{y}_{21}(t)$ is the transmitted output from C_1 to C_2 according to (7.17).

- Due to the interruption of the communication, the local reference signal $y_{s2}(t)$ is adapted by the **decision unit** D_2 (4. Task), i.e,

$$y_{s2}(t) = \begin{cases} \frac{1}{2}(y_1(t) + y_s(t)) & \text{if } |\hat{y}_{d1}(t)| \leq \epsilon \\ y_s(t) & \text{else,} \end{cases} \tag{9.4}$$

where $\tilde{k}_{20}(t) = 1$ for $\mathcal{P}_{C2}(t) = \{0\}$ according to (7.20). Note that for simplicity reasons all elements k_{ij} of the weighting matrix $\boldsymbol{K}$ defining the basic communication graph $\mathcal{G}_B(t) = (\mathcal{V}_C, \mathcal{E}_B, \boldsymbol{K})$ are set to be one, i.e., $k_{ij} = 1$ for all $i, j \in \mathcal{V}_C$ and $j \in \mathcal{P}_{Bi}$.

- The **local control unit** C_0 sends the reference signal $y_s(t)$ to C_1 and C_2 at all times.

The effect of this control strategy is illustrated by the behavior of the controlled agent $\bar{\Sigma}_2$ defined in (7.26) which is undisturbed and has a single input $y_{s2}(t)$:

$$\bar{\Sigma}_2 : \begin{cases} \dot{\boldsymbol{x}}_2(t) = \bar{\boldsymbol{A}}_i \boldsymbol{x}_2(t) + k_{P2}\boldsymbol{b}_2 y_{s2}(t), & \boldsymbol{x}_2(0) = \boldsymbol{x}_{02} \\ y_2(t) = \boldsymbol{c}_2^{\mathrm{T}} \boldsymbol{x}_2(t), \end{cases} \tag{9.5}$$

Since agent Σ_1 is disturbed, according to (9.1) the local reference signal $y_{s2}(t)$ for $\bar{\Sigma}_2$ results to

$$y_{s2}(t) = \begin{cases} \frac{1}{2}(\underbrace{y_{U1}(t) + y_{d1}(t)}_{y_1(t)} + y_s(t)) & \text{if } |\hat{y}_{d1}(t)| \leq \epsilon \\ y_s(t) & \text{else,} \end{cases} \tag{9.6}$$

where the output $y_1(t)$ is divided into the undisturbed behavior represented by $y_{U1}(t)$ and the disturbance impact $y_{d1}(t)$ according to (9.1). Equations (9.5) and (9.6) show that if the relation $\hat{y}_{d1}(t) = y_{d1}(t)$ holds true, the disturbance impact $y_{d1}(t)$ on the controlled agent $\bar{\Sigma}_2$ can be made arbitrarily small by the choice of the switching threshold ϵ since for $|\hat{y}_{d1}(t)| > \epsilon$ the impact $y_{d1}(t)$ vanishes from the local reference signal $y_{s2}(t)$. The following example illustrates that property.

Example 9.1 *Communication interruption within a formation of three robots*

In this example the self-organizing control concept with situation-dependent communication interruption illustrated in Fig. 9.3 is used to bound the disturbance propagation in a formation of three robots as defined in (7.65). The applied basic communication graph $\mathcal{G}_B$ is depicted in Fig. 9.2, where robot Σ_1 is assumed to be disturbed. The follower robots Σ_1 and Σ_2 are both average robots, i.e., $\mathcal{N}_{\text{average}} = \{1, 2\}$ (cf. Table 7.1). The example is divided into two parts. First, the construction of the local control units C_0, C_1 and C_2 is presented. Second, the behavior of the resulting self-organizing control system Σ is investigated.

Parameters of the local control units. Recall that the feedback units F_1 and F_2 use the feedback gain $k_{P1} = k_{P2} = 15$ (cf. Section 7.7). According to the basic communication graph $\mathcal{G}_B$, the sets $\mathcal{F}_{Bi}$ of followers result to: $\mathcal{F}_{B0} = \{1, 2\}$, $\mathcal{F}_{B1} = \{2\}$ and $\mathcal{F}_{B2} = \emptyset$. Since C_2 does not have any follower ($\mathcal{F}_{B2} = \emptyset$), C_2 includes no switching threshold ϵ. For C_1 the switching threshold is set to be $\epsilon = 1\,\text{cm}$.

Behavior of the self-organizing control system Σ. The behavior of the self-organizing controlled robots Σ with the initial states

$$\boldsymbol{x}_{s0} = \boldsymbol{x}_{01} = \boldsymbol{x}_{02} = \begin{pmatrix} 0 \\ 100 \end{pmatrix}$$

and an intermittent disturbance affecting robot Σ_1 is depicted in Fig. 9.4 with the black solid line. The disturbance $d_1(t)$ leads to a deviation between the reference position $y_s(t)$ and the position $s_1(t) = y_1(t)$ of the robots which is indicated by the control error $y_{\Delta 1}(t)$ (left plot in second row). Therefore, also the disturbance impact $y_{d1}(t)$ increases (left plot in lowest row). When the disturbance impact $y_{d1}(t)$ exceeds the switching threshold ϵ the local control unit C_1 interrupts the transmission of the output $y_1(t)$ to C_2. The gray background indicates the time interval in which C_1 interrupts the communication, which leads to a change of local control units from which C_2 receives information indicated by the black bars in the right plot in the lower row. With this interruption the effect of the disturbance $d_1(t)$

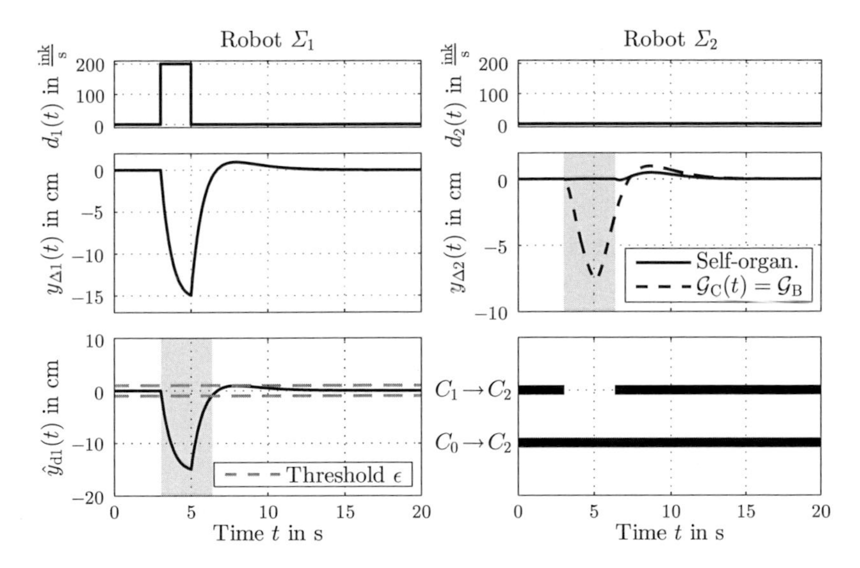

Figure 9.4: Behavior of the robot formation defined in (7.65) with $N = 2$ controlled be the local control units C_i specified in Fig. 9.3.

on the performance of robot Σ_2 is minor which is indicated by the small control error $y_{\Delta 2}(t)$ (right plot in second row). If the disturbance $d_1(t)$ vanishes, in particular, if the disturbance impact $y_{\text{d}1}(t)$ falls below the threshold ϵ, the communication from C_1 to C_1 is activated again and the communication graph $\mathcal{G}_{\text{C}}(t)$ uses the basic communication links within $\mathcal{G}_{\text{B}}$. To illustrate the performance improvement, only the control error $y_{\Delta 2}(t)$ for robot Σ_2 while using the fixed communication graph $\mathcal{G}_{\text{C}}(t) = \mathcal{G}_{\text{B}}$ is depicted (dashed black line in right plot of second row). With the situation-dependent communication interruption the control error $y_{\Delta 2}(t)$ of the self-organizing control system Σ is much smaller then with the fixed communication. Furthermore, the communication interruption is only necessary if a disturbance $d_1(t)$ is present, otherwise the basic communication is reactivated.

Figure 9.5 shows interesting properties of the self-organizing control system Σ. The left plot depicts the maximal value $\max_{t \geq 0} |y_{\Delta 2}(t)| = \max_{t \geq 0} |y_2(t) - y_{\text{s}}(t)|$ of the control error $y_{\Delta 2}(t)$ for different maximal values of the disturbance $d_1(t)$ as in Fig. 9.4 with the threshold $\epsilon = 1\,\text{cm}$ as stated above. The maximal control error $\max_{t \geq 0} |y_{\Delta 2}(t)|$ with the self-organizing controller C is bounded (solid line), where with a fixed communication graph $\mathcal{G}_{\text{C}}(t) = \mathcal{G}_{\text{B}}$ $\max_{t \geq 0} |y_{\Delta 2}(t)|$ (dashed line) it increases linearly with $\max_{t \geq 0} |d_1(t)|$. To achieve this boundedness, the deactivation time increases with $\max_{t \geq 0} |d_1(t)|$ (dot-dashed line). The *deactivation time*, depicted in Fig. 9.5, is the length of the time interval in which C_1 interrupts the transmission of $y_1(t)$ to C_2 which is the length of the gap between the upper black bars in the lower left plot of Fig. 9.4.

The relation between $\max_{t \geq 0} |y_{\Delta 2}(t)|$ and threshold ϵ is depicted in right part of Fig. 9.5. The value for $\max_{t \geq 0} |y_{\Delta 2}(t)|$ increases with switching threshold ϵ until $\epsilon > 15\,\text{cm}$. Thereafter, $\max_{t \geq 0} |y_{\Delta 2}(t)|$ is equal to the value for a fixed communication graph $\mathcal{G}_{\text{C}}(t) = \mathcal{G}_{\text{B}}$ (dashed line) since for large thresholds, ϵ is not exceeded by the disturbance impact $y_{\text{d}1}(t)$ at all. If $\epsilon = 0$ holds, then C_1 interrupts the communication to C_2 at all time. Even for small values of the threshold ϵ there is a huge performance improvement at Σ_2 compared to a fixed communication but only a small time interval in which the com-

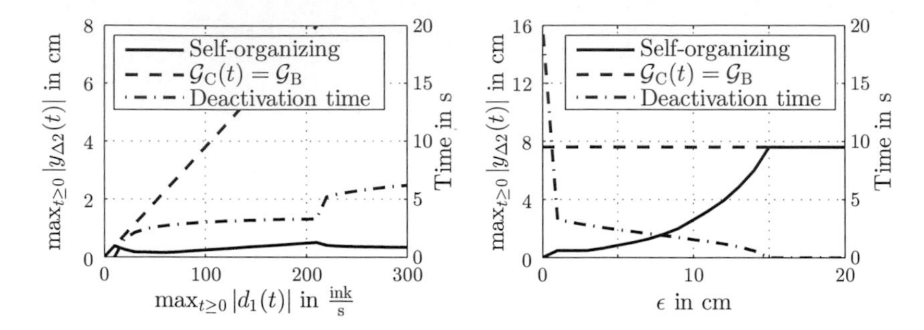

Figure 9.5: Relation between the maximal disturbance $d_1(t)$ at Σ_1 and maximal value of the control error $y_{\Delta 2}(t) = y_2(t) - y_s(t)$ at Σ_2 (left); Relation between the switching threshold ϵ in C_1 and maximal value of $y_{\Delta 2}(t)$ at Σ_2 (right).

munication is interrupted. Hence, the communication structure has to be adapted only for a small time interval to bound the disturbance propagation. This shows that there is a trade-off between the maximal value $\max_{t\geq 0} |y_{\Delta 2}(t)|$ of the control error and the interruption of the communication.

In summary, this example shows that the proposed self-organizing control concept with the interruption of communication links has the following interesting properties:

- The local control units interrupt the communication only if a disturbance affects an agent, which is a situation-dependent communication (cf. Fig. 9.4).
- The disturbance propagation can be adjusted by the choice of the switching threshold ϵ (cf. Fig. 9.5).
- There is a trade-off between the interruption of the communication and the disturbance attenuation (cf. Fig. 9.5 right).

□

Overview of the chapter. In the following sections it is shown that the mentioned properties of the proposed control concept also exist if it is applied to an arbitrary number of agents. Therefore, the content of the next sections is as follows:

- The structure of the self-organizing controller C is extended to an arbitrary number of agents (Section 9.2).
- An analysis of the self-organizing control system Σ yields a condition on the design of the observation unit O_i to reconstruct the disturbance impact y_{di}, a condition on the asymptotic synchronization of the agents and an upper bound on the control error of the undisturbed agents (Section 9.3).

- A method for designing the switching threshold ϵ to guarantee a desired disturbance propagation as claimed in Control aim 7.3 is given (Section 9.4).

- The self-organizing control concept is demonstrated by its application to a large robot formation (Section 9.5).

- The self-organizing properties of the closed-loop system with the situation-dependent communication interruption are evaluated (Section 9.6).

9.2 Structure of the self-organizing controller and the resulting closed-loop system

This section presents the model of the self-organizing controller C and the model of the overall closed-loop system Σ (Section 9.2.3). Therefore, the structure of the local control units C_i, $(\forall i \in \mathcal{V}_{\mathrm{C}})$, is presented (Section 9.2.2) which generate a situation-dependent communication graph $\mathcal{G}_{\mathrm{C}}(t)$ (Section 9.2.1) by interrupting the communication links within a basic communication graph $\mathcal{G}_{\mathrm{B}}$ in case of disturbances affecting the agents.

9.2.1 Communication structure with interruption of communication links

In this section an upper bound and a lower bound for the activation of the communication links is derived according to the definition in (2.25) in Section 2.3. The change of the communication graph $\mathcal{G}_{\mathrm{C}}(t) = (\mathcal{V}_{\mathrm{C}}, \mathcal{E}_{\mathrm{C}}(t), \boldsymbol{K})$ depends on the change of set $\mathcal{E}_{\mathrm{C}}(t)$ communication links defined in (2.17), where $\mathcal{E}_{\mathrm{C}}(t)$ changes due to the situation-dependent interruption of the communication links. According to (9.3), the set $\bar{\mathcal{P}}_{\mathrm{C}i}$ of maximal communicational predecessors and the set $\underline{\mathcal{P}}_{\mathrm{C}i}$ of minimal communicational predecessors is given by

$$\underline{\mathcal{P}}_{\mathrm{C}i} = \mathcal{P}_{\mathrm{B}i} \cap \{0\}, \qquad \forall i \in \mathcal{N} \tag{9.7}$$

$$\bar{\mathcal{P}}_{\mathrm{C}i} = \mathcal{P}_{\mathrm{B}i}, \qquad \forall i \in \mathcal{N}. \tag{9.8}$$

Therefore, the set $\mathcal{P}_{\mathrm{C}i}(t)$ of predecessors is bounded by

$$\underbrace{\mathcal{P}_{\mathrm{B}i} \cap \{0\}}_{\underline{\mathcal{P}}_{\mathrm{C}i}} \subseteq \mathcal{P}_{\mathrm{C}i}(t) \subseteq \underbrace{\mathcal{P}_{\mathrm{B}i}}_{\bar{\mathcal{P}}_{\mathrm{C}i}}, \qquad \forall i \in \mathcal{N}, \quad \forall t \geq 0.$$

The set $\bar{\mathcal{E}}_{\mathrm{C}}$ of maximal communication links and set $\underline{\mathcal{E}}_{\mathrm{C}}$ of minimal communication links follow from (2.21), (2.23), (9.7) and (9.8) to

$$\begin{aligned}\underline{\mathcal{E}}_{\mathrm{C}} &= \{(0 \rightarrow i) \mid i \in \mathcal{F}_{\mathrm{B0}}\}\\ \bar{\mathcal{E}}_{\mathrm{C}} &= \mathcal{E}_{\mathrm{B}}.\end{aligned}$$

Hence, the maximal communication graph $\bar{\mathcal{G}}_{\mathrm{C}} = (\mathcal{V}_{\mathrm{C}}, \bar{\mathcal{E}}_{\mathrm{C}}, \boldsymbol{K})$ defined in (2.19) is identical to the basic communication graph $\mathcal{G}_{\mathrm{B}} = (\mathcal{V}_{\mathrm{C}}, \mathcal{E}_{\mathrm{B}}, \boldsymbol{K})$, i.e.,

$$\bar{\mathcal{G}}_{\mathrm{C}} = \mathcal{G}_{\mathrm{B}}.$$

Note that the communication links $(0 \rightarrow i)$, $(\forall i \in \mathcal{F}_{\mathrm{B0}})$, from the local control unit C_0 of the leader Σ_0 to its basic followers Σ_i always remain since Σ_0 is assumed to be undisturbed (cf. (7.3)).

In summary, the set $\mathcal{E}_{\mathrm{C}}(t)$ of communication links within the communication graph $\mathcal{G}_{\mathrm{C}}(t) = (\mathcal{V}_{\mathrm{C}}, \mathcal{E}_{\mathrm{C}}(t), \boldsymbol{K})$ is bounded:

$$\underbrace{\{(0 \rightarrow i) \mid i \in \mathcal{F}_{\mathrm{B0}}\}}_{\underline{\mathcal{E}}_{\mathrm{C}}} \subseteq \mathcal{E}(t) \subseteq \underbrace{\mathcal{E}_{\mathrm{B}}}_{\bar{\mathcal{E}}_{\mathrm{C}}}, \quad t \geq 0. \tag{9.9}$$

9.2.2 Structure of the local control units

Recall that the local control units C_i include a feedback unit F_i, an observation unit O_i and a decision unit D_i (Fig. 9.6). The feedback unit F_i is used to track the local reference signal $y_{\mathrm{s}i}(t)$. The observation unit O_i detects the current situation of the agent Σ_i indicated by the reconstructed disturbance impact $\hat{y}_{\mathrm{d}i}(t)$ which describes the current impact on $y_i(t)$ by the disturbance $d_i(t)$. The decision unit D_i deactivates the basic communication links depending on the reconstructed disturbance impact $\hat{y}_{\mathrm{d}i}(t)$. The following paragraphs describe the observation unit O_i and the decision unit D_i. The feedback unit F_i is already defined in (7.22) in Section 7.3.2. Finally, the components are combined to define the local control units C_i, $(\forall i \in \mathcal{N})$, of the agents Σ_i and the local control unit C_0 of the leader Σ_0.

Observation unit O_i. The task of the observation unit O_i, $(i \in \mathcal{N})$, is to provide the reconstructed disturbance impact $\hat{y}_{\mathrm{d}i}(t)$ to the decision unit D_i:

$$O_i : \begin{cases} \frac{\mathrm{d}}{\mathrm{d}t}\hat{\boldsymbol{x}}_i(t) = \boldsymbol{A}_i\hat{\boldsymbol{x}}_i(t) + \boldsymbol{b}_i u_i(t) + \boldsymbol{l}_i(y_i(t) - \hat{y}_i(t)), \quad \hat{\boldsymbol{x}}_i(0) = \hat{\boldsymbol{x}}_{0i} \\ \hat{y}_i(t) = \boldsymbol{c}_i^{\mathrm{T}}\hat{\boldsymbol{x}}_i(t) \\ \hat{y}_{\mathrm{d}i}(t) = y_i(t) - \hat{y}_i(t), \end{cases} \tag{9.10}$$

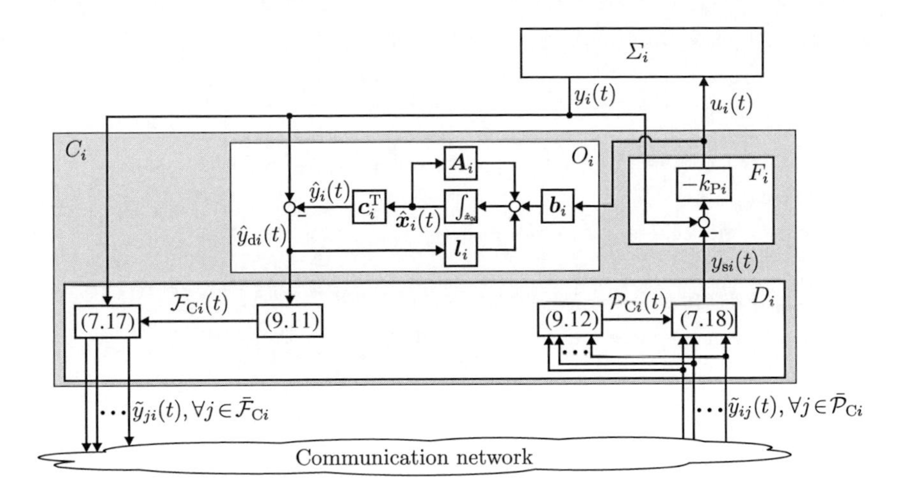

Figure 9.6: Local control unit C_i, ($\forall i \in \mathcal{N}$), designed for a situation-dependent interruption of the communication.

where $\boldsymbol{l}_i$ is the *observer feedback gain*. The reconstructed disturbance impact $\hat{y}_{\mathrm{d}i}(t)$ remains nonzero if the observation unit O_i cannot match its state $\hat{\boldsymbol{x}}_i(t)$ with the state $\boldsymbol{x}_i(t)$ of agent Σ_i due to the disturbance $d_i(t) \neq 0$. This fact is derived in Section 9.3.1.

Decision unit D_i. The decision unit performs three tasks.

First, it determines the local reference signal $y_{\mathrm{s}i}(t)$ for the feedback unit F_i according to (7.18) while using the transmitted outputs $\tilde{y}_{ij}$, ($\forall j \in \mathcal{P}_{\mathrm{B}i}$), and the set of predecessors $\mathcal{P}_i(t)$. Recall that the task of F_i is explained in more detail in Section 7.3.2.

Second, D_i adapts its set $\mathcal{F}_{\mathrm{C}i}(t)$ of followers if the reconstructed disturbance impact $\hat{y}_{\mathrm{d}i}(t)$ exceeds the switching threshold ϵ, i.e.,

$$\mathcal{F}_{\mathrm{C}i}(t) = \begin{cases} \emptyset & \text{if } |\hat{y}_{\mathrm{d}i}(t)| > \epsilon \\ \mathcal{F}_{\mathrm{B}i} & \text{else,} \end{cases} \quad \forall i \in \mathcal{N}. \tag{9.11}$$

In particular, if the relation $|\hat{y}_{\mathrm{d}i}(t)| > \epsilon$ holds true, the transmission of the output $y_i(t)$ from C_i to all C_j, ($\forall j \in \mathcal{F}_{\mathrm{B}i}$), is interrupted. With this, for the transmitted outputs the relation $\tilde{y}_{ji}(t) = \eta$ holds true for all $j \in \mathcal{F}_{\mathrm{B}i}$ according to (7.17).

Third, D_i adapts its current set $\mathcal{P}_{\mathrm{C}i}(t)$ of predecessors depending on the transmitted outputs

$\tilde{y}_{ij}(t)$, $(\forall j \in \mathcal{P}_{\mathrm{B}i})$, received from the local control units C_j, i.e.,

$$\mathcal{P}_{\mathrm{C}i}(t) = \{j \in \mathcal{P}_{\mathrm{B}i} \mid \tilde{y}_{ij}(t) \neq \eta\}, \quad \forall i \in \mathcal{V}_{\mathrm{C}}. \tag{9.12}$$

Hence, if C_j is not transmitting the output $y_i(t)$ anymore, index j is canceled out of the set $\mathcal{P}_{\mathrm{B}i}$ of basic predecessors.

In summary, the decision unit D_i is defined by

$$D_i : \begin{cases} \text{(9.12) generating } \mathcal{P}_{\mathrm{C}i}(t) \text{ using } \tilde{y}_{ij}(t), (\forall j \in \mathcal{P}_{\mathrm{B}i}) \\ \text{(7.18) generating } y_{\mathrm{s}i}(t) \text{ using } \mathcal{P}_{\mathrm{C}i}(t) \text{ and } \tilde{y}_{ij}(t), (\forall j \in \mathcal{P}_{\mathrm{B}i}) \\ \text{(9.11) generating } \mathcal{F}_{\mathrm{C}i}(t) \text{ using } \hat{y}_{\mathrm{d}i}(t) \\ \text{(7.17) generating } \tilde{y}_{ji}(t), (\forall j \in \mathcal{F}_{\mathrm{B}i}), \text{ using } \mathcal{F}_{\mathrm{C}i}(t) \text{ and } y_i(t), \end{cases} \quad \forall i \in \mathcal{N}. \tag{9.13}$$

Local control units C_i of the agents Σ_i. The feedback unit F_i, the observation unit O_i and the decision unit D_i build the local control units

$$C_i : \begin{cases} F_i \text{ defined in (7.22)} \\ O_i \text{ defined in (9.10)} \\ D_i \text{ defined in (9.13),} \end{cases} \qquad \forall i \in \mathcal{N} \tag{9.14}$$

of the agents Σ_i.

Local control unit C_0 of the leader Σ_0. There is no feedback unit and observation unit within the local control unit C_0 of the leader Σ_0. The local control unit

$$C_0 = D_0 : \;\; \text{(7.17) generating } \tilde{y}_{j0}(t), (\forall j \in \mathcal{F}_{\mathrm{B}0}), \text{ using } \mathcal{F}_{\mathrm{C}0}(t) = \mathcal{F}_{\mathrm{B}0} \text{ and } y_{\mathrm{s}}(t) \tag{9.15}$$

includes the decision unit D_0 only. C_0 sends the reference signal $y_{\mathrm{s}}(t)$ to all local control units C_i, $(\forall i \in \mathcal{F}_{\mathrm{B}0})$, included in the set $\mathcal{F}_{\mathrm{B}0}$ of basic followers. Note that the relation $\mathcal{F}_{\mathrm{C}0}(t) = \mathcal{F}_{\mathrm{B}0}$ holds true for all $t \geq 0$.

9.2.3 Self-organizing controller and resulting closed-loop system

In this section the model of the overall self-organizing controller C with situation-dependent interruption of the communication is presented and the resulting model of the overall closed-loop system Σ is derived.

Self-organizing controller C. The self-organizing controller C consisting of the local control units C_i that interrupt the communication to their basic followers to bound the disturbance propagation within the overall closed-loop system results to

$$C\colon\ C_i,\ (\forall i\in\mathcal{V}_{\mathrm{C}}),\ \text{defined in (9.14) and (9.15) with } \mathcal{E}_{\mathrm{C}}(t) \text{ bounded according to (9.9).} \quad (9.16)$$

In particular, $\mathcal{E}_{\mathrm{C}}(t)$ changes due to large disturbance impacts to the agents which are monitored by the observation unit O_i defined in (9.10), where the decision units D_i defined in (9.13) decide whether the impact is so great that communication has to be interrupted or not.

Closed-loop system. The control of the multi-agent system P defined in (7.8) by the self-organizing controller C defined in (9.16) leads to the overall closed-loop system

$$\Sigma:\ \begin{cases} \dot{\boldsymbol{x}}_{\mathrm{os}}(t) = \bar{\boldsymbol{A}}_{\mathrm{os}}(t)\boldsymbol{x}_{\mathrm{os}}(t) + \boldsymbol{G}_{\mathrm{os}}\boldsymbol{d}(t), \quad \boldsymbol{x}_{\mathrm{os}}(0) = \boldsymbol{x}_{\mathrm{os}0} \\ \boldsymbol{y}(t) = \boldsymbol{C}_{\mathrm{os}}\boldsymbol{x}_{\mathrm{os}}(t) \\ \left.\begin{array}{l} O_i,\ (\forall i\in\mathcal{N}) \text{ defined in (9.10)} \\ D_i,\ (\forall i\in\mathcal{N}) \text{ defined in (9.13)} \\ C_0,\ \text{defined in (9.15)} \end{array}\right\} \text{generating } \bar{\boldsymbol{A}}_{\mathrm{os}}(t) \text{ with } \mathcal{P}_{\mathrm{C}i}(t) \text{ defined in (9.12).} \end{cases} \quad (9.17)$$

Note that the general model of the self-organizing control system Σ defined in (7.30) and the model Σ defined in (9.17) with the communication interruption only differ in the structure of the observation unit O_i and of the decision unit D_i.

9.3 Analysis of disturbed self-organizing control system

In this section the behavior of the disturbed self-organizing controlled agents is investigated, where the communication among the corresponding local control units C_i is interrupted in the case of large disturbance effects on the agents. It is shown that the observer feedback gains $\boldsymbol{l}_i$ within the observation units O_i can be chosen such that the reconstructed disturbance impact $\hat{y}_{\mathrm{d}i}(t)$ asymptotically matches with the actual disturbance impact $y_{\mathrm{d}i}(t)$ at the agent (Section 9.3.1). A condition for the asymptotic synchronization of the agents is derived in Theorem 9.1 (Section 9.3.2). The disturbance behavior of the self-organizing control system Σ is analyzed and it is shown that due to the interruption of the communication by the local control units C_i of the disturbed agents Σ_i the propagation of the disturbance is bounded (Section 9.3.3).

9.3.1 Reconstruction of disturbance impact

In this section it is investigated how the effect of the disturbance $d_i(t)$ on the behavior of the controlled agent $\bar{\Sigma}_i$ defined in (7.26) can be reconstructed by using the observation unit O_i. Since $\bar{\Sigma}_i$ has the two inputs $d_i(t)$ and $y_{si}(t)$, its output $y_i(t)$ can be represented by three terms

$$y_i(t) = \underbrace{\boldsymbol{c}_i^{\mathrm{T}} \mathrm{e}^{\bar{\boldsymbol{A}}_i t} \boldsymbol{x}_{0i}}_{y_{0i}(t)} + \underbrace{\int_0^t \boldsymbol{c}_i^{\mathrm{T}} \mathrm{e}^{\bar{\boldsymbol{A}}_i \bar{\tau}} \boldsymbol{b}_i k_{\mathrm{P}i} y_{\mathrm{s}i}(\tau) \mathrm{d}\tau}_{y_{\mathrm{y}i}(t)} + \underbrace{\int_0^t \boldsymbol{c}_i^{\mathrm{T}} \mathrm{e}^{\bar{\boldsymbol{A}}_i \bar{\tau}} \boldsymbol{g}_i d_i(\tau) \mathrm{d}\tau}_{y_{\mathrm{d}i}(t)}, \quad \forall i \in \mathcal{N}, \quad \forall t \geq 0$$

with $\bar{\tau} = t - \tau$. The output $y_i(t)$ includes the free motion $y_{0i}(t)$ of the controlled agent $\bar{\Sigma}_i$, the reference behavior $y_{\mathrm{y}i}(t)$ and the disturbance impact $y_{\mathrm{d}i}(t)$, which are caused by the initial state $\boldsymbol{x}_{0i}$, the reference input $y_{\mathrm{s}i}(t)$ or the disturbance $d_i(t)$, respectively. Note that the free motion $y_{0i}(t)$ together with the reference behavior $y_{\mathrm{y}i}(t)$ build the undisturbed behavior $y_{\mathrm{U}i}(t)$ according to (9.1), i.e., $y_{\mathrm{U}i}(t) = y_{0i}(t) + y_{\mathrm{y}i}(t)$.

It will be shown that the output $\hat{y}_{\mathrm{d}i}(t)$ of the observer O_i asymptotically matches the disturbance impact $y_{\mathrm{d}i}(t)$ of the controlled agent $\bar{\Sigma}_i$ for all initial states $\boldsymbol{x}_{0i}$ of the agent, for all initial states $\hat{\boldsymbol{x}}_{0i}$ of O_i and for all reference signals $y_{\mathrm{s}i}(t)$.

Lemma 9.1 ([95] **Reconstruction of the disturbance effect**) *Consider the observation unit* O_i *defined in* (9.10) *for*

$$\boldsymbol{l}_i = \boldsymbol{b}_i k_{\mathrm{P}i}, \quad i \in \mathcal{N} \tag{9.18}$$

with $k_{\mathrm{P}i}$ *satisfying Assumption 7.5. The reconstructed disturbance impact* $\hat{y}_{\mathrm{d}i}(t)$ *generated by* O_i *asymptotically describes the disturbance impact* $y_{\mathrm{d}i}(t)$ *of the controlled agent* $\bar{\Sigma}_i$ *defined in* (7.26), *i.e.,*

$$\lim_{t\to\infty} |\hat{y}_{\mathrm{d}i}(t) - y_{\mathrm{d}i}(t)| = 0, \quad i \in \mathcal{N}. \tag{9.19}$$

Proof. Consider the state $\boldsymbol{e}_{\mathrm{x}i}(t) = \boldsymbol{x}_i(t) - \hat{\boldsymbol{x}}_i(t)$ which describes the difference between the state $\boldsymbol{x}_i(t)$ of the agent and the state $\hat{\boldsymbol{x}}_i(t)$ of the observation unit O_i, then (7.26) and (9.10) yield

$$O_{\Delta i}: \begin{cases} \dot{\boldsymbol{e}}_{\mathrm{x}i}(t) = (\boldsymbol{A}_i - \boldsymbol{l}_i \boldsymbol{c}_i^{\mathrm{T}}) \boldsymbol{e}_{\mathrm{x}i}(t) + \boldsymbol{g}_i(t) d_i(t), & \boldsymbol{e}_{\mathrm{x}i}(0) = \boldsymbol{x}_{0i} - \hat{\boldsymbol{x}}_{0i} \\ \hat{y}_{\mathrm{d}i}(t) = \boldsymbol{c}_i^{\mathrm{T}} \boldsymbol{e}_{\mathrm{x}i}(t). \end{cases} \tag{9.20}$$

Equations (9.18) and (9.20) lead to

$$\hat{y}_{\mathrm{d}i}(t) = \underbrace{\boldsymbol{c}_i^{\mathrm{T}} \mathrm{e}^{\bar{\boldsymbol{A}}_i t} (\boldsymbol{x}_{0i} - \hat{\boldsymbol{x}}_{0i})}_{r_{0i}(t)} + \underbrace{\int_0^t \boldsymbol{c}_i^{\mathrm{T}} \mathrm{e}^{\bar{\boldsymbol{A}}_i (t-\tau)} \boldsymbol{g}_i d_i(\tau) \mathrm{d}\tau}_{r_{\mathrm{d}i}(t) = y_{\mathrm{d}i}(t)}, \quad i \in \mathcal{N}, \quad t \geq 0. \tag{9.21}$$

Due to Assumption 7.5, the free motion $r_{0i}(t)$ vanishes asymptotically. Since the second addend $r_{\mathrm{d}i}(t)$ is equal to the actual disturbance impact $y_{\mathrm{d}i}(t)$ of $\bar{\Sigma}_i$, the relation in (9.19) holds true. □

Lemma 9.1 shows that the observation unit O_i can be used to reconstruct the effect of the unknown disturbance $d_i(t)$ on the behavior of the controlled agents $\bar{\Sigma}_i$. After the free motion $r_{0i}(t)$ within $\hat{y}_{\mathrm{d}i}(t)$ caused by the unknown state difference $\boldsymbol{x}_{0i} - \hat{\boldsymbol{x}}_{0i}$ has asymptotically vanished, the observation unit O_i delivers the "pure" disturbance effect $y_{\mathrm{d}i}(t)$. This fact holds true no matter what local reference signal $y_{\mathrm{s}i}(t)$ the controlled agent $\bar{\Sigma}_i$ has to track. Hence, the reconstructed disturbance impact $\hat{y}_{\mathrm{d}i}(t)$ generated by the observation unit O_i can be used to decide whether a local control error $e_i(t) = y_{\mathrm{s}i}(t) - y_i(t)$ different from zero, is caused by the local reference signal $y_{\mathrm{s}i}(t)$ or by a disturbance $d_i(t)$ affecting the agent Σ_i. In the first case the observer generates an asymptotically vanishing reconstructed disturbance impact, i.e., $\lim_{t\to\infty} \hat{y}_{\mathrm{d}i}(t) = 0$, whereas in the second case the reconstructed disturbance impact shows the effect of the disturbance on the agent output $y_i(t)$, i.e., $\lim_{t\to\infty} \hat{y}_{\mathrm{d}i}(t) = y_{\mathrm{d}i}(t)$. Furthermore, (9.20) shows that only the disturbance $d_i(t)$ of the same agent Σ_i has an influence on the reconstructed disturbance impact $\hat{y}_{\mathrm{d}i}(t)$. Consequently, the disturbance $d_i(t)$ can only cause an interruption of the communication of the local control unit C_i, but not of other local control units.

In summary, Lemma 9.1 shows that it is reasonable to choose the observer feedback gain $\boldsymbol{l}_i$ according to (9.18).

Assumption 9.1 *The observer feedback gain $\boldsymbol{l}_i$, $(i \subset \mathcal{N})$, is chosen according to (9.18).*

Example 9.1 (cont.) *Communication interruption within a formation of three robots*

Figure 9.7 shows the reconstruction of the disturbance impact $y_{\mathrm{d}1}(t)$ by the observation unit O_1 for the disturbance situation of the robot formation depicted in Fig. 9.4, where the initial state of O_1 is set to be $\hat{\boldsymbol{x}}_{01} = \begin{pmatrix} 0 & 0 \end{pmatrix}^{\mathrm{T}}$. Hence, the difference to the initial state of agent Σ_1 results to

$$\boldsymbol{x}_{01} - \hat{\boldsymbol{x}}_{01} = \begin{pmatrix} 0 \\ 100 \end{pmatrix} - \begin{pmatrix} 0 \\ 0 \end{pmatrix} = \begin{pmatrix} 0 \\ 100 \end{pmatrix}.$$

According to Lemma 9.1 and Assumption 4.3, the observer feedback gain $\boldsymbol{l}_1$ of O_1 is chosen to be $\boldsymbol{l}_1 = \boldsymbol{b}_1 k_{\mathrm{P}1} = \begin{pmatrix} 0.663 & 15 \end{pmatrix}^{\mathrm{T}}$. Figure 9.4 shows that the deviation between the actual disturbance impact $y_{\mathrm{d}1}(t)$ and its reconstructed signal $\hat{y}_{\mathrm{d}1}(t)$ is small. According to (9.21), the deviation only results from the difference $\boldsymbol{x}_{01} - \hat{\boldsymbol{x}}_{01}$ between the initial states which is the free motion $r_{0i}(t)$. In summary, the result in Lemma 9.1 is confirmed since the difference $|\hat{y}_{\mathrm{d}1}(t) - y_{\mathrm{d}1}(t)|$ asymptotically vanishes. □

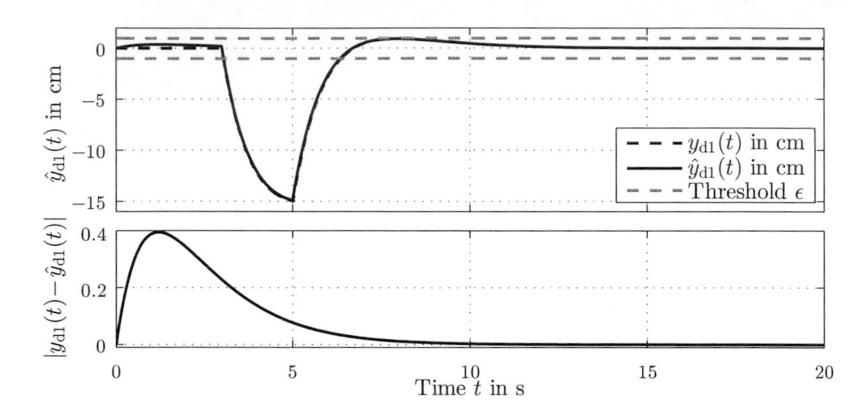

Figure 9.7: Reconstruction of the disturbance impact $y_{d1}(t)$ for the disturbance situation depicted in Fig. 9.4.

9.3.2 Asymptotic synchronization of the agents

This section investigates the asymptotic synchronization of the agents within the self-organizing control system Σ defined in (9.17) (cf. Control aim 7.1). It is shown that if the disturbances affecting the agents Σ_i vanish after a finite time, the agents asymptotically synchronize to the leader Σ_0. Therefore, the conditions in Lemma 7.2 for the general self-organizing control system Σ defined in (7.30) are verified for the specific self-organizing control system Σ with a situation-dependent interruption of the communication.

Theorem 9.1 (Asymptotic synchronization with interruption of communication) *Consider the self-organizing control system Σ defined in* (9.17). *The agents Σ_i, $(\forall i \in \mathcal{N})$, defined in* (7.2) *are asymptotically synchronized to the leader Σ_0 defined in* (7.3) *as desired in Control aim 7.1 if the disturbances $d_i(t)$, $(\forall i \in \mathcal{N})$, affecting the agents Σ_i vanish after a finite time $\bar{t} \in \mathbb{R}_+$, i.e.,*

$$d_i(t) = 0, \quad \forall i \in \mathcal{N}, \quad \forall t \geq \bar{t} \geq 0. \tag{9.22}$$

Proof. The general self-organizing control system Σ defined in (7.30) and the self-organizing control system Σ defined in (9.17) with communication interruption are identical besides the change of the communication. Note that Assumptions 7.1–7.5 also hold for Σ defined in (9.17). The condition (9.22) is identical to the second condition in Lemma 7.2, i.e., (7.49). With this, it has to be shown that the first condition in Lemma 7.2 is fulfilled if the disturbances vanish, i.e., (9.22) holds true. According to (9.21), all reconstructed disturbance impacts $\hat{y}_{di}(t)$, $(\forall i \in \mathcal{N})$, asymptotically vanish, i.e.,

$$\lim_{t\to\infty} |\hat{y}_{\mathrm{d}i}(t)| = 0, \quad \forall i \in \mathcal{N}$$

if (9.22) holds true. Hence, according to the interruption condition (9.11), there exists a finite time $\bar{t}$ for which the set $\mathcal{P}_{\mathrm{C}i}(t)$ of predecessor stays equal to the set $\mathcal{P}_{\mathrm{B}i}$ of basic predecessors. Since $\mathcal{P}_{\mathrm{B}i} \neq \emptyset$ holds true for all $i \in \mathcal{N}$, condition (7.48) in Lemma 7.2 is fulfilled. □

Theorem 9.1 shows that if the disturbances vanish, the interrupted communication links will be activated again and, therefore, the communication graph $\mathcal{G}_{\mathrm{C}}(t)$ gets a spanning tree with the leader Σ_0 as root node (cf. Lemma 7.1). With this, the agents Σ_i can synchronize to the leader Σ_0. Nevertheless, even if the disturbances vanish eventually, the disturbances might lead to large control errors $y_{\Delta i}(t)$ beforehand. The prevention of these errors for the undisturbed agents by situation-dependent communication interruption is investigated in the following section.

Example 9.1 (cont.) *Communication interruption within a formation of three robots*

Figure 9.8 depicts the behavior of the robot formation for two different disturbance scenarios.

Scenario 1 is similar to the situation depicted in Fig. 9.4. The left part of Fig. 9.8 shows that the robots asymptotically track the reference signal $y_{\mathrm{s}}(t)$ of the leader Σ_0 since the disturbance $d_1(t)$ vanishes which is claimed in Theorem 9.1.

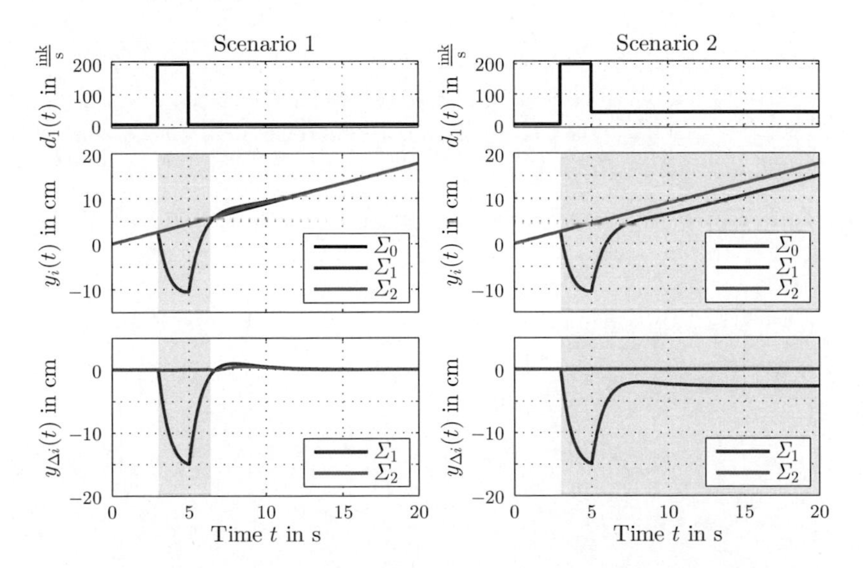

Figure 9.8: Verification of the asymptotic synchronization of the robot formation with communication interruption for two disturbance scenarios.

In **Scenario 2** the disturbance $d_1(t)$ does not vanish which leads to a persistent control error $y_{\Delta 1}(t)$ at the disturbed robot Σ_1 (right part of Fig. 9.8). However, due to the interruption of the communication from C_1 to C_2, the robot Σ_2 still synchronizes with the leader Σ_0. Recall that the time interval in which the communication is interrupted is indicated by the gray area.

In summary, these two scenarios confirm the synchronization condition in Theorem 9.1. □

9.3.3 Disturbance behavior with communication interruption

This section analyzes the disturbance behavior of the self-organizing control system Σ and shows that due to the interruption of the communication by the local control units C_i of the disturbed agents Σ_i to local control units of the other agents the performance can be improved compared to a fixed communication structure. Furthermore, an upper bound on the control error $y_{\Delta i}(t)$, $(\forall i \in \mathcal{U})$, of the undisturbed agents Σ_i is derived which can be used to choose the switching threshold ϵ such that the disturbance propagation is bounded as claimed in Control aim 7.3 (Section 9.4.1). For this, the following situation described in Control aim 7.3 is considered:

- The initial states $\boldsymbol{x}_{0i}$, $(\forall i \in \mathcal{N})$, of the agents Σ_i are considered to match the initial state $\boldsymbol{x}_{s0}$ of the leader Σ_0, i.e.,

$$\boldsymbol{x}_{0i} = \boldsymbol{T}_i \begin{pmatrix} \boldsymbol{x}_{s0} \\ \boldsymbol{0} \end{pmatrix}, \quad \forall i \in \mathcal{N}. \tag{9.23}$$

  Recall that $\boldsymbol{T}_i$ is the transformation matrix from Theorem 7.1. According to (7.43) and (7.44), these initial states guarantee that the agents are synchronized with the leader for all $t \geq 0$ for the situation that no disturbance affects the agents which is required by (7.34) in Control aim 7.3.

- Two groups of agents are considered. The first group includes the *undisturbed agents* Σ_i, $(\forall i \in \mathcal{U})$, with the index i included in the set $\mathcal{U}$ of *undisturbed agents* defined in (7.36). The disturbed agents Σ_j, $(\forall j \in \mathcal{D})$, belong to the second group, where the index j of the agents is included in the *set $\mathcal{D}$ of disturbed agents* defined by $\mathcal{D} = \mathcal{N} \backslash \mathcal{U}$. The scalars

$$\beta_i = \begin{cases} 1 & \text{if } i \in \mathcal{D} \\ 0 & \text{if } i \in \mathcal{U}, \end{cases} \qquad \forall i \in \mathcal{N} \tag{9.24}$$

  illustrate that classification with

$$\boldsymbol{\beta} = \begin{pmatrix} \beta_1 \\ \vdots \\ \beta_N \end{pmatrix}.$$

Recall that according to Lemma 9.1, only the local control units C_i, $(\forall i \in \mathcal{D})$, of the disturbed agents Σ_i possibly interrupt their communication.

To retain the synchronizability of the closed-loop system Σ according to Lemma 7.2 the following assumption is made:

Assumption 9.2 *Despite the communication interruption, the communication graph $\mathcal{G}_{\mathrm{C}}(t)$ retains a spanning tree with the leader Σ_0 as root note. Hence, the following relation holds true:*

$$\mathcal{P}_{\mathrm{C}i}(t) \neq \emptyset, \quad \forall i \in \mathcal{N}, \quad \forall t \geq 0.$$

In order to satisfy this assumption, all agents which have a disturbed agent as predecessor in the communication graph $\mathcal{G}_{\mathrm{C}}(t)$ have to have at least one second predecessor agent.

The following theorem describes an upper bound of the effect of disturbances on all undisturbed agents Σ_i, $(\forall i \in \mathcal{U})$, where the closed-loop system Σ is assumed to be initially synchronized according to (7.34) and (9.23). Furthermore, it is assumed that the free motion $r_{0i}(t)$ has vanished beforehand which permits the assumption that the initial state $\boldsymbol{x}_{0i}$ of the agent Σ_i and the initial state $\hat{\boldsymbol{x}}_{0i}$ of the observation unit O_i are identical, i.e., $\hat{\boldsymbol{x}}_{0i} = \boldsymbol{x}_{0i}$ for all $i \in \mathcal{N}$.

Theorem 9.2 (Bound for disturbance propagation via communication interruption) *Consider the self-organizing control system Σ defined in (9.17), where the initial states $\boldsymbol{x}_{0i}$, $(\forall i \in \mathcal{N})$, fulfill the relation (9.23) and $\hat{\boldsymbol{x}}_{0i} = \boldsymbol{x}_{0i}$ holds true for all $i \in \mathcal{N}$. Furthermore, the agents Σ_j, $(\forall j \in \mathcal{D})$, are assumed to be disturbed. Then the control error $y_{\Delta i}(t) = y_i(t) - y_{\mathrm{s}}(t)$ of the undisturbed agents Σ_i, $(\forall i \in \mathcal{U})$, is upper bounded by*

$$\max_{t \geq 0} |y_{\Delta i}(t)| \leq \bar{y}_{\Delta i} = \boldsymbol{\alpha}_i^{\mathrm{T}} \left(\boldsymbol{I}_N - \boldsymbol{M}_{\mathrm{yy}} \boldsymbol{K}_{\mathrm{d}}\right)^{-1} \boldsymbol{\beta} \epsilon, \quad \forall i \in \mathcal{U} \tag{9.25}$$

with

$$\boldsymbol{K}_{\mathrm{d}} = (k_{\mathrm{d}ij}) \quad \text{with} \quad k_{\mathrm{d}ij} = \begin{cases} \dfrac{k_{ij}}{\sum_{j \in \mathcal{P}_{\mathrm{B}i} \setminus \mathcal{D}} k_{ij}} & \text{if } j \in \mathcal{P}_{\mathrm{B}i} \\ 0 & \text{else,} \end{cases} \quad \forall i, j \in \mathcal{N} \tag{9.26}$$

$$\boldsymbol{\alpha}_i^{\mathrm{T}} = \begin{pmatrix} \alpha_0 & \alpha_1 & \dots & \alpha_N \end{pmatrix} \quad \text{with} \quad \alpha_i = 1, \; \alpha_j = 0, \quad j \neq i, \quad \forall i, j \in \mathcal{N}.$$

Proof. The proof is given in Appendix A.14. □

The theorem shows that the effect of arbitrary disturbances on the undisturbed agents Σ_i, $(\forall i \in \mathcal{U})$, can be bounded by means of the self-organizing controller C. The maximal control error $y_{\Delta i}(t)$ of the undisturbed agents Σ_i, $(\forall i \in \mathcal{D})$, is upper bounded. The bound $\bar{y}_{\Delta i}$ depends on the dynamics of the controlled agents $\bar{\Sigma}_i$, the basic communication graph $\mathcal{G}_{\mathrm{B}}$ and the disturbance situation which are indicated by the matrix $\boldsymbol{M}_{\mathrm{yy}}$, the matrix $\boldsymbol{K}_{\mathrm{d}}$ and the vector

$\boldsymbol{\beta}$, respectively. But the most interesting aspect is that the bound $\bar{y}_{\Delta i}$ can be adjusted by the choice of the threshold ϵ. Note that the transposed vectors $\boldsymbol{\alpha}_i^{\mathrm{T}}$ select the values of the upper bound $\bar{y}_{\Delta i}$ from an overall bound on the control errors (cf. (A.34)).

Since the communication is interrupted by the local control units C_i, $(\forall i \in \mathcal{D})$, of the disturbed agents Σ_i for the time interval in which the effect of the disturbance exceeds a threshold ϵ, the disturbance does not propagate on the other agents. In summary, the following result from Theorem 9.2 is highlighted:

> The maximal control error $y_{\Delta i}(t)$ of the undisturbed agents Σ_i, $(\forall i \in \mathcal{D})$, within an initially synchronized closed-loop system Σ can be arbitrarily adjusted by the choice of the switching threshold ϵ.

Example 9.1 (cont.) *Communication interruption within a formation of three robots*

Consider the disturbance situation in Fig. 9.4. According to (9.25) in Theorem 9.2, the control error $y_{\Delta 2}(t)$ of the undisturbed robot Σ_2 is upper bounded by $\bar{y}_{\Delta 2}$, i.e.,

$$\max_{t \geq 0} |y_{\Delta 2}(t)| \leq \bar{y}_{\Delta 2} = \underbrace{\begin{pmatrix} 0 & 1 \end{pmatrix}}_{\boldsymbol{\alpha}_2^{\mathrm{T}}} \left(\underbrace{\begin{pmatrix} 1 & 0 \\ 0 & 1 \end{pmatrix}}_{\boldsymbol{I}_2} - \underbrace{\begin{pmatrix} 1.27 & 0 \\ 0 & 1.27 \end{pmatrix}}_{\boldsymbol{M}_{\mathrm{yy}}} \underbrace{\begin{pmatrix} 0 & 0 \\ 1 & 0 \end{pmatrix}}_{\boldsymbol{K}_{\mathrm{d}}} \right) \underbrace{\begin{pmatrix} 1 \\ 0 \end{pmatrix}}_{\boldsymbol{\beta}} \underbrace{1}_{\epsilon} = 1.27\,\mathrm{cm}. \quad (9.27)$$

Figure 9.9 shows that $y_{\Delta 2}(t)$ never exceeds the upper bound $\bar{y}_{\Delta 2}$ which is indicated by the gray area. Note that Fig. 9.9 does not show all the details as in Fig. 9.4 since $\hat{y}_{\mathrm{d}1}(t)$ as well as the communication intervals are identical. □

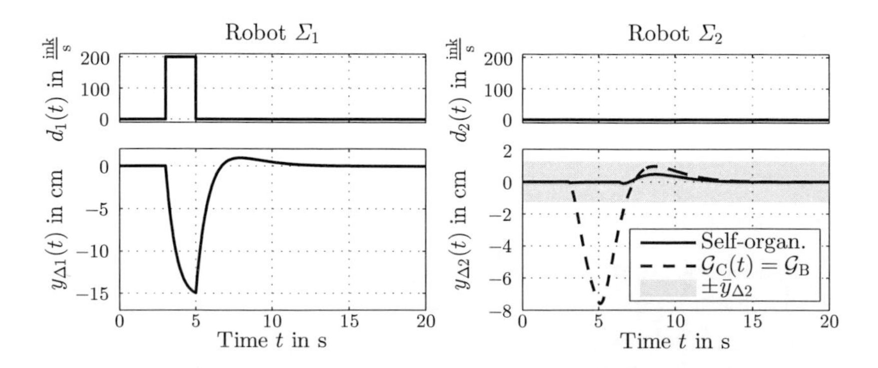

Figure 9.9: Verification of upper bound on the undisturbed robot Σ_2.

9.4 Design of the local control units

In this section methods for choosing the parameters of the local control units C_i are presented. In Section 9.4.1 a condition for choosing the switching threshold ϵ for guaranteeing a desired bound on the disturbance propagation is derived. An algorithm, which summarizes the basic steps for choosing the parameters of C_i, is presented in Section 9.4.2.

9.4.1 Design of the switching threshold to bound the disturbance propagation

This section concerns the design of the switching threshold ϵ for guaranteeing a desired bound on the disturbance propagation as required in Control aim 7.3. Since according to Theorem 9.2 the maximal control error $y_{\Delta i}(t)$ of the undisturbed agents can be arbitrarily adjusted by the switching threshold ϵ, the following theorem presents an appropriate choice for ϵ to guaranteeing a desired maximal control error $\bar{y}^{*}_{\Delta}$ for the undisturbed agents.

Theorem 9.3 (Choice of switching threshold to bound the disturbance propagation) *Consider the self-organizing control system Σ defined in (9.17), where the initial states $\boldsymbol{x}_{0i}$, $(\forall i \in \mathcal{N})$, fulfill the relation (9.23) and $\hat{\boldsymbol{x}}_{0i} = \boldsymbol{x}_{0i}$ holds true for all $i \in \mathcal{N}$. Furthermore, the agents Σ_j, $(\forall j \in \mathcal{D})$, are assumed to be disturbed. Then the control error $y_{\Delta i}(t)$, $(\forall i \in \mathcal{U})$, of the undisturbed agents Σ_i is bounded by the desired maximal control error $\bar{y}^{*}_{\Delta}$ as required in Control aim 7.3 if the switching threshold ϵ is chosen such that the following relation holds:*

$$\epsilon \leq \frac{\bar{y}^{*}_{\Delta}}{\max\limits_{i \in \mathcal{U}} \left\{ \boldsymbol{\alpha}_i^{\mathrm{T}} \left(\boldsymbol{I}_N - \boldsymbol{M}_{\mathrm{yy}} \boldsymbol{K}_{\mathrm{d}} \right)^{-1} \boldsymbol{\beta} \right\}}. \tag{9.28}$$

Proof. According to (9.25), the relation

$$\boldsymbol{\alpha}_i^{\mathrm{T}} \left(\boldsymbol{I}_N - \boldsymbol{M}_{\mathrm{yy}} \boldsymbol{K}_{\mathrm{d}} \right)^{-1} \boldsymbol{\beta} \epsilon \leq \bar{y}^{*}_{\Delta}, \quad \forall i \in \mathcal{U} \tag{9.29}$$

has to be fulfilled, to guarantee Control aim 7.3. Solving (9.29) for ϵ leads to the requirement

$$\epsilon \leq \frac{\bar{y}^{*}_{\Delta}}{\boldsymbol{\alpha}_i^{\mathrm{T}} \left(\boldsymbol{I}_N - \boldsymbol{M}_{\mathrm{yy}} \boldsymbol{K}_{\mathrm{d}} \right)^{-1} \boldsymbol{\beta}}, \quad \forall i \in \mathcal{U}.$$

The usage of the maximal value of the scalar $\boldsymbol{\alpha}_i^{\mathrm{T}} \left(\boldsymbol{I}_N - \boldsymbol{M}_{\mathrm{yy}} \boldsymbol{K}_{\mathrm{d}} \right)^{-1} \boldsymbol{\beta}$ for all $i \in \mathcal{U}$ leads to the relation in (9.28) which completes the proof. □

In Theorem 9.3 the maximal upper bound $\bar{y}_{\Delta i}$ on the control errors $y_{\Delta i}(t)$ from Theorem 9.2 is used to derive a sufficient condition for the choice of the switching threshold ϵ which defines for which disturbance impacts $y_{\mathrm{d}i}(t)$ the communication has to be interrupted (cf. (9.13)). Equation (9.25) shows that the desired maximal control error $\bar{y}_{\Delta}^{*}$ can be chosen arbitrarily small. Furthermore, the choice of the threshold ϵ depends on the disturbance situation described by the set $\mathcal{U}$ of undisturbed agents. The fewer agents are disturbed, the smaller is the denominator in (9.28) since $\boldsymbol{\beta}$ gets more zero entries (cf. (9.24)) and the entries in $\boldsymbol{K}_{\mathrm{d}}$ become smaller (cf. (9.26)). Hence, the fewer agents are disturbed, the greater is the switching threshold ϵ.

Note that for the choice of ϵ overall model information of the controlled agents are necessary since $\boldsymbol{M}_{\mathrm{yy}}$ represents an upper bound on the behavior of all controlled agents $\bar{\Sigma}_i$ and $\boldsymbol{K}_{\mathrm{d}}$ represents an upper bound on the possible communication among the local control units C_i.

Example 9.1 (cont.) *Communication interruption within a formation of three robots*

Consider that the switching threshold ϵ for the robot formation has to be chosen such that the control error $y_{\Delta 2}(t)$ of the undisturbed robot Σ_2 fulfills the relation

$$\max_{t \geq 0} |y_{\Delta 2}(t)| \leq \bar{y}_{\Delta}^{*} = 1\,\mathrm{cm}. \tag{9.30}$$

Recall that $\bar{y}_{\Delta}^{*}$ is the desired maximal control error from Control aim 7.3. According to (9.28) in Theorem 9.3, the solution of (9.27) for ϵ while replacing $\bar{y}_{\Delta 2}$ by $\bar{y}_{\Delta}^{*}$ leads to the switching threshold

$$\epsilon = \frac{\overbrace{1\,\mathrm{cm}}^{\bar{y}_{\Delta}^{*}}}{\underbrace{\begin{pmatrix} 0 & 1 \end{pmatrix}}_{\boldsymbol{\alpha}_2^{\mathrm{T}}} \left(\underbrace{\begin{pmatrix} 1 & 0 \\ 0 & 1 \end{pmatrix}}_{\boldsymbol{I}_2} - \underbrace{\begin{pmatrix} 1.27 & 0 \\ 0 & 1.27 \end{pmatrix}}_{\boldsymbol{M}_{\mathrm{yy}}} \underbrace{\begin{pmatrix} 0 & 0 \\ 1 & 0 \end{pmatrix}}_{\boldsymbol{K}_{\mathrm{d}}} \right) \underbrace{\begin{pmatrix} 1 \\ 0 \end{pmatrix}}_{\boldsymbol{\beta}}} = 0.79\,\mathrm{cm}.$$

which guarantees (9.30).

Figure 9.10 shows that $y_{\Delta 2}(t)$ never exceeds the desired upper bound $\bar{y}_{\Delta}^{*}$ which is indicated by the gray area (right plot of second row). Note that the behavior of the robot formation is similar to the behavior depicted in Fig. 9.4 with $\epsilon = 1\,\mathrm{cm}$. Due to the smaller threshold and the overshoot of the reconstructed disturbance impact $\hat{y}_{\Delta 1}(t)$, there is an additional interruption of the communication from C_1 to C_2 (right plot of lower row). □

9.4.2 Overall design algorithm

In this section the basic steps for choosing the parameters of the local control units C_i interrupting the communication to their followers depending on the actual disturbance impacts are presented.

Generally, it is assumed that there is a given overall system consisting of agents within a leader follower structure which are controlled within a fixed communication structure. The task is to apply the proposed control concept to this given system to bound the disturbance propagation within the resulting self-organizing control system Σ (cf. Control aim 7.3). Hence,

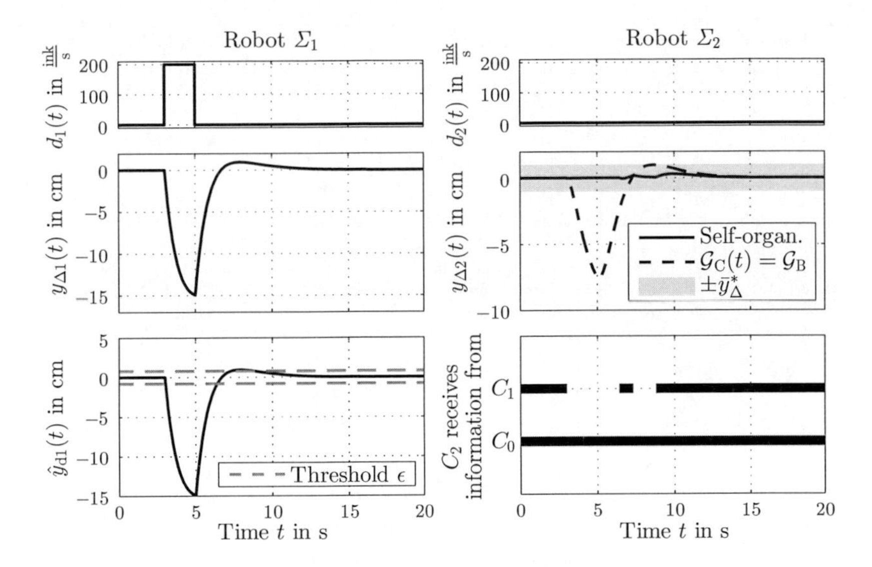

Figure 9.10: Behavior of the robot formation for guaranteeing a desired maximal control error $\bar{y}^*_\Delta = 1\,\mathrm{cm}$ for the undisturbed robot Σ_2.

the following is assumed to be given:

- The model of the controlled agents $\bar{\Sigma}_i$, $(\forall i \in \mathcal{N})$, defined in (7.26) including the gain $k_{\mathrm{P}i}$.
- The basic communication graph $\mathcal{G}_{\mathrm{B}} = (\mathcal{V}_{\mathrm{C}}, \mathcal{E}_{\mathrm{B}}, \boldsymbol{K})$ which defines the set $\mathcal{P}_{\mathrm{B}i}$ of basic predecessors and the set $\mathcal{F}_{\mathrm{B}i}$ of basic followers.
- The set $\mathcal{U}$ of undisturbed agents defined in (7.36).
- The desired maximal control error $\bar{y}^*_\Delta$ defining an acceptable disturbance propagation.

Note that the feedback gains $k_{\mathrm{P}i}$, $(\forall i \in \mathcal{N})$, from the given system retain since $k_{\mathrm{P}i}$ already stabilizes the controlled agents $\bar{\Sigma}_i$ (cf. Assumption 7.5).

Algorithm 9.1 designs the parameters of the local control units C_i such that the resulting self-organizing control system Σ defined in (9.17) fulfills Control aim 7.3. In Step 1 the given feedback gains $k_{\mathrm{P}i}$, $(\forall i \in \mathcal{N})$, are used to construct the feedback units F_i defined in (7.22). With the choice of observer feedback gains $\boldsymbol{l}_i = \boldsymbol{b}_i k_{\mathrm{P}i}$ in Step 2 according to Lemma 9.1 the reconstructed disturbance impact $\hat{y}_{\mathrm{d}i}(t)$ determined by the observation unit O_i defined in (9.10) asymptotically matches with the actual disturbance impact $y_{\mathrm{d}i}(t)$. The last step constructs the decision unit D_i defined in (9.13). The switching threshold ϵ which is equal at every local

Algorithm 9.1: Design local control units C_i, $(\forall i \in \mathcal{N})$, defined in (9.14) which interrupt the communication to bound the disturbance propagation according to Control aim 7.3.

Given: $\bar{y}^*_\Delta$, $\mathcal{G}_\mathrm{B}, \mathcal{U}$, $\bar{\Sigma}_i$ for all $i \in \mathcal{N}$

1. Construct F_i, $(\forall i \in \mathcal{N})$, defined in (7.22) by using the feedback gains $k_{\mathrm{P}i}$ from $\bar{\Sigma}_i$.
2. Construct O_i, $(\forall i \in \mathcal{N})$, defined in (9.10) by choosing the observer feedback gains $\boldsymbol{l}_i$ according to (9.18).
3. Construct D_i, $(\forall i \in \mathcal{N})$, defined in (9.13) by extracting $\mathcal{P}_{\mathrm{B}i}$ and $\mathcal{F}_{\mathrm{B}i}$ from $\mathcal{G}_\mathrm{B}$ as well as by choosing ϵ according to (9.28).

Result: C_i consisting of F_i, O_i and D_i guaranteeing that the control error of the undisturbed agents is upper bounded by $\bar{y}^*_\Delta$ as claimed in Control aim 7.3.

control unit C_i is chosen according to (9.28) in Theorem 9.3. Furthermore, the set $\mathcal{P}_{\mathrm{B}i}$ of basic predecessors and the set $\mathcal{F}_{\mathrm{B}i}$ of basic followers are extracted from the basic communication graph $\mathcal{G}_\mathrm{B}$.

The result of Algorithm 9.1 is that The designed local control units within the self-organizing control system Σ defined in (9.17) guarantee that the control error $y_{\Delta i}(t)$, $(\forall i \in \mathcal{U})$, of the undisturbed agents Σ_i is bounded by the desired maximal control error $\bar{y}^*_\Delta$ (cf. Control aim 7.3).

9.5 Application example: Large robot formation

In this section the proposed self-organizing control concept with a situation-dependent interruption of the communication is applied to the robot formation consisting of 21 robots presented in Section 7.7.3. Section 9.5.1 presents the basic communication graph $\mathcal{G}_\mathrm{B}$, the disturbance situation and the resulting parameters of the local control units C_i. With this, an analysis of the asymptotic synchronization and the disturbance attenuation of the resulting self-organizing controlled system Σ is performed (Section 9.5.2). These results are compared with the actual behavior of the closed-loop system (Section 9.5.3). Section 9.5.4 illustrates the relation between the performance and the communication effort of the self-organizing control system Σ.

9.5.1 Formation problem and parameters of the local control units

Consider a robot formation with 21 robots controlled within the basic communication graph $\mathcal{G}_\mathrm{B}$ depicted in Fig. 9.11, where the robots Σ_5, Σ_9 and Σ_{17} might be affected by disturbances. Hence, the set $\mathcal{D}$ of disturbed agents and the set $\mathcal{U}$ of undisturbed agents result to

$$\mathcal{D} = \{5, 9, 17\}$$
$$\mathcal{U} = \mathcal{N} \backslash \mathcal{D} = \{1, 2, 3, 4, 6, 7, 8, 10, 11, 12, 13, 14, 15, 16, 18, 19, 20\}.$$

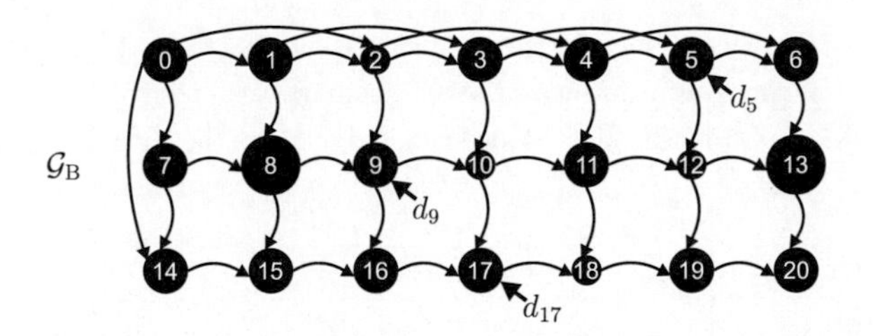

Figure 9.11: Basic communication graph $\mathcal{G}_\mathrm{B}$ for a robot formation of 20 following disturbed robots with individual dynamics.

According to Control aim 7.3, the control error $y_{\Delta i}(t)$, $(\forall i \in \mathcal{U})$, of the undisturbed robots Σ_i should be smaller than one centimeters, i.e,

$$|y_{\Delta i}(t)| \leq \bar{y}_\Delta^* = 1\,\mathrm{cm}, \quad \forall i \in \mathcal{U}, \quad \forall t \geq 0. \tag{9.31}$$

Recall that $\bar{y}_\Delta^*$ is the desired maximal control error.

To fulfill this aim, the local control units C_i defined in (9.14) for the robots Σ_i can be constructed by using Algorithm 9.1. Therefore, the following four steps are performed to parameterize the feedback unit F_i, the observation unit O_i and the decision unit D_i within C_i, where the corresponding parameters are listed in Table 9.1:

1. Recall that all **feedback units** F_i, $(\forall i \in \mathcal{N})$, defined in (7.67) include the gains $k_{\mathrm{P}i} = 15$.

2. The observer gain l_i of the **observation units** O_i, $(\forall i \in \mathcal{N})$, defined in (9.10) are chosen according to Lemma 9.1.

3. The **decision units** D_i, $(\forall i \in \mathcal{N})$, defined in (9.13) are constructed by using the sets $\mathcal{P}_{\mathrm{B}i}$ of basic predecessors and the sets $\mathcal{F}_{\mathrm{B}i}$ of basic followers listed in Table 9.1 which are extracted from the basic communication graph $\mathcal{G}_\mathrm{B}$. Furthermore, the switching threshold ϵ is determined by using (9.28), i.e., $\epsilon = 0.26\,\mathrm{cm}$.

With this, the local control units C_i, $(\forall i \in \mathcal{N})$, for controlling the robots within the formation are completely defined. In the following the robots controlled by these local control units C_i are denoted as self-organizing controlled robots Σ.

9.5.2 Analysis of the behavior of the self-organizing controlled robots

In the following the asymptotic synchronization of the robot formation (Control aim 7.1) and the disturbance attenuation by interrupting the communication (Control aim 7.4) are analyzed.

Table 9.1: Parameters of the local control units C_i defined in (10.14) for controlling the robot formation using the basic communication graph $\mathcal{G}_B$ depicted in Fig. 9.11 by situation-dependent interruption of the communication.

i	$\mathcal{P}_{Bi}$	$\mathcal{F}_{Bi}$	ϵ in cm	k_{Pi}	$\boldsymbol{b}_i$	$\boldsymbol{c}_i^T$	$\boldsymbol{l}_i$	$\boldsymbol{A}_i$
0	$\emptyset$	$\{1,2,7,14\}$	0.26	-	-	-	-	-
1	$\{0\}$	$\{2,3,8\}$	0.26	15	$\begin{pmatrix}0.044 & 1\end{pmatrix}^T$	$\begin{pmatrix}2.5 & 0\end{pmatrix}$	$\begin{pmatrix}0.663 & 15\end{pmatrix}^T$	$\begin{pmatrix}0 & 0.018\\ 0 & 0\end{pmatrix}^T$
2	$\{0,1\}$	$\{3,4,9\}$	0.26	15	$\begin{pmatrix}0.088 & 2\end{pmatrix}^T$	$\begin{pmatrix}2.5 & 0\end{pmatrix}$	$\begin{pmatrix}1.326 & 30\end{pmatrix}^T$	=
3	$\{1,2\}$	$\{4,5,10\}$	0.26	15	$\begin{pmatrix}0.044 & 1\end{pmatrix}^T$	$\begin{pmatrix}2.5 & 0\end{pmatrix}$	$\begin{pmatrix}0.663 & 15\end{pmatrix}^T$	=
4	$\{2,3\}$	$\{5,6,11\}$	0.26	15	$\begin{pmatrix}0.044 & 1\end{pmatrix}^T$	$\begin{pmatrix}2.5 & 0\end{pmatrix}$	$\begin{pmatrix}0.663 & 15\end{pmatrix}^T$	=
5	$\{3,4\}$	$\{6,12\}$	0.26	15	$\begin{pmatrix}0.044 & 1\end{pmatrix}^T$	$\begin{pmatrix}2.5 & 0\end{pmatrix}$	$\begin{pmatrix}0.663 & 15\end{pmatrix}^T$	=
6	$\{4,5\}$	$\{13\}$	0.26	15	$\begin{pmatrix}0.044 & 1\end{pmatrix}^T$	$\begin{pmatrix}2.5 & 0\end{pmatrix}$	$\begin{pmatrix}0.663 & 15\end{pmatrix}^T$	=
7	$\{0\}$	$\{8,14\}$	0.26	15	$\begin{pmatrix}0.044 & 1\end{pmatrix}^T$	$\begin{pmatrix}2.5 & 0\end{pmatrix}$	$\begin{pmatrix}0.663 & 15\end{pmatrix}^T$	=
8	$\{1,7\}$	$\{9,15\}$	0.26	15	$\begin{pmatrix}0.022 & 0.5\end{pmatrix}^T$	$\begin{pmatrix}2.5 & 0\end{pmatrix}$	$\begin{pmatrix}0.332 & 7.5\end{pmatrix}^T$	=
9	$\{2,8\}$	$\{10,16\}$	0.26	15	$\begin{pmatrix}0.044 & 1\end{pmatrix}^T$	$\begin{pmatrix}2.5 & 0\end{pmatrix}$	$\begin{pmatrix}0.663 & 15\end{pmatrix}^T$	=
10	$\{3,9\}$	$\{11,17\}$	0.26	15	$\begin{pmatrix}0.088 & 2\end{pmatrix}^T$	$\begin{pmatrix}2.5 & 0\end{pmatrix}$	$\begin{pmatrix}1.326 & 30\end{pmatrix}^T$	=
11	$\{4,10\}$	$\{12,18\}$	0.26	15	$\begin{pmatrix}0.044 & 1\end{pmatrix}^T$	$\begin{pmatrix}2.5 & 0\end{pmatrix}$	$\begin{pmatrix}0.663 & 15\end{pmatrix}^T$	=
12	$\{5,11\}$	$\{13,19\}$	0.26	15	$\begin{pmatrix}0.088 & 2\end{pmatrix}^T$	$\begin{pmatrix}2.5 & 0\end{pmatrix}$	$\begin{pmatrix}1.326 & 30\end{pmatrix}^T$	=
13	$\{6,12\}$	$\{20\}$	0.26	15	$\begin{pmatrix}0.022 & 0.5\end{pmatrix}^T$	$\begin{pmatrix}2.5 & 0\end{pmatrix}$	$\begin{pmatrix}0.332 & 7.5\end{pmatrix}^T$	=
14	$\{0,7\}$	$\{15\}$	0.26	15	$\begin{pmatrix}0.022 & 0.5\end{pmatrix}^T$	$\begin{pmatrix}2.5 & 0\end{pmatrix}$	$\begin{pmatrix}0.332 & 7.5\end{pmatrix}^T$	=
15	$\{8,14\}$	$\{16\}$	0.26	15	$\begin{pmatrix}0.044 & 1\end{pmatrix}^T$	$\begin{pmatrix}2.5 & 0\end{pmatrix}$	$\begin{pmatrix}0.663 & 15\end{pmatrix}^T$	=
16	$\{9,15\}$	$\{17\}$	0.26	15	$\begin{pmatrix}0.044 & 1\end{pmatrix}^T$	$\begin{pmatrix}2.5 & 0\end{pmatrix}$	$\begin{pmatrix}0.663 & 15\end{pmatrix}^T$	=
17	$\{10,16\}$	$\{18\}$	0.26	15	$\begin{pmatrix}0.044 & 1\end{pmatrix}^T$	$\begin{pmatrix}2.5 & 0\end{pmatrix}$	$\begin{pmatrix}0.663 & 15\end{pmatrix}^T$	=
18	$\{11,17\}$	$\{19\}$	0.26	15	$\begin{pmatrix}0.088 & 2\end{pmatrix}^T$	$\begin{pmatrix}2.5 & 0\end{pmatrix}$	$\begin{pmatrix}1.326 & 30\end{pmatrix}^T$	=
19	$\{12,18\}$	$\{20\}$	0.26	15	$\begin{pmatrix}0.044 & 1\end{pmatrix}^T$	$\begin{pmatrix}2.5 & 0\end{pmatrix}$	$\begin{pmatrix}0.663 & 15\end{pmatrix}^T$	=
20	$\{13,19\}$	$\emptyset$	0.26	15	$\begin{pmatrix}0.044 & 1\end{pmatrix}^T$	$\begin{pmatrix}2.5 & 0\end{pmatrix}$	$\begin{pmatrix}0.663 & 15\end{pmatrix}^T$	=

Asymptotic synchronization. Theorem 9.1 states that the robots can track the current position $s_0(t) = y_s(t)$ of the leading robot Σ_0 asymptotically if there exists a finite time $\bar{t}$ after which all disturbances vanish, i.e., $d_i(t) = 0$ for all $i \in \mathcal{N}$ for all $t \geq \bar{t} \geq 0$. Furthermore, Table 9.1

shows that all local control units that lose a predecessor within their set $\mathcal{P}_{\mathrm{B}i}$ due to the interruption of the communication (highlighted in red in Table 9.1) have at least one other predecessor. Hence, Assumption 9.2 is fulfilled.

Propagation of a disturbance affecting robot Σ_5. Consider that only robot Σ_5 is disturbed, i.e., $\mathcal{D} = \{5\}$, where the resulting communication graph $\mathcal{G}_{\mathrm{C}}(t)$ with the interrupted communication links $(5 \rightarrow 6)$ and $(5 \rightarrow 12)$ is depicted in Fig. 9.13 (a). With this, the propagation of $d_5(t)$ is indicated by

$$\bar{y}_{\Delta 6} = 0.33\,\mathrm{cm} \tag{9.32a}$$

$$\bar{y}_{\Delta 12} = 0.30\,\mathrm{cm} \tag{9.32b}$$

$$\bar{y}_{\Delta 13} = 0.45\,\mathrm{cm} \tag{9.32c}$$

$$\bar{y}_{\Delta 19} = 0.19\,\mathrm{cm} \tag{9.32d}$$

$$\bar{y}_{\Delta 20} = 0.41\,\mathrm{cm} \tag{9.32e}$$

$$\bar{y}_{\Delta i} = 0.00\,\mathrm{cm}, \quad \forall i \in \{1, 2, 3, 4, 7, 8, 9, 10, 11, 14, 15, 16, 17, 18\}. \tag{9.32f}$$

resulting from the analysis in Theorem 9.2. Note that the bounds $\bar{y}_{\Delta i}$ are even smaller than the desired maximal control error of $y^*_{\Delta} = 1\,\mathrm{cm}$ since in this analysis Σ_9 and Σ_{17} are assumed to be undisturbed.

9.5.3 Behavior of the self-organizing controlled robot formation

This section verifies the analysis results in the previous section by evaluating a simulation of the self-organizing controlled robots for a specific disturbance situation.

In Fig. 9.12 the behavior of the robot formation using local control units C_i, $(\forall i \in \mathcal{N})$, specified in Section 9.5.1 is depicted. The topmost plot shows the disturbances $d_i(t)$ affecting the robots Σ_i, where the disturbances affecting Σ_5, Σ_9 and Σ_{17} are colored. The disturbances affecting the other robots are all black. The second plot displays the control error $y_{\Delta i}(t)$, $(\forall i \in \mathcal{U})$, of the undisturbed robots while using the self-organizing controller C (black solid line) and fixed basic communication $\mathcal{G}_{\mathrm{C}}(t) = \mathcal{G}_{\mathrm{B}}$ (gray dashed line). The third plot depicts the control error $y_{\Delta i}(t)$, $(\forall i \in \mathcal{D})$ of the disturbed robots. The transmission of information among the local control units is shown in the lower plot. The black bars indicate at which time the communication links $(i \rightarrow j)$ are in the set $\mathcal{E}_{\mathrm{C}}(t)$ of communicational edges which means that C_i transmits information to C_j. Since there are 210 possible communication links, the lowest plot only shows the links that are activated during runtime. The thin red vertical lines refer to the corresponding communication graphs $\mathcal{G}_{\mathrm{C}}(t)$ for different times shown in Fig. 9.13.

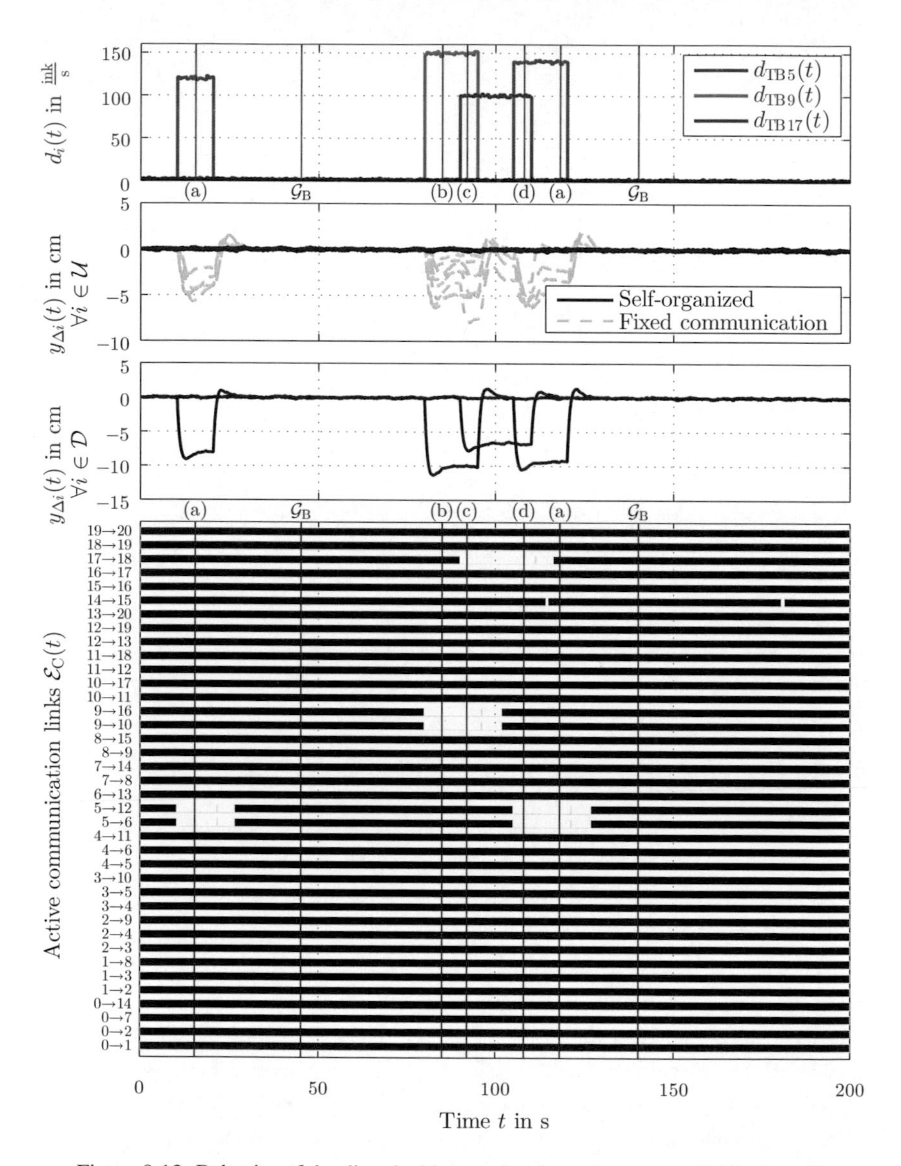

Figure 9.12: Behavior of the disturbed large robot formation controlled by the self-organizing controller C defined in (9.16) with the parameters listed in Table 9.1.

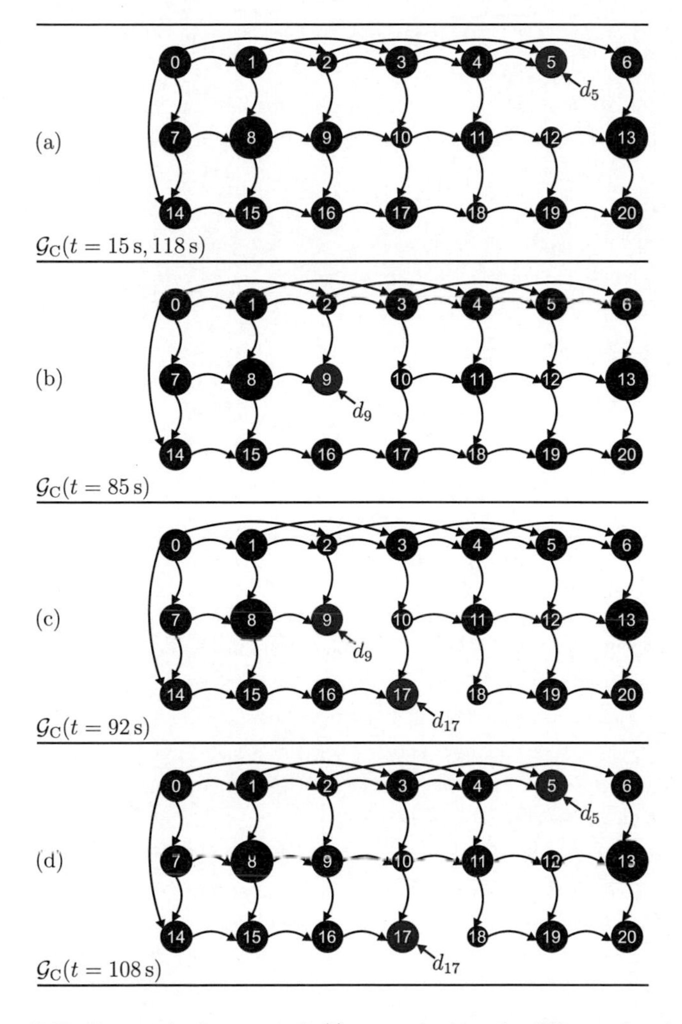

Figure 9.13: Communication graph $\mathcal{G}_C(t)$ at certain times for different disturbance situations in Fig. 9.12 (disturbed robots are colored red).

Figure 9.12 shows that the self-organizing controller C leads to a better performance compared to the usage of a controller with a fixed communication, in particular, the control errors $y_{\Delta i}(t)$, $(\forall i \in \mathcal{U})$, of the undisturbed agents are much smaller while using C (third plot). The performance of the disturbed robots Σ_5, Σ_9 and Σ_{17} is similar for both control concepts since the effect of a disturbance at Σ_i can not be reduced by interrupting the communication to other local control units.

At most times all local control units C_i transmit information to their basic followers (lower plot of Fig. 9.12). The communication is only interrupted if a robot is disturbed significantly. For example, if subsystem Σ_9 is disturbed, the corresponding local control unit C_9 interrupts the communication to C_{10} and C_{16} (Fig. 9.13 (b)). The communication structures in Fig. 9.13 show that it is sufficient to interrupt the communication from the local control units of the disturbed subsystems to bound the disturbance propagation. The local control units of the undisturbed robots do not interrupt their communication.

In summary, the self-organizing controlled robots show characteristic properties of self-organization since the communication graph $\mathcal{G}_{\mathrm{C}}(t)$ adapts to the currently acting disturbances to bound the disturbance propagation within the overall system.

Control aim: Asymptotic synchronization. Figure 9.12 shows that for vanishing disturbances the robots asymptotically track the position of the leading robot.

Control aim: Disturbance attenuation. The control aim in (9.31) is fulfilled (cf. Fig. 9.14). The control errors $y_{\Delta i}(t)$, $(\forall i \in \mathcal{U})$, of the undisturbed agents are smaller than the desired maximal control error of $\bar{y}^*_\Delta = 1\,\mathrm{cm}$ for all $t \geq 0$.

Figure 9.15 depicts the controlled robots with their the maximal values of the control errors $\max_{t\geq 0} |y_{\Delta i}(t)|$, $(\forall i \in \mathcal{U})$, from the behavior of the self-organizing controlled robots Σ shown in Fig. 9.12, where the black bars indicate the self-organizing controller C and the gray bars indicate the controller with a fixed communication $\mathcal{G}_{\mathrm{C}}(t) = \mathcal{G}_{\mathrm{B}}$. Note that the actual amount of the values is not given. This depiction indicates the benefit of the self-organizing control concept. The disturbance propagation with the self-organizing controller C is much smaller than with the fixed communication.

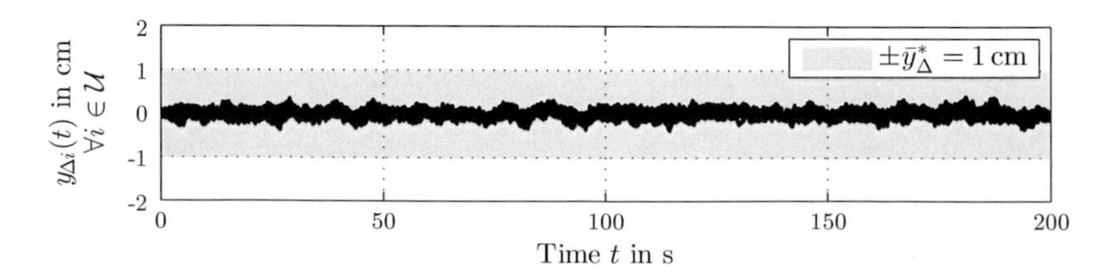

Figure 9.14: Verification of the control aim in (9.31); Control error $y_{\Delta i}(t)$, $(\forall i \in \mathcal{U})$, of the undisturbed robots.

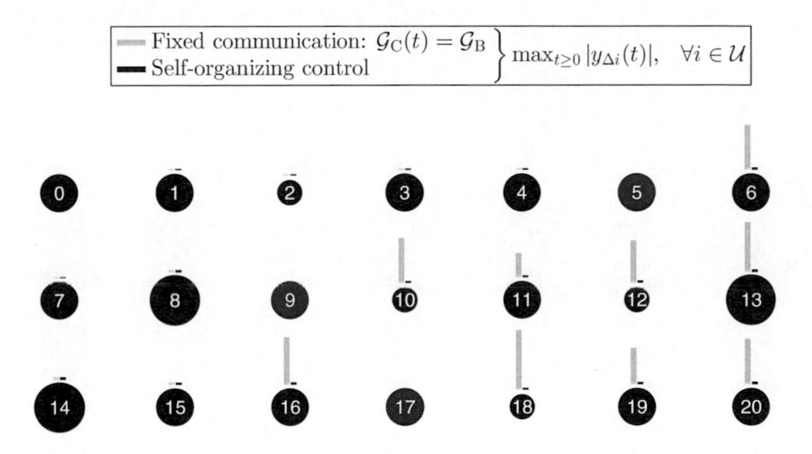

Figure 9.15: Maximal values of the control errors $\max_{t\geq 0} |y_{\Delta i}(t)|$, $\forall i \in \mathcal{U}$, of the undisturbed robots.

Propagation of a disturbance affecting robot Σ_5. For the evaluation of the disturbance propagation of $d_5(t)$ the behavior of the robot formation for the time interval $[0\,\text{s}, 50\,\text{s}]$ is investigated in more detail since in that time interval only robot Σ_5 is disturbed. The analysis result in (9.32) is confirmed, since, e.g.,

$$\max_{t\in[0\,\text{s},50\,\text{s}]} |y_{\Delta 6}(t)| = 0.24\,\text{cm} \leq \bar{y}_{\Delta 6} = 0.33\,\text{cm}$$

$$\max_{t\in[0\,\text{s},50\,\text{s}]} |y_{\Delta 20}(t)| = 0.23\,\text{cm} \leq \bar{y}_{\Delta 20} = 0.41\,\text{cm}.$$

The disturbance propagation of $d_5(t)$ is illustrated in Fig. 9.16 which depicts the robots with their maximal values of the control errors $\max_{t\in[0\,\text{s},50\,\text{s}]} |y_{\Delta i}(t)|$, $(\forall i \in \mathcal{U})$, from the behavior of the self-organizing controlled robots Σ shown in Fig. 9.12. With a fixed communication the disturbance $d_5(t)$ also has a great influence on the robots that are close to Σ_5. With the self-organizing controller C this influence nearly vanishes.

9.5.4 Trade-off between performance and communication effort

Consider different self-organizing controllers C designed for a certain desired maximal control error $\bar{y}^*_\Delta$ for the undisturbed robots that lead to different self-organizing control systems Σ. Figure 9.17 compares the performance of the different Σ indicated by $\bar{y}^*_\Delta$ and the communication effort indicated by the deactivation time $t_{\Sigma\text{D}}$ for the same disturbance situation as depicted in Fig. 9.12. The time $t_{\Sigma\text{D}}$ is the sum of the lengths of intervals for which communication links

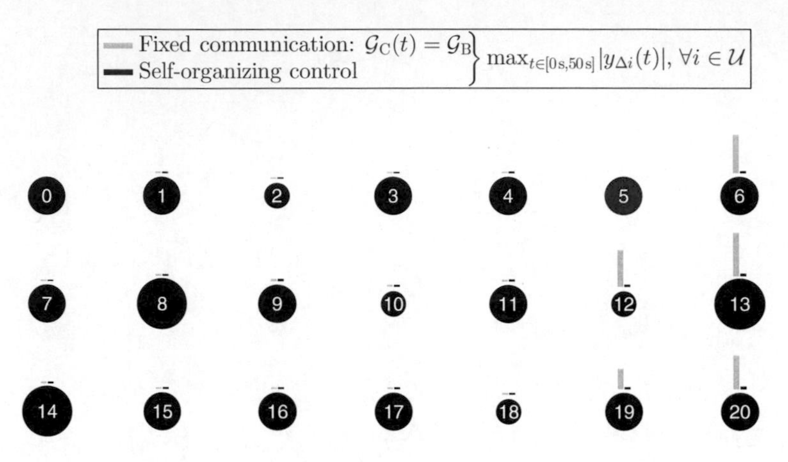

Figure 9.16: Maximal values of the control errors $\max_{t\in[0\,s,50\,s]} |y_{\Delta i}(t)|,\ \forall i \in \mathcal{U}$, of the undisturbed robots for the time interval $[0\,s, 50\,s]$ in which only robot Σ_5 is disturbed.

$(i \rightarrow j) \in \mathcal{E}_B$ within $\mathcal{G}_B$ are deactivated which is the length of all gaps within the black bars in the lower plot of Fig. 9.12 in a row. For example with $\bar{y}^*_\Delta = 1\,\text{cm}$, the overall deactivation time is $t_{\Sigma D} \approx 160$ s (cf. Fig. 9.12). Figure 9.17 shows that even for small values of $\bar{y}^*_\Delta$, there is a huge reduction of the communication deactivation compared to a permanent deactivation of the communication links of the robots that might be disturbed. Hence, the desired maximal control error $\bar{y}^*_\Delta = 1\,\text{cm}$ claimed in (9.31) ensures a good compromise between a minor change of the basic communication and a reasonable disturbance attenuation within the overall system.

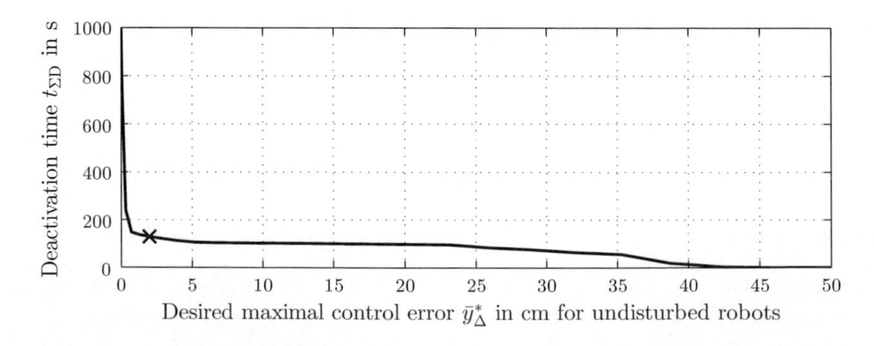

Figure 9.17: Deactivation time for certain self-organizing systems Σ of the form (9.17) resulting from different desired maximal control errors $\bar{y}^*_\Delta$.

9.6 Summary and evaluation of the proposed control concept

The main properties of the self-organizing control concept with situation-dependent interruption of the communication for the control of a multi-agent system within a leader-follower structure can be summarized as follows:

- The concept bounds the disturbance propagation within the resulting closed-loop system.
- The maximal control error of the undisturbed subsystems can be arbitrarily adjusted by the choice of the switching threshold ϵ (Theorem 9.2 and Theorem 9.3).
- Despite the interruption of the communication, the agents can asymptotically synchronize with the leader after the disturbances have vanished (Theorem 9.1).

In summary, the proposed self-organizing control concept is applicable for multi-agent systems with significant disturbances effecting the agents which leads to a bad overall synchronization performance.

Properties of self-organization. The claimed properties of a self-organizing control system in Section 1.2 are verified for the proposed control concept as follows:

- **Flexibility.** The communication structure is adapting to the currently acting disturbances. In particular, the communication is only interrupted for disturbances with a significant impact on the behavior of the agents.
- **Scalability.** The reconstruction of the disturbance impact by the observation unit O_i can be performed by using local model information only. For designing the switching threshold ϵ to guarantee a desired maximal control error of the undisturbed agents global model information is used. Furthermore, the local control units C_i need information about the basic communication graph $\mathcal{G}_\mathrm{B}$, i.e.,

$$\left.\begin{array}{r} \mathcal{G}_\mathrm{B} \Rightarrow \mathcal{P}_{\mathrm{B}i} \text{ and } \mathcal{F}_{\mathrm{B}i} \\ \bar{\Sigma}_i \\ \epsilon \end{array}\right\} \text{Parameters of } C_i,\ \forall i \in \mathcal{N}.$$

  Hence, the complexity of the local control units scales linearly with the number of subsystems. For the analysis of the behavior of the resulting closed-loop system global model information is needed.

- **Fault tolerance.** The proposed control concept is not analyzed w.r.t. faults within the agents or local control units. Since links within the basic communication graph $\mathcal{G}_\mathrm{B}$ are switched off over time, faulty links within $\mathcal{G}_{\mathrm{B}}$ are very crucial for not losing the connectivity among the agents.

In summary, the proposed self-organizing control concept shows flexibility and a good scalability.

Note that Assumption 9.2 requires that there exists a spanning tree with the leader as root despite the deactivation of the communication links. If this requirement can non be fulfilled, the combination of the concept presented in this chapter with the control concept that reconfigures the communication structure in case of faulty communication links presented in Chapter 10 can solve this problem. This combination is discussed in more detail in Chapter 11.

10 Preservation of the performance by local reconfiguration of the faulty communication structure

This chapter introduces a control concept for preserving the desired basic synchronization performance in case of faulty/missing communication links. The local control units perform a local reconfiguration of the communication structure by compensating the missing links by admissible new communication links. Therefore, in most cases an asymptotic synchronization to the leader can still be achieved despite the faults and the maximal basic control error remains for the fault-free agents. The concept is applied to the control of a robot formation.

Chapter contents

10.1 Local reconfiguration of the communication structure

Consider the situation that the local control units C_i, that communicate within the basic communication graph $\mathcal{G}_{\rm B}$, i.e., $\mathcal{G}_{\rm C}(0) = \mathcal{G}_{\rm B}$, might have faulty transmission units (Fig. 10.1). Due to such a fault, the corresponding local control unit C_j, $(j \in \mathcal{N}_{\rm F}(t))$, can not transmit any information. $\mathcal{N}_{\rm F}(t)$ is called *set of faulty agents* despite the fact that only the corresponding local control units are faulty. Hence, the output $y_j(t)$ can not be transmitted to the desired followers C_i, $(i \in \mathcal{F}_{{\rm B}j})$. The affected local control units C_i have to compensate the missing output $y_j(t)$ by an admissible new output $y_l(t)$ from an other local control unit C_l, $(l < i)$, to preserve an asymptotic synchronization with the leader Σ_0 as well as a desired synchronization performance. To avoid an unsolvable problem, the local control unit C_0 of the leader Σ_0 is assumed to be fault-free at any time. Based in this problem formulation, the main **aim** of the control concept that will be presented in this chapter can be summarized as follows:

> Develop a self-organizing controller C with a situation-dependent reconfiguration of the communication structure in case of a failure of specific communication links to preserve the synchronization performance of the fault-free communication (cf. Control aim 7.4).

To get a fault-tolerant overall closed-loop system that guarantees asymptotic synchronization and a desired performance of the fault-free agents Σ_i, $(\forall i \in \mathcal{N} \backslash \mathcal{N}_{\rm F}(t))$, the local control units C_i perform two common tasks of a fault-tolerant system by using locally available information:

- *Diagnosis of the faulty agents* by evaluating the transmitted information.
- *Reconfiguration of the communication structure* by replacing the failed communication links $(j \rightarrow i)$, $(\forall j \in \mathcal{P}_{{\rm B}i})$, with admissible new communication links $(l \rightarrow i)$, $(\forall l \in \mathcal{P}_{{\rm S}i})$, see Fig. 10.1.

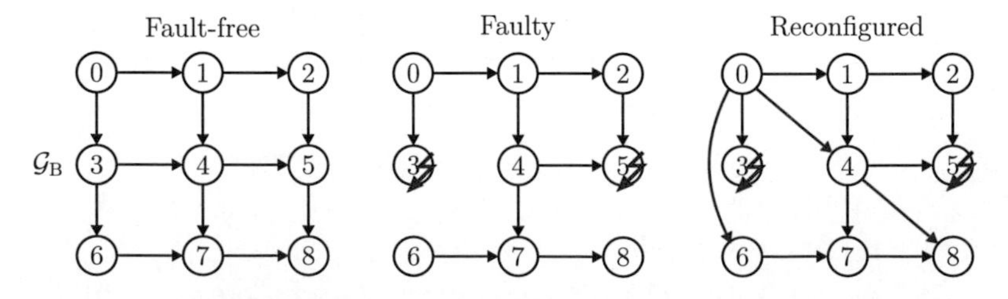

Figure 10.1: Idea of the situation-dependent reconfiguration of the communication structure.

The *set* $\mathcal{P}_{\mathrm{S}i}(t)$ *of substitutional predecessors* includes the index of the agent Σ_l whose output $y_l(t)$ is chosen to compensate the missing output $y_j(t)$, $(j \in \mathcal{P}_{\mathrm{B}i} \cap \mathcal{N}_{\mathrm{F}}(t))$, where the index j is included in the set $\mathcal{P}_{\mathrm{B}i}$ of basic predecessors. With this, the local reference signal $y_{\mathrm{s}i}(t)$ of the affected local control unit C_i results to

$$y_{\mathrm{s}i}(t) = \sum_{j \in \mathcal{P}_{\mathrm{C}i}(t)} \tilde{k}_{ij}(t) y_j(t), \qquad \forall i \in \mathcal{N}, \tag{10.1}$$

with $\mathcal{P}_{\mathrm{C}i}(t) = \{\mathcal{P}_{\mathrm{B}i} \cup \mathcal{P}_{\mathrm{S}i}(t)\} \backslash \mathcal{N}_{\mathrm{F}}(t)$. The vanishing indices $j \in \mathcal{P}_{\mathrm{B}i} \cap \mathcal{N}_{\mathrm{F}}(t)$ are compensated by the set of substitutional predecessors

$$\mathcal{P}_{\mathrm{S}i}(t) = \left\{ \arg\min_{l \in \mathcal{V}_{\mathrm{C}} \backslash \mathcal{N}_{\mathrm{F}}(t), l<i} \{c_{il} \mid \bar{y}_{\Delta\mathrm{R}il}(t) \leq \bar{y}_{\Delta\mathrm{B}i}\} \right\}, \quad i \in \mathcal{N}. \tag{10.2}$$

An output $y_j(t)$ is chosen such that the maximal control error $y_{\Delta\mathrm{R}il}(t)$ while using the reconfigured communication structure is smaller than or equal to the basic maximal control error $\bar{y}_{\Delta\mathrm{B}i}$ while using the basic communication graph $\mathcal{G}_{\mathrm{B}}$. Furthermore, the resulting communication cost c_{ij} of the new communication link $(j \to i)$ is minimal.

Based on these considerations, the **idea** of the self-organizing control concept presented in this chapter can be summarized as follows:

> Whenever a faulty communication link leads to a communication structure which does not guarantee the same basic maximal control error $\bar{y}_{\Delta\mathrm{B}i}$ while using the basic communication graph $\mathcal{G}_{\mathrm{B}}$, the corresponding local control unit C_i compensates the missing predecessor C_j, $(j \in \mathcal{P}_{\mathrm{B}i} \cap \mathcal{N}_{\mathrm{F}}(t))$ by choosing an admissible new predecessor C_l, $(l \in \mathcal{P}_{\mathrm{S}i})$ with a minimal communication cost c_{il} to recover the desired bound $\bar{y}_{\Delta\mathrm{B}i}$.

Figure 10.1 illustrates the idea of the situation-dependent reconfiguration of the communication structure. If the local control unit C_5 of agent Σ_5 gets faulty, the communication link $(5 \to 8)$ vanishes and C_8 does not receive $y_5(t)$ anymore. Due to this disconnection, C_8 recognizes that C_5 is faulty and choses C_4 as a new predecessor to preserve its performance.

Equations (10.1) and (10.2) characterize the fundamental operating principle of the local control units C_i, $(\forall i \in \mathcal{N})$, (Fig. 10.1), whereas their actual realization is different from this operation. The diagnosis of the faulty agents is performed locally. Furthermore, for the choice of the set $\mathcal{P}_{\mathrm{S}i}$ of substitutional predecessors the calculation of the maximal control error $\bar{y}_{\Delta\mathrm{R}ij}$ while using the corresponding reconfigured communication structure has to be specified. In summary, the basic tasks of the local control units C_i, $(\forall i \in \mathcal{N})$, can be defined as follows:

1. As explained in Section 7.3, the feedback unit F_i tracks the local reference signal $y_{\mathrm{s}i}(t)$.

2. The observation unit O_i detects the currently faulty agents.

3. The decision unit D_i choses new predecessors C_l, $(l \in \mathcal{P}_{\mathrm{S}i})$ to compensate the missing outputs of the faulty agents.

4. The decision unit D_i adapts the local reference signal $y_{\mathrm{s}i}(t)$ depending on the received outputs $y_j(t)$, $(\forall j \in \mathcal{P}_{\mathrm{C}i}(t))$.

In the following the structure of the local control units that fulfill these four tasks is illustrated for multi-agent system with two followers.

Structure of the self-organizing controller for two followers. The concerned basic communication graph $\mathcal{G}_{\mathrm{B}}$ for controlling the multi-agent system P defined in (7.8) consisting of the leader Σ_0 and two follower agents Σ_1 and Σ_2 is depicted in Fig. 10.2. Furthermore, Fig. 10.2 shows the possible communication graph $\mathcal{G}_{\mathrm{C}}(t)$ for considering that the transmission unit of C_1 might be faulty.

Figure 10.3 depicts the resulting structure of the self-organizing control system Σ with a situation-dependent reconfiguration of the communication structure. In case that the transmission unit (abbreviated TU) of C_1 is faulty, the components of the local control units C_0, C_1 and C_2 work as follows:

- The **feedback units** F_i, $(i = 1, 2)$, generate the inputs $u_i(t)$ to the agents Σ_i by comparing the corresponding output $y_i(t)$ with the local reference signal $y_{\mathrm{s}i}(t)$ to track the local reference signal $y_{\mathrm{s}i}(t)$ (1. Task). Recall that F_i is defined in (7.22).

- The **observation unit** O_2 determines the local set $\mathcal{P}_{\mathrm{F}2}(t)$ of faulty agents using the transmitted output

$$\tilde{y}_{21}(t) = \begin{cases} \eta & \text{if transmission unit (TU) of } C_1 \text{ is faulty} \\ y_1(t) & \text{else} \end{cases}$$

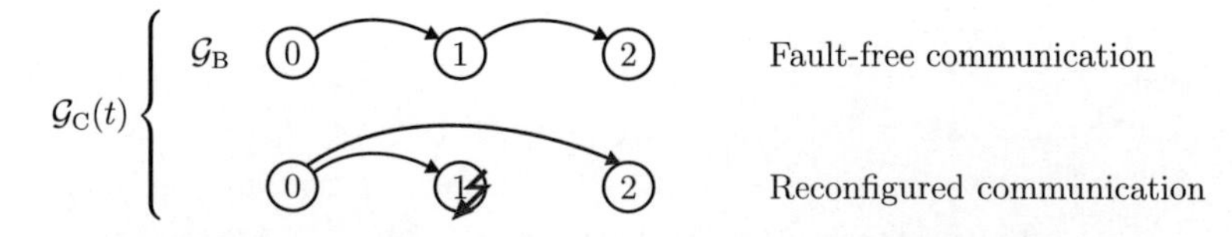

Figure 10.2: Basic communication graph $\mathcal{G}_{\mathrm{B}}$ for the self-organizing control of a multi-agent system with two followers and reconfigured communication graph $\mathcal{G}_{\mathrm{C}}(t)$ for a fault affecting C_1.

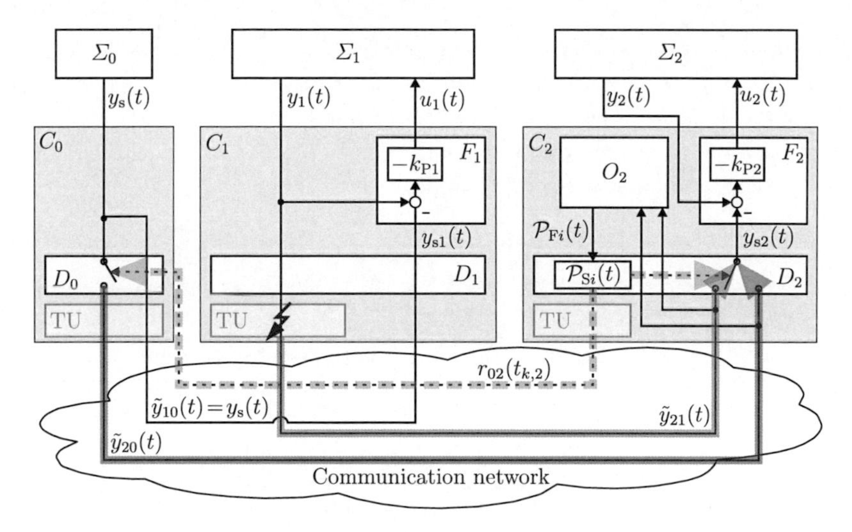

Figure 10.3: Self-organizing control concept with situation-dependent reconfiguration of the communication structure for two followers.

which usually equals the output $y_1(t)$ transmitted by C_1 (2. Task). However, if C_1 becomes faulty, the transmission of $y_1(t)$ is canceled which is indicated by the symbol η.

- The **decision unit** D_2 sends a request $r_{02}(t_{k,2})$ at the time instance $t_{k,2}$, ($k \in \mathcal{N}$), to the local control unit C_0 if $\mathcal{P}_{\mathrm{F}2}(t) = \{1\}$ holds true, whereupon C_0 sends the reference signal $y_{\mathrm{s}}(t)$ to C_2 (3. Task).

- The **decision unit** D_2 in the local control unit C_2 uses the reference signal $y_{\mathrm{s}}(t)$ to adapt its local reference signal $y_{\mathrm{s}2}(t)$ (4. Task), i.e.,

$$y_{\mathrm{s}2}(t) = \begin{cases} y_{\mathrm{s}}(t) & \text{if } \mathcal{P}_{\mathrm{F}2}(t) = \{1\} \\ y_1(t) & \text{else.} \end{cases} \tag{10.3}$$

- The **local control unit** C_0 sends the reference signal $y_{\mathrm{s}}(t)$ to C_1 at all times.

The effect of this control strategy is illustrated by the behavior of the controlled agent $\bar{\Sigma}_2$ defined in (7.26) which is assumed to be fault-free:

$$\bar{\Sigma}_2 : \begin{cases} \dot{\boldsymbol{x}}_2(t) = \bar{\boldsymbol{A}}_i \boldsymbol{x}_2(t) + k_{\mathrm{P}2} \boldsymbol{b}_2 y_{\mathrm{s}2}(t), & \boldsymbol{x}_2(0) = \boldsymbol{x}_{02} \\ y_2(t) = \boldsymbol{c}_2^{\mathrm{T}} \boldsymbol{x}_2(t). \end{cases} \tag{10.4}$$

Since agent C_1 is faulty, according to (10.3) the local reference signal $y_{s2}(t)$ always includes a signal which guarantees asymptotic synchronization to the leader Σ_0. $y_{s2}(t)$ even is identical to the reference signal $y_s(t)$ which leads to the lowest possible control error $y_{\Delta 2}(t)$. Since C_0 is the only choice of C_2 as a new predecessor, no real decision process is necessary. For a more complex reconfiguration problem see Section 10.4.

Example 10.1 *Communication reconfiguration within a formation of three robots*

In this example the self-organizing control concept with situation-dependent reconfiguration of the communication structure illustrated in Fig. 10.3 is used to preserve the synchronization performance within a formation of three robots which are defined in (7.65). The applied basic communication graph $\mathcal{G}_B$ is depicted in Fig. 10.2, where the local control unit C_1 of robot Σ_1 is assumed to become faulty at some time. The follower robots Σ_1 and Σ_2 are both average robots, i.e., $\mathcal{N}_{\text{average}} = \{1, 2\}$ (cf. Table 7.1). The example is divided into two parts. First, the construction of the local control units C_0, C_1 and C_2 is presented. Second, the behavior of the resulting self-organizing control system Σ is investigated.

Parameters of the local control units. Recall that the feedback units F_1 and F_2 use the feedback gain $k_{P1} = k_{P2} = 15$ (cf. Section 7.7). According to the basic communication graph $\mathcal{G}_B$, the sets $\mathcal{P}_{Bi}$ of basic predecessors result to: $\mathcal{P}_{B0} = \emptyset$, $\mathcal{P}_{B1} = \{0\}$ and $\mathcal{P}_{B2} = \{1\}$.

Behavior of the self-organizing control system Σ. The behavior of the self-organizing controlled robots Σ with initial states

$$\boldsymbol{x}_{s0} = \boldsymbol{x}_{01} = \boldsymbol{x}_{02} = \begin{pmatrix} 0 \\ 100 \end{pmatrix},$$

a stepwise change of the reference velocity $\bar{v}(t)$ and an intermittent fault affecting the local control unit C_1 of robot Σ_1 is depicted in Fig. 10.4 with solid lines. The change of the reference velocity $\bar{v}(t)$ leads to a deviation between the reference position $y_s(t)$ and the position $y_i(t)$ of the robots (third plot) which is indicated by the control errors $y_{\Delta 1}(t)$ and $y_{\Delta 2}(t)$ in the fourth row of plots. The fault at C_1 represented by the set of faulty agents $\mathcal{N}_F(t) = \{1\}$ leads to a reconfigured communication structure. In particular, C_2 receives the reference signal $y_s(t)$ from C_0 instead of the output $y_1(t)$ from C_1 indicated by the black bars in the right plot in the lowest row. The local reconfiguration within C_2 is performed in three basic steps: First, the fault is diagnosed by recognizing that C_2 receives no information from C_1 which results in a set of locally faulty agents, i.e., $\mathcal{P}_{F2}(t) = \{1\}$. Second, C_2 choses C_0 as its substitutional predecessor by including 0 into the set $\mathcal{P}_{S2}(t)$ of substitutional predecessors, i.e., $\mathcal{P}_{S2}(t) = \{0\}$. Third, C_2 updates its set of predecessors to $\mathcal{P}_{C2}(t) = \{0\}$ by sending a request $r_{02}(t_{1,2})$ to C_0. With this reconfiguration the control error $y_{\Delta 2}(t)$ is even smaller than in the fault-free case (dashed black line) since the C_2 receives the actual reference position $s_0(t) = y_s(t)$ from the leading robot Σ_0. The drawback is that the communication cost c_{20} is higher than c_{12} within the basic communication graph $\mathcal{G}_B$.

When the fault vanishes ($\mathcal{N}_F(t) = \emptyset$) at $t = 32$ s the reconfiguration of the communication structure is made undone. Since C_2 receives the output $y_i(t)$ of C_1 again, the sets $\mathcal{P}_{F2}$ and $\mathcal{P}_{S2}$ become empty. Therefore, the set of predecessors changes to $\mathcal{P}_{C2} = \{1\}$ which has to be transmitted to C_0 by $r_{02}(t_{2,2})$. Thereafter, the control error $y_{\Delta 2}(t)$ is similar to the one in the fault-free case.

The red dashed dotted line in the third plot shows the position $s_2(t) = y_2(t)$ of robot Σ_2 without reconfiguration of the communication structure. Robot Σ_2 remains at its position at which the fault occurs since the local reference signal $y_{s2}(t)$ remains constant as long as $\mathcal{P}_{C1}(t) = \emptyset$ holds true (cf. (7.18)). This shows that the presented control concept leads to an enormous improvement of the synchronization performance. Even if robot Σ_2 can track the reference position again, after the fault vanishes, in the meantime its control error $y_{\Delta 2}(t)$ is very large. Furthermore, if the fault does not vanish, robot Σ_2 could never track the reference position $y_s(t)$ again. Recall that this improvement has been achieved by

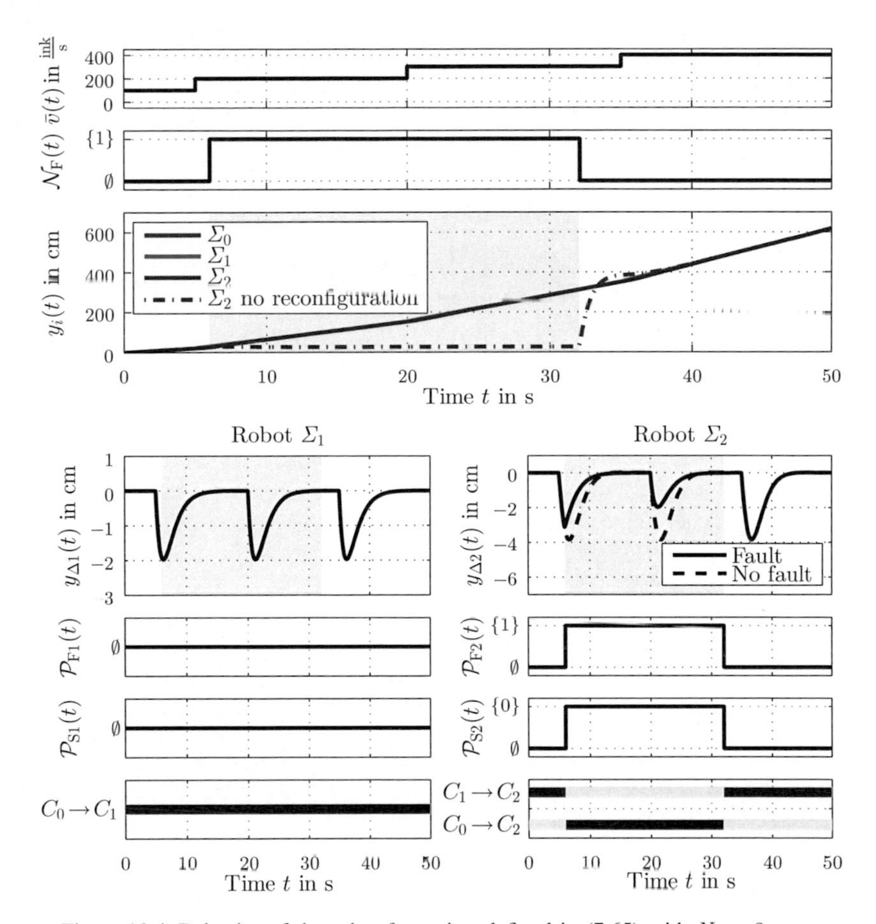

Figure 10.4: Behavior of the robot formation defined in (7.65) with $N = 2$ controlled be the local control units C_i specified in Fig. 10.3.

decentralized fault diagnosis and decentralized reconfiguration.

In summary, this example shows that the proposed self-organizing control concept with reconfiguration of communication links has the following interesting properties:

- The local control units reconfigure the communication structure only if a fault affects the local control units, which is a situation-dependent communication (cf. lower row of plots in Fig. 10.4).
- The synchronization performance is preserved after the reconfiguration (cf. fourth row of plots in Fig. 10.4).
- If a fault vanishes, the basic communication is used again (cf. lowest row of plots in Fig. 10.4).

□

Overview of the chapter. In the following sections it is shown that the mentioned properties of the proposed control concept also exist if it is applied to an arbitrary number of agents. Therefore, the content of the next sections is as follows:

- The structure of the self-organizing controller C is extended to an arbitrary number of agents (Section 10.2).
- An analysis of the self-organizing control system Σ yields a condition on the asymptotic synchronization of the agents and a result which agents can retain their synchronization performance (Section 10.3).
- The self-organizing control concept is demonstrated by its application to a large robot formation (Section 10.4).
- The self-organizing properties of the closed-loop system with the situation-dependent reconfiguration of the communication structure are evaluated (Section 10.5).

10.2 Structure of the self-organizing control system

This section presents the model of the fault-tolerant self-organizing control system Σ (Section 10.2.4). Therefore, an upper and a lower bound for the situation-dependent reconfigured communication links is derived (Section 10.2.1). Then the structure of the local control units C_i, $(\forall i \in \mathcal{V}_{\mathrm{C}})$, is presented (Section 10.2.2) which replace faulty predecessors by admissible new predecessors. The aggregation of the local control units C_i, $(\forall i \in \mathcal{V}_{\mathrm{C}})$, leads to the overall self-organizing controller C introduced in Section 10.2.3.

10.2.1 Communication structure with compensation of faulty links

In this section an upper bound and a lower bound for the activation of the communication links is derived according to definition (2.25) in Section 2.3. The change of the communication graph $\mathcal{G}_{\mathrm{C}}(t) = (\mathcal{V}_{\mathrm{C}}, \mathcal{E}_{\mathrm{C}}(t), \boldsymbol{K})$ depends on the set $\mathcal{E}_{\mathrm{C}}(t)$ of communication links defined in (2.17), where $\mathcal{E}_{\mathrm{C}}(t)$ changes due to the situation-dependent reconfiguration of the communication links. According to (10.1), the set $\bar{\mathcal{P}}_{\mathrm{C}i}$ of maximal communicational predecessors and the set $\underline{\mathcal{P}}_{\mathrm{C}i}$ of minimal communicational predecessors is given by

$$\underline{\mathcal{P}}_{\mathrm{C}i} = \mathcal{P}_{\mathrm{B}i} \cap \{0\}, \qquad \forall i \in \mathcal{N} \tag{10.5}$$

$$\bar{\mathcal{P}}_{\mathrm{C}i} = \{0, 1, \ldots, i-1\}, \quad \forall i \in \mathcal{N}. \tag{10.6}$$

Therefore, the set $\mathcal{P}_{\mathrm{C}i}(t)$ of predecessors is bounded by

$$\underbrace{\mathcal{P}_{\mathrm{B}i} \cap \{0\}}_{\underline{\mathcal{P}}_{\mathrm{C}i}} \subseteq \mathcal{P}_{\mathrm{C}i}(t) \subseteq \underbrace{\{0, 1, \ldots, i-1\}}_{\bar{\mathcal{P}}_{\mathrm{C}i}}, \qquad \forall i \in \mathcal{N}, \quad \forall t \geq 0.$$

The set $\bar{\mathcal{E}}_{\mathrm{C}}$ of maximal communication links and set $\underline{\mathcal{E}}_{\mathrm{C}}$ of minimal communication links follow from (2.21), (2.23), (10.5) and (10.6) to

$$\underline{\mathcal{E}}_{\mathrm{C}} = \{(0 \to i) \mid i \in \mathcal{F}_{\mathrm{B}0}\}$$
$$\bar{\mathcal{E}}_{\mathrm{C}} = \{(j \to i) \mid j \in \mathcal{V}_{\mathrm{C}}, j < i\}.$$

Hence, the maximal communication graph $\bar{\mathcal{G}}_{\mathrm{C}} = (\mathcal{V}_{\mathrm{C}}, \bar{\mathcal{E}}_{\mathrm{C}}, \boldsymbol{K})$ defined in (2.19) includes all possible communication links. Note that the communication links $(0 \to i)$, $(\forall i \in \mathcal{F}_{\mathrm{B}0})$, from the local control unit C_0 of the leader Σ_0 to its basic followers Σ_i are guaranteed to remain since C_0 is assumed to be fault-free.

In summary, the set $\mathcal{E}_{\mathrm{C}}(t)$ of communication links within the communication graph $\mathcal{G}_{\mathrm{C}}(t) = (\mathcal{V}_{\mathrm{C}}, \mathcal{E}_{\mathrm{C}}(t), \boldsymbol{K})$ is bounded, i.e.,

$$\underbrace{\{(0 \to i) \mid i \in \mathcal{F}_{\mathrm{B}0}\}}_{\underline{\mathcal{E}}_{\mathrm{C}}} \subseteq \mathcal{E}(t) \subseteq \underbrace{\{(j \to i) \mid j \in \mathcal{V}_{\mathrm{C}}, j < i\}}_{\bar{\mathcal{E}}_{\mathrm{C}}}, \quad t \geq 0. \qquad (10.7)$$

10.2.2 Structure of the local control units

Recall that the local control units C_i include a feedback unit F_i, an observation unit O_i and a decision unit D_i (Fig. 10.5). The feedback unit F_i is used to track the local reference signal $y_{\mathrm{s}i}(t)$. The observation unit O_i detects the current situation of the agents Σ_i indicated by the set $\mathcal{P}_{\mathrm{F}i}(t)$ of faulty agents which are locally diagnosed at C_i. The decision unit D_i handles the incoming and outgoing information and determines the *set* $\mathcal{P}_{\mathrm{S}i}(t)$ *of substitutional predecessors* which replace the faulty predecessors included in set $\mathcal{P}_{\mathrm{F}i}(t)$. Hence, the observation unit O_i performs the diagnosis of the faulty agents and the decision unit D_i performs the reconfiguration of the communication structures. The following paragraphs describe the observation unit O_i and the decision unit D_i. The feedback unit F_i is already defined in (7.22) in Section 7.3.2. Finally, the components are combined to define the local control units C_i, $(\forall i \in \mathcal{N})$, of the agents Σ_i and the local control unit C_0 of the leader Σ_0.

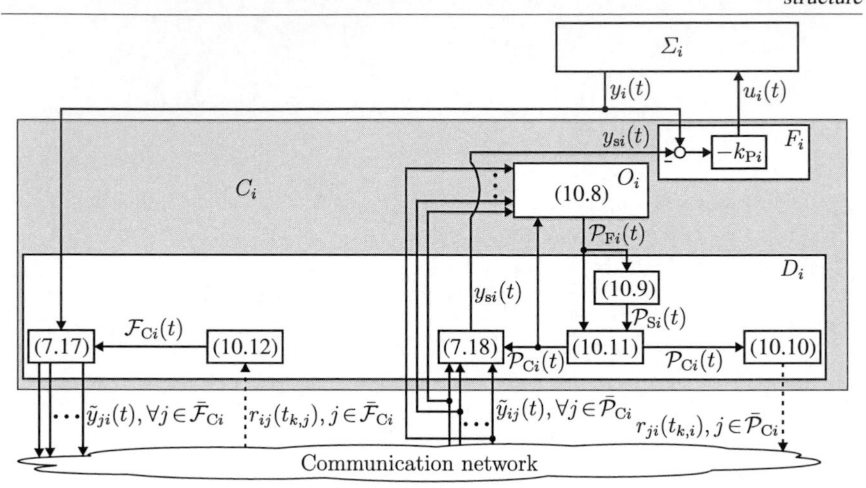

Figure 10.5: Local control unit C_i, $(\forall i \in \mathcal{N})$, designed for a situation-dependent reconfiguration of the communication structure.

Observation unit O_i. The task of the observation unit O_i, $(i \in \mathcal{N})$ is to perform the diagnosis by determining the *set* $\mathcal{P}_{Fi}(t)$ *faulty predecessors*

$$O_i : \mathcal{P}_{Fi}(t) = \begin{cases} \mathcal{P}_{Fi}(t^-) \cup \{j\} & \text{if } \tilde{y}_{ij}(t) = \eta, j \in \mathcal{P}_{Ci}(t) \\ \mathcal{P}_{Fi}(t^-) \backslash \{j\} & \text{if } \tilde{y}_{ij}(t) \neq \eta, j \in \mathcal{P}_{Fi}(t^-) \\ \mathcal{P}_{Fi}(t^-) & \text{else} \end{cases} \tag{10.8}$$

for the decision unit D_i. Therefore, O_i uses the transmitted outputs $\tilde{y}_{ij}(t)$, $(\forall j \in \bar{\mathcal{P}}_{Ci})$, defined in (7.17). If C_i receives no information from its desired predecessor C_j, $(j \in \mathcal{P}_{Ci}(t))$, which is denoted by the symbol η for $\tilde{y}_{ij}(t)$, the index j is added to $\mathcal{P}_{Fi}(t)$. Note that the symbol t^- denotes the time instance directly before the time t with $\mathcal{P}_{Fi}(t^-) = \emptyset$ for $t = 0$.

Decision unit D_i. The decision unit performs four tasks.

First, it determines the local reference signal $y_{si}(t)$ for the feedback unit F_i according to (7.18) while using the transmitted outputs $\tilde{y}_{ij}$, $(\forall j \in \mathcal{P}_{Bi})$ and the set of predecessors $\mathcal{P}_{Ci}(t)$. Recall that the task of F_i is explained in more detail in Section 7.3.2.

Second, D_i determines the set of substitutional predecessors

$$\mathcal{P}_{\mathrm{S}i}(t) = \begin{cases} \mathcal{P}_{\mathrm{S}i}(t^-) & \text{if } \left(\mathcal{P}_{\mathrm{F}i}(t) = \mathcal{P}_{\mathrm{F}i}(t^-)\right) \wedge (i \in \mathcal{N}_{\mathrm{F}}(t)) & (10.9a) \\ \emptyset & \text{if } (\bar{y}_{\Delta\mathrm{R}ii}(t) \leq \bar{y}^*_{\Delta\mathrm{B}i}) \wedge (\mathcal{P}_{\mathrm{B}i} \backslash \mathcal{P}_{\mathrm{F}i}(t) \neq \emptyset) & (10.9b) \\ \mathcal{P}_{\mathrm{N}i}(t) & \text{else,} & (10.9c) \end{cases}$$

for all $i \in \mathcal{N}$ with $\mathcal{P}_{\mathrm{S}i}(t^-) = \emptyset$ for $t = 0$ and

$$\begin{aligned} \mathcal{P}_{\mathrm{N}i}(t) &= \left\{ \underset{j \in \mathcal{V}_{\mathrm{C}} \backslash \{\mathcal{P}_{\mathrm{B}i} \cup \mathcal{P}_{\mathrm{F}i}(t)\}, j<i}{\arg\min} \{c_{ij} \mid \bar{y}_{\Delta\mathrm{R}ij}(t) \leq \bar{y}^*_{\Delta\mathrm{B}i}\} \right\}, && i \in \mathcal{N} \\ \bar{y}_{\Delta\mathrm{R}ij}(t) &= m_{\mathrm{y}\epsilon\mathrm{B}i} + m_{\mathrm{yy}i} \sum_{j \in \mathcal{P}_{\mathrm{CR}ij}(t)} \frac{k_{ij}}{\sum_{l \in \mathcal{P}_{\mathrm{CR}ij}(t)} k_{il}} \bar{y}^*_{\Delta\mathrm{B}j}, && i \in \mathcal{N}, \quad j \leq i, \quad j \in \mathcal{V}_{\mathrm{C}} \\ \mathcal{P}_{\mathrm{CR}ij}(t) &= \{\mathcal{P}_{\mathrm{B}i} \backslash \mathcal{P}_{\mathrm{F}i}(t)\} \cup \{j\}, && i \in \mathcal{N}, \quad j < i, \quad j \in \mathcal{V}_{\mathrm{C}} \\ \mathcal{P}_{\mathrm{CR}ii}(t) &= \mathcal{P}_{\mathrm{B}i} \backslash \mathcal{P}_{\mathrm{F}i}(t), && i \in \mathcal{N}. \end{aligned}$$

There exist three cases for the determination of $\mathcal{P}_{\mathrm{S}i}(t)$. The set $\mathcal{P}_{\mathrm{S}i}(t)$ remains unchanged if the set $\mathcal{P}_{\mathrm{F}i}(t)$ of faulty agents stays constant (cf. (10.9a)). If not all basic predecessors of C_i are faulty, i.e., $\mathcal{P}_{\mathrm{B}i} \backslash \mathcal{P}_{\mathrm{F}i}(t) \neq \emptyset$ and the actual upper bound $\bar{y}_{\Delta\mathrm{R}ii}(t)$ on the control error $y_{\Delta i}(t)$ without compensating the faulty predecessors ($\mathcal{P}_{\mathrm{CR}ii} = \mathcal{P}_{\mathrm{B}i} \backslash \mathcal{P}_{\mathrm{F}i}(t)$) is smaller or equal to the maximal basic control error $y^*_{\Delta\mathrm{B}i}$, then $\mathcal{P}_{\mathrm{S}i}(t)$ is set to be empty (cf. (10.9b)). For any other case $\mathcal{P}_{\mathrm{S}i}(t)$ is the predecessor C_j with the smallest communication cost c_{ij} which guarantees the same bound $y^*_{\Delta\mathrm{B}i}$ as with the basic communication graph $\mathcal{G}_{\mathrm{B}}$ (cf. (10.9c)).

Third, D_i sends an *information request* $r_{ji}(t_{k,i})$, ($k \in \mathbb{N}_0$), at the *request time instance* $t_{k,i}$ to the local control unit C_j, ($j \in \bar{\mathcal{P}}_{\mathrm{C}i}$), if its set $\mathcal{P}_{\mathrm{C}i}(t)$ of predecessors is reconfigured. An information request is defined by

$$r_{ji}(t_{k,i}) = \begin{cases} 1 & \text{if } j \in \mathcal{P}_{\mathrm{C}i}(t_{k,i}) \backslash \mathcal{P}_{\mathrm{C}i}(t_{k-1,i}) \\ 0 & \text{if } j \in \mathcal{P}_{\mathrm{C}i}(t_{k-1,i}) \backslash \mathcal{P}_{\mathrm{C}i}(t_{k,i}) \\ \eta & \text{else,} \end{cases} \qquad j \in \bar{\mathcal{P}}_{\mathrm{C}i}, \quad k > 0, \quad i \in \mathcal{N} \qquad (10.10a)$$

and the request time instance by

$$\begin{aligned} t_{0,i} &= 0 && (10.10b) \\ t_{k+1,i} &= \inf\{t > t_{k,i} \mid \mathcal{P}_{\mathrm{C}i}(t) \neq \mathcal{P}_{\mathrm{C}i}(t_{k,i})\}. && (10.10c) \end{aligned}$$

The set $\mathcal{P}_{\mathrm{C}i}(t)$ of predecessors is reconfigured if a newly faulty predecessor included in $\mathcal{P}_{\mathrm{F}i}(t)$ has to be compensated by an admissible new predecessor included in the set $\mathcal{P}_{\mathrm{S}i}(t)$ of substitutional predecessors, i.e., $\mathcal{P}_{\mathrm{S}i}(t) \neq \emptyset$. Hence, the set of predecessors results to

$$\mathcal{P}_{\mathrm{C}i}(t) = \{\mathcal{P}_{\mathrm{B}i} \cup \mathcal{P}_{\mathrm{S}i}(t)\} \backslash \mathcal{P}_{\mathrm{F}i}(t). \tag{10.11}$$

Fourth, D_i sends the output $y_i(t)$ to its current followers C_j, $(\forall j \in \mathcal{F}_{\mathrm{C}i}(t))$ with the transmitted output $\tilde{y}_{ji}(t)$ according to (7.17) explained in Section 7.3.2. The set of followers

$$\mathcal{F}_{\mathrm{C}i}(t) = \begin{cases} \mathcal{F}_{\mathrm{C}i}(t^-) \cup \{j\} & \text{if } (\exists k \in \mathbb{N}, j \in \bar{\mathcal{F}}_{\mathrm{C}i}) \;\; t = t_{k,j} \;\wedge\; r_{ij}(t_{k,j}) = 1 \\ \mathcal{F}_{\mathrm{C}i}(t^-) \backslash \{j\} & \text{if } (\exists k \in \mathbb{N}, j \in \bar{\mathcal{F}}_{\mathrm{C}i}) \;\; t = t_{k,j} \;\wedge\; r_{ij}(t_{k,j}) = 0 \\ \mathcal{F}_{\mathrm{C}i}(t^-) & \text{else,} \end{cases} \qquad \forall i \in \mathcal{V}_{\mathrm{C}} \tag{10.12}$$

changes if there is an information request $r_{ij}(t_{k,j})$, $(j \in \bar{\mathcal{F}}_{\mathrm{C}i})$, from the local control unit C_j at the request time instance $t_{k,j}$. Note that $\mathcal{F}_{\mathrm{C}i}(t^-) = \mathcal{F}_{\mathrm{B}i}$ for $t = 0$.

In summary, the decision unit D_i is defined by

$$D_i : \begin{cases} \text{(10.11) generating } \mathcal{P}_{\mathrm{C}i}(t) \text{ using } \mathcal{P}_{\mathrm{S}i}(t) \text{ determined by (10.9)} \\ \text{(7.18) generating } y_{\mathrm{s}i}(t) \text{ using } \mathcal{P}_{\mathrm{C}i}(t) \text{ and } \tilde{y}_{ij}(t), (\forall j \in \bar{\mathcal{P}}_{\mathrm{C}i}) \\ \text{(10.10) generating } r_{ji}(t_{k,i}), (j \in \bar{\mathcal{P}}_{\mathrm{C}i}), \text{ using } \mathcal{P}_{\mathrm{C}i}(t) \\ \text{(10.12) generating } \mathcal{F}_{\mathrm{C}i}(t) \text{ using } r_{ij}(t_{k,j}), (j \in \bar{\mathcal{F}}_{\mathrm{C}i}) \\ \text{(7.17) generating } \tilde{y}_{ji}(t), (\forall j \in \bar{\mathcal{F}}_{\mathrm{C}i}), \text{ using } \mathcal{F}_{\mathrm{C}i}(t) \text{ and } y_i(t), \end{cases} \qquad \forall i \in \mathcal{N}. \tag{10.13}$$

Local control units C_i of the followers. The feedback unit F_i, the observation unit O_i and the decision unit D_i build the local control units

$$C_i : \begin{cases} F_i \text{ defined in (7.22)} \\ O_i \text{ defined in (10.8)} \\ D_i \text{ defined in (10.13),} \end{cases} \qquad \forall i \in \mathcal{N} \tag{10.14}$$

of the agents Σ_i.

Local control unit C_0 of the leader Σ_0. There is no feedback unit and observation unit within the local control unit C_0 of the leader Σ_0. The local control unit

$$C_0 = D_0 : \begin{cases} \text{(10.12) generating } \mathcal{F}_{\mathrm{C}0}(t) \text{ using } r_{0j}(t_{k,j}), (j \in \bar{\mathcal{F}}_{\mathrm{C}0}) \\ \text{(7.17) generating } \tilde{y}_{j0}(t), (\forall j \in \bar{\mathcal{F}}_{\mathrm{C}0}), \text{ using } \mathcal{F}_{\mathrm{C}0}(t) \text{ and } y_0(t) = y_{\mathrm{s}}(t) \end{cases} \tag{10.15}$$

Algorithm 10.1: Construct local control units C_i defined in (10.14) which locally reconfigure the communication structure.

Given: $\mathcal{G}_{\rm B} = (\mathcal{V}_{\rm C}, \mathcal{E}_{\rm B}, \boldsymbol{K})$, $\bar{\boldsymbol{y}}_{\Delta{\rm B}}$, $m_{{\rm y}\epsilon{\rm B}i}$, $m_{{\rm yy}i}$, $k_{{\rm P}i}$ and c_{ij} for all $i, j \in \mathcal{V}_{\rm C}$

1 Construct F_i, $(\forall i \in \mathcal{N})$, defined in (7.22) by using the feedback gains $k_{{\rm P}i}$.
2 Construct O_i, $(\forall i \in \mathcal{N})$, defined in (10.8) without any parametrization.
3 Construct D_i, $(\forall i \in \mathcal{N})$, defined in (10.13) by using $\mathcal{P}_{{\rm B}i}$, $\mathcal{F}_{{\rm B}i}$, $m_{{\rm y}\epsilon{\rm B}i}$, $m_{{\rm yy}i}$ and $\bar{y}_{\Delta{\rm B}i}$ as well as $\bar{y}_{\Delta{\rm B}j}$, c_{ij} and k_{ij} for $j = 0, 1, \ldots, i-1$. Extract $\mathcal{P}_{{\rm B}i}$, $(\forall i \in \mathcal{N})$, and $\mathcal{F}_{{\rm B}i}$ from $\mathcal{G}_{\rm B}$.

Result: C_i, $(\forall i \in \mathcal{N})$, consisting of F_i, O_i and D_i.

includes the decision unit D_0 only. C_0 handles the information requests $r_{0i}(t_{k,i})$, $(i \in \bar{\mathcal{F}}_{\rm C0})$, from the local control units C_i of the agents Σ_i.

Construction of the local control units. Algorithm 10.1 summarizes the construction of the local control units C_i defined in (10.14) for a local reconfiguration of the communication structure, where the following parameters are used:

- The basic communication graph $\mathcal{G}_{\rm B}$ is used to initialize the communication among the local control units C_i.
- The overall maximal basic control error $\bar{\boldsymbol{y}}_{\Delta{\rm B}}$ defined in (7.59) is used to chose an adequate communication link which guarantees a similar control error as in the fault-free case.
- The bounds $m_{{\rm y}\epsilon{\rm B}i}$ and $m_{{\rm yy}i}$ on the dynamics of controlled subsystems $\bar{\Sigma}_i$, $(\forall i \in \mathcal{N})$.
- The feedback gains $k_{{\rm P}i}$, $(\forall i \in \mathcal{N})$, are used within the feedback units F_i.
- The communication costs c_{ij}, $(\forall i, j \in \mathcal{V}_{\rm C})$, are used to decide about a reconfigured communication link with the smallest communication cost as required in Control aim 7.4.

Note that the local control units C_i, $(\forall i \in \mathcal{N})$, do not need any model information about the other agents Σ_j, $(j \in \mathcal{N}, j \neq i)$.

In Step 1 of Algorithm 10.1 the given feedback gains $k_{{\rm P}i}$, $(\forall i \in \mathcal{N})$, are used to construct the feedback units F_i defined in (7.22). For the construction of the observation unit O_i defined in (10.8) no parameters are needed (Step 2). The decision unit D_i defined in (10.13) is constructed in Step 3. Therefore, the maximal basic control errors $\bar{y}_{\Delta{\rm B}j}$, $(j = 0, 1, \ldots, i-1)$, the communication costs c_{ij}, the weightings k_{ij} of the communication links and the bounds $m_{{\rm y}\epsilon{\rm B}i}$ and $m_{{\rm yy}i}$ on the dynamics of $\bar{\Sigma}_i$ are used for the choice of the set $\mathcal{P}_{{\rm S}i}(t)$ of substitutional predecessors according to (10.9). Furthermore, the set $\mathcal{P}_{{\rm B}i}$ of basic predecessors and the set $\mathcal{F}_{{\rm B}i}$ of basic followers are extracted from the basic communication graph $\mathcal{G}_{\rm B}$.

10.2.3 Fault-tolerant self-organizing controller

The self-organizing controller C in which the local control units C_i locally reconfigure the communication links to preserve the basic performance within the overall closed-loop system results to

$$C\colon\ C_i,\ (\forall i \in \mathcal{V}_{\mathrm{C}}),\ \text{defined in (10.14) and (10.15) with } \mathcal{E}_{\mathrm{C}}(t) \text{ bounded according to (10.7).} \tag{10.16}$$

In particular, $\mathcal{E}_{\mathrm{C}}(t)$ changes due to faulty agents which are monitored by the observation unit O_i defined in (10.8), whereas the decision units D_i defined in (10.13) decide how to compensate the vanished predecessors by admissible new predecessors.

10.2.4 Overall fault-tolerant self-organizing control system

The control of the multi-agent system P defined in (7.8) by the self-organizing controller C defined in (10.16) leads to the overall fault-tolerant closed-loop system

$$\Sigma:\begin{cases} \dot{\boldsymbol{x}}_{\mathrm{os}}(t) = \bar{\boldsymbol{A}}_{\mathrm{os}}(t)\boldsymbol{x}_{\mathrm{os}}(t) + \boldsymbol{G}_{\mathrm{os}}\boldsymbol{d}(t), \quad \boldsymbol{x}_{\mathrm{os}}(0) = \boldsymbol{x}_{\mathrm{os}0} \\ \boldsymbol{y}(t) = \boldsymbol{C}_{\mathrm{os}}\boldsymbol{x}_{\mathrm{os}}(t) \\ \left.\begin{array}{l} O_i,\ (\forall i \in \mathcal{N}) \text{ defined in (10.8)} \\ D_i,\ (\forall i \in \mathcal{N}) \text{ defined in (10.13)} \\ C_0,\ \text{defined in (10.15)} \end{array}\right\} \text{generating } \bar{\boldsymbol{A}}_{\mathrm{os}}(t) \text{ with } \mathcal{P}_{\mathrm{C}i}(t) \text{ defined in (10.11).} \end{cases} \tag{10.17}$$

Note that the general model of the self-organizing control system Σ defined in (7.30) and the model Σ defined in (10.17) with the local reconfiguration of communication links only differ in the structure of the observation unit O_i and the decision unit D_i.

10.3 Analysis of the overall fault-tolerant self-organizing control system

In this section the behavior of the self-organizing controlled agents affected by faults within the communication among the local control units is investigated, where the communication structure is locally reconfigured if the fault causes a violation of the performance requirements. A condition for the asymptotic synchronization of the agents is derived in Theorem 10.1 (Section 10.3.1). The behavior of the self-organizing control system Σ after the reconfiguration is analyzed and it is shown that due to the reconfiguration of the communication structure the performance of the fault-free agents remains unchanged compared to a fault-free communication (Section 10.3.2).

10.3.1 Asymptotic synchronization for time-varying faults

This section investigates the asymptotic synchronization of the agents within the self-organizing control system Σ defined in (10.17) (cf. Control aim 7.1). It is shown, that despite the loss of communication links the agents can asymptotically synchronize to the leader Σ_0 if the disconnected agents are not faulty. Therefore, the conditions in Lemma 7.2 for the general self-organizing control system Σ defined in (7.30) are verified for a specific self-organizing control system Σ with a situation-dependent reconfiguration of the communication links.

Theorem 10.1 (Asymptotic synchronization with local reconfiguration of the communication) *Consider the self-organizing control system Σ defined in* (10.17) *for $d_i(t) = 0$ for all $i \in \mathcal{N}$ and all $t \geq 0$. The agents Σ_i, $(\forall i \in \mathcal{N})$, defined in* (7.2) *are asymptotically synchronized to the leader Σ_0 defined in* (7.3) *as desired in Control aim 7.1 if there exists a finite time $\bar{t} \in \mathbb{R}_+$ after which the condition*

$$\{\mathcal{P}_{\mathrm{B}i} \cup \mathcal{P}_{\mathrm{S}i}(t)\} \backslash \mathcal{N}_{\mathrm{F}}(t) \neq \emptyset, \quad \forall i \in \mathcal{N}_{\mathrm{F}}(t), \quad \forall t \geq \bar{t} \geq 0 \tag{10.18}$$

is fulfilled.

Proof. The general self-organizing control system Σ defined in (7.30) and the self-organizing control system Σ defined in (10.17) with local reconfiguration of the communication structure are identical besides the change of the communication. Note that Assumptions 7.1 and 7.3–7.5 also hold for Σ defined in (10.17). Hence, the two conditions Lemma 7.2 have to be fulfilled to proof the asymptotic synchronization of the agents:

1. According to (10.9) and (10.11), the vanishing predecessors $\mathcal{N}_{\mathrm{F}}(t)$ in the set $\mathcal{P}_{\mathrm{C}i}(t)$ of predecessor are always replaced by admissible predecessors $\mathcal{P}_{\mathrm{S}i}(t)$ as long as C_i can transmit information, i.e., $i \notin \mathcal{N}_{\mathrm{F}}(t)$, since C_0 is assumed to be fault-free at all times. Hence, the relation $\mathcal{P}_{\mathrm{C}i}(t) = \emptyset$ can only hold true if C_i is faulty, i.e., $i \in \mathcal{N}_{\mathrm{F}}(t)$, and all possible predecessors C_j, $(\forall j \in \mathcal{P}_{\mathrm{B}i} \cup \mathcal{P}_{\mathrm{S}i}(t))$, are also faulty, i.e., $\mathcal{P}_{\mathrm{B}i} \cup \mathcal{P}_{\mathrm{S}i}(t) \subseteq \mathcal{N}_{\mathrm{F}}(t)$.

2. The second condition in Lemma 7.2 is fulfilled since $\boldsymbol{d}(t) = \mathbf{0}$ holds true for all $t \geq 0$.

□

Theorem 10.1 shows that the property of asymptotic synchronization does not get lost immediately if an agent Σ_i gets faulty. Only if all predecessors, which were active at the time at which

Σ_i gets faulty, are faulty or become faulty, the agents can not synchronize to the leader Σ_0 anymore. As soon as either Σ_i or one of its predecessors becomes fault-free, then Σ_i gets connected with the leader Σ_0 again. Hence, the following corollary directly follows from Theorem 10.1.

Corollary 10.1 (Asymptotic synchronization with local reconfiguration of the communication for vanishing faults) *Consider the self-organizing control system Σ defined in (10.17) for $d_i(t) = 0$ for all $i \in \mathcal{N}$ and all $t \geq 0$. The agents Σ_i, $(\forall i \in \mathcal{N})$, defined in (7.2) are asymptotically synchronized to the leader Σ_0 defined in (7.3) as desired in Control aim 7.1 if there exists a finite time $\bar{t} \in \mathbb{R}_+$ after which the condition*

$$\mathcal{N}_{\mathrm{F}}(t) = \emptyset, \quad \forall t \geq \bar{t} \geq 0 \tag{10.19}$$

is fulfilled.

Proof. Relation (10.19) implies (10.18) since for $\mathcal{N}_{\mathrm{F}}(t) = \emptyset$ there is no index i for which the relation $\{\mathcal{P}_{\mathrm{B}i} \cup \mathcal{P}_{\mathrm{S}i}(t)\} \backslash \mathcal{N}_{\mathrm{F}}(t) \neq \emptyset$ has to be tested. □

Corollary 10.1 states that if all agents become fault-free again, an asymptotic synchronization is guaranteed. Nevertheless, even if the faults vanish eventually, the control errors $y_{\Delta i}(t)$ might be very large beforehand. An analysis of these errors for the situation-dependent reconfiguration of the communication structure is performed in the following section.

Example 10.1 (cont.) *Communication reconfiguration within a formation of three robots*

Figure 10.6 depicts the behavior of the robot formation for two different fault scenarios. Note that Σ_1 always has the same synchronization performance since C_1 always receives information from C_0 which is assumed to be fault-free at all times.

Scenario 1 is similar to the situation depicted in Fig. 10.4, but additionally also C_2 becomes faulty. The left part of Fig. 10.6 shows that the robots asymptotically track the reference signal $y_{\mathrm{s}}(t)$ of the leader Σ_0 even if C_1 and C_2 are simultaneously faulty. Since at first C_1 is faulty ($\mathcal{N}_{\mathrm{F}}(t) = \mathcal{P}_{\mathrm{F2}}(t) = \{1\}$), C_2 choses C_0 as its new predecessor. With this, a faulty transmission unit at C_2 has no effect on the asymptotic synchronization of Σ_2. According to Theorem 10.1, this result is expected since

$$\{\mathcal{P}_{\mathrm{B2}} \cup \mathcal{P}_{\mathrm{S2}}(t)\} \backslash \mathcal{N}_{\mathrm{F}}(t) = \{0\} \neq \emptyset, \quad \forall t \geq 6\,\mathrm{s}.$$

holds true.

In **Scenario 2** at first C_2 is faulty and then C_1 (right part of Fig. 10.6). Since C_2 can not transmit information to C_0, when C_1 becomes faulty, robot Σ_2 loses connection to the overall formation. But since the fault in C_2 vanishes, Σ_2 can track the reference position $s_0(t)$ again after time $t \geq 20\,\mathrm{s}$. This result is expected since

$$\{\mathcal{P}_{\mathrm{B2}} \cup \mathcal{P}_{\mathrm{S2}}(t)\} \backslash \mathcal{N}_{\mathrm{F}}(t) = \{0\} \neq \emptyset, \quad \forall t \geq 20\,\mathrm{s}.$$

is fulfilled according to Theorem 10.1.

In summary, these two scenarios confirm the synchronization condition in Theorem 10.1. □

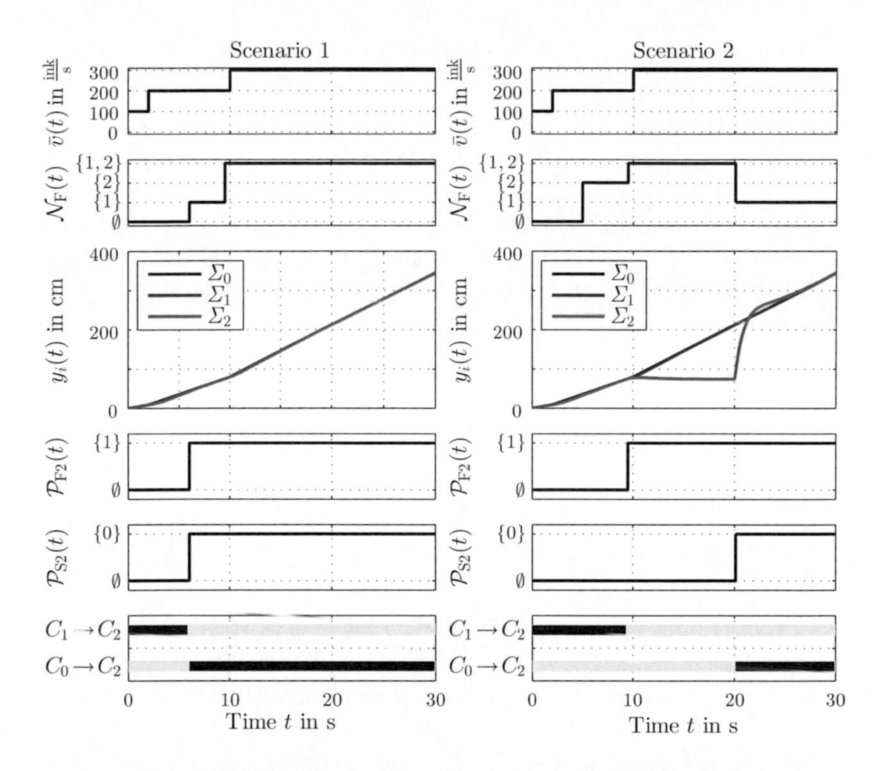

Figure 10.6: Verification of the asymptotic synchronization of the robot formation with local reconfiguration of the communication structure for two different fault scenarios.

10.3.2 Performance after reconfiguration of the communication structure

This section investigates the behavior of the self-organizing controlled agents after a reconfiguration of the communication structure due to faults affecting the local control units. The local reconfiguration procedure of the local control units C_i is assumed to be completed and the set $\mathcal{N}_{\text{F}}$ of faulty agents remains constant, i.e., $\mathcal{N}_{\text{F}}(t) = \mathcal{N}_{\text{F}}^*$ for all $t \geq 0$. Hence, the relations $\mathcal{P}_{\text{F}i}(t) = \mathcal{P}_{\text{F}i}^*$ and $\mathcal{P}_{\text{S}i}(t) = \mathcal{P}_{\text{S}i}^*$ hold true for all $t \geq 0$. Furthermore, the agents are assumed to be initially synchronized with the leader Σ_0.

Consider Control aim 7.4 with the situation that the transformed initial states of all agents Σ_i, $(\forall i \in \mathcal{N})$, are identical according to (7.52) indicating that initially the outputs $y_i(t)$, $(\forall i \in \mathcal{N})$, of the agents are identical as well (7.32). Furthermore, the initial state $\boldsymbol{x}_{\text{s}0}$ of the leader Σ_0 is assumed to be different from the initial state $\boldsymbol{x}_{\text{q}0}$ according to (7.31) and (7.53). With this, the initial state $\tilde{\boldsymbol{x}}^*_{\Delta 0}$ defined in (7.54) represents the difference between the initial state $\boldsymbol{x}_{\mathrm{s}0}$ of the leader Σ_0 defined in (7.3) and the transformed initial state $\tilde{\boldsymbol{x}}_{0i}$ of the agents in (7.52).

The following theorem indicates for which agents the maximal control error remains unchanged compared to the fault-free communication within the basic communication graph $\mathcal{G}_{\mathrm{B}}$.

Theorem 10.2 (Maximal control error with local reconfiguration of the communication structure) *Consider the self-organizing control system Σ defined in* (10.17) *with the initial situation in* (7.52) *and* (7.53). *The control errors $y_{\Delta i}(t)$ of the fault-free agent Σ_i, ($\forall i \in \mathcal{N} \backslash \mathcal{N}_{\mathrm{F}}$), are bounded by the maximal basic control error $\bar{y}_{\Delta \mathrm{B}i}$ and the substitutional communication links are chosen such that the local communication cost is minimal as required in Control aim 7.4:*

$$\max_{t \geq 0} |y_{\Delta i}(t)| \leq \bar{y}_{\Delta \mathrm{B}i}, \quad \forall i \in \mathcal{N} \backslash \mathcal{N}_0, \quad \forall t \geq 0 \tag{10.20}$$

with $\mathcal{N}_0 = \{i \in \mathcal{N}^*_{\mathrm{F}} \mid \mathcal{P}^*_{\mathrm{F}i} = \mathcal{P}_{\mathrm{F}i}(t^-_{\mathrm{F}i})\}$. $t^-_{\mathrm{F}i}$ is the time instance directly before the time $t_{\mathrm{F}i} \leq 0$ at which C_i becomes faulty.

Proof. The choice of the set $\mathcal{P}_{\mathrm{S}i}$ of substitutional predecessors in (10.9) guarantees that the control error $y_{\Delta i}(t)$ with the reconfigured communication is always smaller than or equal to the maximal basic control error $y_{\Delta \mathrm{B}i}$ whenever the corresponding agent Σ_i is not faulty, i.e., $i \notin \mathcal{N}_{\mathrm{F}}(t)$, since (10.9b) and (10.9c) check whether the maximal control error $\bar{y}_{\Delta \mathrm{R}ij}(t)$ with the current set $\mathcal{P}_{\mathrm{CR}ij}(t)$ of reconfigured predecessors fulfills the relation $\bar{y}_{\Delta \mathrm{R}ij}(t) \leq \bar{y}_{\Delta \mathrm{B}i}$. If a new predecessor is necessary (10.9c) also guarantee a substitutional communication link $(j \to i)$ with a minimal communication cost as required in (7.38). Hence, if the set of local faulty agents remains constant for the time at which C_i is faulty, i.e., $\mathcal{P}^*_{\mathrm{F}i} = \mathcal{P}_{\mathrm{F}i}(t^-_{\mathrm{F}i})$, the reconfiguration is successful even if C_i is faulty. With this, Control aim 7.4 is satisfied. □

Theorem 10.2 shows that not only the fault-free agents Σ_i, ($\forall i \in \mathcal{N} \backslash \mathcal{N}_{\mathrm{F}}$), keep the desired maximal control error compared to the fault-free communication, i.e., $\mathcal{G}_{\mathrm{C}}(t) = \mathcal{G}_{\mathrm{B}}$ for all $t \geq 0$. If after a fault in C_i there is no fault within desired substitutional predecessors C_j, ($j \in \mathcal{P}^*_{\mathrm{S}i}$), the reconfigured performance remains unchanged. With this, the following corollary directly follows from Theorem 10.2.

Corollary 10.2 (Maximal control error with local reconfiguration of the communication structure of the fault-free agents) *Consider the self-organizing control system Σ defined in* (10.17) *with the initial situation in* (7.52) *and* (7.53). *The control errors $y_{\Delta i}(t)$ of the fault-free agent Σ_i, ($\forall i \in \mathcal{N} \backslash \mathcal{N}_{\mathrm{F}}$), are bounded by the maximal basic control error $\bar{y}_{\Delta \mathrm{B}i}$ and the substitutional communication links are chosen such that the local communication cost is minimal as* *required in Control aim 7.4:*

$$\max_{t\geq 0} |y_{\Delta i}(t)| \leq \bar{y}_{\Delta \mathrm{B}i}, \quad \forall i \in \mathcal{N}\backslash\mathcal{N}_{\mathrm{F}}^{*}, \quad \forall t \geq 0. \tag{10.21}$$

Proof. Relation (10.20) implies (10.21) since $\mathcal{N}_0 \subseteq \mathcal{N}_{\mathrm{F}}^{*}$ holds true. □

Corollary 10.2 states that at least all agents Σ_i, $(\forall i \in \mathcal{N})$, with a fault-free local control unit C_i preserve their synchronization performance.

Example 10.1 (cont.) *Communication reconfiguration within a formation of three robots*

Consider the fault situations in Fig. 10.6.

For **Scenario 1** according to (10.20) in Theorem 10.2 the control error $y_{\Delta 2}(t)$ of robot Σ_2 is always upper bounded by $\bar{y}_{\Delta \mathrm{B}2} = 5.36\,\mathrm{cm}$ for a stepwise change of the reference velocity $\bar{v}(t)$ by $100\frac{\mathrm{inc}}{\mathrm{s}}$ (left part of Fig. 10.7) since

$$\mathcal{P}_{\mathrm{F}2}(t) = \{1\}, \quad \forall t \geq 6\,\mathrm{s}$$

holds true for analyzing the time $t \geq 10\,\mathrm{s}$ with $\mathcal{P}_{\mathrm{F}2}^{*} = \{1\}$ and $\mathcal{N}_{\mathrm{F}}^{*} = \{1, 2\}$.

Since for **Scenario 2** the set of local faulty predecessors changes from $\mathcal{P}_{\mathrm{F}2}(t) = \emptyset$ to $\mathcal{P}_{\mathrm{F}2}(t) = \{1\}$ after time $t = 5\,\mathrm{s}$ at which C_2 becomes faulty, the maximal basic control error $\bar{y}_{\Delta \mathrm{B}2} = 5.36\,\mathrm{cm}$ is exceeded for a stepwise change of the reference velocity $\bar{v}(t)$ by $100\,\frac{\mathrm{inc}}{\mathrm{s}}$ (cf. right part of Fig. 10.7). □

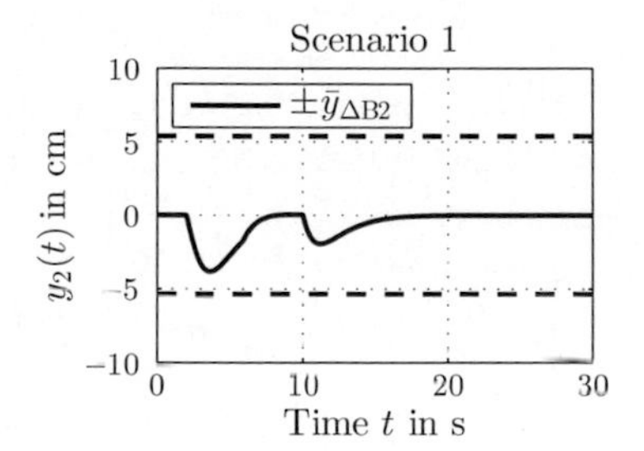

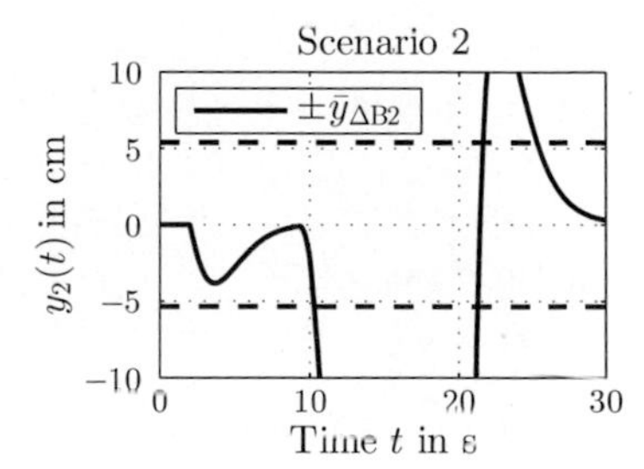

Figure 10.7: Verification of maximal control error $y_{\Delta 2}(t)$ for robot Σ_2.

10.4 Application example: Large robot formation

In this section the proposed self-organizing control concept with a situation-dependent reconfiguration of the communication structure is applied to a robot formation consisting of 21 robots presented in Section 7.7.3. Section 10.4.1 introduces the basic communication graph $\mathcal{G}_{\mathrm{B}}$, the fault situation and the resulting parameters of the local control units C_i. With this, an analysis of the asymptotic synchronization and the adherence to the basic performance of the resulting self-organizing controlled system is performed (Section 10.4.2). These results are compared with the actual behavior of the closed-loop system (Section 10.4.3).

10.4.1 Formation problem and parameters of the local control units

Consider a robot formation with 21 robots controlled within the basic communication graph $\mathcal{G}_{\mathrm{B}}$ depicted in Fig. 10.8. According to Control aim 7.4, the local control errors $y_{\Delta i}(t)$, $(\forall i \in \mathcal{N})$, of the robots Σ_i should not exceed the maximal basic control errors $y_{\Delta \mathrm{B}i}$ listed in Table 10.1 for a stepwise change of the reference velocity $\bar{v}(t)$ by $100\frac{\mathrm{inc}}{\mathrm{s}}$, despite a faulty transmission unit within the local control units C_2, C_4, C_5, C_9 and C_{17}. Recall that the maximal basic control errors are determined as presented in Lemma 7.3.

The local control units C_i defined in (10.14) for the robots Σ_i can be constructed with Algorithm 10.1. Hence, the following four steps are performed to define the feedback unit F_i, the observation unit O_i and the decision unit D_i within C_i:

1. Recall that all **feedback units** F_i, $(\forall i \in \mathcal{N})$, defined in (7.67) have the same feedback gain $k_{\mathrm{P}i} = 15$.

2. The **observation units** O_i, $(\forall i \in \mathcal{N})$, defined in (10.8) can be constructed without any parametrization.

3. The **decision units** D_i, $(\forall i \in \mathcal{N})$, defined in (10.13) are constructed by using $\mathcal{P}_{\mathrm{B}i}$, $\mathcal{F}_{\mathrm{B}i}$ and $\bar{y}_{\Delta \mathrm{B}j}$, $(j = 0, 1, \ldots, i)$, listed in Table 10.1. Furthermore, the communication costs c_{ij}, $(j = 0, 1, \ldots, i-1)$, listed in Table B.16 and the elements k_{ij}, $(j = 0, 1, \ldots, i-1)$, of the weighting matrix $\boldsymbol{K}$ are needed. Recall that all elements k_{ij}, $(\forall i, j \in \mathcal{V}_{\mathrm{C}}, j < i)$, are set to be one, i.e., $k_{ij} = 1$ for all $i, j \in \mathcal{V}_{\mathrm{C}}$ and $j < i$.

4. Recall that the sets $\mathcal{P}_{\mathrm{B}i}$, $(\forall i \in \mathcal{N})$, of basic predecessors and the sets $\mathcal{F}_{\mathrm{B}i}$ of basic followers listed in Table 10.1 are extracted from the basic communication graph $\mathcal{G}_{\mathrm{B}}$ depicted in Fig. 10.8.

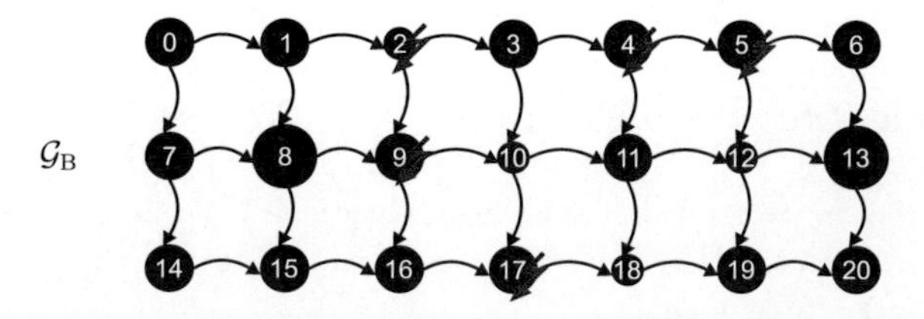

Figure 10.8: Basic communication graph $\mathcal{G}_{\mathrm{B}}$ for a robot formation of 21 robots with individual dynamics affected by faulty transmission units.

Table 10.1: Parameters of the local control units C_i defined in (10.14) for controlling the robot formation using the basic communication graph $\mathcal{G}_{\mathrm{B}}$ depicted in Fig. 10.8 by local reconfiguration of the communication in case of faulty transmission units.

i	$\mathcal{P}_{\mathrm{B}i}$	$\mathcal{F}_{\mathrm{B}i}$	Faulty	Reconf.	$y_{\Delta \mathrm{B}i}$ in cm
0	$\emptyset$	$\{1,7\}$			0.00
1	$\{0\}$	$\{2,8\}$			1.97
2	$\{1\}$	$\{3,9\}$	x		3.38
3	$\{2\}$	$\{4,10\}$		x	6.23
4	$\{3\}$	$\{5,11\}$	x		9.84
5	$\{4\}$	$\{6,12\}$	x	x	14.39
6	$\{5\}$	$\{13\}$		x	20.14
7	$\{0\}$	$\{8,14\}$			1.97
8	$\{1,7\}$	$\{9,15\}$			6.25
9	$\{2,8\}$	$\{10,16\}$	x	x	8.05
10	$\{3,9\}$	$\{11,17\}$		x	9.37
11	$\{4,10\}$	$\{12,18\}$		x	14.09
12	$\{5,11\}$	$\{13,19\}$		x	17.60
13	$\{6,12\}$	$\{20\}$			30.06
14	$\{7\}$	$\{15\}$			6.25
15	$\{8,14\}$	$\{16\}$			9.86
16	$\{9,15\}$	$\{17\}$		x	13.27
17	$\{10,16\}$	$\{18\}$	x		16.26
18	$\{11,17\}$	$\{19\}$		x	18.68
19	$\{12,18\}$	$\{20\}$			24.87
20	$\{13,19\}$	$\emptyset$			36.63

With this, the local control units C_i, $(\forall i \in \mathcal{N})$, for controlling the robots within the formation are completely defined. In the following the robots controlled by these local control units C_i are denoted as self-organizing controlled robots Σ.

10.4.2 Analysis of the behavior of the self-organizing controlled robot formation

In the following the asymptotic synchronization of the robots (Control aim 7.1) and the preservation of the maximal control error (Control aim 7.4) is analyzed. Table 10.1 shows that all local control units that lose a predecessor within its set $\mathcal{P}_{\mathrm{B}i}$ (highlighted in red in Table 10.1) except C_5 can compensate the missing information by local reconfiguration of their communication structure since they are not affected by a fault. Hence, the only critical situation is a fault within C_5 whose only basic predecessor C_4 also might be faulty. There exist different scenarios, e.g.:

1. If either C_4 or C_5 stays fault-free after some time $\bar{t}$, then
   - asymptotic synchronization of all robots is guaranteed since condition (10.18) in Theorem 10.1 holds true.
   - the control error $y_{\Delta 5}(t)$ might exceed the maximal basic control error $\bar{y}_{\Delta \mathrm{B}5}$ since the set $\mathcal{P}_{\mathrm{F}5}(t)$ of local faulty predecessors might change after the time instance $t_{\mathrm{F}5}^-$ directly before the time $t_{\mathrm{F}5}$ at which C_5 becomes faulty (cf. Theorem 10.2).
2. If C_2 and C_4 stay faulty before C_5 becomes faulty, then
   - asymptotic synchronization of all robots is guaranteed since condition (10.18) in Theorem 10.1 holds true.
   - the control error $y_{\Delta 5}(t)$ is smaller than the maximal basic control error $\bar{y}_{\Delta \mathrm{B}5}$ since the set $\mathcal{P}_{\mathrm{F}5}(t)$ of local faulty predecessors does not change after the time instance $t_{\mathrm{F}5}^-$ as required in Theorem 10.2.
3. If C_5 is faulty and stays faulty before C_4 becomes and stays faulty, then
   - asymptotic synchronization of all robots is not guaranteed since condition (10.18) in Theorem 10.1 does not hold true.
   - the control error $y_{\Delta 5}(t)$ might exceed the maximal basic control error $\bar{y}_{\Delta \mathrm{B}5}$ since the set $\mathcal{P}_{\mathrm{F}5}(t)$ of local faulty predecessors changes after the time instance $t_{\mathrm{F}5}^-$ (cf. Theorem 10.2).

There are multiple other cases depending on the times at which C_2, C_4 and C_5 become faulty or become fault-free again.

In summary, only if C_4 becomes and remains faulty before C_5 becomes and remains faulty, then there is no asymptotic synchronization of robot Σ_5 (cf. Theorem 10.1). Furthermore, all robots keep their synchronization performance except perhaps Σ_5, i.e., the local control errors $y_{\Delta i}(t)$, $(\forall i \in \mathcal{N} \backslash \{5\})$, are smaller than the maximal basic control errors $\bar{y}_{\Delta \mathrm{B}i}$.

10.4.3 Behavior of the self-organizing controlled robot formation

This section verifies the analysis results in the previous section by evaluating a simulation of the self-organizing controlled robots for a specific fault situation.

In Fig. 10.9 the behavior of the robot formation using the local control units C_i, $(\forall i \in \mathcal{N})$, specified in Section 10.4.1 is depicted. The topmost plot shows the reference velocity $\bar{v}(t)$

of the leading robot Σ_0. The second plot depicts the course of the set $\mathcal{N}_{\mathrm{F}}(t)$ of faulty agents which changes over time. The third plot displays the position $y_i(t) = s_i(t)$ of the robots while using the self-organizing controller C (solid line) and a corresponding controller with a fixed communication graph $\mathcal{G}_{\mathrm{C}}(t) = \mathcal{G}_{\mathrm{B}}$ with no reconfiguration of the communication structure (red dashed dotted line). The fourth plot shows the control error $y_{\Delta i}(t)$ of the robots using the self-organizing controller C in case of faults (black solid line) and no faults (gray dashed line) affecting the local control units C_i. The transmission of information among the local control units is shown in the lowest plot. The black bars indicate at which time the communication links $(i \rightarrow j)$ are in the set $\mathcal{E}_{\mathrm{C}}(t)$ of communicational edges which means that C_i transmits information to C_j. Note that the new communication links resulting from the reconfiguration are highlighted in red. Since there are 210 possible communication links, the lowest plot only shows the links that are activated during runtime. The thin red vertical lines refer to the corresponding communication graphs $\mathcal{G}_{\mathrm{C}}(t)$ for different times shown in Fig. 10.10.

Figure 10.10 shows that the faults are compensated locally by an adaption of the communication links of affected agents only. If, for example, C_2 is faulty, only C_3 and C_9 adapt their predecessors to retain their desired synchronization performance (Figure 10.10 (a)). The other local control units do not even know that C_2 is faulty. For a fault at C_{17} there is no need to receive the position of another robot by C_{18} since position $s_{11}(t)$ is sufficient to retain the desired control error (Figure 10.10 (e)). An important property of the used local control units is that for vanishing faults, the reconfiguration of the communication graph $\mathcal{G}_{\mathrm{C}}(t)$ is made undone to achieve the basic performance again.

Verification of the control aims. Figure 10.9 shows that the self-organizing controller C leads to a much better behavior compared to the usage of a controller with no reconfiguration of the communication structure. In particular, the robots asymptotically synchronize to the position of the leading robot (third plot). Furthermore, Fig. 10.11 shows that the maximal control error $\max_{t\geq 0} |y_{\Delta i}(t)|$ of all robots is similar to its value for no faults affecting the local control units. As a matter of fact the control errors $y_{\Delta i}(t)$ are smaller than in the case of no faults since generally the substitutional predecessors are closer to the leading robot and, therefore, have a better performance. The drawback is that generally the communication cost from such agents is larger.

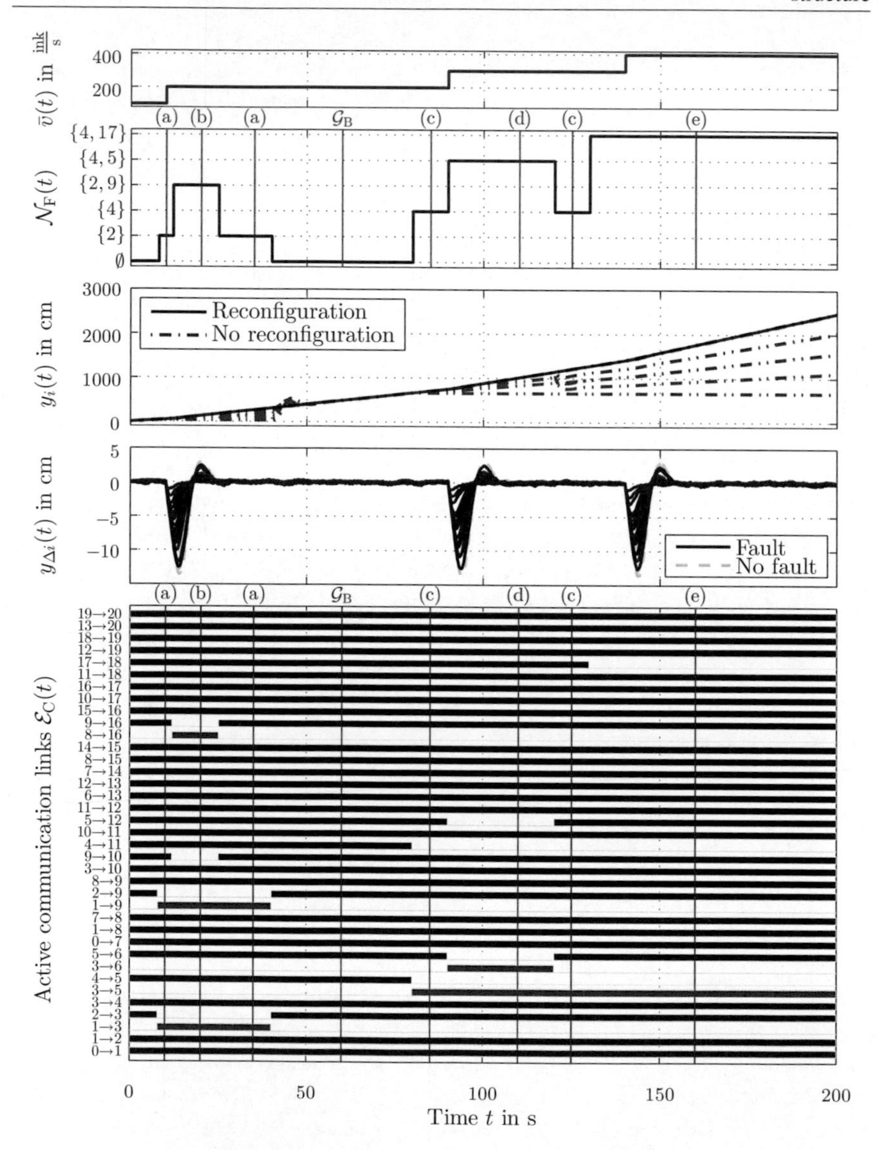

Figure 10.9: Behavior of the robot formation controlled by the self-organizing controller C defined in (10.16) with the parameters presented in Section 10.4.1.

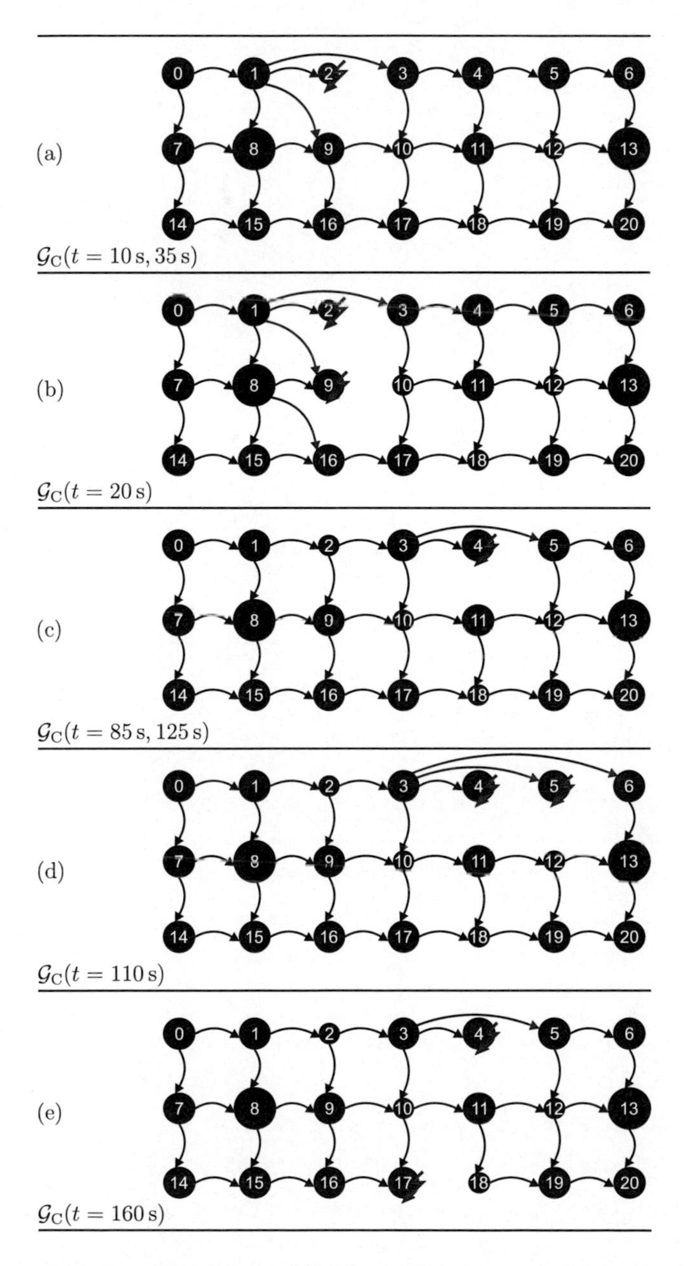

Figure 10.10: Communication graph $\mathcal{G}_\mathrm{C}(t)$ at certain times for different fault situations in Fig. 10.9 (faulty agents have a red lightning).

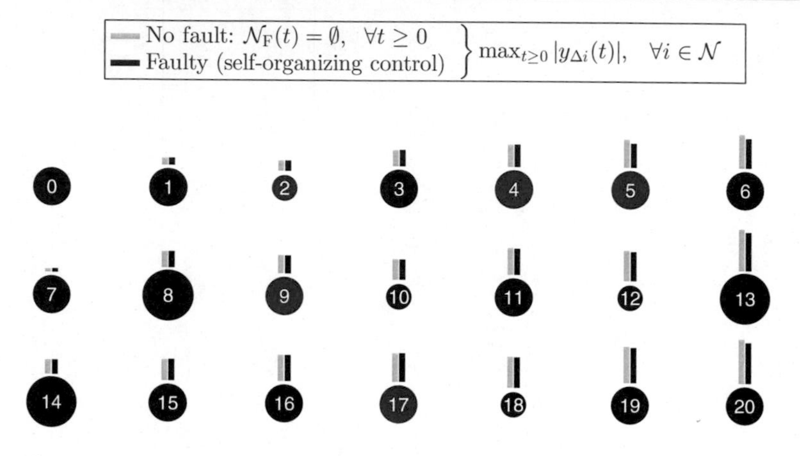

Figure 10.11: Maximal control error $\max_{t\geq 0} |y_{\Delta i}(t)|$ of the self-organizing controlled robots in case of faults (black line) and no faults (gray line) affecting the local control units C_i w.r.t the behavior depicted in Fig. 10.9.

10.5 Summary and evaluation of the proposed control concept

The main properties of the self-organizing control concept with situation-dependent reconfiguration of the communication structure for the control of multi-agent systems affected by faults within the communication can be summarized as follows:

- The compensation of missing communication links by admissible new links based on a local reconfiguration by the local control units enables to retain the desired basic synchronization performance of the closed-loop system.
- The maximal basic control error remains for the fault-free agents (Theorem 10.2 and Corollary 10.2).
- In many cases asymptotic synchronization to the leader can also be achieved by agents with faulty local control units. (Theorem 10.1 and Corollary 10.1).

In summary, the proposed self-organizing control concept is applicable for multi-agent systems which are expected to lose communication links, but shall not use redundant communication links all over the time.

Properties of self-organization. The claimed properties of a self-organizing control system in Section 1.2 for the proposed control concept are verified as follows:

- **Flexibility.** The communication structure is adapting to the current faults affecting the local control units. The adaption takes place in the part of the communication graph which is affected by the fault. No overall redesign of the communication structure is performed. For vanishing faults the reconfiguration of the communication graph is made undone, to keep the properties of the basic communication whenever this is possible.

- **Scalability.** The local control units C_i do not include any model information from other agents. For the reconfiguration, the maximal basic control error $\bar{y}_{\Delta \mathrm{B}j}$, $(j \leq i)$ of possible new predecessors as well as the communication cost c_{ij} to receive information from this predecessors have to be known. Furthermore, the local control units C_i need information about the basic communication graph $\mathcal{G}_\mathrm{B}$ and the bounds $m_{\mathrm{yy}i}$ and $m_{\mathrm{y}\epsilon\mathrm{B}i}$ on dynamics of the controlled agent $\bar{\Sigma}_i$, i.e.,

$$\left.\begin{array}{r} \mathcal{G}_\mathrm{B} \Rightarrow \mathcal{P}_{\mathrm{B}i} \text{ and } \mathcal{F}_{\mathrm{B}i} \\ \bar{\Sigma}_i \Rightarrow m_{\mathrm{yy}i} \text{ and } m_{\mathrm{y}\epsilon\mathrm{B}i} \\ \bar{y}_{\Delta\mathrm{B}j}, j \leq i \\ c_{ij}, j < i \end{array}\right\} \text{Parameters of } C_i, \forall i \in \mathcal{N}.$$

  Hence, the complexity of the local control units scales linearly with the number of subsystems.

- **Fault tolerance.** The self-organizing controller leads to a closed-loop system which prevents the propagation of local faults within the overall system. If the transmission unit of a specific local control unit becomes faulty, a local reconfiguration of the communication structure is sufficient to preserve the desired performance of the overall system. Hence, the resulting closed-loop system is a fault-tolerant system.

In summary, the proposed self-organizing control concept shows all desired self-organizing properties.

Note that the presented structure of the local control units can be combined with the concept of situation-dependent information request (Chapter 8) or the concept of situation-dependent communication interruption due to disturbance (Chapter 9). This combination is discussed in more detail in Chapter 11.

11 Comparison of the control concepts for multi-agent systems

This section compares the main properties of the three control concepts for multi-agent systems presented in Chapters 8–10. Table 11.1 gives an overview about the most important aspects.

Main aim and basic idea. The concepts have different aims. The concept with the situation-dependent request of information aims to guarantee a good synchronization performance and a low communication effort (cf. Chapter 8). In the following this concept is denoted as *first concept*. The aim of the concept presented in Chapter 9 is to bound the disturbance propagation within the overall system. This concept is denoted as *second concept*. The *third concept* presented in Chapter 10 aims to preserve the synchronization performance while using the basic communication graph $\mathcal{G}_{\rm B}$ despite the vanishing of communication links due to faulty transmission units within the local control units C_i.

Due to the different aims, every concept has a different basic idea to achieve these aims. The calculation of the local reference signal $y_{{\rm s}i}(t)$ which has to be tracked by the controlled agents $\bar{\Sigma}_i$ is equal for every concept. The main difference between the concepts is the determination of the set $\mathcal{P}_{{\rm C}i}(t)$ of predecessors. Within the first concept the local control unit C_i **requests** the output $y_j(t)$ from its adjusted predecessors C_j, $(\forall j \in \mathcal{P}_{{\rm A}i})$, whenever its local control error exceeds the switching threshold ϵ_i in order to improve its local reference signal $y_{{\rm s}i}(t)$ compared to the basic communication. The local control units C_j of the second concept **interrupt** the transmission of the output $y_j(t)$ to its follower C_i, $(j \in \mathcal{P}_{{\rm B}i})$, whenever the disturbance impact $y_{{\rm d}j}(t)$ exceeds the threshold ϵ in order to bound the propagation of the disturbance $d_j(t)$ within the overall system. For the third concept the local control units C_i **compensate** missing/faulty predecessors C_j, $(\mathcal{N}_{\rm F}(t))$, by admissible new predecessors C_p, $(p \in \mathcal{P}_{{\rm S}i}(t))$, whenever the remaining predecessors are not sufficient to preserve the desired basic synchronization performance.

Hence, the basis of all three concepts is a basic communication graph $\mathcal{G}_{\rm B}$ which is generally assumed to be given (Fig. 11.1). The first concept adds new communication links to $\mathcal{G}_{\rm B}$, whenever this is appropriate to improve the synchronization performance. The second concept switches off basic communication links in $\mathcal{G}_{\rm B}$, whenever it is appropriate to bound the propagation of a disturbance. Compared to the first two concepts, in the third concept the local control units are forced to add new communication links, whenever a missing/faulty communication link occurs.

Table 11.1: Comparison of the three control concepts for multi-agent systems.

Concept	1. Performance improvement by requesting additional information	2. Disturbance attenuation by communication interruption	3. Performance preservation by local reconfiguration of the communication structure
Idea	Situation-dependent usage of communication links from a graph with a good synchronization performance.	Situation-dependent interruption of the transmission of $y_i(t)$ for large disturbance impacts $y_{\mathrm{d}i}(t)$ affecting agent Σ_i.	Situation-dependent compensation of missing/faulty communication links by admissible new communication links.
Feedback unit F_i	Tracks local reference signal $y_{\mathrm{s}i}(t)$: $u_i(t) = -k_{\mathrm{P}i}(y_i(t) - y_{\mathrm{s}i}(t))$		
Observation unit O_i	Determines local control error $e_i(t) = y_{\mathrm{s}i}(t) - y_i(t)$	Reconstructs disturbance impact $y_{\mathrm{d}i}(t)$ on agent Σ_i by using an observer	Detects faulty predecessors $\mathcal{P}_{\mathrm{F}i}(t)$ by evaluating received information
Decision unit D_i	Adapts set of predecessors		
	$\mathcal{P}_{\mathrm{C}i}(t) = \begin{cases} \mathcal{P}_{\mathrm{A}i} & \text{if } \lvert e_i(t)\rvert \geq \epsilon_i \\ \mathcal{P}_{\mathrm{B}i} & \text{else,} \end{cases}$	$\mathcal{P}_{\mathrm{C}i}(t) = \mathcal{P}_{\mathrm{B}i} \backslash \{j \mid \lvert y_{\mathrm{d}j}(t)\rvert > \epsilon\}$	$\mathcal{P}_{\mathrm{C}i}(t) = \{\mathcal{P}_{\mathrm{B}i} \cup \mathcal{P}_{\mathrm{S}i}(t)\} \backslash \mathcal{N}_{\mathrm{F}}(t)$
	Adapts local reference signal $y_{\mathrm{s}i}(t)$ depending on the received outputs $y_j(t)$, $(j \in \mathcal{P}_{\mathrm{C}i}(t))$: $y_{\mathrm{s}i}(t) = \sum_{j\in\mathcal{P}_{\mathrm{C}i}(t)} \tilde{k}_{ij}(t) y_j(t)$		
$\underline{\mathcal{E}}_{\mathrm{C}} =$	$\mathcal{E}_{\mathrm{B}} \cap \mathcal{E}_{\mathrm{A}}$	$\{(0 \to i) \mid i \in \mathcal{F}_{\mathrm{B0}}\}$	
$\bar{\mathcal{E}}_{\mathrm{C}} =$	$\mathcal{E}_{\mathrm{B}} \cup \mathcal{E}_{\mathrm{A}}$	$\mathcal{E}_{\mathrm{B}}$	$\{(j \to i) \mid i, j \in \mathcal{V}_{\mathrm{C}},\ j < i\}$
Asymptotic synchronization	Is always fulfilled (Theorem 8.1)	If $d_i(t) = 0,\ \forall i \in \mathcal{N},\ \forall t \geq \bar{t} \geq 0$ (Theorem 9.1)	If $\{\mathcal{P}_{\mathrm{B}i} \cup \mathcal{P}_{\mathrm{S}i}(t)\} \backslash \mathcal{N}_{\mathrm{F}}(t) \neq \emptyset$, $\forall i \in \mathcal{N}_{\mathrm{F}}(t),\ \forall t \geq \bar{t} \geq 0$ (Theorem 10.1)
Control error	Adjustable by $\mathcal{G}_{\mathrm{A}}$ and ϵ_i (Theorem 8.2, Algorithm 8.2)	Arbitrary adjustable for undisturbed agents (Theorem 9.2, Theorem 9.3)	Basic error remains for fault-free agents (Theorem 10.2, Corollary 10.2)
Properties of self-organization	-Flexibility ✓ -Scalability✓ -Fault-tolerance	-Flexibility ✓ -Scalability✓ -Fault-tolerance	-Flexibility ✓ -Scalability✓ -Fault-tolerance✓

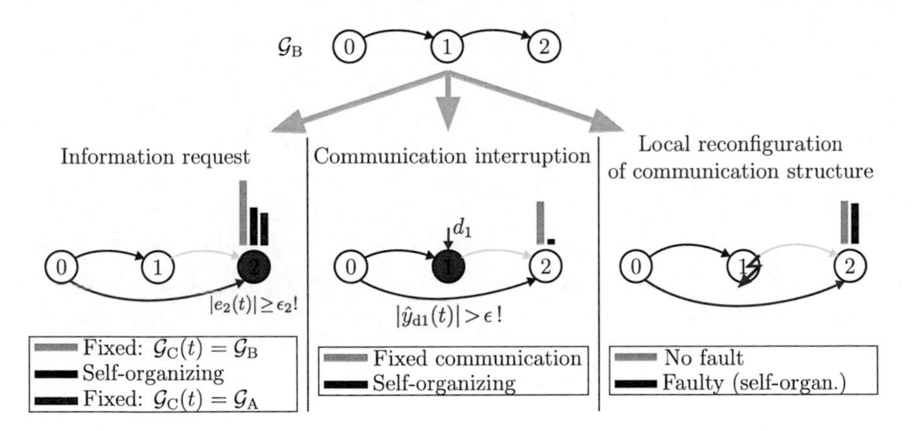

Figure 11.1: Comparison of the main aims and the basic ideas of the control concepts for multi-agent systems, where the bars indicate maximal control error $\max_{t \geq 0} |y_{\Delta 2}(t)|$ w.r.t. Example 8.1, Example 9.1 and Example 10.1.

With this, the components of the local control units have different tasks which are compared in the following.

Tasks of the local control units. The feedback units F_i, $(\forall i \in \mathcal{N})$, defined in (7.22) are identical in all three concepts, in which their task is to track the local reference signal $y_{si}(t)$ determined by the decision unit D_i.

Since the situation to be detected is different for all three desired aims, the observation units O_i have different tasks. For the first concept, O_i just calculates the local control error $e_i(t) = y_i(t) - y_{si}(t)$ which is a measure for the actual control error $y_{\Delta i}(t)$. Within the second concept O_i reconstructs the actual disturbance impact $y_{di}(t)$ by using an observer with a specific choice of the observer gain, i.e., $\boldsymbol{l}_i = k_{Pi}\boldsymbol{b}_i$. The observation unit O_i within the third concept detects the currently faulty predecessors $\mathcal{P}_{Fi}(t) \subseteq \mathcal{N}_F(t)$.

The decision units D_i request information – first concept – or interrupt the communication – second concept–, whenever the local control error $e_i(t)$ or the reconstructed disturbance impact $\hat{y}_{di}(t)$ exceeds a switching threshold, respectively. For the third concept, there exists no switching threshold since the change of the set $\mathcal{P}_{Fi}(t)$ faulty predecessors is discrete. The local performance requirement is checked for every change of $\mathcal{P}_{Fi}(t)$ and, if necessary, new predecessors are chosen.

Communication structure. Recall that the set $\mathcal{E}_C(t)$ of communication links within the communication graph $\mathcal{G}_C(t)$ has an upper limit and a lower limit for all presented control concepts (cf. (2.25) in Section 2.3). The lower limit, namely the set $\underline{\mathcal{E}}_{\mathrm{C}}$ of minimal communication links within the minimal communication graph $\underline{\mathcal{G}}_{\mathrm{C}} = (\mathcal{V}_{\mathrm{C}}, \underline{\mathcal{E}}_{\mathrm{C}})$, is identical for the second and the third concept, i.e., $\underline{\mathcal{E}}_{\mathrm{C}} = \{(0 \to i) \mid i \in \mathcal{F}_{\mathrm{B0}}\}$. This set $\underline{\mathcal{E}}_{\mathrm{C}}$ shows that in the worst case only the communication links with the leader Σ_0 remain unchanged since Σ_0 is assumed to be undisturbed and fault-free. The lower limit $\underline{\mathcal{E}}_{\mathrm{C}}$ for the first concept is the intersection of the set $\mathcal{E}_{\mathrm{B}}$ of basic communication links and the set $\mathcal{E}_{\mathrm{A}}$ of adjusted communication links, i.e., $\underline{\mathcal{E}}_{\mathrm{C}} = \mathcal{E}_{\mathrm{B}} \cap \mathcal{E}_{\mathrm{A}}$. Furthermore, the corresponding upper limit, that is, the set $\bar{\mathcal{E}}_{\mathrm{C}}$ of maximal communication links within the maximal communication graph $\bar{\mathcal{G}}_{\mathrm{C}} = (\mathcal{V}_{\mathrm{C}}, \bar{\mathcal{E}}_{\mathrm{C}})$, is the union of both sets $\bar{\mathcal{E}}_{\mathrm{C}} = \mathcal{E}_{\mathrm{B}} \cup \mathcal{E}_{\mathrm{A}}$. Since for the second concept communication links within $\mathcal{E}_{\mathrm{B}}$ can only vanish due to the communication interruption, $\bar{\mathcal{E}}_{\mathrm{C}}$ is equal to $\mathcal{E}_{\mathrm{B}}$. For the reconfiguration of the communication structure within the third concept all possible communication links can be used, i.e., $\bar{\mathcal{E}}_{\mathrm{C}} = \{(j \to i) \mid i, j \in \mathcal{V}_{\mathrm{C}},\ j < i\}$.

Comparison of the main results. The following properties of the self-organizing system Σ resulting from the three control concepts are compared:

- **Asymptotic synchronization** (Control aim 7.1): There are no additional requirements on the ability of asymptotic synchronization compared to the control within a fixed communication structure. For the first concept, asymptotic synchronization of the agents Σ_i with the leader is always fulfilled, where all agents are assumed to be undisturbed (Theorem 8.1). Asymptotic synchronization with a fixed communication structure can not be achieved if some agent Σ_i is affected by a non-vanishing disturbance $d_i(t)$. The same requirement holds for the second concept (Theorem 9.1). Hence, this requirement is weak. The requirement for the third concept is weak since for certain missing/faulty communication links no asymptotic synchronization is guaranteed without reconfiguration of the communication structure (Theorem 10.1).

- **Maximal control error** (Control aim 7.2, Control aim 7.3, Control aim 7.4): For the first concept the maximal control error of the agents can be adjusted by the choice of the adjusted communication graph $\mathcal{G}_{\mathrm{A}}$ and the switching thresholds ϵ_i, $(\forall i \in \mathcal{N})$ (Theorem 8.2), where Algorithm 8.2 presents a method for designing $\mathcal{G}_{\mathrm{A}}$ and ϵ_i. Due to large disturbances affecting the agents Σ_i within the second concept, a desired maximal control $\bar{y}^*_\Delta$ error can only be guaranteed for the undisturbed agents Σ_j, $(\forall i \in \mathcal{U})$. $\bar{y}^*_\Delta$ can be arbitrarily adjusted by the choice of the threshold ϵ (Theorem 9.2, Theorem 9.3). Despite missing/faulty communication links, concept three with the reconfiguration of the communication structure can preserve the maximal basic control error $\bar{y}_{\Delta \mathrm{B}i}(t)$ of the agents with a the fault-free local control unit. Fig. 11.1 illustrates these properties for the maximal control error $\max_{t \geq 0} |y_{\Delta 2}(t)|$ w.r.t. Example 8.1, Example 9.1 and Example 10.1.

Properties of self-organization.

- **Flexibility**: All three concepts are adapting the communication among the local control units to the current situation to guarantee a desired performance. These situations are the current local control error $e_i(t)$ (first concept), the current difference impact $y_{\mathrm{d}i}(t)$ (second concept) and for the third concept the currently faulty local control units C_i, $(\forall i \in \mathcal{N}_{\mathrm{F}})$.
- **Scalability**: The concepts are applicable for systems with a large number of subsystem (complex systems) since the local control units C_i include local model information only. Hence, new agents can be easily integrated into a present system. The complexity of the certain local control units scales linearly with the number of agents for all three concepts.
- **Fault-tolerance**: The first and the second concept have no fault-tolerant properties. The third concept prevents the propagation of local faults within the overall system. If the transmission unit of a specific local control unit becomes faulty, a local reconfiguration of the communication structure is sufficient to retain the desired performance of the overall system.

In summary, all three concepts show characteristic properties of self-organizing systems where the first and second concept have no fault-tolerant properties. To obtain these properties, the first and second concept can be combined with the third concept which is briefly discussed in the following.

Combination of the concepts. Consider that the concept performing information requests to improve the synchronization performance (first concept) and the concept performing communication interruptions to bound the disturbance propagation (second concept) shall also compensate missing/faulty communication links. In the following the combination of the first concept and the second concept with the concept performing a reconfiguration of the communication structure (third concept) is discussed:

- **Information requests & reconfiguration of communication structure**: The combination of the concept which requests information and the concept with a local reconfiguration of the communication structure is presented in [7]. Since the local control units switch between the links within the basic communication graph $\mathcal{G}_{\mathrm{B}}$ and the adjusted communication graph $\mathcal{G}_{\mathrm{A}}$, both graphs have to be reconfigured in case of a missing/faulty communication link (Fig. 11.2). Hence, $\mathcal{G}_{\mathrm{B}}(t)$ and $\mathcal{G}_{\mathrm{A}}(t)$ become time-variant. In Fig. 11.2 a red arrow indicates a reconfigured communication link and a green arrow an information request. Each local control unit C_i has to detect its faulty predecessors which depend on the current request of information. Furthermore, C_i has to perform two parallel reconfigurations, one for $\mathcal{G}_{\mathrm{B}}(t)$ and one for $\mathcal{G}_{\mathrm{A}}(t)$. Note that a faulty predecessor does not necessarily lead to a reconfiguration of $\mathcal{G}_{\mathrm{B}}(t)$ and $\mathcal{G}_{\mathrm{A}}(t)$ since the set $\mathcal{P}_{\mathrm{B}i}$ of basic predecessors and the set $\mathcal{P}_{\mathrm{A}i}$ of adjusted predecessors might include different indices.

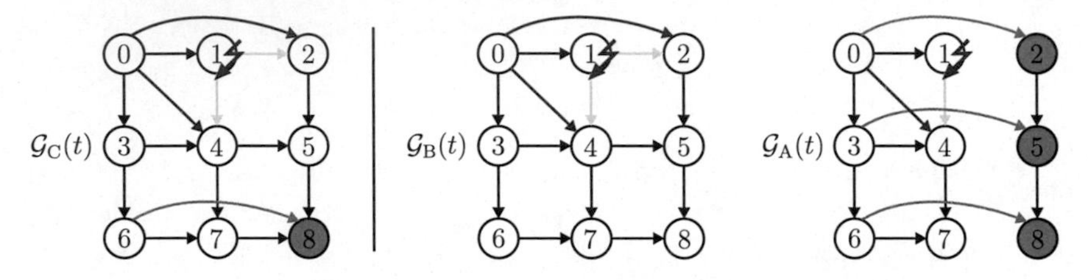

Figure 11.2: Combination of the concept performing information requests with the concept of reconfiguration of the communication structure.

- **Communication deactivation & reconfiguration of communication structure**: The combination of the concept of interrupting the communication due to large disturbance impacts and the concept of a local reconfiguration of the communication structure solve two problems simultaneously (Fig. 11.3). First, faulty communication links are compensated by admissible new communication links to preserve the synchronization performance while using the fault-free basic communication graph $\mathcal{G}_{\mathrm{B}}$. Second, since the deactivation of the communication to bound the disturbance propagation is similar to faulty communication links, the reconfiguration of the communication structure makes no difference between a faulty link or an interrupted link. Hence, due to the reconfiguration, the agents are always connected with the leader Σ_0 no matter whether a disturbance occurs or not. Hence, asymptotic synchronization is guaranteed.

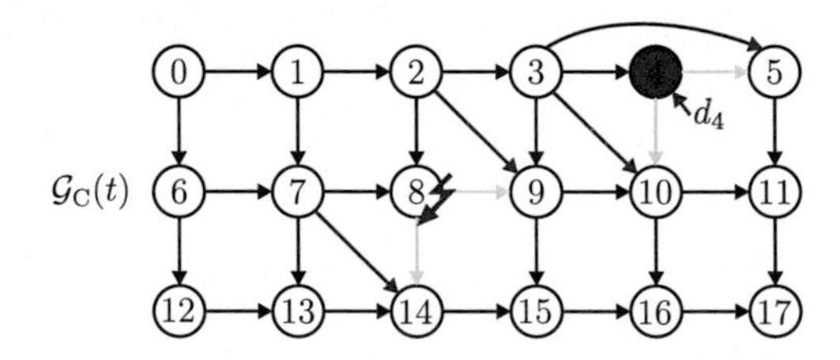

Figure 11.3: Combination of the concept performing communication interruptions with the concept of reconfiguration of the communication structure.

The opportunity to combine the concepts is a result of the strict allocation of the local control units C_i into a feedback unit F_i, an observation unit O_i and a decision units D_i, which therefore can be easily combined with the components of other concepts.

Part III

Conclusion

12 Summary and comparison of the developed control concepts

A general structure for local control units C_i that perform a situation-dependent information exchange in order to guarantee a desired behavior of the overall closed-loop system has been developed. The local control units C_i are divided into three components: the feedback unit F_i realizes a local feedback to guarantee basic performance requirements, the observation unit O_i detects the current situation of the subsystems and the decision unit D_i decides about the communication exchange based on locally available information. Hence, there exists no central control unit. Five concepts which are aimed to control physically interconnected systems as well as multi-agent systems have been presented, where situations like disturbances or faults affecting the subsystems and set-point changes are investigated.

General comparison of the developed control concepts. The control concepts are developed to adapt the communication among the local control units C_i with respect to specific situations (Fig. 12.1):

- **Disturbances** require different strategies for adapting the communication structure to guarantee a desired performance. For controlling physically interconnected subsystems the local control units C_i of the disturbed subsystems Σ_i transmit information to local control units of neighboring subsystems to either bound the disturbance propagation (Chapter 4) or to mimic the behavior of a corresponding centralized control system (Chapter 5). For controlling multi-agent systems the communication to basic followers is interrupted to bound the disturbance propagation (Chapter 9).

- The propagation of the effect of a **fault** within physically interconnected systems can be bounded by the transmission of locally determined coupling inputs to the local control units of neighboring subsystems in order to compensate large coupling impacts caused by the fault (Chapter 4). Vanishing communication links within multi-agent systems due to faults within a transmission unit are compensated by choosing appropriate new links at the local control units C_i which have lost required information (Chapter 10).

- In order to improve the synchronization performance of a multi-agent system with a basic communication in case of a **set-point change** at the leader, the local control units C_i request additional information by using communication links from an adjusted communication structure (Chapter 8). A set-point change or a disturbance forces a subsystem out

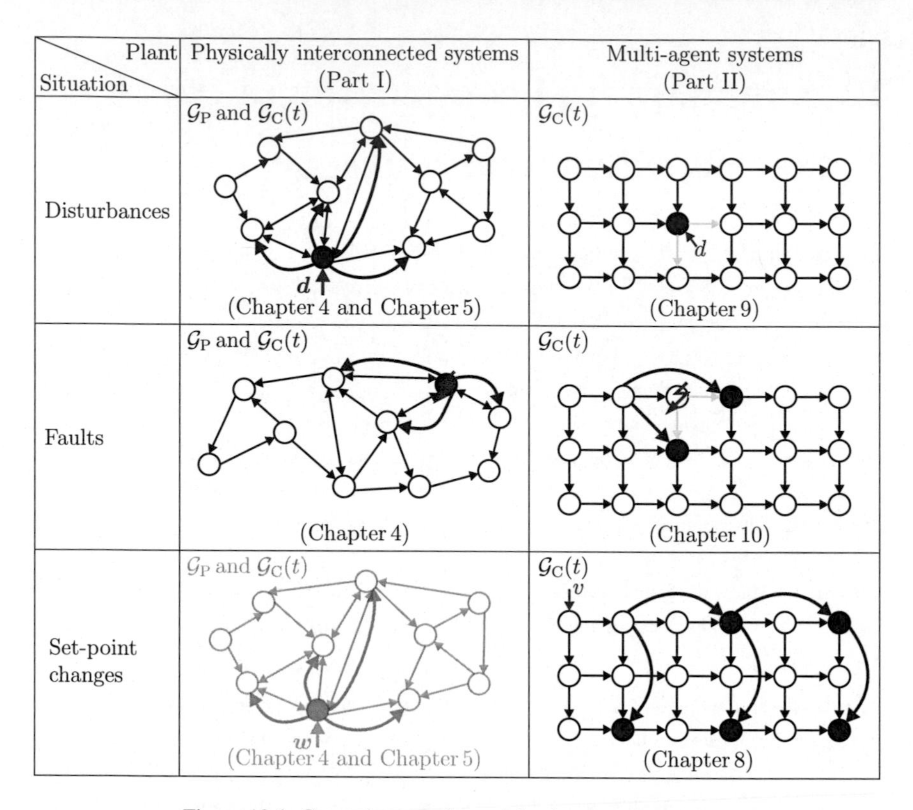

Figure 12.1: General comparison of the developed concepts.

of its equilibrium point, hence, the concepts in Chapter 4 and Chapter 5 can also be used to bound the propagation of the effect of a set-point change on the neighboring subsystems (Chapter 4) or to mimic the behavior of a corresponding centralized control system in case of a set-point change (Chapter 5). Recall that this situation is not investigated in this thesis.

This overview shows that different situation require the use of different concepts. Furthermore, the concepts vary for the two different types of plants within the same situation.

In summary, the following result shall be highlighted:

> For all concepts the local control units C_i only exchange information at times in which a disturbance, a fault or a set-point change is present. At other times a basic communication is sufficient.

Components of the local control units C_i. To detect the different situations and to decide about an appropriate communication among the local control units to guarantee a desired performance of the closed-loop system, the components of the local control units for the specific control concepts are different (Table 12.1):

- The **feedback units** F_i within the two control concepts for physically interconnected systems have the same structure. The local control input $\boldsymbol{u}_i(t)$ for the coupled subsystems is composed of two inputs: a permanent static local state feedback and the communicated control input $\tilde{\boldsymbol{u}}_i(t)$ determined by the decision unit D_i. For controlling multi-agent systems the feedback unit F_i within all concepts includes a tracking controller for tracking the local reference signal $y_{\mathrm{s}i}(t)$ determined by the corresponding decision unit D_i.
- The **observation units** O_i for controlling physically interconnected systems detect an actual or a desired effect on the neighboring subsystems which are the coupling inputs $\hat{\boldsymbol{s}}_{ji}(t)$ of its subsystem to neighboring subsystems (Chapter 4) or the local control inputs $\hat{\boldsymbol{u}}_{ji}(t)$ desired for the other subsystems (Chapter 5). Two observation units O_i for controlling multi-agent systems derive information about the actual situation of their own controlled subsystem which are the local control error $e_i(t)$ (Chapter 8) or the reconstructed disturbance impact $\hat{y}_{\mathrm{d}i}(t)$ (Chapter 9), respectively. The other observation unit detects faulty predecessors $\mathcal{P}_{\mathrm{F}i}(t)$ by evaluating the received information (Chapter 10).
- All **decision units** D_i perform two tasks:
  - The first task is to decide when to adapt the communicated information. The execution of this task is different in all five concepts. For controlling physically interconnected systems the decision process is similar but the transmitted information is different. If either the coupling inputs $\hat{\boldsymbol{s}}_{ji}(t)$ (Chapter 4) or the local control inputs $\hat{\boldsymbol{u}}_{ji}(t)$ (Chapter 5) determined by the corresponding observation unit O_i exceed a threshold ϵ_{ji} or $\boldsymbol{\epsilon}_{ji}$ the current value of $\hat{\boldsymbol{s}}_{ji}(t)$ or $\hat{\boldsymbol{u}}_{ji}(t)$ is transmitted from C_i to C_j, respectively. Since the concerned situation is different for all three developed control concepts for multi-agent systems, the decisions about adapting the set $\mathcal{P}_{\mathrm{C}i}(t)$ of predecessors also vary: if the local control errors $e_i(t)$ exceeds the threshold ϵ_i the set $\mathcal{P}_{\mathrm{A}i}$ of adjusted predecessors is used instead of the set $\mathcal{P}_{\mathrm{B}i}$ of basic predecessors (Chapter 8), predecessors with large disturbance impacts ($\hat{y}_{\mathrm{d}i}(t) > \epsilon$) vanish from the set $\mathcal{P}_{\mathrm{B}i}$ of basic predecessors (Chapter 9) and faulty predecessors $\mathcal{N}_{\mathrm{F}}(t)$ are replaced by substitutional predecessors $\mathcal{P}_{\mathrm{S}i}(t)$ to reconstruct the communication structure (Chapter 10).
  - The second task for controlling multi-agent systems is identical for all three control concepts which is the determination of the local reference signal $y_{\mathrm{s}i}(t)$ for the

feedback unit based on the incoming outputs from the other agents (Chapters 8–10). Both decision units D_i for controlling physically interconnected systems determine the communicated control input $\tilde{\boldsymbol{u}}_i(t)$ but which includes different information: a signal including the determined coupling inputs $\hat{\boldsymbol{s}}_{ij}(t)$ to compensate the actual coupling inputs $\boldsymbol{s}_{ij}(t)$ (Chapter 4) or an additional control input which supports the local feedback by using the local control inputs $\hat{\boldsymbol{u}}_{ij}(t)$ transmitted by the other local control units (Chapter 5).

In summary, the overview shows that classical feedback components from literature (feedback unit F_i) are joint with newly developed components (observation unit O_i and decision unit D_i) to obtain a situation-dependent information exchange among the local control units C_i. Furthermore, there are some tasks of the components which are general task and do not depend on the actual situation of the subsystems and other tasks which are tailored for the actual situation.

Main results. There are two concepts that bound the disturbance propagation, one for physically interconnected systems (Chapter 4) and one for multi-agents systems (Chapter 9), where the maximal error of the undisturbed subsystems/agents can be arbitrarily adjusted by the choice of the corresponding switching thresholds. The maximal error is either the maximal value for the performance output (Theorem 4.5, Theorem 4.6) or the maximal value for the control error (Theorem 9.2, Theorem 9.3).

Two concepts try to achieve the performance of a specific controller, one for physically interconnected systems (Chapter 5) and one for multi-agents systems (Chapter 8). For one concept the deviation to the behavior of a given centralized control system Σ_{C} can be arbitrarily adjusted by the choice of the switching thresholds (Theorem 5.1, Theorem 5.5). For the other concept the control errors can be adjusted by the choice of an adjusted communication graph $\mathcal{G}_{\mathrm{A}}$ and the switching thresholds ϵ_i (Theorem 8.2, Algorithm 8.2).

Another concept for multi-agent systems (Chapter 10) can preserve the desired maximal value of a basic control error of the fault-free agents (Theorem 10.2, Corollary 10.2).

Evaluation of the desired self-organizing properties. It has been shown that all concepts adapt their communication structure depending on the current situation of the subsystems. Hence, all concepts have the desired flexibility property. Furthermore, the local control units only use local model information and, therefore, the scalability property is given for all concepts. The concepts in Chapter 9 and Chapter 10 show a fault-tolerant behavior, where the thesis discusses approaches to combine these concepts with the other concepts to obtain fault-tolerance for all concepts.

Table 12.1: Overall comparison of the four control concepts presented in this thesis.

Plant	Physically interconnected systems (Part I)		Multi-agent systems (Part II)		
Concept	Disturbance attenuation by compensating physical couplings (Chapter 4)	Mimicry of a centralized controller (Chapter 5)	Performance improvement by requesting additional information (Chapter 8)	Disturbance attenuation by communication interruption (Chapter 9)	Performance preservation by local reconfiguration of the communication structure (Chapter 10)
Feedback unit F_i	Decentralized static state feedback with additional communicated control input $\tilde{\boldsymbol{u}}_i(t)$: $\boldsymbol{u}_i(t) = -\boldsymbol{K}_{\mathrm{C}i}\boldsymbol{x}_i(t) - \tilde{\boldsymbol{u}}_i(t)$		Tracks local reference signal $y_{\mathrm{s}i}(t)$: $u_i(t) = -k_{\mathrm{P}i}(y_i(t) - y_{\mathrm{s}i}(t))$		
Observation unit O_i	Determines physical coupling inputs $\hat{\boldsymbol{s}}_{ji}(t) = \boldsymbol{L}_{ji}\boldsymbol{C}_{\mathrm{z}i}\boldsymbol{x}_i(t)$	Determines local control signals $\hat{\boldsymbol{u}}_{ji}(t) = \boldsymbol{K}_{\mathrm{C}ji}\boldsymbol{x}_i(t)$	Determines local control error $e_i(t) = y_{\mathrm{s}i}(t) - y_i(t)$	Reconstructs disturbance impact $y_{\mathrm{d}i}(t)$ on agent Σ_i by using an observer	Detects faulty predecessors $\mathcal{P}_{\mathrm{F}i}(t)$ by evaluating received information
Decision unit D_i	Sends coupling input $\hat{\boldsymbol{s}}_{ji}(t)$ to C_j if $\Vert\hat{\boldsymbol{s}}_{ji}(t)\Vert \geq \epsilon_{ji}$	Sends local control signal $\hat{\boldsymbol{u}}_{ji}(t)$ to C_j if $\lvert\hat{\boldsymbol{u}}_{ji}(t)\rvert \overset{\exists}{\geq} \boldsymbol{\epsilon}_{ji}$	Adapts set of predecessors: $\mathcal{P}_{\mathrm{C}i}(t) =$ $\begin{cases} \mathcal{P}_{\mathrm{A}i} & \text{if } \lvert e_i(t)\rvert \geq \epsilon_i \\ \mathcal{P}_{\mathrm{B}i} & \text{else,} \end{cases}$	$\mathcal{P}_{\mathrm{B}i} \backslash \{j \mid \lvert y_{\mathrm{d}j}(t)\rvert > \epsilon\}$	$\{\mathcal{P}_{\mathrm{B}i} \cup \mathcal{P}_{\mathrm{S}i}(t)\} \backslash \mathcal{N}_{\mathrm{F}}(t)$
	Calculates communicated control input $\tilde{\boldsymbol{u}}_i(t) =$ $\boldsymbol{K}_{\mathrm{D}i}\sum_{j\in\mathcal{P}_{\mathrm{C}i}(t)}\hat{\boldsymbol{s}}_{ij}(t)$	$\sum_{j\in\mathcal{P}_{\mathrm{C}i}(t)}\hat{\boldsymbol{u}}_{ij}(t)$	Determines local reference signal: $y_{\mathrm{s}i}(t) = \sum_{j\in\mathcal{P}_{\mathrm{C}i}(t)}\tilde{k}_{ij}(t)y_j(t)$		
Main result	Max. error of undisturbed subsystems is arbitrarily adjustable (Theorem 4.5 and Theorem 4.6)	Difference behavior to Σ_{C} is arbitrarily adjustable (Theorem 5.1 and Theorem 5.5)	Max. control errors are adjustable by choice of $\mathcal{G}_{\mathrm{A}}$ and ϵ_i (Theorem 8.2 and Algorithm 8.2)	Max. control error of undisturbed agents is arbitrarily adjustable (Theorem 9.2 and Theorem 9.3)	Max. basic control error remains for fault-free agents (Theorem 10.2 and Corollary 10.2)
Properties of self-organization	-Flexibility ✓ -Scalability ✓ -Fault-tolerance ✓	-Flexibility ✓ -Scalability ✓ -Fault-tolerance	-Flexibility ✓ -Scalability ✓ -Fault-tolerance	-Flexibility ✓ -Scalability ✓ -Fault-tolerance	-Flexibility ✓ -Scalability ✓ -Fault-tolerance ✓

Trade-off between communication effort/change and performance. The concepts in Chapters 4, 5 and 8 show that there is a trade-off between a low communication effort (or a low communication change Chapter 9) and a small deviation to a corresponding best possible performance (Fig. 12.2). Taken only a small deviation from the best possible performance into account, e.g., 10%, leads to an enormous reduction of the communication effort. For example, if a small propagation of a disturbance to undisturbed subsystems is tolerable, the communication effort among the local control units decreases rapidly compared to a complete disturbance compensation with a permanent communication (cf. Fig. 4.5 in Chapter 4). Similar effects can be observed also for the concepts in Chapters 5, 8 and 9.

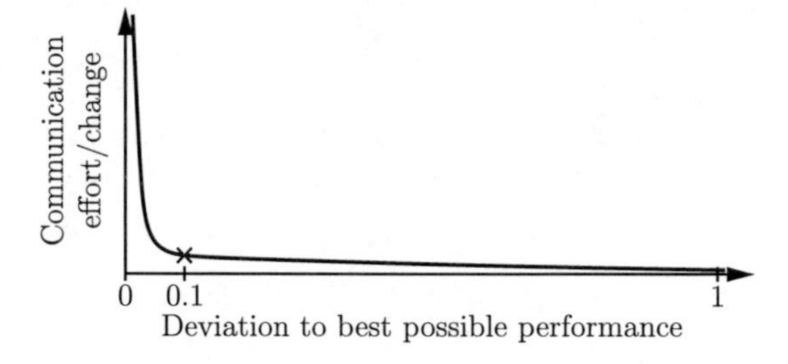

Figure 12.2: Trade-off between communication effort/change and performance.

13 Outlook

The presented control methods that generate a self-organizing behavior could be extended in different directions:

- **Evolution of the communication structure**: For most concepts in this thesis the possible communication links to be used are given or designed a priori. The exception is the concept in Chapter 10 which reconfigures the communication structure in order to maintain the synchronization performance in case of faulty communication links. A reasonable extension, e.g., for multi-agent systems would be a concept with an online evolution of the communication structure. The agents could, e.g., start with a basic communication structure and local performance requirement. During an external stimulation like a set-point change, the corresponding local control units would test specific communication links in a systematic way until their local performance requirements are achieved.

- **Online identification of the coupling strength**: In both concepts for physically interconnected systems the coupling strength among the systems is assumed to be known. In some application there even might not exist an exact value for these coupling strengths but at least a structure which describes the interconnection relations among the subsystems. A concept could be developed that either does not need the actual values of the coupling strengths or which identifies the values by an external stimulation (e.g. disturbances) to keep using the proposed methods.

- **More general communication structures**: The concepts for multi-agent systems in this thesis concern cycle-free communication structures which is appropriate for many technical systems. Nevertheless, the extension of the proposed concepts to more general communication structures with cycles as well as from a leader-follower structure to a leaderless structure would be appropriate.

- **Extensions to nonlinear systems**: The proposed concepts concern linear time-invariant systems to be controlled for physically interconnected systems as well as for multi-agent systems. An interesting topic would be the adaption of the concepts to nonlinear system which are necessary to describe the behavior of real technical systems. A good starting point would be the investigation of systems classes that feature special aspects of nonlinearities like Wiener systems and Hammerstein systems that describe static nonlinearities in the output or input, respectively.

- **Combination of concepts**: In Chapter 6 and Chapter 11 possible combinations of the proposed concepts are discussed for physically interconnected systems and multi-agent systems, respectively, where [7] presents a combination of the concepts in Chapter 8 and Chapter 10. The actual realization of the suggested combinations would combine the benefits of the concepts to obtain closed-loop systems with all desired properties of self-organization.

Bibliography

Contributions of the author

[1] R. Schuh and H. Aschemann. "Nonlinear control of a condensation turbine with steam extraction". In: *Proc. 16th Int. Conf. Methods and Models in Automation and Robotics*. 2011, pp. 236–241.

[2] R. Schuh and J. Lunze. "A switching control strategy for disturbed networked systems with similar dynamics". In: *Proc. 4th IFAC Workshop on Distributed Estimation and Control in Networked Systems*. 2013, pp. 168–173.

[3] R. Schuh and J. Lunze. "Stability analysis of networked systems with similar dynamics". In: *Proc. 12th European Control Conf.* 2013, pp. 4359–4364.

[4] R. Schuh and J. Lunze. "Control Theory of Digitally Networked Dynamic Systems". In: ed. by J. Lunze. Springer, 2014. Chap. Multi-agent Systems: Autonomy and cooperation in networked systems, pp. 305–323.

[5] R. Schuh and J. Lunze. "Self-organizing control of unidirectionally coupled heterogeneous agents with information request". In: *Proc. 53rd IEEE Conf. on Decision and Contr.* 2014, pp. 2453–2460.

[6] R. Schuh and J. Lunze. "Design of the communication structure of a self-organizing networked controller for heterogeneous agents". In: *Proc. 14th European Control Conf.* 2015, pp. 2199–2206.

[7] R. Schuh and J. Lunze. "Fault-tolerant self-organizing control for command tracking of heterogeneous agents". In: *Proc. 5th IFAC Workshop on Distributed Estimation and Control in Networked Systems*. 2015, pp. 118–125.

[8] R. Schuh and J. Lunze. "Self-organizing control of physically interconnected systems for disturbance attenuation". In: *Proc. 14th European Control Conf.* 2015, pp. 2191–2198.

[9] R. Schuh and J. Lunze. "Self-organizing distributed control with situation-dependent communication". In: *Proc. 15th European Control Conf.* 2016, pp. 222–229.

[10] R. Schuh and J. Lunze. "Self-organizing output feedback control of physically interconnected systems for disturbance attenuation". In: *Proc. 2016 American Cont. Conf.* 2016, pp. 3358–3365.

Supervised theses

[11] A. Feiler. “Entwicklung eines Beschleunigungsmesssystems für Stranggusskokillen”. Diploma thesis. Ruhr-Universität Bochum, 2012.

[12] D. Johnen. “Erstellung einer Simulationsumgebung zur Analyse gekoppelter Teilsysteme mit ähnlicher Dynamik”. Bachelor thesis. Ruhr-Universität Bochum, 2013.

[13] K. Kavak. “Erprobung einer sich selbstorganisierenden Regelung mit Informationsanforderung zur Positionierung von Fahrzeugen”. Bachelor thesis. Ruhr-Universität Bochum, 2014.

[14] B. Kilic. “Erprobung einer sich selbstorganisierenden Regelung für physikalisch gekoppelte Systeme”. Master thesis. Ruhr-Universität Bochum, 2014.

[15] B. Kroninger. “Entwurf und Analyse einer dezentralen Regelung physikalisch gekoppelter Teilsysteme mit ähnlicher Dynamik unter Variation eines Nominalmodells”. Bachelor thesis. Ruhr-Universität Bochum, 2012.

[16] K. Lamers. “Erprobung einer sich selbstorganisierenden Regelung zur Positionierung von autonomen Fahrzeugen”. Master thesis. Ruhr-Universität Bochum, 2014.

[17] H. Rong. “Autonome und kooperative Zustandsbeobachtung vernetzter Systeme”. Master thesis. Ruhr-Universität Bochum, 2013.

[18] D. Schwolgin. “Entwurf einer vernetzenden Regelung für Multiagentensysteme an der verfahrenstechnischen Anlage VERA”. Bachelor thesis. Ruhr-Universität Bochum, 2013.

[19] A. Váš. “Erprobung einer sich selbstorganisierenden Regelung an der verfahrenstechnischen Anlage VERA”. Master thesis. Ruhr-Universität Bochum, 2014.

Further literature

[20] B. D. O. Anderson and J. B. Moore. *Optimal control: linear quadratic methods*. Courier Corporation, 2007.

[21] W. R. Ashby. “Principles of the self-organizing system”. In: *Facets of Systems Science*. Springer, 1991, pp. 521–536.

[22] K. J. Aström and R. M. Murray. *Feedback Systems: An Introduction for Scientists and Engineers*. Princeton University Press, 2010.

[23] S. M. Azizi and K. Khorasani. “Cooperative actuator fault accommodation in formation flight of unmanned vehicles using absolute measurements”. In: *IET Control Theory Applications* 6.18 (2012), pp. 2805–2819.

[24] A. Bemporad, M. Heemels, and M. Johansson. *Networked Control Systems*. Springer, 2010.

[25] A. Berman and R. J. Plemmons. *Nonnegative Matrices in the Mathematical Sciences*. Academic Press, 1979.

[26] M. Blanke, M. Kinnaert, J. Lunze, and M. Staroswiecki. *Diagnosis and Fault-Tolerant Control*. Springer, 2016.

[27] S. Bodenburg, V. Kraus, and J. Lunze. "A design method for plug-and-play control of large-scale systems with a varying number of subsystems". In: *Proc. 2016 American Control Conf.* 2016, pp. 5314–5321.

[28] S. Bodenburg and J. Lunze. "Plug-and-play control of interconnected systems with a changing number of subsystems". In: *Proc. 14th European Control Conf.* 2015, pp. 3520–3527.

[29] S. Bodenburg and J. Lunze. "Plug-and-play reconfiguration of locally interconnected systems with limited model information". In: *Proc. 5th IFAC Workshop on Distributed Estimation and Control in Networked Systems*. 2015, pp. 20–27.

[30] F. Borrelli and T. Keviczky. "Distributed LQR design for dynamically decoupled systems". In: *Proc. 45th IEEE Conf. on Decision and Control*. 2006, pp. 5639–5644.

[31] F. Borrelli and T. Keviczky. "Distributed LQR design for identical dynamically decoupled systems". In: *IEEE Trans. Automat. Control* 53.8 (2008), pp. 1901–1912.

[32] S. Boyd and L. Vandenberghe. *Convex Optimization*. Cambridge, United Kingdom: Cambridge University Press, 2003.

[33] G. Chen and Y. Song. "Fault-tolerant output synchronisation control of multi-vehicle systems". In: *IET Control Theory Applications* 8.8 (2014), pp. 574–584.

[34] J. Chen and R. J. Patton. *Robust Model-based Fault Diagnosis for Dynamic Systems*. Vol. 3. Springer Science & Business Media, 2012.

[35] R. Dai and M. Mesbahi. "Optimal topology design for dynamic networks". In: *Proc. Joint 50th IEEE Conf. Decision and Control and European Control Conf.* 2011, pp. 1280–1285.

[36] B. d'Andréa-Novel and M.. De Lara. *Control Theory for Engineers*. Springer, 2013.

[37] G. B. Dantzig and M. N. Thapa. *Linear programming 1: Introduction*. Springer Science & Business Media, 2006.

[38] P. DeLellis and M. Di Bernardo. "Synchronization of complex networks via hybrid adaptive coupling and evolving topologies". In: *Analysis and Control of Chaotic Systems* 2.1 (2009), pp. 50–55.

[39] P. DeLellis, M. Di Bernardo, and F. Garofalo. "Synchronization of complex networks through local adaptive coupling". In: *Chaos* 18.3 (2008).

[40] P. DeLellis, M. Di Bernardo, and F. Garofalo. "Novel decentralized adaptive strategies for the synchronization of complex networks". In: *Automatica* 45.5 (2009), pp. 1312–1318.

[41] P. DeLellis, M. Di Bernardo, F. Garofalo, and M. Porfiri. "Evolution of complex networks via edge snapping". In: *Trans. Cir. Sys. Part I* 57.8 (2010), pp. 2132–2143.

[42] P. DeLellis, M. Di Bernardo, T. E. Gorochowski, and G. Russo. "Synchronization and control of complex networks via contraction, adaptation and evolution". In: *Circuits and Systems Magazine, IEEE* 10.3 (2010), pp. 64–82.

[43] P. DeLellis, M. Di Bernardo, F. Sorrentino, and A. Tierno. "Adaptive synchronization of complex networks". In: *Int. J. of Computer Mathematics* 85.8 (2008), pp. 1189–1218.

[44] P. DeLellis, M. diBernardo, and M. Porfiri. "Achieving consensus and synchronization by adapting the network topology". In: *Proc. 18th IFAC World Congress* 18.2 (2011), pp. 1243–1248.

[45] O. Demir and J. Lunze. "Autonomie und Kooperation in seriell gekoppelten Systemen". In: *Automatisierungstechnik* 4 (2010), pp. 217–226.

[46] O. Demir and J. Lunze. "A decomposition approach to decentralized and distributed control of spatially interconnected systems". In: *Proc. 18th IFAC World Congress*. 2011.

[47] O. Demir and J. Lunze. "Networked control of symmetrically interconnected systems". In: *Proc. 19th Mediterranean Conf. on Control and Automation*. 2011, pp. 1014–1021.

[48] O. Demir and J. Lunze. "Cooperative control of multi-agent systems with event-based communication". In: *Proc. 2012 American Control Conf.* 2012.

[49] O. Demir and J. Lunze. "Event-based synchronisation of multi-agent systems". In: *Proc. 4th IFAC Conf. on Analysis and Design of Hybrid Systems*. 2012.

[50] D. V. Dimarogonas, E. Frazzoli, and K. H. Johansson. "Distributed event-triggered control for multi-agent systems". In: *IEEE Trans. Automat. Control* 57.5 (2012), pp. 1291–1297.

[51] M. C. F. Donkers and W. P. M. H. Heemels. "Output-based event-triggered control with guaranteed $\mathcal{L}_\infty$-gain and improved and decentralized event-triggering". In: *IEEE Trans. Automat. Control* 57.6 (2012), pp. 1362–1376.

[52] Y. Fan, G. Feng, J. Wang, and C. Song. "Distributed event-triggered control of multi-agent systems with combinational measurements". In: *Automatica* 49.2 (2013), pp. 671–675.

[53] M. Fardad and M. R. Jovanović. "On the design of optimal structured and sparse feedback gains via sequential convex programming". In: *Proc. 2014 American Control Conf.* 2014, pp. 2426–2431.

[54] M. Fardad, F. Lin, and M. R. Jovanović. "On the optimal design of structured feedback gains for interconnected systems". In: *Proc. Joint 48th IEEE Conf. on Decision and Control and 28th Chinese Contr. Conf.* 2009, pp. 978–983.

[55] J. A. Fax and R. M. Murray. "Information flow and cooperative control of vehicle formations". In: *IEEE Trans. Automat. Control* 49.9 (2004), pp. 1465–1476.

[56] P. A. Fuhrmann and U. Helmke. *The Mathematics of Networks of Linear Systems.* Springer, 2015.

[57] E. Garcia and P. J. Antsaklis. "Decentralized model-based event-triggered control of networked systems". In: *Proc. 2012 American Control Conf.* 2012, pp. 6485–6490.

[58] E. Garcia, Y. Cao, H. Yu, P. Antsaklis, and D. Casbeer. "Decentralised event-triggered cooperative control with limited communication". In: *Int. J. Control* 86.9 (2013), pp. 1479–1488.

[59] H. F. Grip, T. Yang, A. Saberi, and A. A. Stoorvogel. "Output synchronization for heterogeneous networks of non-introspective agents". In: *Automatica* 48.10 (2012), pp. 2444–2453.

[60] D. Groß, M. Jilg, and O. Stursberg. "Design of distributed controllers and communication topologies considering link failures". In: *Proc. 12th European Control Conf.* 2013, pp. 3288–3292.

[61] D. Groß and O. Stursberg. "Optimized distributed control and network topology design for interconnected systems". In: *Proc. Joint 50th IEEE Conf. Decision and Control and European Contr. Conf.* 2011, pp. 8112–8117.

[62] M. Guinaldo, D. V. Dimarogonas, K. H. Johansson, J. Sanchez, and S. Dormido. "Distributed event-based control for interconnected linear systems". In: *Proc. 50th IEEE Conf. on Decision and Control.* 2011, pp. 2553–2558.

[63] A. Gusrialdi and S. Hirche. "Performance-oriented communication topology design for large-scale interconnected systems". In: *Proc. 49th IEEE Conf. on Decision and Control.* 2010, pp. 5707–5713.

[64] A. Gusrialdi and S. Hirche. "Communication topology design for large-scale interconnected systems with time delay". In: *Proc. 2011 American Contr. Conf.* 2011, pp. 4508–4513.

[65] J.P. Hespanha, P. Naghshtabrizi, and Y. Xu. "A survey of recent results in networked control systems". In: *Proc. IEEE* 95.1 (2007), pp. 138–162.

[66] L. Hogben, ed. *Handbook of Linear Algebra*. CRC Press, 2006.

[67] H. Huang, C. Yu, A. Gusrialdi, and S. Hirche. "Topology design for distributed formation control towards optimal convergence rate". In: *Proc. 2012 American Control Conf.* 2012, pp. 3895–3900.

[68] O. C. Imer, S. Yüksel, and T. Başar. "Optimal control of LTI systems over unreliable communication links". In: *Automatica* 42.9 (2006), pp. 1429–1439.

[69] A. Jadbabaie, J. Lin, and A. Morse. "Coordination of groups of mobile autonomous agents using nearest neighbor rules". In: *IEEE Trans. Automat. Control* 48.6 (2003), pp. 988–1001.

[70] M. Jamshidi. *Large-Scale Systems: Modeling and Control*. North-Holland, 1983.

[71] M. Jilg and O. Stursberg. "Hierarchical Distributed Control for Interconnected Systems". In: *Proc. 13th IFAC Symposium on Large Scale Complex Systems*. 2013, pp. 419–425.

[72] M. Jilg and O. Stursberg. "Optimized distributed control and topology design for hierarchically interconnected systems". In: *Proc. 12th European Control Conf.* 2013, pp. 4340–4346.

[73] M. Jilg and O. Stursberg. "Hierarchical Distributed Control of a Class of Interconnected Jump Semi-Markov Linear Systems". In: *Proc. 15th European Control Conf.* 2016, pp. 236–242.

[74] M. Jilg, J. Tonne, and O. Stursberg. "Design of distributed $\mathcal{H}_2$-optimized controllers considering stochastic communication link failures". In: *Proc. 2015 American Control Conf.* 2015, pp. 3540–3545.

[75] M. R. Jovanović and N. K. Dhingra. "Controller architectures: Tradeoffs between performance and structure". In: *Eur. J. Control* 30 (2016). 15th European Contr. Conf., pp. 76–91.

[76] R. E. Kalman. "Mathematical description of linear dynamical systems". In: *Journal of the Society for Industrial & Applied Mathematics, Series A: Control* 1.2 (1963), pp. 152–192.

[77] C. Kambhampati, R. J. Patton, and F. J. Uppal. "Reconfiguration in networked control systems: Fault tolerant control and plug-and-play". In: *Proc. Fault Detection, Supervision and Safety of Technical Processes*. 2006.

[78] S. Kar, S. Aldosari, and J. M. F. Moura. "Topology for distributed inference on graphs". In: *IEEE Trans. on Signal Processing* 56.6 (2008), pp. 2609–2613.

[79] S. Kar and J. M. F. Moura. "Sensor Networks With Random Links: Topology Design for Distributed Consensus". In: *IEEE Trans. on Signal Processing* 56.7 (2008), pp. 3315–3326.

[80] P. L. Kempker, A. C. M. Ran, and J. H. van Schuppen. "A linear-quadratic coordination control procedure with event-based communication". In: *Proc. IFAC Conference on Analysis and Design of Hybrid Systems*. 2012, pp. 13–18.

[81] H. K. Khalil. *Nonlinear Systems*. Prentice Hall, 2002.

[82] P. P. Khargonekar, I. R. Petersen, and K. Zhou. "Robust stabilization of uncertain linear systems: quadratic stabilizability and H infin; control theory". In: *IEEE Trans. Automat. Control* 35.3 (1990), pp. 356–361.

[83] H. Kim, H. Shim, and J. H. Seo. "Output consensus of heterogeneous uncertain linear multi-agent systems". In: *IEEE Trans. Automat. Control* 56 (2011), pp. 200–206.

[84] A. J. Laub. *Matrix Analysis for Scientists and Engineers*. Society for Industrial and Applied Mathematics, 2004.

[85] D. Liberzon. *Switching in Systems and Control*. Systems and Control. Birkhauser, 2003.

[86] F. Lin, M. Fardad, and M. R. Jovanović. "Augmented lagrangian approach to design of structured optimal state feedback gains". In: *IEEE Trans. Automat. Control* 56.12 (2011), pp. 2923–2929.

[87] F. Lin, M. Fardad, and M. R. Jovanović. "Design of optimal sparse feedback gains via the alternating direction method of multipliers". In: *IEEE Trans. Automat. Control* 58.9 (2013), pp. 2426–2431.

[88] J. Lunze. *Feedback Control of Large-Scale Systems*. Prentice Hall, 1992.

[89] J. Lunze. "An internal-model principle for the synchronisation of autonomous agents with individual dynamics". In: *Proc. Joint 50th IEEE Conf. on Decision and Control and European Control Conf.* 2011.

[90] J. Lunze. "Synchronization of heterogeneous agents". In: *IEEE Trans. Automat. Control* 47 (2012), pp. 2885–2890.

[91] J. Lunze. "A method for designing the communication structure of networked controllers". In: *Int. J. Control* 86.9 (2013), pp. 1489–1502.

[92] J. Lunze. "Synchronisation of autonomous agents with minimum communication". In: *Proc. 52nd IEEE Conf. on Decision and Contr.* 2013, pp. 5415–5420.

[93] J. Lunze, ed. *Control Theory of Digitally Networked Dynamic Systems*. Springer, 2014.

[94] J. Lunze. *Regelungstechnik 2 (8. Auflage)*. Springer-Verlag, 2014.

[95] J. Lunze. "Self-organising disturbance attenuation for synchronised agents with individual dynamics". In: *Int. J. Control* 88.3 (2015), pp. 469–483.

[96] P. Massioni and M. Verhaegen. "Distributed control for identical dynamically coupled systems: A decomposition approach". In: *IEEE Trans. Automat. Control* 54.1 (2009), pp. 124–135.

[97] M. Mazo and P. Tabuada. "Decentralized event-triggered control over wireless sensor/actuator networks". In: *IEEE Trans. Automat. Control* 56.10 (2011), pp. 2456–2461.

[98] X. Meng and T. Chen. "Event based agreement protocols for multi-agent networks". In: *Automatica* 49.7 (2013), pp. 2125–2132.

[99] P. P. Menon and C. Edwards. "Decentralised static output feedback stabilisation and synchronisation of networks". In: *Automatica* 45.12 (2009), pp. 2910–2916.

[100] L. Moreau. "Stability of multi-agent systems with time-dependent communication links". In: *IEEE Trans. Automat. Control* 50 (2005), pp. 169–182.

[101] A. Mosebach and J. Lunze. "Synchronization of multi-agent systems with similar dynamics". In: *Proc. 4th IFAC Workshop on Distributed Estimation and Control in Networked Systems*. 2013, pp. 102–109.

[102] A. Mosebach, J. Lunze, and C. Kampmeyer. "Transient behavior of synchronized agents: A design method for dynamic networked controllers". In: *Proc. 14th European Control Conf.* 2015, pp. 203–210.

[103] R. Olfati-Saber. "Ultrafast consensus in small-world networks". In: *Proc. 2005 American Contr. Conf.* 2005, pp. 2371–2378.

[104] R. Olfati-Saber. "Flocking for multi-agent dynamic systems: Algorithms and theory". In: *IEEE Trans. Automat. Control* 51 (2006), pp. 401–420.

[105] R. Olfati-Saber and R. Murray. "Consensus problems in networks of agents with switching topology and time-delays". In: *IEEE Trans. Automat. Control* 49 (Sept. 2004), pp. 1520–1533.

[106] R. J. Patton, C. Kambhampati, A. Casavola, P. Zhang, S. Ding, and D. Sauter. "A generic strategy for fault-tolerance in control systems distributed over a network". In: *Eur. J. Control* 13.2 (2007), pp. 280–296.

[107] C. De Persis, R. Sailer, and F. Wirth. "Parsimonious Event-Triggered Distributed Control: A Zeno Free Approach". In: *Automatica* (2012), pp. 2116–2124.

[108] R. J. Plemmons. "M-matrix characterizations. I–nonsingular M-matrices." In: *Linear Algebra and its Applications* 18.2 (1977), pp. 175–188.

[109] A. Popov and H. Werner. "Robust stability of a multi-agent system under arbitrary and time-varying communication topologies and communication delays." In: *IEEE Trans. Automat. Control* 57.9 (2012), pp. 2343–2347.

[110] R. Postoyan, P. Tabuada, D. Nešić, and A. Anta. "Event-triggered and self-triggered stabilization of distributed networked control systems". In: *Proc. Joint 50th IEEE Conf. Decision and Control and European Control Conf.* 2011, pp. 2565–2570.

[111] M. Rafiee and A. M. Bayen. "Optimal network topology design in multi-agent systems for efficient average consensus". In: *Proc. 49th IEEE Conf. on Decision and Control.* 2010, pp. 3877–3883.

[112] W. Ren and R. W. Beard. "Consensus seeking in multiagent systems under dynamically changing interaction topologies". In: *IEEE Trans. Automat. Control* 50 (2005), pp. 655–661.

[113] C. W. Reynolds. "Flocks, Herds, and Schools: A Distributed Behavioral Model". In: *Computer Graphics*. 1987, pp. 25–34.

[114] S. Riverso, M. Farina, and G. Ferrari-Trecate. "Plug-and-Play Decentralized Model Predictive Control for Linear Systems". In: *IEEE Trans. Automat. Control* 58.10 (2013), pp. 2608–2614.

[115] S. J. Russell and P. Norvig. *Artificial Intelligence: A Modern Approach.* Prentice Hall series in artificial intelligence. Prentice Hall, 2010.

[116] I. Saboori and Khorasani K. "Actuator fault accommodation strategy for a team of multi-agent systems subject to switching topology". In: *Automatica* 62 (2015), pp. 200–207.

[117] L. Scardovi and R. Sepulchre. "Synchronization in networks of identical linear systems". In: *Automatica* 45.11 (2009), pp. 2557–2562.

[118] K. Schenk and J. Lunze. "Fault-tolerant control of networked systems with re-distribution of control tasks in case of faults". In: *Proc. 3rd International Conference on Control and Fault-Tolerant Systems*. 2016, pp. 723–729.

[119] G. Seyboth, D. V. Dimarogonas, and K. H. Johansson. "Event-based broadcasting for multi-agent average consensus". In: *Automatica* 49.1 (2013), pp. 245–252.

[120] D. D. Šiljak. *Large-Scale Dynamic Systems: Stability and Structure*. North-Holland, 1978.

[121] D. D. Šiljak. *Decentralized Control of Complex Systems*. Elsevier Science, 1991.

[122] E. D. Sontag. *Mathematical Control Theory: Deterministic Finite Dimensional Systems (2nd Edition)*. Springer, New York, 1998.

[123] E. D. Sontag and Y. Wang. "New characterizations of input-to-state stability". In: *IEEE Trans. Automat. Control* 41.9 (1996), pp. 1283–1294.

[124] C. Stöcker. "Event-based state-feedback control of physically interconnected systems". PhD thesis. Ruhr University Bochum, 2014.

[125] J. Stoustrup. "Plug & play control: Control technology towards new challenges". In: *Proc. 10th European Control Conf.* 2009, pp. 1668–1683.

[126] H. Su, X. Wang, and Z. Lin. "Flocking of Multi-Agents With a Virtual Leader". In: *IEEE Trans. Automat. Control* 54.2 (2009), pp. 293–307.

[127] H. G. Tanner, A. Jadbabaie, and G. J. Pappas. "Stable flocking of mobile agents, Part I: Fixed topology". In: *Proc. 42nd IEEE Conf. on Decision and Control*. Vol. 2. 2003, pp. 2010–2015.

[128] H. G. Tanner, A. Jadbabaie, and G. J. Pappas. "Stable flocking of mobile agents, Part II: Dynamic topology". In: *Proc. 42nd IEEE Conf. on Decision and Control*. Vol. 2. 2003, pp. 2016–2021.

[129] A. Teixeira, J. Araújo, H. Sandberg, and K. H. Johansson. "Distributed actuator reconfiguration in networked control systems". In: *Proc. IFAC Workshop on Distributed Estimation and Control in Networked Systems*. 2013, pp. 61–68.

[130] D. Vey, S. Hugging, S. Bodenburg, and J. Lunze. "Control reconfiguration of physically interconnected systems by decentralized virtual actuators". In: *Proc. 9th IFAC Symposium on Fault Detection, Supervision and Safety for Technical Processes*. 2015, pp. 360–367.

[131] T. Vicsek, A. Czirók, E. Ben-Jacob, I. Cohen, and O. Shochet. "Novel Type of Phase Transition in a System of Self-Driven Particles". In: *Phys. Rev. Lett.* 75 (6 1995), pp. 1226–1229.

[132] F. Y. Wang and D. Liu, eds. *Networked control systems*. Springer, 2008.

[133] X. Wang and M. Lemmon. "Event-triggering in distributed networked control systems". In: *IEEE Trans. Automat. Control* 56 (2011), pp. 586–601.

[134] P. Wieland, R. Sepulchre, and F. Allgöwer. "An internal model principle is necessary and sufficient for linear output synchronization". In: *Automatica* 47.5 (2011), pp. 1068–1074.

[135] Z. Wu and R. Wang. "The consensus in multi-agent system with speed-optimized network". In: *International Journal of Modern Physics B* 23.10 (2009), pp. 2339–2348.

[136] L. Xiao and S. Boyd. "Fast Linear Iterations for Distributed Averaging". In: *Systems and Control Letters* 53 (2004), pp. 65–78.

[137] D. Xue and S. Hirche. "Event-triggered consensus of heterogeneous multi-agent systems with double-integrator dynamics". In: *Proc. 12th European Control Conference*. 2013, pp. 1162–1167.

[138] H. Yang, M. Staroswiecki, B. Jiang, and J. Liu. "Fault tolerant cooperative control for a class of nonlinear multi-agent systems". In: *Systems & Control Letters* 60.4 (2011), pp. 271–277.

[139] X. Yin, D. Yue, and S. Hu. "Distributed event-triggered control of discrete-time heterogeneous multi-agent systems". In: *J. Franklin Institute* 350.3 (2013), pp. 651–669.

[140] J. K. Yook, D. M. Tilbury, and N. R. Soparkar. "Trading computation for bandwidth: Reducing communication in distributed control systems using state estimators". In: *IEEE Trans. Control Syst. Technol.* 10.4 (2002), pp. 503–518.

[141] W. Yu, P. DeLellis, G. Chen, M. Di Bernardo, and J. Kurths. "Distributed Adaptive Control of Synchronization in Complex Networks". In: *IEEE Trans. Automat. Control* 57.8 (2012), pp. 2153–2158.

[142] H. Zhang, F. L. Lewis, and A. Das. "Optimal Design for Synchronization of Cooperative Systems: State Feedback, Observer and Output Feedback". In: *IEEE Trans. Automat. Control* 56.8 (2011), pp. 1948–1952.

[143] W. Zhu, Z. Jiang, and G. Feng. "Event-based consensus of multi-agent systems with general linear models". In: *Automatica* 50.2 (2014), pp. 552–558.

Appendix

A Proofs

A.1 Proof of Lemma 4.3

Proof. From (4.24) it follows that

$$\|\boldsymbol{v}_i(t)\| = \left\| \boldsymbol{C}_{\text{v}i}\mathrm{e}^{\bar{\boldsymbol{A}}_i t}\boldsymbol{x}_{0i} + \int_0^t \boldsymbol{C}_{\text{v}i}\mathrm{e}^{\bar{\boldsymbol{A}}_i(t-\tau)}\boldsymbol{G}_i\boldsymbol{d}_i(\tau)\mathrm{d}\tau + \int_0^t \boldsymbol{C}_{\text{v}i}\mathrm{e}^{\bar{\boldsymbol{A}}_i(t-\tau)}\boldsymbol{E}_i\tilde{\boldsymbol{s}}_{\Delta i}(\tau)\mathrm{d}\tau \right\|$$

holds for all $t \geq 0$ and all $i \in \mathcal{N}$. According to (4.13) and (4.3) the relation $\max_{t\geq 0} ||\tilde{\boldsymbol{s}}_{\Delta i}(t)|| \leq \sum_{l\in\bar{\mathcal{P}}_{\text{C}i}} \epsilon_{il}$, $(\forall i \in \mathcal{N})$, holds and with Assumption 3.1 it follows that

$$\begin{aligned}
\|\boldsymbol{v}_i(t)\| \leq & \left\|\boldsymbol{C}_{\text{v}i}\mathrm{e}^{\bar{\boldsymbol{A}}_i t}\right\| \|\boldsymbol{x}_{0i}\| + \int_0^\infty \left\|\boldsymbol{C}_{\text{v}i}\mathrm{e}^{\bar{\boldsymbol{A}}_i \tau}\boldsymbol{G}_i\right\| \mathrm{d}\tau \max_{t\geq 0} \|\boldsymbol{d}_i(t)\| \\
& + \int_0^\infty \left\|\boldsymbol{C}_{\text{v}i}\mathrm{e}^{\bar{\boldsymbol{A}}_i \tau}\boldsymbol{E}_i\right\| \mathrm{d}\tau \max_{t\geq 0} \|\tilde{\boldsymbol{s}}_{\Delta i}(t)\|, \\
\leq & \left\|\boldsymbol{C}_{\text{v}i}\mathrm{e}^{\bar{\boldsymbol{A}}_i t}\right\| \|\boldsymbol{x}_{0i}\| + m_{\text{vd}i}\bar{d}_i + m_{\text{vs}i}\sum\nolimits_{l\in\bar{\mathcal{P}}_{\text{C}i}} \epsilon_{il}.
\end{aligned}$$

From (4.25) it follows that

$$\|\hat{\boldsymbol{s}}_{ji}(t)\| = \left\| \boldsymbol{L}_{ji}\boldsymbol{C}_{\text{z}i}\mathrm{e}^{\bar{\boldsymbol{A}}_i t}\boldsymbol{x}_{0i} + \int_0^t \boldsymbol{L}_{ji}\boldsymbol{C}_{\text{z}i}\mathrm{e}^{\bar{\boldsymbol{A}}_i \bar{\tau}}\boldsymbol{G}_i\boldsymbol{d}_i(\tau)\mathrm{d}\tau + \int_0^t \boldsymbol{L}_{ji}\boldsymbol{C}_{\text{z}i}\mathrm{e}^{\bar{\boldsymbol{A}}_i \bar{\tau}}\boldsymbol{E}_i\tilde{\boldsymbol{s}}_{\Delta i}(\tau)\mathrm{d}\tau \right\| \quad \text{(A.1)}$$

holds for all $t \geq 0$, all $j \in \bar{\mathcal{F}}_{\text{C}i}$ and all $i \in \mathcal{N}$. According to (4.13) and (4.3) the relation $\max_{t\geq 0} ||\tilde{\boldsymbol{s}}_{\Delta i}(t)|| \leq \sum_{l\in\bar{\mathcal{P}}_{\text{C}i}} \epsilon_{il}$, $(\forall i \in \mathcal{N})$, holds and with Assumption 3.1 it follows that

$$\begin{aligned}
\|\hat{\boldsymbol{s}}_{ji}(t)\| \leq & \left\|\boldsymbol{L}_{ji}\boldsymbol{C}_{\text{z}i}\mathrm{e}^{\bar{\boldsymbol{A}}_i t}\right\| \|\boldsymbol{x}_{0i}\| + \int_0^\infty \left\|\boldsymbol{L}_{ji}\boldsymbol{C}_{\text{z}i}\mathrm{e}^{\bar{\boldsymbol{A}}_i \tau}\boldsymbol{G}_i\right\| \mathrm{d}\tau \max_{t\geq 0} \|\boldsymbol{d}_i(t)\| \\
& + \int_0^\infty \left\|\boldsymbol{L}_{ji}\boldsymbol{C}_{\text{z}i}\mathrm{e}^{\bar{\boldsymbol{A}}_i \tau}\boldsymbol{E}_i\right\| \mathrm{d}\tau \max_{t\geq 0} \|\tilde{\boldsymbol{s}}_{\Delta i}(t)\|, \\
\leq & \left\|\boldsymbol{L}_{ji}\boldsymbol{C}_{\text{z}i}\mathrm{e}^{\bar{\boldsymbol{A}}_i t}\right\| \|\boldsymbol{x}_{0i}\| + m_{\text{sd}ji}\bar{d}_i + m_{\text{ys}ji}\sum\nolimits_{l\in\bar{\mathcal{P}}_{\text{C}i}} \epsilon_{il}, \quad \forall j \in \bar{\mathcal{F}}_{\text{C}i}, \quad \forall i \in \mathcal{N}.
\end{aligned}$$

According to Assumption 4.1 the matrices $\bar{\boldsymbol{A}}_i$, $(\forall i \in \mathcal{N})$, are Hurwitz matrices, hence, the scalars $m_{\text{vd}i}$, $m_{\text{vs}i}$, $m_{\text{sd}ji}$ and $m_{\text{ss}ji}$ are finite which completes the proof. □

A.2 Proof of Theorem 4.3

Proof. The proof is divided into two parts. In the first part it is shown that the condition (4.42) guarantees that the communication stays deactivated after a certain time. The second part proves that condition (4.42) also implies the asymptotic stability of the decentralized control system Σ_{D} defined in (4.38).

Due to (4.40), the overall closed-loop system Σ is stable if the decentralized controlled system Σ_{D} is stable and the entire communication among the local control units is deactivated after some time $\bar{t}$. The first condition in guarantees the asymptotic stability of Σ_{D}. According to Lemma 4.5, the second condition guarantees that there exists an *overall deactivation time*

$$t_{\mathrm{D}} = \max\{t_{\mathrm{D}ji}, \forall j \in \bar{\mathcal{F}}_{\mathrm{C}i}, \forall i \in \mathcal{N}\}$$

after which $\mathcal{G}_{\mathrm{C}}(t) = \emptyset$ holds true which completes the first part of the proof.

The decentralized controlled subsystems $\Sigma_{\mathrm{D}i}$, $(\forall i \in \mathcal{N})$ result from (3.1) and (4.11) to

$$\Sigma_{\mathrm{D}i}: \begin{cases} \dot{\boldsymbol{x}}_i(t) = \bar{\boldsymbol{A}}_i \boldsymbol{x}_i(t) + \boldsymbol{E}_i \boldsymbol{s}_i(t) + \boldsymbol{G}_i \boldsymbol{d}_i(t), & \boldsymbol{x}_i(0) = \boldsymbol{x}_{0i} \\ \boldsymbol{y}_i(t) = \boldsymbol{C}_i \boldsymbol{x}_i(t) \\ \boldsymbol{z}_i(t) = \boldsymbol{C}_{\mathrm{z}i} \boldsymbol{x}_i(t). \end{cases} \tag{A.2}$$

Hence, the overall decentralized control system Σ_{D} results input-output representation of the decentralized controlled subsystems Σ_i, $(\forall i \in \mathcal{N})$, with $\boldsymbol{x}_{0i} = \boldsymbol{0}$ while including the physical interconnection

$$\Sigma_{\mathrm{D}}: \begin{cases} \boldsymbol{y}_i(t) = \boldsymbol{G}_{\mathrm{ys}i} * \boldsymbol{s}_i + \boldsymbol{G}_{\mathrm{yd}i} * \boldsymbol{d}_i, & \forall i \in \mathcal{N} \\ \boldsymbol{s}_{ji}(t) = \boldsymbol{G}_{\mathrm{ss}ji} * \boldsymbol{s}_i + \boldsymbol{G}_{\mathrm{sd}ji} * \boldsymbol{d}_i, & \forall i, j \in \mathcal{N}, \quad i \neq j \\ \boldsymbol{s}_i(t) = \sum_{l \in \mathcal{N}} \boldsymbol{s}_{il}(t) = \sum_{l \in \mathcal{N}} \boldsymbol{L}_{il} \boldsymbol{z}_l(l), & \forall i \in \mathcal{N} \end{cases} \tag{A.3}$$

with

$$\begin{aligned} \boldsymbol{G}_{\mathrm{ys}i}(t) &= \boldsymbol{C}_i \mathrm{e}^{\bar{\boldsymbol{A}}_i t} \boldsymbol{E}_i, \quad \forall i \in \mathcal{N} \\ \boldsymbol{G}_{\mathrm{yd}i}(t) &= \boldsymbol{C}_i \mathrm{e}^{\bar{\boldsymbol{A}}_i t} \boldsymbol{G}_i, \quad \forall i \in \mathcal{N} \\ \boldsymbol{G}_{\mathrm{ss}ji}(t) &= \boldsymbol{L}_{ji} \boldsymbol{C}_{\mathrm{z}i} \mathrm{e}^{\bar{\boldsymbol{A}}_i t} \boldsymbol{E}_i, \quad \forall i, j \in \mathcal{N}, \quad i \neq j \\ \boldsymbol{G}_{\mathrm{sd}ji}(t) &= \boldsymbol{L}_{ji} \boldsymbol{C}_{\mathrm{z}i} \mathrm{e}^{\bar{\boldsymbol{A}}_i t} \boldsymbol{G}_i, \quad \forall i, j \in \mathcal{N}, \quad i \neq j. \end{aligned}$$

The system

$$\Sigma_{\mathrm{RD}}:\begin{cases} r_{\mathrm{y}i}(t) = v_{\mathrm{ys}i} * \|\boldsymbol{s}_i\| + v_{\mathrm{yd}i} * r_{\mathrm{d}i} \geq \|\boldsymbol{y}_i(t)\| \quad \forall i \in \mathcal{N} \\ r_{\mathrm{s}ji}(t) = v_{\mathrm{ss}ji} * \|\boldsymbol{s}_i\| + v_{\mathrm{sd}ji} * r_{\mathrm{d}i} \geq \|\boldsymbol{s}_{il}(t)\|, \quad \forall i,j \in \mathcal{N}, \quad i \neq j \\ \|\boldsymbol{s}_i(t)\| \leq \sum_{l \in \mathcal{N}} \|\boldsymbol{s}_{il}(t)\| \leq \sum_{l \in \mathcal{N}} r_{\mathrm{s}il}(t), \quad \forall i \in \mathcal{N} \end{cases} \tag{A.4}$$

with $r_{\mathrm{d}i}(t) := \|\boldsymbol{d}_i(t)\|$, $v_{\mathrm{ys}i}(t) := \|\boldsymbol{G}_{\mathrm{ys}i}(t)\|$, $v_{\mathrm{yd}i}(t) := \|\boldsymbol{G}_{\mathrm{yd}i}(t)\|$, $v_{\mathrm{ss}ji}(t) := \|\boldsymbol{G}_{\mathrm{ss}ji}(t)\|$ and $v_{\mathrm{ys}ji}(t) := \|\boldsymbol{G}_{\mathrm{sd}i}(t)\|$ is a compression system for Σ_{D} defined in (A.3) whose behavior is an upper bound on the behavior of Σ_{D}. With

$$\boldsymbol{r}_{\mathrm{y}}(t) := \begin{pmatrix} r_{\mathrm{y}1}(t) \\ \vdots \\ r_{\mathrm{y}N}(t) \end{pmatrix}, \quad \boldsymbol{r}_{\mathrm{s}}(t) := \begin{pmatrix} \sum_{l \in \mathcal{N}} r_{\mathrm{s}1l}(t) \\ \vdots \\ \sum_{l \in \mathcal{N}} r_{\mathrm{s}Nl}(t) \end{pmatrix}, \quad \boldsymbol{r}_{\mathrm{d}}(t) := \begin{pmatrix} r_{\mathrm{d}1}(t) \\ \vdots \\ r_{\mathrm{d}N}(t) \end{pmatrix}$$

the comparison system Σ_{RD} can be written as

$$\Sigma_{\mathrm{RD}}:\begin{cases} \boldsymbol{r}_{\mathrm{y}}(t) = \boldsymbol{V}_{\mathrm{ys}} * \boldsymbol{r}_{\mathrm{s}} + \boldsymbol{V}_{\mathrm{yd}} * \boldsymbol{r}_{\mathrm{d}} \\ \boldsymbol{r}_{\mathrm{s}}(t) = \boldsymbol{V}_{\mathrm{ss}} * \boldsymbol{r}_{\mathrm{s}} + \boldsymbol{V}_{\mathrm{sd}} * \boldsymbol{r}_{\mathrm{d}} \end{cases} \tag{A.5}$$

with

$$\begin{aligned} \boldsymbol{V}_{\mathrm{ys}} &:= \mathrm{diag}(v_{\mathrm{ys}1} \dots v_{\mathrm{ys}N}) \\ \boldsymbol{V}_{\mathrm{yd}} &:= \mathrm{diag}(v_{\mathrm{yd}1} \dots v_{\mathrm{yd}N}) \\ \boldsymbol{V}_{\mathrm{sd}} &:= \mathrm{diag}(v_{\mathrm{sd}1} \dots v_{\mathrm{sd}N}) \\ \boldsymbol{V}_{\mathrm{ss}} &:= \begin{pmatrix} 0 & v_{\mathrm{ss}12} & \cdots & v_{\mathrm{ss}1N} \\ v_{\mathrm{ss}21} & 0 & \ddots & \vdots \\ \vdots & \ddots & \ddots & v_{\mathrm{ss}N-1,N} \\ v_{\mathrm{ss}N1} & \cdots & v_{\mathrm{ss}N,N-1} & 0 \end{pmatrix}. \end{aligned}$$

With this, the comparison system Σ_{RD} can be written as

$$\Sigma_{\mathrm{RD}}: \boldsymbol{r}_{\mathrm{y}}(t) = \boldsymbol{V} * \boldsymbol{r}_{\mathrm{d}}$$

with

$$\boldsymbol{V}(t) := \boldsymbol{V}_{\mathrm{yd}} + \boldsymbol{V}_{\mathrm{ys}} * \bar{\boldsymbol{V}}_{\mathrm{sd}}$$

$$\bar{V}_{sd}(t) := V_{sd} + V_{ss} * \bar{V}_{sd}.$$

If the comparison system Σ_{RD} is asymptotically stable, then the decentralized control Σ_D system is asymptotically stable as well since the behavior of Σ_{RD} is an upper bound on the behavior of Σ_D. According to Theorem 2.6, Σ_{RD} is asymptotically stable if the relation

$$\int_0^\infty \|V(t)\| \, dt < \infty \tag{A.6}$$

holds true. Condition (A.6) is fulfilled if the matrix $M := \int_0^\infty V(t) dt$ is finite, with

$$\begin{aligned}
M &\leq M_{yd} + M_{ys} * \bar{M}_{sd} \\
\bar{M}_{sd} &\leq M_{sd} + M_{ss} * \bar{M}_{sd} \\
M_{yd} &:= \int_0^\infty V_{yd}(t) dt \\
M_{ys} &:= \int_0^\infty V_{ys}(t) dt \\
\bar{M}_{sd} &:= \int_0^\infty \bar{V}_{sd}(t) dt \\
M_{ss} &:= \int_0^\infty V_{ss}(t) dt
\end{aligned}$$

Since $\bar{A}_i$ are Hurwitz matrices according to Assumption 4.1 and $\bar{M}_{sd} \leq (I - M_{ss})^{-1} M_{sd}$, the matrix M is finite. If the matrix $S := I - M_{ss}$ is an M-Matrix, then S^{-1} exists and is non-singular. Hence, Σ_D is asymptotically stable if S is an M-Matrix. In the following it is shown that condition (4.42) implies that the matrix S is an M-Matrix. Condition (4.42) implies

$$\epsilon \geq M_{ss}\epsilon, \quad \forall \epsilon \geq 0.$$

With this, the relation

$$\epsilon (I - M_{ss}) = \epsilon S > 0, \quad \forall \epsilon \geq 0 \tag{A.7}$$

holds true. Due to (2.7), the relation (A.7) implies that S is a nonsingular M-Matrix, where the relation (A.7) is implied by (4.42), which completes the proof. □

A.3 Proof of Proposition 4.2

Proof. From Lemma 4.3 it follows that

$$\max_{t\geq 0} \|\hat{\boldsymbol{s}}_{hi}(t)\| \leq m_{\mathrm{sd}hi}\bar{d}_i + m_{\mathrm{sd}hi} \sum_{l\in\bar{\mathcal{P}}_{\mathrm{C}i}} \epsilon_{il}, \quad \forall h \in \bar{\mathcal{F}}_{\mathrm{C}i}, \quad \forall i \in \mathcal{N} \tag{A.8}$$

holds true. According to the definition of the decision unit D_i in (4.13), there is only communication from C_i, $(i \in \mathcal{N})$, to C_h, $(h \in \bar{\mathcal{F}}_{\mathrm{C}i})$, if the relation $\|\hat{\boldsymbol{s}}_{hi}(t)\| \geq \epsilon_{hi}$ holds. Distinguishing between the disturbed subsystems Σ_j, $(\forall j \in \mathcal{D})$, and the undisturbed subsystems Σ_p, $(\forall p \in \mathcal{U})$, leads to the bounds $\bar{\boldsymbol{s}}_{\mathrm{D}hi}(\mathcal{D})$ for $\hat{\boldsymbol{s}}_{hi}(t)$, $(\forall i \in \mathcal{N})$. Hence, the set $\bar{\mathcal{F}}_{\mathrm{CD}i}(\mathcal{D})$ includes all followers of C_i whose switching threshold ϵ_{hi} might be exceeded and, therefore, C_i sends information to C_h. According to (A.8), it is guaranteed that all other possible followers C_l, $(\forall l \in \bar{\mathcal{F}}_{\mathrm{C}i}\backslash\bar{\mathcal{F}}_{\mathrm{CD}i}(\mathcal{D}))$, do not receive information from C_i, $(i \in \mathcal{N})$, which completes the proof. □

A.4 Proof of Proposition 4.3

Proof. From the model of the extended subsystem $\bar{\Sigma}^{*}_{\mathrm{F}i}$, $(i\in\mathcal{N})$, that might be faulty defined in (4.76) for $x_{\mathrm{F}0i} = \mathbf{0}$ it follows that

$$\boldsymbol{x}_{\mathrm{F}i}(t) = \mathrm{e}^{\bar{\boldsymbol{A}}_{\mathrm{F}i}t}\boldsymbol{x}_{\mathrm{F}0i} + \int_0^t \mathrm{e}^{\bar{\boldsymbol{A}}_{\mathrm{F}i}(t-\tau)}\boldsymbol{G}_i\boldsymbol{d}_i(\tau)\mathrm{d}\tau + \int_0^t \mathrm{e}^{\bar{\boldsymbol{A}}_{\mathrm{F}i}(t-\tau)}\boldsymbol{E}_i\tilde{\boldsymbol{s}}_{\mathrm{F}\Delta i}(\tau)\mathrm{d}\tau$$

for all $i \in \mathcal{N}$ and $t \geq 0$. With this, it follows that

$$\|\boldsymbol{x}_{\mathrm{F}i}(t)\| = \left\| \mathrm{e}^{\bar{\boldsymbol{A}}_{\mathrm{F}i}t}\boldsymbol{x}_{\mathrm{F}0i} + \int_0^t \mathrm{e}^{\bar{\boldsymbol{A}}_{\mathrm{F}i}(t-\tau)}\boldsymbol{G}_i\boldsymbol{d}_i(\tau)\mathrm{d}\tau + \int_0^t \mathrm{e}^{\bar{\boldsymbol{A}}_{\mathrm{F}i}(t-\tau)}\boldsymbol{E}_i\tilde{\boldsymbol{s}}_{\mathrm{F}\Delta i}(\tau)\mathrm{d}\tau \right\|$$

for all $i \in \mathcal{N}$ and $t \geq 0$. According to Lemma 4.1 and Assumption 3.1, it follows that

$$\begin{aligned}
\|\boldsymbol{x}_{\mathrm{F}i}(t)\| &\leq \left\|\mathrm{e}^{\bar{\boldsymbol{A}}_{\mathrm{F}i}t}\boldsymbol{x}_{\mathrm{F}0i}\right\| + \int_0^\infty \left\|\mathrm{e}^{\bar{\boldsymbol{A}}_{\mathrm{F}i}\tau}\boldsymbol{G}_i\right\| \mathrm{d}\tau \max_{t\geq 0}\|\boldsymbol{d}_i(t)\| + \int_0^\infty \left\|\mathrm{e}^{\bar{\boldsymbol{A}}_{\mathrm{F}i}\tau}\boldsymbol{E}_i\right\| \mathrm{d}\tau \max_{t\geq 0}\|\tilde{\boldsymbol{s}}_{\mathrm{F}\Delta i}(t)\| \\
&\leq \left\|\mathrm{e}^{\bar{\boldsymbol{A}}_{\mathrm{F}i}t}\boldsymbol{x}_{\mathrm{F}0i}\right\| + m_{\mathrm{Fxd}i}\bar{d}_i + m_{\mathrm{Fxs}i}\sum\nolimits_{j\in\bar{\mathcal{P}}_{\mathrm{C}i}} \epsilon_{ij}.
\end{aligned}$$

According to Assumption 4.1 and (4.73), the matrices $\bar{\boldsymbol{A}}_{\mathrm{F}j} = \bar{\boldsymbol{A}}_j$, $(\forall j \in \mathcal{N}\backslash\mathcal{N}_{\mathrm{F}})$, are Hurwitz matrices and the scalars $m_{\mathrm{Fxd}j} = m_{\mathrm{xd}j}$ as well as $m_{\mathrm{Fxs}j} = m_{\mathrm{xs}i}$ are identical and finite. Hence, $\bar{x}_{\mathrm{F}j} = \bar{x}_j$ holds true for all $j \in \mathcal{N}\backslash\mathcal{N}_{\mathrm{F}}$. Since $\bar{\boldsymbol{A}}_{\mathrm{F}k}$, $(\forall k \in \mathcal{N}_{\mathrm{F}}\backslash\mathcal{H}_{\mathrm{F}})$, are Hurwitz matrices, the scalars $m_{\mathrm{Fxd}k}$ and $m_{\mathrm{Fxs}k}$ are finite which leads to the modified bounds $\bar{x}_{\mathrm{F}k}$ in (4.79) compared to the nominal bounds $\bar{x}_k$. The matrices $\bar{\boldsymbol{A}}_{\mathrm{F}h}$, $(\forall h \in \mathcal{H}_{\mathrm{F}})$, are not Hurwitz matrices, hence, the scalars $m_{\mathrm{Fxd}h}$ and $m_{\mathrm{Fxs}h}$ are not finite. With this, the states $x_{\mathrm{F}h}(t)$, $(\forall h \in \mathcal{H}_{\mathrm{F}})$, are not bounded for $t \geq 0$ which completes the proof. □

A.5 Proof of Corollary 5.1

Proof. From the model Σ_Δ in (5.25) it follows that

$$\boldsymbol{x}_{\Delta i}(t) = \check{\boldsymbol{C}}_{\mathrm{x}i} \mathrm{e}^{\boldsymbol{A}_{\mathrm{C}} t} \boldsymbol{x}_{\Delta 0} + \int_0^t \check{\boldsymbol{C}}_{\mathrm{x}i} \mathrm{e}^{\boldsymbol{A}_{\mathrm{C}}(t-\tau)} \boldsymbol{B} \boldsymbol{u}_\Delta(\tau) \mathrm{d}\tau, \quad \forall i \in \mathcal{N}, \quad \forall t \geq 0.$$

With this, from Lemma 5.1, and $\boldsymbol{x}_{\Delta 0} = \boldsymbol{0}$ it follows that

$$\begin{aligned} \max_{t \geq 0} |\boldsymbol{x}_{\Delta i}(t)| &\leq \int_0^\infty \left| \check{\boldsymbol{C}}_{\mathrm{x}i} \mathrm{e}^{\boldsymbol{A}_{\mathrm{C}} \tau} \boldsymbol{B} \right| \mathrm{d}\tau \max_{t \geq 0} |\boldsymbol{u}_\Delta(t)| \\ &\leq \int_0^\infty \left| \check{\boldsymbol{C}}_{\mathrm{x}i} \mathrm{e}^{\boldsymbol{A}_{\mathrm{C}} \tau} \boldsymbol{B} \right| \mathrm{d}\tau \cdot \boldsymbol{\epsilon} \end{aligned}$$

holds for all $i \in \mathcal{N}$. Since $\boldsymbol{A}_{\mathrm{C}}$ is a Hurwitz matrix according to Assumption 5.1, the matrices $\boldsymbol{M}_{\mathrm{xu}i}$, $(\forall i \in \mathcal{N})$ are finite.

From the model Σ_Δ in (5.25) it follows that

$$\boldsymbol{v}_{\Delta i}(t) = \check{\boldsymbol{C}}_{\mathrm{v}i} \mathrm{e}^{\boldsymbol{A}_{\mathrm{C}} t} \boldsymbol{x}_{\Delta 0} + \int_0^t \check{\boldsymbol{C}}_{\mathrm{v}i} \mathrm{e}^{\boldsymbol{A}_{\mathrm{C}}(t-\tau)} \boldsymbol{B} \boldsymbol{u}_\Delta(\tau) \mathrm{d}\tau, \quad \forall i \in \mathcal{N}, \quad \forall t \geq 0. \tag{A.9}$$

With this, from Lemma 5.1, and $\boldsymbol{x}_{\Delta 0} = \boldsymbol{0}$ it follows that

$$\begin{aligned} \max_{t \geq 0} |\boldsymbol{v}_{\Delta i}(t)| &\leq \int_0^\infty \left| \check{\boldsymbol{C}}_{\mathrm{v}i} \mathrm{e}^{\boldsymbol{A}_{\mathrm{C}} \tau} \boldsymbol{B} \right| \mathrm{d}\tau \max_{t \geq 0} |\boldsymbol{u}_\Delta(t)| \\ &\leq \int_0^\infty \left| \check{\boldsymbol{C}}_{\mathrm{v}i} \mathrm{e}^{\boldsymbol{A}_{\mathrm{C}} \tau} \boldsymbol{B} \right| \mathrm{d}\tau \cdot \boldsymbol{\epsilon} \end{aligned}$$

holds for all $i \in \mathcal{N}$. Since $\boldsymbol{A}_{\mathrm{C}}$ is a Hurwitz matrix according to Assumption 5.1, the matrices $\boldsymbol{M}_{\mathrm{vu}i}$, $(\forall i \in \mathcal{N})$, are finite.

From the definition of the observation unit O_i in (5.14) and the model Σ_Δ defined in (5.25) it follows that

$$\tilde{\boldsymbol{u}}_{\Delta ji}(t) = \check{\boldsymbol{C}}_{\mathrm{u}ji} \mathrm{e}^{\boldsymbol{A}_{\mathrm{C}} t} \boldsymbol{x}_{\Delta 0} + \int_0^t \check{\boldsymbol{C}}_{\mathrm{u}ji} \mathrm{e}^{\boldsymbol{A}_{\mathrm{C}}(t-\tau)} \boldsymbol{B} \boldsymbol{u}_\Delta(\tau) \mathrm{d}\tau, \quad \forall i \in \mathcal{N}, \quad \forall t \geq 0. \tag{A.10}$$

With this, from Lemma 5.1, and $\boldsymbol{x}_{\Delta 0} = \boldsymbol{0}$ it follows that

$$\begin{aligned}\max_{t\geq 0} |\tilde{\boldsymbol{u}}_{\Delta ji}(t)| &\leq \int_0^\infty \left|\check{\boldsymbol{C}}_{\mathrm{u}ji}\mathrm{e}^{\boldsymbol{A}_\mathrm{C}\tau}\boldsymbol{B}\right| \mathrm{d}\tau \max_{t\geq 0} |\boldsymbol{u}_\Delta(t)| \\ &\leq \int_0^\infty \left|\check{\boldsymbol{C}}_{\mathrm{u}ji}\mathrm{e}^{\boldsymbol{A}_\mathrm{C}\tau}\boldsymbol{B}\right| \mathrm{d}\tau \cdot \boldsymbol{\epsilon}\end{aligned}$$

holds for all $i, j \in \mathcal{N}$ and $i \neq j$. Since $\boldsymbol{A}_\mathrm{C}$ is a Hurwitz matrix according to Assumption 5.1, the matrices $\boldsymbol{M}_{\mathrm{uu}ji}$, $(\forall i, j \in \mathcal{N}, i \neq j)$ are finite which completes the proof. □

A.6 Proof of Lemma 5.2

Proof. From $\boldsymbol{x}_i(t) = \boldsymbol{x}_{\mathrm{C}i}(t) + \boldsymbol{x}_{\Delta i}(t)$, $\boldsymbol{x}_0 = \boldsymbol{x}_{\mathrm{C}0}$, (5.19), (5.21) and (5.25) it follows that

$$\boldsymbol{x}_i(t) = \check{\boldsymbol{C}}_{\mathrm{x}i}\mathrm{e}^{\boldsymbol{A}_\mathrm{C}t}\boldsymbol{x}_0 + \int_0^t \check{\boldsymbol{C}}_{\mathrm{x}i}\mathrm{e}^{\boldsymbol{A}_\mathrm{C}(t-\tau)}\boldsymbol{G}\boldsymbol{d}(\tau)\mathrm{d}\tau + \int_0^t \check{\boldsymbol{C}}_{\mathrm{x}i}\mathrm{e}^{\boldsymbol{A}_\mathrm{C}(t-\tau)}\boldsymbol{B}\boldsymbol{u}_\Delta(\tau)\mathrm{d}\tau$$

holds true for all $i \in \mathcal{N}$ and $t \geq 0$. With this, Assumption 3.1 and Lemma 5.1 yield

$$\begin{aligned}\max_{t\geq 0} |\boldsymbol{x}_i(t)| &\leq \left|\check{\boldsymbol{C}}_{\mathrm{x}i}\mathrm{e}^{\boldsymbol{A}_\mathrm{C}t}\right| |\boldsymbol{x}_0| + \int_0^\infty \left|\check{\boldsymbol{C}}_{\mathrm{x}i}\mathrm{e}^{\boldsymbol{A}_\mathrm{C}\tau}\boldsymbol{G}\right| \mathrm{d}\tau \max_{t\geq 0} |\boldsymbol{d}(t)| + \int_0^\infty \left|\check{\boldsymbol{C}}_{\mathrm{x}i}\mathrm{e}^{\boldsymbol{A}_\mathrm{C}\tau}\boldsymbol{B}\right| \mathrm{d}\tau \max_{t\geq 0} |\boldsymbol{u}_\Delta(t)| \\ &\leq \left|\check{\boldsymbol{C}}_{\prime yi}\mathrm{e}^{\boldsymbol{A}_\mathrm{C}t}\right| |\boldsymbol{x}_0| + \int_0^\infty \left|\check{\boldsymbol{C}}_{\mathrm{x}i}\mathrm{e}^{\boldsymbol{A}_\mathrm{C}\tau}\boldsymbol{G}\right| \mathrm{d}\tau \cdot \bar{\boldsymbol{d}} + \int_0^\infty \left|\check{\boldsymbol{C}}_{\mathrm{x}i}\mathrm{e}^{\boldsymbol{A}_\mathrm{C}\tau}\boldsymbol{B}\right| \mathrm{d}\tau \cdot \boldsymbol{\epsilon}\end{aligned}$$

holds for all $i \in \mathcal{N}$. Since $\boldsymbol{A}_\mathrm{C}$ is a Hurwitz matrix according to Assumption 5.1, the matrices $\boldsymbol{M}_{\mathrm{xd}i}$, $(\forall i \in \mathcal{N})$, and $\boldsymbol{M}_{\mathrm{xu}i}$ are finite.

From $\boldsymbol{v}_i(t) = \boldsymbol{v}_{\mathrm{C}i}(t) + \boldsymbol{v}_{\Delta i}(t)$, $\boldsymbol{x}_0 = \boldsymbol{x}_{\mathrm{C}0}$, (5.19), (5.21) and (5.25) it follows that

$$\boldsymbol{v}_i(t) = \check{\boldsymbol{C}}_{\mathrm{v}i}\mathrm{e}^{\boldsymbol{A}_\mathrm{C}t}\boldsymbol{x}_0 + \int_0^t \check{\boldsymbol{C}}_{\mathrm{v}i}\mathrm{e}^{\boldsymbol{A}_\mathrm{C}(t-\tau)}\boldsymbol{G}\boldsymbol{d}(\tau)\mathrm{d}\tau + \int_0^t \check{\boldsymbol{C}}_{\mathrm{v}i}\mathrm{e}^{\boldsymbol{A}_\mathrm{C}(t-\tau)}\boldsymbol{B}\boldsymbol{u}_\Delta(\tau)\mathrm{d}\tau$$

holds true for all $i \in \mathcal{N}$ and $t \geq 0$. With this, Assumption 3.1 and Lemma 5.1 yield

$$\max_{t\geq 0} |\boldsymbol{v}_i(t)| \leq \left|\check{\boldsymbol{C}}_{\mathrm{v}i}\mathrm{e}^{\boldsymbol{A}_\mathrm{C}t}\right| |\boldsymbol{x}_0| + \int_0^\infty \left|\check{\boldsymbol{C}}_{\mathrm{v}i}\mathrm{e}^{\boldsymbol{A}_\mathrm{C}\tau}\boldsymbol{G}\right| \mathrm{d}\tau \cdot \bar{\boldsymbol{d}} + \int_0^\infty \left|\check{\boldsymbol{C}}_{\mathrm{v}i}\mathrm{e}^{\boldsymbol{A}_\mathrm{C}\tau}\boldsymbol{B}\right| \mathrm{d}\tau \cdot \boldsymbol{\epsilon}$$

holds for all $i \in \mathcal{N}$. Since $\boldsymbol{A}_\mathrm{C}$ is a Hurwitz matrix according to Assumption 5.1, the matrices $\boldsymbol{M}_{\mathrm{vd}i}$, $(\forall i \in \mathcal{N})$, and $\boldsymbol{M}_{\mathrm{vu}i}$ are finite.

From $\hat{\boldsymbol{u}}_{ji}(t) = \hat{\boldsymbol{u}}_{\mathrm{C}ji}(t) + \hat{\boldsymbol{u}}_{\Delta ji}(t)$, $\boldsymbol{x}_0 = \boldsymbol{x}_{\mathrm{C}0}$, (5.19), (5.21) and (5.25) it follows that

$$\hat{\boldsymbol{u}}_{ji}(t) = \check{\boldsymbol{C}}_{\mathrm{u}ji}\mathrm{e}^{\boldsymbol{A}_\mathrm{C}t}\boldsymbol{x}_0 + \int_0^t \check{\boldsymbol{C}}_{\mathrm{u}ji}\mathrm{e}^{\boldsymbol{A}_\mathrm{C}(t-\tau)}\boldsymbol{G}\boldsymbol{d}(\tau)\mathrm{d}\tau + \int_0^t \check{\boldsymbol{C}}_{\mathrm{u}ji}\mathrm{e}^{\boldsymbol{A}_\mathrm{C}(t-\tau)}\boldsymbol{B}\boldsymbol{u}_\Delta(\tau)\mathrm{d}\tau \tag{A.11}$$

holds true for all $i \in \mathcal{N}$, all $j \in \bar{\mathcal{F}}_{\mathrm{C}i}$ and $t \geq 0$. With this, Assumption 3.1 and Lemma 5.1 yield

$$\max_{t\geq 0} |\hat{\boldsymbol{u}}_{ji}| \leq \left|\check{\boldsymbol{C}}_{\mathrm{u}ji}\mathrm{e}^{\boldsymbol{A}_\mathrm{C}t}\right| |\boldsymbol{x}_0| + \int_0^\infty \left|\check{\boldsymbol{C}}_{\mathrm{u}ji}\mathrm{e}^{\boldsymbol{A}_\mathrm{C}\tau}\boldsymbol{G}\right| \mathrm{d}\tau \cdot \bar{\boldsymbol{d}} + \int_0^\infty \left|\check{\boldsymbol{C}}_{\mathrm{u}ji}\mathrm{e}^{\boldsymbol{A}_\mathrm{C}\tau}\boldsymbol{B}\right| \mathrm{d}\tau \cdot \boldsymbol{\epsilon}$$

holds for all $i \in \mathcal{N}$ and all $j \in \bar{\mathcal{F}}_{\mathrm{C}i}$. Since $\boldsymbol{A}_\mathrm{C}$ is a Hurwitz matrix according to Assumption 5.1, the matrices $\boldsymbol{M}_{\mathrm{ud}ji}$, $(\forall i \in \mathcal{N}, \forall j \in \bar{\mathcal{F}}_{\mathrm{C}i})$, and $\boldsymbol{M}_{\mathrm{uu}ji}$ are finite. □

A.7 Proof of Lemma 5.3

Proof. According to (A.11), Lemma 5.1 and Assumption 3.1, the relation

$$|\hat{\boldsymbol{u}}_{ji}(t)| \leq \left|\check{\boldsymbol{C}}_{\mathrm{u}ji}\mathrm{e}^{\boldsymbol{A}_\mathrm{C}t}\right| |\boldsymbol{x}_0| + \int_0^t \left|\check{\boldsymbol{C}}_{\mathrm{u}ji}\mathrm{e}^{\boldsymbol{A}_\mathrm{C}\tau}\boldsymbol{G}\right| \mathrm{d}\tau\bar{\boldsymbol{d}} + \int_0^t \left|\check{\boldsymbol{C}}_{\mathrm{u}ji}\mathrm{e}^{\boldsymbol{A}_\mathrm{C}\tau}\boldsymbol{B}\right| \mathrm{d}\tau\boldsymbol{\epsilon} = \bar{\boldsymbol{u}}_{\mathrm{T}ji}(t) \tag{A.12}$$

holds true for all $t \geq 0$, all $j \in \bar{\mathcal{F}}_{\mathrm{C}i}$ and all $i \in \mathcal{N}$. Hence, only if $\bar{\boldsymbol{u}}_{\mathrm{T}ji}(t)$ exceeds $\boldsymbol{\epsilon}_{ji}$, the relation (5.42) can be fulfilled which completes the proof. □

A.8 Proof of Lemma 5.4

Proof. Since $\boldsymbol{A}_\mathrm{C}$ is a Hurwitz matrix (cf. Assumption 5.1) and according to (5.38), the relation

$$\lim_{t\to\infty} |\hat{\boldsymbol{u}}_{ji}(t)| \leq \boldsymbol{M}_{\mathrm{uu}ji}\boldsymbol{\epsilon} + \boldsymbol{M}_{\mathrm{ud}ji}\bar{\boldsymbol{d}}$$

holds true for all $t \geq 0$, all $j \in \bar{\mathcal{F}}_{\mathrm{C}i}$ and all $i \in \mathcal{N}$. Note that $\lim_{t\to\infty} \left|\check{\boldsymbol{C}}_{\mathrm{u}ji}\mathrm{e}^{\boldsymbol{A}_\mathrm{C}t}\right| |\boldsymbol{x}_0| = \boldsymbol{0}$ since $\boldsymbol{A}_\mathrm{C}$ is a Hurwitz matrix. Hence, if the relation in (5.47) holds true, the condition for deactivating the communication in (5.43) is fulfilled. Consider (A.12), then the time for which $\bar{\boldsymbol{u}}_{\mathrm{T}ji}(t)$ stays below the threshold $\boldsymbol{\epsilon}_{ji}$ is upper bounded by $\bar{t}_{\mathrm{D}ji}$. □

A.9 Proof of Proposition 5.3

Proof. From Lemma 5.2 it follows that

$$\max_{t\geq 0} \|\hat{\boldsymbol{u}}_{hi}(t)\| \leq \boldsymbol{M}_{\mathrm{ud}hi}\bar{\boldsymbol{d}}^* + \boldsymbol{M}_{\mathrm{uu}hi}\boldsymbol{\epsilon}, \quad \forall h \in \bar{\mathcal{F}}_{\mathrm{C}i}, \quad \forall i \in \mathcal{N} \tag{A.13}$$

holds true. According to the definition of the decision unit in (5.15), there only is communication from C_i, $(i \in \mathcal{N})$, to C_h, $(h \in \bar{\mathcal{F}}_{\mathrm{C}i})$, if the relation $\|\hat{\boldsymbol{u}}_{hi}(t)\| \overset{\exists}{\geq} \boldsymbol{\epsilon}_{hi}$ holds which leads to the bounds $\bar{\boldsymbol{u}}_{\mathrm{D}hi}(\mathcal{D})$ for $\hat{\boldsymbol{u}}_{hi}(t)$, $(\forall i \in \mathcal{N})$. Hence, the set $\bar{\mathcal{F}}_{\mathrm{CD}i}(\mathcal{D})$ includes all followers of C_i whose switching threshold $\boldsymbol{\epsilon}_{ih}$ might be exceeded and, therefore, C_i sends information to C_h which completes the proof. □

A.10 Proof of Lemma 7.1

Proof. The relation

$$(\exists\,\mathrm{Path}(0 \to 1) \in \mathcal{G}_{\mathrm{C}}(t), \forall t \geq 0) \wedge (\exists\,\mathrm{Path}(0 \to j) \in \mathcal{G}_{\mathrm{C}}(t), \forall i \in \mathcal{N}\backslash\{1\}, \forall t \geq 0) \tag{A.14}$$

is equivalent to the requirement in (7.15). Since

$$(\mathcal{P}_{\mathrm{C}1}(t) = \{0\} \neq \emptyset, \forall t \geq 0) \Leftrightarrow ((0 \to 1) \in \mathcal{E}_{\mathrm{C}}(t), \forall t \geq 0) \Leftrightarrow (\exists\,\mathrm{Path}(0 \to 1) \in \mathcal{G}_{\mathrm{C}}(t), \forall t \geq 0) \tag{A.15}$$

holds true according to (7.13), agent Σ_1 is always connected with the leader Σ_0 if and only if $\mathcal{P}_{\mathrm{C}1}(t) \neq \emptyset$, $t \geq 0$. The right part of (A.14) is equivalent to

$$(\exists\,(0 \to i) \in \mathcal{E}_{\mathrm{C}}(t), 0 \in \mathcal{P}_{\mathrm{C}i}(t)) \vee (\exists\,\mathrm{Path}(0 \to j) \in \mathcal{G}_{\mathrm{C}}(t), j \in \mathcal{P}_{\mathrm{C}i}(t)\backslash\{0\}) \tag{A.16}$$

for all $i \in \mathcal{N}\backslash\{1\}$ and all $t \geq 0$. The left part of (A.16) holds true if and only if $\mathcal{P}_{\mathrm{C}i}(t) \neq \emptyset$, $(t \geq 0)$, since otherwise there would be no edge $(0 \to i)$. The equivalence of the right part of (A.16) with $\mathcal{P}_{\mathrm{C}i}(t) \neq \emptyset$, $(\forall i \in \mathcal{N}, t \geq 0)$, is proven by induction. Consider the *base case* with $i = 2$. The relation

$$\exists\,\mathrm{Path}(0 \to j) \in \mathcal{G}_{\mathrm{C}}(t), \quad j \in \mathcal{P}_{\mathrm{C}2}(t)\backslash\{0\}, \quad \forall t \geq 0$$

is fulfilled if and only if $\mathcal{P}_{\mathrm{C}i}(t) \neq \emptyset$, $(i = 1, 2)$, for all $t \geq 0$. The relation $j \in \mathcal{P}_{\mathrm{C}2}(t)\backslash\{0\}$ and $j \in \mathcal{P}_{\mathrm{C}2}(t) \neq \emptyset$ is equivalent to $j \in \mathcal{P}_{\mathrm{C}2}(t) = 1$ for all $t \geq 0$. According to (A.16), there is a path from 0 to 1 if and only if $\mathcal{P}_{\mathrm{C}1}(t) \neq \emptyset$ for all $t \geq 0$. This completes the proof of the base case. For the *induction step* $i = n+1$ the relation

$$\exists\,\mathrm{Path}(0 \to j) \in \mathcal{G}_\mathrm{C}(t), \quad j \in \mathcal{P}_{\mathrm{C}n+1}(t)\backslash\{0\}, \quad i = 2, \dots n{+}1, \quad \forall t \geq 0,$$

has to be fulfilled if and only if $\mathcal{P}_{\mathrm{C}i}(t) \neq \emptyset$ holds true for $i = 1, \dots, n+1$ and all $t \geq 0$. Therefore, the *induction hypothesis* is:

$$\exists\,\mathrm{Path}(0 \to j) \in \mathcal{G}_\mathrm{C}(t), \quad j \in \mathcal{P}_{\mathrm{C}n}(t)\backslash\{0\}, \quad i = 2, \dots n, \quad \forall t \geq 0,$$

is fulfilled if and only if $\mathcal{P}_{\mathrm{C}i}(t) \neq \emptyset$ holds true for $i = 1, \dots, n$ and all $t \geq 0$. Relation (7.12) yields

$$\mathcal{P}_{\mathrm{C}n+1}(t) \subseteq \{j \in \mathcal{V}_\mathrm{C} \mid j \leq n\}.$$

Hence, if and only if $\mathcal{P}_{\mathrm{C}i}(t) \neq \emptyset$ holds true for $i = 1, \dots, n+1$ and all $t \geq 0$ there is an edge from j to $n+1$ for $j < n+1$. Due to the *induction hypothesis*, there exists a path from 0 to j for all $i = 1, \dots, n$ if and only if $\mathcal{P}_{\mathrm{C}i}(t) \neq \emptyset$ holds true for $i = 1, \dots, n$ and all $t \geq 0$. Due to both facts, there exists a path from the leader Σ_0 to Σ_{n+1}. With this, the induction proof for the right part of (A.16) is completed which completes the overall proof. □

A.11 Proof of Lemma 7.2

Proof. The proof is divided into five parts:

1. Conditions on the construction of the transformed controlled difference agents $\tilde{\tilde{\Sigma}}_{\Delta i}$ are investigated.

2. The behavior of $\tilde{\tilde{\Sigma}}_{\Delta i}$ is analyzed for $t \in [0, \bar{t}]$ in which the agents are disturbed and not all agents might have a predecessor.

3. Conditions on the construction of the overall transformed difference system $\tilde{\Sigma}_\Delta$ are investigated.

4. It is shown that under the stated assumptions and presented conditions the undisturbed overall transformed difference system $\tilde{\Sigma}_\Delta$ is asymptotically stable which guarantees that the control error $\boldsymbol{y}_\Delta(t)$ vanishes for $t \to \infty$.

5. The sufficiency of the conditions is explained.

Construction of the transformed controlled difference agents $\tilde{\tilde{\Sigma}}_{\Delta i}$. The transformed controlled difference subsystems $\tilde{\tilde{\Sigma}}_{\Delta i}$ defined in (7.43) result from the transformed controlled agents $\tilde{\tilde{\Sigma}}_i$ defined in (7.27) and the leader Σ_0 defined in (7.3). Due to Theorem 7.1 and Assumption 7.4, the transformed controlled agents $\tilde{\tilde{\Sigma}}_i$ can be constructed which implies that $\tilde{\tilde{\Sigma}}_{\Delta i}$ can be constructed.

Behavior of $\tilde{\tilde{\Sigma}}_{\Delta i}$ for $t \in [0, \bar{t}]$. Consider the model of $\tilde{\tilde{\Sigma}}_{\Delta i}$ defined in (7.43). The system matrix $\tilde{\tilde{\boldsymbol{A}}}_i$ of $\tilde{\tilde{\Sigma}}_{\Delta i}$ is a Hurwitz matrix since $\bar{\boldsymbol{A}}_i$ is a Hurwitz matrix according to Assumption 7.5. The transformation of $\bar{\boldsymbol{A}}_i$ in Theorem 7.1 to $\tilde{\tilde{\boldsymbol{A}}}_i$ does not change the eigenvalues of $\tilde{\tilde{\boldsymbol{A}}}_i$ compared to the eigenvalues of $\bar{\boldsymbol{A}}_i$. The inputs in $\tilde{\tilde{\Sigma}}_{\Delta i}$ are bounded for $t \in [0, \bar{t}]$. The disturbance $d_i(t)$ is bounded according to Assumption 7.2. The local reference error $y_{\Delta \mathrm{s}i}$ is also bounded since $y_{\mathrm{s}}(t)$ and $y_j(t)$, $(\forall j \in \mathcal{P}_{\mathrm{C}i}(t))$, can only increase exponentially according to (7.3) and (7.28). In summary, even if the agents Σ_i are disturbed or might not have a predecessor, the differences states $\tilde{\boldsymbol{x}}_{\Delta i}(t)$ are upper bounded, i.e., $\|\tilde{\boldsymbol{x}}_{\Delta i}(t)\| < \infty$ for $t \in [0, \bar{t}]$. Hence, for the investigation of the behavior of $\tilde{\Sigma}_{\Delta}$ for $t \geq \bar{t}$ the corresponding initial state is always upper bounded.

Construction of the overall transformed difference system $\tilde{\Sigma}_{\Delta}$. According to Lemma 7.1, the well-known requirement of a spanning tree with the leader as root node to guarantee asymptotic synchronization of the agents is equivalent to the condition in (7.48). This requirement for synchronization can also be seen in (7.44), since the local reference error $y_{\Delta \mathrm{s}i}(t)$, $(i \in \mathcal{N})$, can not be expressed by the control errors $y_{\Delta j}(t)$ for $\mathcal{P}_{\mathrm{C}i}(t) = \emptyset$. If $\mathcal{P}_{\mathrm{C}i}(t) = \emptyset$, $(i \in \mathcal{N})$, holds true, then $\tilde{\Sigma}_{\Delta}$ would have the reference signal $y_{\mathrm{s}}(t)$ as an input. Therefore, generally $\tilde{\Sigma}_{\Delta}$ would not be asymptotically stable since $y_{\mathrm{s}}(t)$ does not vanish for $\boldsymbol{x}_{\mathrm{s}0} \neq \boldsymbol{0}$ according to Assumption 7.3. Hence, the overall transformed difference system $\tilde{\Sigma}_{\Delta}$ of the form (7.45) can be constructed if condition (7.48) holds true.

Behavior of the undisturbed overall transformed difference system $\tilde{\Sigma}_{\Delta}$ for $t \geq \bar{t}$. If $\tilde{\Sigma}_{\Delta}$ defined in (7.45) is disturbed, generally $\tilde{\Sigma}_{\Delta}$ is not asymptotically stable. Hence, condition (7.49) is required. With this, it is sufficient to prove that the undisturbed overall transformed difference system $\tilde{\Sigma}_{\Delta}$ given by

$$\tilde{\Sigma}_{\Delta}: \begin{cases} \dot{\tilde{\boldsymbol{x}}}_{\Delta}(t) = \tilde{\tilde{\boldsymbol{A}}}(t)\tilde{\boldsymbol{x}}_{\Delta}(t), \quad \tilde{\boldsymbol{x}}_{\Delta}(0) = \tilde{\boldsymbol{x}}_{\Delta 0} \\ \boldsymbol{y}_{\Delta}(t) = \tilde{\boldsymbol{C}}\tilde{\boldsymbol{x}}_{\Delta}(t) \end{cases} \tag{A.17}$$

is asymptotically stable for a bounded initial state $\tilde{\boldsymbol{x}}_{\Delta 0}$. This guarantees that control error $\boldsymbol{y}_{\Delta}(t)$ vanishes for $t \to \infty$. Hence, the main task is to prove the asymptotic stability of $\tilde{\Sigma}_{\Delta}$ defined in (A.17) with the switching system matrix $\tilde{\tilde{\boldsymbol{A}}}(t)$. $\tilde{\Sigma}_{\Delta}$ is asymptotically stable, if

$$\|\tilde{\boldsymbol{x}}_{\Delta i}(t)\| \leq \gamma \mathrm{e}^{-\mu t}, \quad \forall i \in \mathcal{N}, \quad \forall t \geq 0 \tag{A.18}$$

hold true for some $\gamma, \mu > 0$. To prove that there exist some $\gamma, \mu > 0$ that guarantee (A.18) is made by induction. Consider the *base step* with $i = 1$. According to (A.17) and Assumption 7.5, the relation

$$\|\tilde{\boldsymbol{x}}_{\Delta 1}(t)\| \leq \left\|\mathrm{e}^{\tilde{\bar{\boldsymbol{A}}}_1 t}\right\| \|\tilde{\boldsymbol{x}}_{\Delta 01}\| \leq \gamma_1 \mathrm{e}^{-\mu_1 t}, \quad \forall i \in \mathcal{N}, \quad \forall t \geq 0$$

holds true for some $\gamma \geq \gamma_1 > 0$ and some $\mu \geq \mu_1 > 0$. Consider the *induction step* $i = n+1$. Consider the *induction hypothesis*

$$\|\tilde{\boldsymbol{x}}_{\Delta i}(t)\| \leq \left\|\mathrm{e}^{\tilde{\bar{\boldsymbol{A}}}_i t}\right\| \|\tilde{\boldsymbol{x}}_{\Delta 0,i}\| \leq \gamma_i \mathrm{e}^{-\mu_i t}, \quad i = 1, \dots n, \quad \forall t \geq 0 \tag{A.19}$$

for some $\gamma \geq \gamma_i > 0$ and $\mu \geq \mu_i > 0$. With this, Assumption 7.5 and equations (A.17), (7.44) yield that the relation

$$\begin{aligned} \|\tilde{\boldsymbol{x}}_{\Delta n+1}(t)\| \leq & \left\|\mathrm{e}^{\tilde{\bar{\boldsymbol{A}}}_{n+1} t}\right\| \|\tilde{\boldsymbol{x}}_{\Delta 0,n+1}\| \\ & + \sum_{j \in \mathcal{P}_{\mathrm{C}n+1}(t)} \int_0^t \underbrace{\left\|\mathrm{e}^{\tilde{\bar{\boldsymbol{A}}}_{n+1} t} \tilde{\boldsymbol{b}}_{n+1} k_{\mathrm{P}n+1}\right\|}_{\leq \bar{\gamma}_{n+1} \mathrm{e}^{-\bar{\mu}_{n+1} t}} \underbrace{\left\|\tilde{\boldsymbol{c}}_j^{\mathrm{T}} \tilde{k}_{n+1,j}(\tau)\right\|}_{<\infty} \underbrace{\|\tilde{\boldsymbol{x}}_{\Delta j}(\tau)\|}_{\leq \gamma_j \mathrm{e}^{-\mu_j t}} \mathrm{d}\tau \\ \leq & \gamma_{n+1} \mathrm{e}^{-\mu_{n+1} t} \end{aligned} \tag{A.20}$$

holds true for all $t \geq 0$, some $\gamma \geq \gamma_{n+1} > 0$, some $\mu \geq \mu_{n+1} > 0$, some $\gamma_{n+1} \geq \bar{\gamma}_{n+1} > 0$ and some $\mu_{n+1} \geq \bar{\mu}_{n+1} > 0$. According to (7.12) and the induction hypothesis in (A.18), the relation $\|\tilde{\boldsymbol{x}}_{\Delta j}(\tau)\| \leq \gamma_j \mathrm{e}^{-\mu_j t}$ holds true for all $j \in \mathcal{P}_{\mathrm{C}n+1}(t)$. Since the elements $\tilde{k}_{n+1,j}(\tau)$ are finite according to (7.19) for all $\tau > 0$ and arbitrary switching, the element $\left\|\tilde{\boldsymbol{c}}_j^{\mathrm{T}} \tilde{k}_{n+1,j}(\tau)\right\|$ is also upper bounded. Therewith, the relation (A.20) holds true which completes the induction.
Sufficiency of the conditions. For investigating the sufficiency of condition (7.48) consider a leader Σ_0 with one agent Σ_1. Agent Σ_1 can still be synchronized to Σ_0 even if condition (7.48) does not hold true, since Σ_1 can asymptotically track $y_{\mathrm{s}}(t)$ also for a periodic switching of the communication link $(0 \rightarrow 1)$. Therefore, the convergence to $y_{\mathrm{s}}(t)$ during the activation of $(0 \rightarrow 1)$ must be larger than the divergence during the deactivation of $(0 \rightarrow 1)$. Condition (7.49) is sufficient but not necessary since, e.g., the controlled agents $\bar{\Sigma}_i$ might have the ability to compensate a constant disturbance $d_i(t)$ and, therefore, still synchronize to the leader Σ_0. □

A.12 Proof of Theorem 8.2

Proof. Keeping Fig. 8.7 in mind, the proof is divided into three steps:

1. An upper bound for the control error $\max_{t \in T_{2,i}} |y_{\Delta i}(t)|$ in the time interval $t \in T_{2,i} = [t_{\mathrm{R}i}, \bar{t}_{\mathrm{R}i}]$ is determined which depends on the upper bound of the difference state $|\tilde{\boldsymbol{x}}_{\Delta i}(t_{\mathrm{R}i})|$

at time $t_{\mathrm{R}i}$.

2. A maximum value $\max_{t\in T_{1,i}} |\tilde{\boldsymbol{x}}_{\Delta i}(t)|$ for the difference state in the time interval $t \in T_{1,i} = [0, t_{\mathrm{R}i}]$ is calculated that depends on the switching time $t_{\mathrm{R}i}$.

3. The results of the previous steps are composed to get an upper bound for the overall control error $\max_{t\geq 0} |\boldsymbol{y}_{\Delta}(t)|$.

Control error $y_{\Delta i}(t)$ in the time interval $T_{2,i}$. According to (7.55), the control error $y_{\Delta i}(t)$ (7.33) in the time interval $t \in T_{2i}$ is bounded by

$$\max_{t\in T_{2,i}} |y_{\Delta i}(t)| \leq \left|\tilde{\boldsymbol{c}}_i^{\mathrm{T}} \mathrm{e}^{\tilde{\tilde{\boldsymbol{A}}}_i(t-t_{\mathrm{R}i})}\right| |\tilde{\boldsymbol{x}}_{\Delta i}(t_{\mathrm{R}i})| + \int_{t_{\mathrm{R}i}}^{\infty} \left|\tilde{\boldsymbol{c}}_i^{\mathrm{T}} \mathrm{e}^{\tilde{\tilde{\boldsymbol{A}}}_i\tau} k_{\mathrm{P}i}\tilde{\boldsymbol{b}}_i\right| \mathrm{d}\tau \max_{t\in T_{2,i}} |y_{\Delta \mathrm{s}i}(t)|$$

which leads to a maximum error

$$\max_{t\in T_{2,i}} |y_{\Delta i}(t)| \leq \boldsymbol{m}_{\mathrm{y}0i}^{\mathrm{T}} \max_{t\in T_{1i}} |\tilde{\boldsymbol{x}}_{\Delta i}(t)| + m_{\mathrm{yy}i} \max_{t\in T_{2i}} |y_{\Delta \mathrm{s}i}(t)| \tag{A.21}$$

since

$$|\tilde{\boldsymbol{x}}_{\Delta i}(t_{\mathrm{R}i})| \leq \max_{t\in T_{1i}} |\tilde{\boldsymbol{x}}_{\Delta i}(t)|$$

holds.

With this and (7.44) an upper bound for $y_{\Delta i}(t)$ is given by

$$\max_{t\in T_{2i}} |y_{\Delta i}(t)| \leq \boldsymbol{m}_{\mathrm{y}0i}^{\mathrm{T}} \max_{t\in T_{1i}} |\tilde{\boldsymbol{x}}_{\Delta i}(t)| + m_{\mathrm{yy}i} \sum_{j\in\mathcal{P}_{\mathrm{A}i}} k_{\mathrm{A}ij} \max_{t\in T_{2i}} |y_{\Delta j}(t)| \tag{A.22}$$

since $\max_{t\in T_{2i}} \tilde{k}_{ij}(t) \leq k_{\mathrm{A}ij}$.

State deviation $\tilde{x}_{\Delta i}(t)$ in the time interval T_{1i}. The model (7.43) can be rewritten as

$$\tilde{\tilde{\Sigma}}_{\Delta i}: \begin{cases} \frac{\mathrm{d}}{\mathrm{d}t}\tilde{\boldsymbol{x}}_{\Delta i}(t) = \tilde{\boldsymbol{A}}_i\tilde{\boldsymbol{x}}_{\Delta i}(t) + k_{\mathrm{P}i}\tilde{\boldsymbol{b}}_i e_i(t), & \tilde{\boldsymbol{x}}_{\Delta i}(0) = \tilde{\boldsymbol{x}}_{\Delta 0i} \\ y_{\Delta i}(t) = \tilde{\boldsymbol{c}}_i^{\mathrm{T}}\tilde{\boldsymbol{x}}_{\Delta i}(t) \end{cases} \tag{A.23}$$

with the local control error $e_i(t)$ as input which leads to

$$\tilde{\boldsymbol{x}}_{\Delta i}(t) = \mathrm{e}^{\tilde{\boldsymbol{A}}_i t}\tilde{\boldsymbol{x}}_{\Delta 0i} + \int_0^{t_{\mathrm{R}i}} \mathrm{e}^{\tilde{\boldsymbol{A}}_i(t-\tau)} k_{\mathrm{P}i}\tilde{\boldsymbol{b}}_i e_i(\tau)\mathrm{d}\tau, \quad \forall i\in\mathcal{N}, \quad t\in T_{1,i}. \tag{A.24}$$

Note that this model uses the uncontrolled system matrix $\tilde{\boldsymbol{A}}_i$. Therefore, a bound for $\tilde{\boldsymbol{x}}_{\Delta i}(t)$ is given as

$$\max_{t\in T_{1,i}} |\tilde{\boldsymbol{x}}_{\Delta i}(t)| = \left|\mathrm{e}^{\tilde{\boldsymbol{A}}_i t}\right| |\tilde{\boldsymbol{x}}_{\Delta 0i}| + \int_0^{t_{\mathrm{R}i}} \left|\mathrm{e}^{\tilde{\boldsymbol{A}}_i(t-\tau)} k_{\mathrm{P}i}\tilde{\boldsymbol{b}}_i\right| \mathrm{d}\tau \max_{t\in T_{1,i}} |e_i(t)|, \quad \forall i \in \mathcal{N}$$

which results to the maximum error

$$\max_{t\in T_{1i}} |\tilde{\boldsymbol{x}}_{\Delta i}(t)| \leq \boldsymbol{m}_{\mathrm{x}0i} \cdot \bar{\boldsymbol{x}}_{\Delta 0} + \boldsymbol{m}_{\mathrm{xe}i} \cdot \epsilon_i \tag{A.25}$$

since Assumption 8.2 and the switching condition in (8.11) guarantees that $e_i(t)$ does not exceed the switching threshold ϵ_i in the time interval $t \in T_{1i}$, i.e.,

$$\max_{t\in T_{1i}} |e_i(t)| \leq \epsilon_i, \quad \forall i \in \mathcal{N}$$

and $\tilde{\boldsymbol{x}}_{\Delta 0i} = \bar{\boldsymbol{x}}_{\Delta 0}$ holds according to (7.54). Due to Assumption 8.2, an upper bound for the switching time $t_{\mathrm{R}i}$ is given by $t_{\mathrm{B}i}$ which can be determined as stated in Lemma 8.1.

Composition. Equation (A.22), (A.25) and Assumption 8.2 yield

$$\max_{t\geq 0} |\boldsymbol{y}_{\Delta}(t)| \leq \boldsymbol{M}_{\mathrm{yy}} \boldsymbol{K}_{\mathrm{A}}^{*} \max_{t\geq 0} |\boldsymbol{y}_{\Delta}(t)| + \boldsymbol{m}_{\mathrm{y}0}. \tag{A.26}$$

Solving (A.26) for $\max_{t\geq 0} |\boldsymbol{y}_{\Delta}(t)|$ leads to (8.23). The inverse matrix $(\boldsymbol{I} - \boldsymbol{M}_{\mathrm{yy}}\boldsymbol{K}_{\mathrm{A}}^{*})^{-1}$ exists, because $\boldsymbol{K}_{\mathrm{A}}^{*}$ is a lower triangular matrix with vanishing diagonal elements. Hence, the matrix $\boldsymbol{I} - \boldsymbol{M}_{\mathrm{yy}}\boldsymbol{K}_{\mathrm{A}}^{*}$ has N identical eigenvalues with the value 1, i.e., $\lambda_i(\boldsymbol{I} - \boldsymbol{M}_{\mathrm{yy}}\boldsymbol{K}_{\mathrm{A}}^{*}) = 1$ for all $i \in \mathcal{N}$. Furthermore, the matrix $\boldsymbol{I} - \boldsymbol{M}_{\mathrm{yy}}\boldsymbol{K}_{\mathrm{A}}^{*}$ is an M-Matrix and, hence, its inverse a non-negative matrix. Consequently, the inequality sign remains when multiplying with the inverse matrix. □

A.13 Proof of Proposition 8.1

Proof. With (8.21) and (8.27), the relation (8.28) can be written as

$$m_{\mathrm{y}0\mathrm{A}i} + m_{\mathrm{yy}i} \sum_{j\in\mathcal{P}_{\mathrm{A}i}} k_{\mathrm{A}ij}\bar{y}_{\Delta\mathrm{A}j} \geq m_{\mathrm{y}0\mathrm{A}i} + m_{\mathrm{yy}i}\bar{\tilde{y}}_{\Delta\mathrm{A}l}, \quad l \in \tilde{\mathcal{P}}_{\mathrm{A}i}, \quad \forall i \in \mathcal{N}$$

which leads to

$$\sum_{j\in\mathcal{P}_{\mathrm{A}i}} k_{\mathrm{A}ij}\bar{y}_{\Delta\mathrm{A}j} \geq \bar{\tilde{y}}_{\Delta\mathrm{A}l}, \quad l \in \tilde{\mathcal{P}}_{\mathrm{A}i}, \quad \forall i \in \mathcal{N} \tag{A.27}$$

where $\tilde{\mathcal{P}}_{\mathrm{A}i}$, $(\forall i \in \mathcal{N})$, are the sets of predecessors within the communication graph $\tilde{\mathcal{G}}_{\mathrm{A}}$. In the following (A.27) is used to prove (8.28) in Proposition 8.1 by induction. The *base case* is to prove that (A.27) holds for $i = 1$ which leads to

$$\sum_{j \in \mathcal{P}_{\mathrm{A}1}} k_{\mathrm{A}1j} \bar{y}_{\Delta j} \geq \bar{y}^{*}_{\Delta l}, \quad l \in \tilde{\mathcal{P}}_{\mathrm{A}i}. \tag{A.28}$$

Since $\mathcal{P}_{\mathrm{A}1} = \tilde{\mathcal{P}}_{\mathrm{A}1} = \{0\}$ and $\bar{y}_{\Delta \mathrm{A}0} = \tilde{\bar{y}}_{\Delta \mathrm{A}0} = 0$, relation (A.28) holds. For the *induction step* $i = n + 1$ the relation

$$\sum_{j \in \mathcal{P}_{\mathrm{A}n+1}} k_{\mathrm{A}n+1,j} \bar{y}_{\Delta \mathrm{A}j} \geq \tilde{\bar{y}}_{\Delta \mathrm{A}l}, \quad l \in \tilde{\mathcal{P}}_{\mathrm{A}n+1}. \tag{A.29}$$

has to be fulfilled under the *induction hypothesis*

$$\sum_{j \in \mathcal{P}_{\mathrm{A}p}} k_{\mathrm{A}p,j} \bar{y}_{\Delta \mathrm{A}j} \geq \tilde{\bar{y}}_{\Delta \mathrm{A}l}, \quad l \in \tilde{\mathcal{P}}_{\mathrm{A}p}, \quad p \leq n. \tag{A.30}$$

If $\tilde{\mathcal{P}}_{\mathrm{A}l}$ with $l \in \tilde{\mathcal{P}}_{\mathrm{A}n+1}$ is chosen such that

$$\tilde{\mathcal{P}}_{\mathrm{A}l} = \{\arg\min_{p \in \mathcal{P}_{\mathrm{A}l}} (\tilde{\bar{y}}_{\Delta \mathrm{A}p})\}$$

holds true, then relation (A.29) holds true according to the induction hypothesis in (A.30) which completes the proof. □

A.14 Proof of Theorem 9.2

Proof. Consider the difference signal

$$\tilde{y}_{\Delta i}(t) = \begin{cases} y_{\Delta i}(t) & \text{if } |\hat{y}_{\mathrm{d}i}(t)| \leq \epsilon \\ 0 & \text{else} \end{cases} \tag{A.31}$$

where the zero in the second line indicates that the difference signal $\tilde{y}_{\Delta i}(t)$ does not have any meaning as long as the communication from C_i is interrupted. Equations (7.18) and (7.20) yield

$$y_{\Delta \mathrm{s}i}(t) = \sum_{j \in \mathcal{P}_{\mathrm{C}i}(t)} \tilde{k}_{ij}(t) \tilde{y}_{\Delta j}(t) \tag{A.32}$$

and, furthermore, with (9.26) the relation

$$\begin{aligned}\max_{t\geq 0}|y_{\Delta \mathrm{s}i}(t)| &\leq \max_{t\geq 0}\sum_{j\in\mathcal{P}_{\mathrm{C}i}(t)}\tilde{k}_{ij}(t)\,|\tilde{y}_{\Delta j}(t)| \\ &\leq \sum_{j\in\mathcal{P}_{\mathrm{B}i}} k_{\mathrm{d}ij}(t)\max_{t\geq 0}|\tilde{y}_{\Delta j}(t)|\,.\end{aligned}$$

holds true. According to (7.43) and (9.23), the relation

$$\begin{aligned}|y_{\Delta i}| &= \left|\int_0^t \boldsymbol{c}_i^{\mathrm{T}}\mathrm{e}^{\bar{\boldsymbol{A}}_i(t-\tau)}\boldsymbol{b}_i k_{\mathrm{P}i} y_{\mathrm{s}i}(\tau)\mathrm{d}\tau + y_{\mathrm{d}i}(t)\right| \\ &\leq m_{\mathrm{yy}i}\cdot\max_{t\geq 0}|y_{\Delta \mathrm{s}i}(t)| + |y_{\mathrm{d}i}(t)|\end{aligned}$$

holds true for all $i\in\mathcal{N}$ and all $t\geq 0$ with $m_{\mathrm{yy}i}$ from Lemma 7.3. Hence, for (9.21) and (A.31) with $\hat{\boldsymbol{x}}_{0i}=\boldsymbol{x}_{0i}$ one gets

$$\max_{t\geq 0}|\tilde{y}_{\Delta i}(t)| \leq \begin{cases} m_{\mathrm{yy}i}\cdot\max_{t\geq 0}|y_{\Delta \mathrm{s}i}(t)| + \beta_i\epsilon & \text{if } |\hat{y}_{\mathrm{d}i}(t)|\leq\epsilon \\ 0 & \text{else,}\end{cases} \quad \forall i\in\mathcal{N}. \tag{A.33}$$

Equations (A.32) and (A.33) result to

$$\begin{aligned}\max_{t\geq 0}|\tilde{y}_{\Delta i}(t)| &\leq m_{\mathrm{yy}i}\cdot\max_{t\geq 0}|y_{\Delta \mathrm{s}i}(t)| + \beta_i\epsilon \\ &\leq m_{\mathrm{yy}i}\cdot\sum_{j\in\mathcal{P}_{\mathrm{B}i}} k_{\mathrm{d}ij}(t)\max_{t\geq 0}|\tilde{y}_{\Delta j}(t)| + \beta_i\epsilon\end{aligned}$$

for all $i\in\mathcal{N}$ and all $t\geq 0$ and in vector notation

$$\begin{pmatrix}\max_{t\geq 0}|\tilde{y}_{\Delta 1}(t)| \\ \vdots \\ \max_{t\geq 0}|\tilde{y}_{\Delta N}(t)|\end{pmatrix} \leq \boldsymbol{M}_{\mathrm{yy}}\boldsymbol{K}_{\mathrm{d}}\begin{pmatrix}\max_{t\geq 0}|\tilde{y}_{\Delta 1}(t)| \\ \vdots \\ \max_{t\geq 0}|\tilde{y}_{\Delta N}(t)|\end{pmatrix} + \begin{pmatrix}\beta_1 \\ \vdots \\ \beta_N\end{pmatrix}\epsilon, \quad t\geq 0. \tag{A.34}$$

This inequality can be reformulated as

$$\begin{pmatrix}\max_{t\geq 0}|\tilde{y}_{\Delta 1}(t)| \\ \vdots \\ \max_{t\geq 0}|\tilde{y}_{\Delta N}(t)|\end{pmatrix} \leq (\boldsymbol{I}_N - \boldsymbol{M}_{\mathrm{yy}}\boldsymbol{K}_{\mathrm{d}})^{-1}\begin{pmatrix}\beta_1 \\ \vdots \\ \beta_N\end{pmatrix}\epsilon, \quad t\geq 0.$$

The inverse matrix $(\boldsymbol{I}-\boldsymbol{M}_{\mathrm{yy}}\boldsymbol{K}_{\mathrm{d}})^{-1}$ exists, because $\boldsymbol{K}_{\mathrm{d}}$ is a lower triangular matrix with vanishing diagonal elements. Hence, the matrix $\boldsymbol{I}-\boldsymbol{M}_{\mathrm{yy}}\boldsymbol{K}_{\mathrm{d}}$ has N identical eigenvalues with the value

1, i.e., $\lambda_i(\boldsymbol{I}-\boldsymbol{M}_{\rm yy}\boldsymbol{K}_{\rm d}) = 1$ for all $i \in \mathcal{N}$. Furthermore, the matrix $\boldsymbol{I}-\boldsymbol{M}_{\rm yy}\boldsymbol{K}_{\rm d}$ is an M-Matrix and, hence, its inverse a non-negative matrix. Consequently, the inequality sign remains when multiplying with the inverse matrix. Since for the undisturbed agents Σ_i, $(\forall i \in \mathcal{U})$, the relation

$$y_{\Delta i}(t) = \tilde{y}_{\Delta i}(t), \quad \forall i \in \mathcal{U}$$

holds true, the extraction of the particular control errors $y_{\Delta i}(t)$ by $\boldsymbol{\alpha}_i^{\rm T}$ leads to (9.25) which completes the proof. □

B Further information on the application examples

Table B.1: Analysis of the asymptotic stability for the overall coupled tanks.

j	i	ϵ_{ji}	$m_{\mathrm{ss}ji}\sum_{l\in\bar{\mathcal{P}}_{\mathrm{C}i}}\epsilon_{il}$	$\epsilon_{ji} > m_{\mathrm{ss}ji}\sum_{l\in\bar{\mathcal{P}}_{\mathrm{C}i}}\epsilon_{il}$
1	2	0.1001	0.0640	True
2	3	0.1001	0.0220	True
6	3	0.0499	0.0220	True
7	3	0.0499	0.0220	True
4	5	0.1001	0.0640	True
3	6	0.0499	0.0220	True
5	6	0.1001	0.0220	True
8	6	0.0499	0.0220	True
3	7	0.0499	0.0220	True
6	8	0.0499	0.0220	True
7	9	0.0499	0.0176	True
8	9	0.0499	0.0176	True
10	9	0.0492	0.0176	True
11	9	0.0492	0.0176	True
9	10	0.0492	0.0176	True
12	10	0.0499	0.0176	True
9	11	0.0492	0.0176	True
16	11	0.0499	0.0176	True
10	12	0.0492	0.0220	True
13	12	0.1001	0.0220	True
14	12	0.0499	0.0220	True
12	14	0.0499	0.0220	True
15	14	0.1001	0.0220	True
17	14	0.0499	0.0220	True
11	16	0.0492	0.0220	True
17	16	0.0499	0.0220	True
14	17	0.0499	0.0220	True
16	17	0.0499	0.0220	True
18	17	0.1001	0.0220	True
21	17	0.1001	0.0220	True
19	18	0.1001	0.0640	True
20	19	0.1001	0.0640	True
22	21	0.1001	0.0640	True
23	22	0.1001	0.0640	True

Table B.2: Analysis of the communication activation of the self-organizing controlled tanks for $\mathcal{D} = \{17\}$ by Proposition 4.2.

i	$\bar{\mathcal{F}}_{\mathrm{C}i}$	$\bar{\mathcal{F}}_{\mathrm{CD}i}$	h	ϵ_{hi}	$\bar{u}_{\mathrm{D}hi}$	$\epsilon_{hi} \leq \bar{s}_{\mathrm{D}hi}$
1	$\emptyset$	$\emptyset$				
2	$\{1\}$	$\emptyset$	1	0.1001	0.0640	False
3	$\{2,6,7\}$	$\emptyset$	2	0.1001	0.0220	False
			6	0.0499	0.0220	False
			7	0.0499	0.0220	False
4	$\emptyset$	$\emptyset$				
5	$\{4\}$	$\emptyset$	4	0.1001	0.0640	False
6	$\{3,5,8\}$	$\emptyset$	3	0.0499	0.0220	False
			5	0.1001	0.0220	False
			8	0.0499	0.0220	False
7	$\{3\}$	$\emptyset$	3	0.0499	0.0220	False
8	$\{6\}$	$\emptyset$	6	0.0499	0.0220	False
9	$\{7,8,10,11\}$	$\emptyset$	7	0.0499	0.0176	False
			8	0.0499	0.0176	False
			10	0.0492	0.0176	False
			11	0.0492	0.0176	False
10	$\{9,12\}$	$\emptyset$	9	0.0492	0.0176	False
			12	0.0499	0.0176	False
11	$\{9,16\}$	$\emptyset$	9	0.0492	0.0176	False
			16	0.0499	0.0176	False
12	$\{10,13,14\}$	$\emptyset$	10	0.0492	0.0220	False
			13	0.1001	0.0220	False
			14	0.0499	0.0220	False
13	$\emptyset$	$\emptyset$				
14	$\{12,15,17\}$	$\emptyset$	12	0.0499	0.0220	False
			15	0.1001	0.0220	False
			17	0.0499	0.0220	False
15	$\emptyset$	$\emptyset$				
16	$\{11,17\}$	$\emptyset$	11	0.0492	0.0220	False
			17	0.0499	0.0220	False
17	$\{14,16,18,21\}$	$\{14,16,18,21\}$	14	0.0499	0.1765	True
			16	0.0499	0.1765	True
			18	0.1001	0.1765	True
			21	0.1001	0.1765	True
18	$\{19\}$	$\emptyset$	19	0.1001	0.0640	False
19	$\{20\}$	$\emptyset$	20	0.1001	0.0640	False
20	$\emptyset$	$\emptyset$				
21	$\{22\}$	$\emptyset$	22	0.1001	0.0640	False
22	$\{23\}$	$\emptyset$	23	0.1001	0.0640	False
23	$\emptyset$	$\emptyset$				

Table B.3: Analysis of the communication activation for $\mathcal{D} = \mathcal{N}$.

i	$\bar{\mathcal{F}}_{\mathrm{C}i}$	$\bar{\mathcal{F}}_{\mathrm{CD}i}$	h	ϵ_{hi}	$\bar{s}_{\mathrm{D}hi}$	$\epsilon_{hi} \leq \bar{s}_{\mathrm{D}hi}$
1	$\emptyset$	$\emptyset$				
2	$\{1\}$	$\emptyset$	1	0.1001	0.0640	False
3	$\{2,6,7\}$	$\emptyset$	2	0.1001	0.0220	False
			6	0.0499	0.0220	False
			7	0.0499	0.0220	False
4	$\emptyset$	$\emptyset$				
5	$\{4\}$	$\emptyset$	4	0.1001	0.0640	False
6	$\{3,5,8\}$	$\{3,5,8\}$	3	0.0499	0.1544	True
			5	0.1001	0.1544	True
			8	0.0499	0.1544	True
7	$\{3\}$	$\emptyset$	3	0.0499	0.0220	False
8	$\{6\}$	$\emptyset$	6	0.0499	0.0220	False
9	$\{7,8,10,11\}$	$\{7,8,10,11\}$	7	0.0499	0.1787	True
			8	0.0499	0.1787	True
			10	0.0492	0.1787	True
			11	0.0492	0.1787	True
10	$\{9,12\}$	$\emptyset$	9	0.0492	0.0176	False
			12	0.0499	0.0176	False
11	$\{9,16\}$	$\emptyset$	9	0.0492	0.0176	False
			16	0.0499	0.0176	False
12	$\{10,13,14\}$	$\emptyset$	10	0.0492	0.0220	False
			13	0.1001	0.0220	False
			14	0.0499	0.0220	False
13	$\emptyset$	$\emptyset$				
14	$\{12,15,17\}$	$\emptyset$	12	0.0499	0.0220	False
			15	0.1001	0.0220	False
			17	0.0499	0.0220	False
15	$\emptyset$	$\emptyset$				
16	$\{11,17\}$	$\emptyset$	11	0.0492	0.0220	False
			17	0.0499	0.0220	False
17	$\{14,16,18,21\}$	$\{14,16,18,21\}$	14	0.0499	0.1765	True
			16	0.0499	0.1765	True
			18	0.1001	0.1765	True
			21	0.1001	0.1765	True
18	$\{19\}$	$\emptyset$	19	0.1001	0.0640	False
19	$\{20\}$	$\emptyset$	20	0.1001	0.0640	False
20	$\emptyset$	$\emptyset$				
21	$\{22\}$	$\emptyset$	22	0.1001	0.0640	False
22	$\{23\}$	$\emptyset$	23	0.1001	0.0640	False
23	$\emptyset$	$\emptyset$				

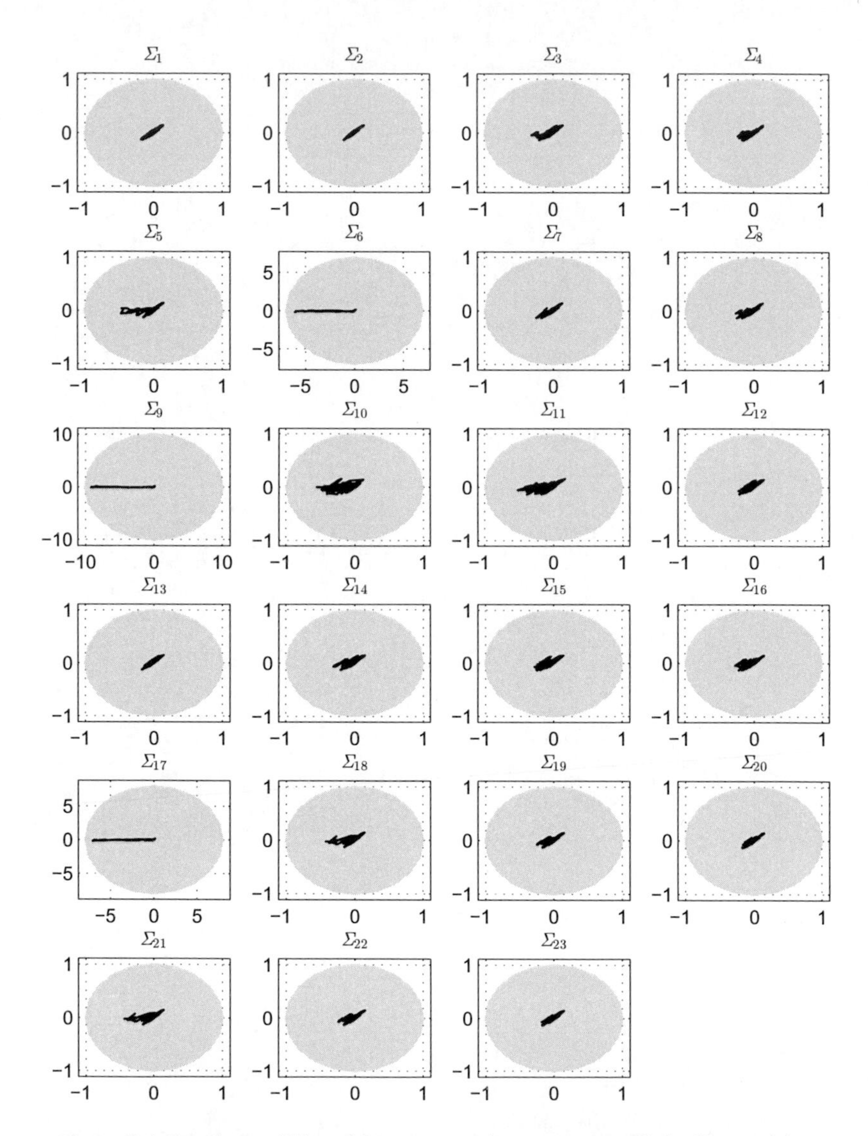

Figure B.1: Practical stability of the water supply system controlled self-organizing controller defined in (4.15).

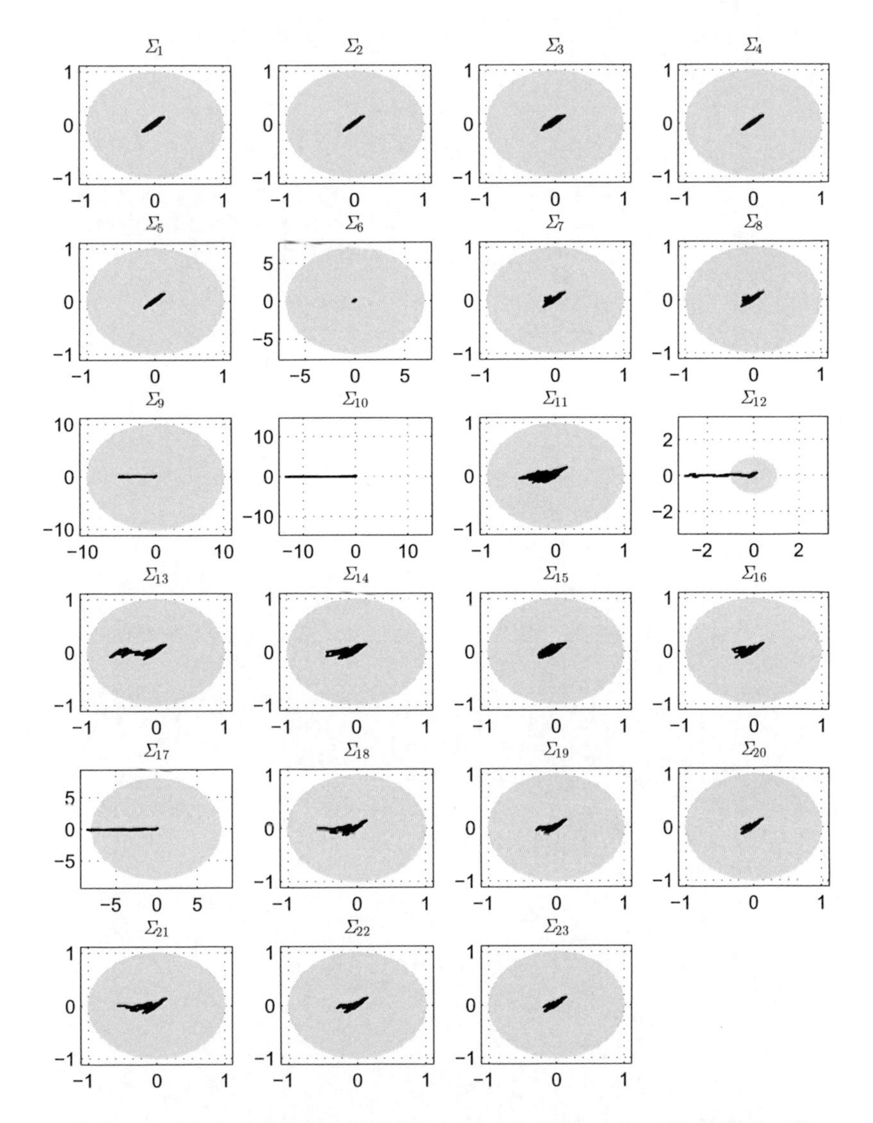

Figure B.2: Practical stability of the faulty water supply system controlled self-organizing controller defined in (4.15).

Table B.4: Parameters α_{ij}, ($\forall i, j \in \mathcal{N}, i \neq j$), for weighting the communication links.

$i \backslash j$	1	2	3	4	5	6	7	8	9	10	11	12	13	14	15	16	17	18	19	20	21	22	23
1		50.00	10.00	1.00	1.00	5.00	5.00	1.00	1.00	1.00	1.00	1.00	1.00	1.00	1.00	1.00	1.00	1.00	1.00	1.00	1.00	1.00	1.00
2	50.00		50.00	1.00	1.00	1.00	1.00	1.00	1.00	1.00	1.00	1.00	1.00	1.00	1.00	1.00	1.00	1.00	1.00	1.00	1.00	1.00	1.00
3	5.00	50.00		1.00	5.00	50.00	50.00	5.00	5.00	1.00	1.00	1.00	1.00	1.00	1.00	1.00	1.00	1.00	1.00	1.00	1.00	1.00	1.00
4	1.00	1.00	1.00		50.00	10.00	1.00	1.00	1.00	1.00	1.00	1.00	1.00	1.00	1.00	1.00	1.00	1.00	1.00	1.00	1.00	1.00	1.00
5	1.00	1.00	5.00	50.00		50.00	1.00	5.00	1.00	1.00	1.00	1.00	1.00	1.00	1.00	1.00	1.00	1.00	1.00	1.00	1.00	1.00	1.00
6	1.00	5.00	50.00	5.00	50.00		5.00	50.00	5.00	1.00	1.00	1.00	1.00	1.00	1.00	1.00	1.00	1.00	1.00	1.00	1.00	1.00	1.00
7	1.00	1.00	50.00	1.00	1.00	5.00		5.00	50.00	1.00	1.00	1.00	1.00	1.00	1.00	1.00	1.00	1.00	1.00	1.00	1.00	1.00	1.00
8	1.00	1.00	5.00	1.00	1.00	50.00	5.00		50.00	1.00	1.00	1.00	1.00	1.00	1.00	1.00	1.00	1.00	1.00	1.00	1.00	1.00	1.00
9	1.00	1.00	8.00	1.00	1.00	8.00	33.00	33.00		50.00	50.00	8.00	1.00	1.00	1.00	8.00	1.00	1.00	1.00	1.00	1.00	1.00	1.00
10	1.00	1.00	1.00	1.00	1.00	1.00	1.00	1.00	33.00		5.00	45.00	2.00	6.00	1.00	1.00	1.00	1.00	1.00	1.00	1.00	1.00	1.00
11	1.00	1.00	1.00	1.00	1.00	1.00	1.00	1.00	33.00	6.00		1.00	1.00	1.00	1.00	45.00	6.00	1.00	1.00	1.00	1.00	1.00	1.00
12	1.00	1.00	1.00	1.00	1.00	1.00	1.00	1.00	3.00	40.00	1.00		30.00	40.00	1.00	1.00	6.00	1.00	1.00	1.00	1.00	1.00	1.00
13	1.00	1.00	1.00	1.00	1.00	1.00	1.00	1.00	1.00	10.00	1.00	50.00		10.00	1.00	1.00	1.00	1.00	1.00	1.00	1.00	1.00	1.00
14	1.00	1.00	1.00	1.00	1.00	1.00	1.00	1.00	1.00	5.00	1.00	50.00	5.00		50.00	5.00	50.00	5.00	1.00	1.00	1.00	1.00	1.00
15	1.00	1.00	1.00	1.00	1.00	1.00	1.00	1.00	1.00	1.00	1.00	10.00	1.00	50.00		1.00	10.00	1.00	1.00	1.00	1.00	1.00	1.00
16	1.00	1.00	1.00	1.00	1.00	1.00	1.00	1.00	5.00	1.00	50.00	1.00	1.00	5.00	1.00		50.00	5.00	1.00	1.00	5.00	1.00	1.00
17	1.00	1.00	1.00	1.00	1.00	1.00	1.00	1.00	1.00	1.00	3.00	6.00	1.00	50.00	3.00	50.00		30.00	5.00	1.00	30.00	5.00	1.00
18	1.00	1.00	1.00	1.00	1.00	1.00	1.00	1.00	1.00	1.00	1.00	1.00	1.00	5.00	1.00	5.00	50.00		50.00	10.00	5.00	1.00	1.00
19	1.00	1.00	1.00	1.00	1.00	1.00	1.00	1.00	1.00	1.00	1.00	1.00	1.00	1.00	1.00	1.00	10.00	60.00		55.00	1.00	1.00	1.00
20	1.00	1.00	1.00	1.00	1.00	1.00	1.00	1.00	1.00	1.00	1.00	1.00	1.00	1.00	1.00	1.00	1.00	10.00	50.00		1.00	1.00	1.00
21	1.00	1.00	1.00	1.00	1.00	1.00	1.00	1.00	1.00	1.00	1.00	1.00	1.00	5.00	1.00	5.00	50.00	5.00	1.00	1.00		55.00	10.00
22	1.00	1.00	1.00	1.00	1.00	1.00	1.00	1.00	1.00	1.00	1.00	1.00	1.00	1.00	1.00	1.00	10.00	1.00	1.00	1.00	60.00		55.00
23	1.00	1.00	1.00	1.00	1.00	1.00	1.00	1.00	1.00	1.00	1.00	1.00	1.00	1.00	1.00	1.00	1.00	1.00	1.00	1.00	10.00	55.00	

Table B.5: Elements $\boldsymbol{K}_{\text{C}ij}$, ($\forall i \in \mathcal{N}, \forall j \in \{1, \ldots, 12\}$) of the central feedback gain matrix $\boldsymbol{K}_{\text{C}}$ for controlling the water supply system.

$i\backslash j$	1	2	3	4	5	6	7	8	9	10	11	12
1	$\begin{pmatrix} 3.50 & 0.12 \\ 0.43 & 9.01 \end{pmatrix}$	$\begin{pmatrix} -0.00 & 0.05 \\ -0.03 & 2.75 \end{pmatrix}$	$\begin{pmatrix} -0.00 & 0.00 \\ 0.00 & 0.15 \end{pmatrix}$	$\begin{pmatrix} -0.00 & -0.00 \\ -0.00 & -0.00 \end{pmatrix}$	$\begin{pmatrix} 0.00 & -0.00 \\ 0.00 & 0.00 \end{pmatrix}$	$\begin{pmatrix} -0.00 & 0.00 \\ 0.00 & 0.00 \end{pmatrix}$	$\begin{pmatrix} -0.00 & 0.00 \\ 0.00 & 0.00 \end{pmatrix}$	$\begin{pmatrix} -0.00 & -0.00 \\ -0.00 & -0.00 \end{pmatrix}$	$\begin{pmatrix} 0.00 & -0.00 \\ -0.00 & -0.00 \end{pmatrix}$	$\begin{pmatrix} 0.00 & -0.00 \\ -0.00 & -0.00 \end{pmatrix}$	$\begin{pmatrix} 0.00 & -0.00 \\ -0.00 & -0.00 \end{pmatrix}$	$\begin{pmatrix} 0.00 & 0.00 \\ 0.00 & 0.00 \end{pmatrix}$
2	$\begin{pmatrix} -0.00 & -0.01 \\ 0.16 & 2.75 \end{pmatrix}$	$\begin{pmatrix} 3.50 & 0.12 \\ 0.43 & 10.63 \end{pmatrix}$	$\begin{pmatrix} -0.00 & 0.01 \\ -0.00 & 1.01 \end{pmatrix}$	$\begin{pmatrix} 0.00 & 0.00 \\ -0.00 & 0.00 \end{pmatrix}$	$\begin{pmatrix} -0.00 & -0.00 \\ -0.00 & -0.00 \end{pmatrix}$	$\begin{pmatrix} -0.00 & 0.00 \\ -0.00 & 0.00 \end{pmatrix}$	$\begin{pmatrix} -0.00 & 0.00 \\ -0.00 & 0.00 \end{pmatrix}$	$\begin{pmatrix} 0.00 & -0.00 \\ 0.00 & -0.01 \end{pmatrix}$	$\begin{pmatrix} 0.00 & -0.00 \\ 0.00 & -0.00 \end{pmatrix}$	$\begin{pmatrix} 0.00 & 0.00 \\ 0.00 & 0.00 \end{pmatrix}$	$\begin{pmatrix} 0.00 & 0.00 \\ 0.00 & 0.00 \end{pmatrix}$	$\begin{pmatrix} -0.00 & 0.00 \\ -0.00 & 0.00 \end{pmatrix}$
3	$\begin{pmatrix} -0.00 & 0.00 \\ 0.01 & 0.17 \end{pmatrix}$	$\begin{pmatrix} -0.00 & -0.00 \\ 0.06 & 1.18 \end{pmatrix}$	$\begin{pmatrix} 3.84 & 0.06 \\ 0.18 & 10.39 \end{pmatrix}$	$\begin{pmatrix} -0.00 & 0.00 \\ 0.00 & 0.00 \end{pmatrix}$	$\begin{pmatrix} -0.00 & -0.00 \\ 0.00 & 0.00 \end{pmatrix}$	$\begin{pmatrix} -0.00 & 0.01 \\ 0.02 & 2.15 \end{pmatrix}$	$\begin{pmatrix} -0.00 & 0.01 \\ 0.02 & 2.14 \end{pmatrix}$	$\begin{pmatrix} -0.00 & -0.00 \\ -0.00 & 0.22 \end{pmatrix}$	$\begin{pmatrix} 0.00 & 0.00 \\ -0.00 & 0.12 \end{pmatrix}$	$\begin{pmatrix} 0.00 & 0.00 \\ -0.00 & 0.00 \end{pmatrix}$	$\begin{pmatrix} 0.00 & 0.00 \\ -0.00 & 0.00 \end{pmatrix}$	$\begin{pmatrix} -0.00 & 0.00 \\ 0.00 & -0.00 \end{pmatrix}$
4	$\begin{pmatrix} -0.00 & -0.00 \\ -0.00 & -0.00 \end{pmatrix}$	$\begin{pmatrix} 0.00 & -0.00 \\ 0.00 & 0.00 \end{pmatrix}$	$\begin{pmatrix} -0.00 & 0.00 \\ 0.00 & 0.00 \end{pmatrix}$	$\begin{pmatrix} 3.50 & 0.12 \\ 0.43 & 9.01 \end{pmatrix}$	$\begin{pmatrix} -0.00 & 0.05 \\ -0.03 & 2.75 \end{pmatrix}$	$\begin{pmatrix} -0.00 & 0.00 \\ 0.00 & 0.15 \end{pmatrix}$	$\begin{pmatrix} -0.00 & -0.00 \\ -0.00 & -0.00 \end{pmatrix}$	$\begin{pmatrix} -0.00 & 0.00 \\ 0.00 & 0.00 \end{pmatrix}$	$\begin{pmatrix} 0.00 & -0.00 \\ -0.00 & -0.00 \end{pmatrix}$	$\begin{pmatrix} 0.00 & -0.00 \\ -0.00 & -0.00 \end{pmatrix}$	$\begin{pmatrix} 0.00 & -0.00 \\ -0.00 & -0.00 \end{pmatrix}$	$\begin{pmatrix} 0.00 & 0.00 \\ 0.00 & 0.00 \end{pmatrix}$
5	$\begin{pmatrix} 0.00 & 0.00 \\ -0.00 & 0.00 \end{pmatrix}$	$\begin{pmatrix} -0.00 & -0.00 \\ -0.00 & -0.00 \end{pmatrix}$	$\begin{pmatrix} -0.00 & 0.00 \\ -0.00 & 0.00 \end{pmatrix}$	$\begin{pmatrix} -0.00 & -0.01 \\ 0.16 & 2.75 \end{pmatrix}$	$\begin{pmatrix} 3.50 & 0.12 \\ 0.43 & 10.63 \end{pmatrix}$	$\begin{pmatrix} -0.00 & 0.01 \\ -0.00 & 1.01 \end{pmatrix}$	$\begin{pmatrix} 0.00 & -0.00 \\ 0.00 & -0.01 \end{pmatrix}$	$\begin{pmatrix} -0.00 & 0.00 \\ -0.00 & 0.00 \end{pmatrix}$	$\begin{pmatrix} 0.00 & -0.00 \\ 0.00 & -0.00 \end{pmatrix}$	$\begin{pmatrix} 0.00 & 0.00 \\ 0.00 & 0.00 \end{pmatrix}$	$\begin{pmatrix} 0.00 & 0.00 \\ 0.00 & 0.00 \end{pmatrix}$	$\begin{pmatrix} -0.00 & 0.00 \\ -0.00 & 0.00 \end{pmatrix}$
6	$\begin{pmatrix} -0.00 & 0.00 \\ 0.00 & 0.00 \end{pmatrix}$	$\begin{pmatrix} -0.00 & -0.00 \\ 0.00 & 0.00 \end{pmatrix}$	$\begin{pmatrix} -0.00 & 0.01 \\ 0.02 & 2.15 \end{pmatrix}$	$\begin{pmatrix} -0.00 & 0.00 \\ 0.01 & 0.17 \end{pmatrix}$	$\begin{pmatrix} -0.00 & -0.00 \\ 0.06 & 1.18 \end{pmatrix}$	$\begin{pmatrix} 3.84 & 0.06 \\ 0.18 & 10.39 \end{pmatrix}$	$\begin{pmatrix} -0.00 & -0.00 \\ -0.00 & 0.22 \end{pmatrix}$	$\begin{pmatrix} -0.00 & 0.01 \\ 0.02 & 2.14 \end{pmatrix}$	$\begin{pmatrix} 0.00 & 0.00 \\ -0.00 & 0.12 \end{pmatrix}$	$\begin{pmatrix} 0.00 & 0.00 \\ -0.00 & 0.00 \end{pmatrix}$	$\begin{pmatrix} 0.00 & 0.00 \\ -0.00 & 0.00 \end{pmatrix}$	$\begin{pmatrix} -0.00 & 0.00 \\ 0.00 & -0.00 \end{pmatrix}$
7	$\begin{pmatrix} -0.00 & 0.00 \\ 0.00 & 0.00 \end{pmatrix}$	$\begin{pmatrix} -0.00 & -0.00 \\ 0.00 & 0.00 \end{pmatrix}$	$\begin{pmatrix} -0.00 & 0.01 \\ 0.02 & 2.14 \end{pmatrix}$	$\begin{pmatrix} -0.00 & -0.00 \\ -0.00 & -0.00 \end{pmatrix}$	$\begin{pmatrix} 0.00 & 0.00 \\ -0.00 & -0.01 \end{pmatrix}$	$\begin{pmatrix} -0.00 & -0.00 \\ -0.00 & 0.22 \end{pmatrix}$	$\begin{pmatrix} 3.84 & 0.06 \\ 0.19 & 9.93 \end{pmatrix}$	$\begin{pmatrix} -0.00 & -0.00 \\ -0.00 & -0.06 \end{pmatrix}$	$\begin{pmatrix} -0.00 & 0.00 \\ -0.00 & 0.52 \end{pmatrix}$	$\begin{pmatrix} -0.00 & -0.00 \\ -0.00 & -0.00 \end{pmatrix}$	$\begin{pmatrix} -0.00 & -0.00 \\ -0.00 & -0.00 \end{pmatrix}$	$\begin{pmatrix} 0.00 & -0.00 \\ 0.00 & -0.01 \end{pmatrix}$
8	$\begin{pmatrix} -0.00 & -0.00 \\ -0.00 & -0.00 \end{pmatrix}$	$\begin{pmatrix} 0.00 & 0.00 \\ -0.00 & -0.01 \end{pmatrix}$	$\begin{pmatrix} -0.00 & -0.00 \\ -0.00 & 0.22 \end{pmatrix}$	$\begin{pmatrix} -0.00 & 0.00 \\ 0.00 & 0.00 \end{pmatrix}$	$\begin{pmatrix} -0.00 & -0.00 \\ 0.00 & 0.00 \end{pmatrix}$	$\begin{pmatrix} -0.00 & 0.01 \\ 0.02 & 2.14 \end{pmatrix}$	$\begin{pmatrix} -0.00 & -0.00 \\ -0.00 & -0.06 \end{pmatrix}$	$\begin{pmatrix} 3.84 & 0.06 \\ 0.19 & 9.93 \end{pmatrix}$	$\begin{pmatrix} -0.00 & 0.00 \\ -0.00 & 0.52 \end{pmatrix}$	$\begin{pmatrix} -0.00 & -0.00 \\ -0.00 & -0.00 \end{pmatrix}$	$\begin{pmatrix} -0.00 & -0.00 \\ -0.00 & -0.00 \end{pmatrix}$	$\begin{pmatrix} 0.00 & -0.00 \\ 0.00 & -0.01 \end{pmatrix}$
9	$\begin{pmatrix} 0.00 & -0.00 \\ -0.00 & -0.00 \end{pmatrix}$	$\begin{pmatrix} 0.00 & 0.00 \\ -0.00 & -0.01 \end{pmatrix}$	$\begin{pmatrix} 0.00 & -0.00 \\ 0.00 & 0.27 \end{pmatrix}$	$\begin{pmatrix} 0.00 & -0.00 \\ -0.00 & -0.00 \end{pmatrix}$	$\begin{pmatrix} 0.00 & 0.00 \\ -0.00 & -0.01 \end{pmatrix}$	$\begin{pmatrix} 0.00 & -0.00 \\ 0.00 & 0.27 \end{pmatrix}$	$\begin{pmatrix} -0.00 & -0.00 \\ 0.02 & 1.15 \end{pmatrix}$	$\begin{pmatrix} -0.00 & -0.00 \\ 0.02 & 1.15 \end{pmatrix}$	$\begin{pmatrix} 4.01 & 0.02 \\ 0.06 & 10.40 \end{pmatrix}$	$\begin{pmatrix} -0.00 & 0.00 \\ 0.00 & 1.75 \end{pmatrix}$	$\begin{pmatrix} -0.00 & 0.00 \\ 0.00 & 1.75 \end{pmatrix}$	$\begin{pmatrix} -0.00 & -0.00 \\ -0.00 & 0.25 \end{pmatrix}$
10	$\begin{pmatrix} 0.00 & -0.00 \\ -0.00 & -0.00 \end{pmatrix}$	$\begin{pmatrix} 0.00 & 0.00 \\ 0.00 & 0.00 \end{pmatrix}$	$\begin{pmatrix} 0.00 & -0.00 \\ 0.00 & 0.00 \end{pmatrix}$	$\begin{pmatrix} 0.00 & -0.00 \\ -0.00 & -0.00 \end{pmatrix}$	$\begin{pmatrix} 0.00 & 0.00 \\ 0.00 & 0.00 \end{pmatrix}$	$\begin{pmatrix} 0.00 & -0.00 \\ 0.00 & 0.00 \end{pmatrix}$	$\begin{pmatrix} -0.00 & -0.00 \\ -0.00 & -0.00 \end{pmatrix}$	$\begin{pmatrix} -0.00 & -0.00 \\ -0.00 & -0.00 \end{pmatrix}$	$\begin{pmatrix} -0.00 & 0.00 \\ 0.00 & 1.75 \end{pmatrix}$	$\begin{pmatrix} 4.01 & 0.02 \\ 0.06 & 10.23 \end{pmatrix}$	$\begin{pmatrix} -0.00 & -0.00 \\ -0.00 & 0.15 \end{pmatrix}$	$\begin{pmatrix} -0.00 & 0.00 \\ 0.02 & 2.70 \end{pmatrix}$
11	$\begin{pmatrix} 0.00 & -0.00 \\ -0.00 & -0.00 \end{pmatrix}$	$\begin{pmatrix} 0.00 & 0.00 \\ 0.00 & 0.00 \end{pmatrix}$	$\begin{pmatrix} 0.00 & -0.00 \\ 0.00 & 0.00 \end{pmatrix}$	$\begin{pmatrix} 0.00 & -0.00 \\ -0.00 & -0.00 \end{pmatrix}$	$\begin{pmatrix} 0.00 & 0.00 \\ 0.00 & 0.00 \end{pmatrix}$	$\begin{pmatrix} 0.00 & -0.00 \\ 0.00 & 0.00 \end{pmatrix}$	$\begin{pmatrix} -0.00 & -0.00 \\ -0.00 & -0.00 \end{pmatrix}$	$\begin{pmatrix} -0.00 & -0.00 \\ -0.00 & -0.00 \end{pmatrix}$	$\begin{pmatrix} -0.00 & 0.00 \\ 0.00 & 1.75 \end{pmatrix}$	$\begin{pmatrix} -0.00 & -0.00 \\ -0.00 & 0.15 \end{pmatrix}$	$\begin{pmatrix} 4.01 & 0.02 \\ 0.06 & 10.23 \end{pmatrix}$	$\begin{pmatrix} 0.00 & -0.00 \\ -0.00 & -0.01 \end{pmatrix}$
12	$\begin{pmatrix} 0.00 & 0.00 \\ 0.00 & 0.00 \end{pmatrix}$	$\begin{pmatrix} -0.00 & -0.00 \\ 0.00 & 0.00 \end{pmatrix}$	$\begin{pmatrix} -0.00 & 0.00 \\ 0.00 & -0.00 \end{pmatrix}$	$\begin{pmatrix} 0.00 & 0.00 \\ 0.00 & 0.00 \end{pmatrix}$	$\begin{pmatrix} -0.00 & -0.00 \\ 0.00 & 0.00 \end{pmatrix}$	$\begin{pmatrix} -0.00 & 0.00 \\ 0.00 & -0.00 \end{pmatrix}$	$\begin{pmatrix} 0.00 & 0.00 \\ -0.00 & -0.01 \end{pmatrix}$	$\begin{pmatrix} 0.00 & 0.00 \\ -0.00 & -0.01 \end{pmatrix}$	$\begin{pmatrix} -0.00 & -0.00 \\ -0.00 & 0.11 \end{pmatrix}$	$\begin{pmatrix} -0.00 & 0.00 \\ 0.00 & 1.23 \end{pmatrix}$	$\begin{pmatrix} 0.00 & -0.00 \\ -0.00 & -0.00 \end{pmatrix}$	$\begin{pmatrix} 3.84 & 0.06 \\ 0.18 & 10.37 \end{pmatrix}$
13	$\begin{pmatrix} 0.00 & 0.00 \\ 0.00 & 0.00 \end{pmatrix}$	$\begin{pmatrix} -0.00 & -0.00 \\ -0.00 & -0.00 \end{pmatrix}$	$\begin{pmatrix} -0.00 & 0.00 \\ -0.00 & 0.00 \end{pmatrix}$	$\begin{pmatrix} 0.00 & 0.00 \\ 0.00 & 0.00 \end{pmatrix}$	$\begin{pmatrix} -0.00 & -0.00 \\ -0.00 & -0.00 \end{pmatrix}$	$\begin{pmatrix} -0.00 & 0.00 \\ -0.00 & 0.00 \end{pmatrix}$	$\begin{pmatrix} 0.00 & 0.00 \\ 0.00 & 0.00 \end{pmatrix}$	$\begin{pmatrix} 0.00 & 0.00 \\ 0.00 & 0.00 \end{pmatrix}$	$\begin{pmatrix} 0.00 & -0.00 \\ 0.00 & -0.00 \end{pmatrix}$	$\begin{pmatrix} -0.00 & 0.00 \\ -0.00 & 0.02 \end{pmatrix}$	$\begin{pmatrix} 0.00 & -0.00 \\ 0.00 & -0.00 \end{pmatrix}$	$\begin{pmatrix} -0.00 & 0.02 \\ -0.00 & 0.92 \end{pmatrix}$
14	$\begin{pmatrix} 0.00 & 0.00 \\ -0.00 & 0.00 \end{pmatrix}$	$\begin{pmatrix} -0.00 & -0.00 \\ -0.00 & -0.00 \end{pmatrix}$	$\begin{pmatrix} -0.00 & 0.00 \\ -0.00 & 0.00 \end{pmatrix}$	$\begin{pmatrix} 0.00 & 0.00 \\ -0.00 & 0.00 \end{pmatrix}$	$\begin{pmatrix} -0.00 & -0.00 \\ -0.00 & -0.00 \end{pmatrix}$	$\begin{pmatrix} -0.00 & 0.00 \\ -0.00 & 0.00 \end{pmatrix}$	$\begin{pmatrix} 0.00 & 0.00 \\ 0.00 & 0.00 \end{pmatrix}$	$\begin{pmatrix} 0.00 & 0.00 \\ 0.00 & 0.00 \end{pmatrix}$	$\begin{pmatrix} 0.00 & -0.00 \\ -0.00 & -0.00 \end{pmatrix}$	$\begin{pmatrix} -0.00 & 0.00 \\ -0.00 & 0.12 \end{pmatrix}$	$\begin{pmatrix} 0.00 & -0.00 \\ -0.00 & 0.00 \end{pmatrix}$	$\begin{pmatrix} -0.00 & 0.01 \\ 0.02 & 2.14 \end{pmatrix}$
15	$\begin{pmatrix} -0.00 & -0.00 \\ -0.00 & -0.00 \end{pmatrix}$	$\begin{pmatrix} 0.00 & 0.00 \\ 0.00 & 0.00 \end{pmatrix}$	$\begin{pmatrix} 0.00 & -0.00 \\ 0.00 & -0.00 \end{pmatrix}$	$\begin{pmatrix} -0.00 & -0.00 \\ -0.00 & -0.00 \end{pmatrix}$	$\begin{pmatrix} 0.00 & 0.00 \\ 0.00 & 0.00 \end{pmatrix}$	$\begin{pmatrix} 0.00 & -0.00 \\ 0.00 & -0.00 \end{pmatrix}$	$\begin{pmatrix} -0.00 & -0.00 \\ -0.00 & -0.00 \end{pmatrix}$	$\begin{pmatrix} -0.00 & -0.00 \\ -0.00 & -0.00 \end{pmatrix}$	$\begin{pmatrix} 0.00 & 0.00 \\ 0.00 & 0.00 \end{pmatrix}$	$\begin{pmatrix} 0.00 & -0.00 \\ 0.00 & -0.00 \end{pmatrix}$	$\begin{pmatrix} 0.00 & -0.00 \\ 0.00 & -0.00 \end{pmatrix}$	$\begin{pmatrix} -0.00 & 0.00 \\ -0.00 & 0.00 \end{pmatrix}$
16	$\begin{pmatrix} 0.00 & 0.00 \\ 0.00 & -0.00 \end{pmatrix}$	$\begin{pmatrix} -0.00 & -0.00 \\ 0.00 & 0.00 \end{pmatrix}$	$\begin{pmatrix} -0.00 & 0.00 \\ 0.00 & -0.00 \end{pmatrix}$	$\begin{pmatrix} 0.00 & 0.00 \\ 0.00 & -0.00 \end{pmatrix}$	$\begin{pmatrix} -0.00 & -0.00 \\ 0.00 & 0.00 \end{pmatrix}$	$\begin{pmatrix} -0.00 & 0.00 \\ 0.00 & -0.00 \end{pmatrix}$	$\begin{pmatrix} 0.00 & 0.00 \\ -0.00 & -0.01 \end{pmatrix}$	$\begin{pmatrix} 0.00 & 0.00 \\ -0.00 & -0.01 \end{pmatrix}$	$\begin{pmatrix} -0.00 & -0.00 \\ -0.00 & 0.11 \end{pmatrix}$	$\begin{pmatrix} 0.00 & -0.00 \\ -0.00 & -0.00 \end{pmatrix}$	$\begin{pmatrix} -0.00 & 0.00 \\ 0.00 & 1.23 \end{pmatrix}$	$\begin{pmatrix} 0.00 & -0.00 \\ -0.00 & -0.00 \end{pmatrix}$
17	$\begin{pmatrix} 0.00 & 0.00 \\ -0.00 & 0.00 \end{pmatrix}$	$\begin{pmatrix} -0.00 & -0.00 \\ -0.00 & -0.00 \end{pmatrix}$	$\begin{pmatrix} -0.00 & 0.00 \\ -0.00 & 0.00 \end{pmatrix}$	$\begin{pmatrix} 0.00 & 0.00 \\ -0.00 & 0.00 \end{pmatrix}$	$\begin{pmatrix} -0.00 & -0.00 \\ -0.00 & -0.00 \end{pmatrix}$	$\begin{pmatrix} -0.00 & 0.00 \\ -0.00 & 0.00 \end{pmatrix}$	$\begin{pmatrix} 0.00 & 0.00 \\ 0.00 & 0.00 \end{pmatrix}$	$\begin{pmatrix} 0.00 & 0.00 \\ 0.00 & 0.00 \end{pmatrix}$	$\begin{pmatrix} 0.00 & -0.00 \\ -0.00 & -0.00 \end{pmatrix}$	$\begin{pmatrix} 0.00 & -0.00 \\ -0.00 & 0.00 \end{pmatrix}$	$\begin{pmatrix} -0.00 & 0.00 \\ -0.00 & 0.11 \end{pmatrix}$	$\begin{pmatrix} -0.00 & -0.00 \\ -0.00 & 0.22 \end{pmatrix}$
18	$\begin{pmatrix} -0.00 & -0.00 \\ -0.00 & -0.00 \end{pmatrix}$	$\begin{pmatrix} 0.00 & 0.00 \\ 0.00 & 0.00 \end{pmatrix}$	$\begin{pmatrix} 0.00 & -0.00 \\ 0.00 & -0.00 \end{pmatrix}$	$\begin{pmatrix} -0.00 & -0.00 \\ -0.00 & -0.00 \end{pmatrix}$	$\begin{pmatrix} 0.00 & 0.00 \\ 0.00 & 0.00 \end{pmatrix}$	$\begin{pmatrix} 0.00 & -0.00 \\ 0.00 & -0.00 \end{pmatrix}$	$\begin{pmatrix} -0.00 & -0.00 \\ -0.00 & -0.00 \end{pmatrix}$	$\begin{pmatrix} -0.00 & -0.00 \\ -0.00 & -0.00 \end{pmatrix}$	$\begin{pmatrix} 0.00 & 0.00 \\ 0.00 & 0.00 \end{pmatrix}$	$\begin{pmatrix} 0.00 & -0.00 \\ 0.00 & -0.00 \end{pmatrix}$	$\begin{pmatrix} 0.00 & -0.00 \\ 0.00 & -0.00 \end{pmatrix}$	$\begin{pmatrix} 0.00 & -0.00 \\ 0.00 & -0.01 \end{pmatrix}$
19	$\begin{pmatrix} 0.00 & 0.00 \\ 0.00 & 0.00 \end{pmatrix}$	$\begin{pmatrix} -0.00 & -0.00 \\ -0.00 & -0.00 \end{pmatrix}$	$\begin{pmatrix} -0.00 & -0.00 \\ -0.00 & 0.00 \end{pmatrix}$	$\begin{pmatrix} 0.00 & 0.00 \\ 0.00 & 0.00 \end{pmatrix}$	$\begin{pmatrix} -0.00 & -0.00 \\ -0.00 & -0.00 \end{pmatrix}$	$\begin{pmatrix} -0.00 & -0.00 \\ -0.00 & 0.00 \end{pmatrix}$	$\begin{pmatrix} 0.00 & -0.00 \\ 0.00 & 0.00 \end{pmatrix}$	$\begin{pmatrix} 0.00 & -0.00 \\ 0.00 & 0.00 \end{pmatrix}$	$\begin{pmatrix} 0.00 & 0.00 \\ -0.00 & -0.00 \end{pmatrix}$	$\begin{pmatrix} 0.00 & -0.00 \\ -0.00 & 0.00 \end{pmatrix}$	$\begin{pmatrix} 0.00 & -0.00 \\ -0.00 & -0.00 \end{pmatrix}$	$\begin{pmatrix} -0.00 & -0.00 \\ -0.00 & -0.00 \end{pmatrix}$
20	$\begin{pmatrix} 0.00 & 0.00 \\ -0.00 & 0.00 \end{pmatrix}$	$\begin{pmatrix} -0.00 & -0.00 \\ -0.00 & -0.00 \end{pmatrix}$	$\begin{pmatrix} -0.00 & 0.00 \\ -0.00 & 0.00 \end{pmatrix}$	$\begin{pmatrix} 0.00 & 0.00 \\ -0.00 & 0.00 \end{pmatrix}$	$\begin{pmatrix} -0.00 & -0.00 \\ -0.00 & -0.00 \end{pmatrix}$	$\begin{pmatrix} -0.00 & 0.00 \\ -0.00 & 0.00 \end{pmatrix}$	$\begin{pmatrix} 0.00 & 0.00 \\ 0.00 & 0.00 \end{pmatrix}$	$\begin{pmatrix} 0.00 & 0.00 \\ 0.00 & 0.00 \end{pmatrix}$	$\begin{pmatrix} -0.00 & -0.00 \\ -0.00 & -0.00 \end{pmatrix}$	$\begin{pmatrix} -0.00 & 0.00 \\ -0.00 & 0.00 \end{pmatrix}$	$\begin{pmatrix} -0.00 & 0.00 \\ -0.00 & 0.00 \end{pmatrix}$	$\begin{pmatrix} -0.00 & 0.00 \\ -0.00 & 0.00 \end{pmatrix}$
21	$\begin{pmatrix} -0.00 & -0.00 \\ -0.00 & -0.00 \end{pmatrix}$	$\begin{pmatrix} 0.00 & 0.00 \\ 0.00 & 0.00 \end{pmatrix}$	$\begin{pmatrix} 0.00 & -0.00 \\ 0.00 & -0.00 \end{pmatrix}$	$\begin{pmatrix} -0.00 & -0.00 \\ -0.00 & -0.00 \end{pmatrix}$	$\begin{pmatrix} 0.00 & 0.00 \\ 0.00 & 0.00 \end{pmatrix}$	$\begin{pmatrix} 0.00 & -0.00 \\ 0.00 & -0.00 \end{pmatrix}$	$\begin{pmatrix} -0.00 & -0.00 \\ -0.00 & -0.00 \end{pmatrix}$	$\begin{pmatrix} -0.00 & -0.00 \\ -0.00 & -0.00 \end{pmatrix}$	$\begin{pmatrix} 0.00 & 0.00 \\ 0.00 & 0.00 \end{pmatrix}$	$\begin{pmatrix} 0.00 & -0.00 \\ 0.00 & -0.00 \end{pmatrix}$	$\begin{pmatrix} 0.00 & -0.00 \\ 0.00 & -0.00 \end{pmatrix}$	$\begin{pmatrix} 0.00 & -0.00 \\ 0.00 & -0.01 \end{pmatrix}$
22	$\begin{pmatrix} 0.00 & 0.00 \\ 0.00 & 0.00 \end{pmatrix}$	$\begin{pmatrix} -0.00 & -0.00 \\ -0.00 & -0.00 \end{pmatrix}$	$\begin{pmatrix} -0.00 & -0.00 \\ -0.00 & 0.00 \end{pmatrix}$	$\begin{pmatrix} 0.00 & 0.00 \\ 0.00 & 0.00 \end{pmatrix}$	$\begin{pmatrix} -0.00 & -0.00 \\ -0.00 & -0.00 \end{pmatrix}$	$\begin{pmatrix} -0.00 & -0.00 \\ -0.00 & 0.00 \end{pmatrix}$	$\begin{pmatrix} 0.00 & -0.00 \\ 0.00 & 0.00 \end{pmatrix}$	$\begin{pmatrix} 0.00 & -0.00 \\ 0.00 & 0.00 \end{pmatrix}$	$\begin{pmatrix} 0.00 & 0.00 \\ -0.00 & -0.00 \end{pmatrix}$	$\begin{pmatrix} 0.00 & -0.00 \\ -0.00 & 0.00 \end{pmatrix}$	$\begin{pmatrix} 0.00 & -0.00 \\ -0.00 & -0.00 \end{pmatrix}$	$\begin{pmatrix} -0.00 & -0.00 \\ -0.00 & -0.00 \end{pmatrix}$
23	$\begin{pmatrix} 0.00 & 0.00 \\ -0.00 & 0.00 \end{pmatrix}$	$\begin{pmatrix} -0.00 & -0.00 \\ -0.00 & -0.00 \end{pmatrix}$	$\begin{pmatrix} -0.00 & 0.00 \\ -0.00 & 0.00 \end{pmatrix}$	$\begin{pmatrix} 0.00 & 0.00 \\ -0.00 & 0.00 \end{pmatrix}$	$\begin{pmatrix} -0.00 & -0.00 \\ -0.00 & -0.00 \end{pmatrix}$	$\begin{pmatrix} -0.00 & 0.00 \\ -0.00 & 0.00 \end{pmatrix}$	$\begin{pmatrix} 0.00 & 0.00 \\ 0.00 & 0.00 \end{pmatrix}$	$\begin{pmatrix} 0.00 & 0.00 \\ 0.00 & 0.00 \end{pmatrix}$	$\begin{pmatrix} -0.00 & -0.00 \\ -0.00 & -0.00 \end{pmatrix}$	$\begin{pmatrix} -0.00 & 0.00 \\ -0.00 & 0.00 \end{pmatrix}$	$\begin{pmatrix} -0.00 & 0.00 \\ -0.00 & 0.00 \end{pmatrix}$	$\begin{pmatrix} -0.00 & 0.00 \\ -0.00 & 0.00 \end{pmatrix}$

Table B.6: Elements $\boldsymbol{K}_{\mathrm{C}ij}$, $(\forall i \in \mathcal{N}, \forall j \in \{13, \ldots, 23\})$ of the central feedback gain matrix $\boldsymbol{K}_{\mathrm{C}}$ for controlling the water supply system.

$i \backslash j$	13	14	15	16	17	18	19	20	21	22	23
1	$\begin{pmatrix} 0.00 & 0.00 \\ 0.00 & 0.00 \end{pmatrix}$	$\begin{pmatrix} 0.00 & -0.00 \\ 0.00 & 0.00 \end{pmatrix}$	$\begin{pmatrix} -0.00 & -0.00 \\ -0.00 & -0.00 \end{pmatrix}$	$\begin{pmatrix} 0.00 & 0.00 \\ 0.00 & -0.00 \end{pmatrix}$	$\begin{pmatrix} 0.00 & -0.00 \\ 0.00 & 0.00 \end{pmatrix}$	$\begin{pmatrix} -0.00 & -0.00 \\ -0.00 & -0.00 \end{pmatrix}$	$\begin{pmatrix} 0.00 & 0.00 \\ 0.00 & 0.00 \end{pmatrix}$	$\begin{pmatrix} 0.00 & -0.00 \\ 0.00 & 0.00 \end{pmatrix}$	$\begin{pmatrix} -0.00 & -0.00 \\ -0.00 & -0.00 \end{pmatrix}$	$\begin{pmatrix} 0.00 & 0.00 \\ 0.00 & 0.00 \end{pmatrix}$	$\begin{pmatrix} 0.00 & -0.00 \\ 0.00 & 0.00 \end{pmatrix}$
2	$\begin{pmatrix} -0.00 & -0.00 \\ -0.00 & -0.00 \end{pmatrix}$	$\begin{pmatrix} -0.00 & -0.00 \\ -0.00 & -0.00 \end{pmatrix}$	$\begin{pmatrix} 0.00 & 0.00 \\ 0.00 & 0.00 \end{pmatrix}$	$\begin{pmatrix} -0.00 & 0.00 \\ -0.00 & 0.00 \end{pmatrix}$	$\begin{pmatrix} -0.00 & -0.00 \\ -0.00 & -0.00 \end{pmatrix}$	$\begin{pmatrix} 0.00 & 0.00 \\ 0.00 & 0.00 \end{pmatrix}$	$\begin{pmatrix} -0.00 & -0.00 \\ -0.00 & -0.00 \end{pmatrix}$	$\begin{pmatrix} -0.00 & -0.00 \\ -0.00 & -0.00 \end{pmatrix}$	$\begin{pmatrix} 0.00 & 0.00 \\ 0.00 & 0.00 \end{pmatrix}$	$\begin{pmatrix} -0.00 & -0.00 \\ -0.00 & -0.00 \end{pmatrix}$	$\begin{pmatrix} -0.00 & -0.00 \\ -0.00 & -0.00 \end{pmatrix}$
3	$\begin{pmatrix} -0.00 & -0.00 \\ 0.00 & 0.00 \end{pmatrix}$	$\begin{pmatrix} -0.00 & -0.00 \\ 0.00 & 0.00 \end{pmatrix}$	$\begin{pmatrix} 0.00 & 0.00 \\ -0.00 & -0.00 \end{pmatrix}$	$\begin{pmatrix} -0.00 & 0.00 \\ 0.00 & -0.00 \end{pmatrix}$	$\begin{pmatrix} -0.00 & -0.00 \\ 0.00 & 0.00 \end{pmatrix}$	$\begin{pmatrix} 0.00 & 0.00 \\ -0.00 & -0.00 \end{pmatrix}$	$\begin{pmatrix} -0.00 & -0.00 \\ -0.00 & 0.00 \end{pmatrix}$	$\begin{pmatrix} -0.00 & -0.00 \\ 0.00 & 0.00 \end{pmatrix}$	$\begin{pmatrix} 0.00 & 0.00 \\ -0.00 & -0.00 \end{pmatrix}$	$\begin{pmatrix} -0.00 & -0.00 \\ -0.00 & 0.00 \end{pmatrix}$	$\begin{pmatrix} -0.00 & -0.00 \\ 0.00 & 0.00 \end{pmatrix}$
4	$\begin{pmatrix} 0.00 & 0.00 \\ 0.00 & 0.00 \end{pmatrix}$	$\begin{pmatrix} 0.00 & -0.00 \\ 0.00 & 0.00 \end{pmatrix}$	$\begin{pmatrix} -0.00 & -0.00 \\ -0.00 & -0.00 \end{pmatrix}$	$\begin{pmatrix} 0.00 & 0.00 \\ 0.00 & -0.00 \end{pmatrix}$	$\begin{pmatrix} 0.00 & -0.00 \\ 0.00 & 0.00 \end{pmatrix}$	$\begin{pmatrix} -0.00 & -0.00 \\ -0.00 & -0.00 \end{pmatrix}$	$\begin{pmatrix} 0.00 & 0.00 \\ 0.00 & 0.00 \end{pmatrix}$	$\begin{pmatrix} 0.00 & -0.00 \\ 0.00 & 0.00 \end{pmatrix}$	$\begin{pmatrix} -0.00 & -0.00 \\ -0.00 & -0.00 \end{pmatrix}$	$\begin{pmatrix} 0.00 & 0.00 \\ 0.00 & 0.00 \end{pmatrix}$	$\begin{pmatrix} 0.00 & -0.00 \\ 0.00 & 0.00 \end{pmatrix}$
5	$\begin{pmatrix} -0.00 & -0.00 \\ -0.00 & -0.00 \end{pmatrix}$	$\begin{pmatrix} -0.00 & -0.00 \\ -0.00 & -0.00 \end{pmatrix}$	$\begin{pmatrix} 0.00 & 0.00 \\ 0.00 & 0.00 \end{pmatrix}$	$\begin{pmatrix} -0.00 & 0.00 \\ -0.00 & 0.00 \end{pmatrix}$	$\begin{pmatrix} -0.00 & -0.00 \\ -0.00 & -0.00 \end{pmatrix}$	$\begin{pmatrix} 0.00 & 0.00 \\ 0.00 & 0.00 \end{pmatrix}$	$\begin{pmatrix} -0.00 & -0.00 \\ -0.00 & -0.00 \end{pmatrix}$	$\begin{pmatrix} -0.00 & -0.00 \\ -0.00 & -0.00 \end{pmatrix}$	$\begin{pmatrix} 0.00 & 0.00 \\ 0.00 & 0.00 \end{pmatrix}$	$\begin{pmatrix} -0.00 & -0.00 \\ -0.00 & -0.00 \end{pmatrix}$	$\begin{pmatrix} -0.00 & -0.00 \\ -0.00 & -0.00 \end{pmatrix}$
6	$\begin{pmatrix} -0.00 & -0.00 \\ 0.00 & 0.00 \end{pmatrix}$	$\begin{pmatrix} -0.00 & -0.00 \\ 0.00 & 0.00 \end{pmatrix}$	$\begin{pmatrix} 0.00 & 0.00 \\ -0.00 & -0.00 \end{pmatrix}$	$\begin{pmatrix} -0.00 & 0.00 \\ 0.00 & -0.00 \end{pmatrix}$	$\begin{pmatrix} -0.00 & -0.00 \\ 0.00 & 0.00 \end{pmatrix}$	$\begin{pmatrix} 0.00 & 0.00 \\ -0.00 & -0.00 \end{pmatrix}$	$\begin{pmatrix} -0.00 & -0.00 \\ -0.00 & 0.00 \end{pmatrix}$	$\begin{pmatrix} -0.00 & -0.00 \\ 0.00 & 0.00 \end{pmatrix}$	$\begin{pmatrix} 0.00 & 0.00 \\ -0.00 & -0.00 \end{pmatrix}$	$\begin{pmatrix} -0.00 & -0.00 \\ -0.00 & 0.00 \end{pmatrix}$	$\begin{pmatrix} -0.00 & -0.00 \\ 0.00 & 0.00 \end{pmatrix}$
7	$\begin{pmatrix} 0.00 & 0.00 \\ 0.00 & 0.00 \end{pmatrix}$	$\begin{pmatrix} 0.00 & 0.00 \\ 0.00 & 0.00 \end{pmatrix}$	$\begin{pmatrix} -0.00 & -0.00 \\ -0.00 & -0.00 \end{pmatrix}$	$\begin{pmatrix} 0.00 & -0.00 \\ 0.00 & -0.01 \end{pmatrix}$	$\begin{pmatrix} 0.00 & 0.00 \\ 0.00 & 0.00 \end{pmatrix}$	$\begin{pmatrix} -0.00 & -0.00 \\ -0.00 & -0.00 \end{pmatrix}$	$\begin{pmatrix} 0.00 & 0.00 \\ -0.00 & 0.00 \end{pmatrix}$	$\begin{pmatrix} 0.00 & 0.00 \\ 0.00 & 0.00 \end{pmatrix}$	$\begin{pmatrix} -0.00 & -0.00 \\ -0.00 & -0.00 \end{pmatrix}$	$\begin{pmatrix} 0.00 & 0.00 \\ -0.00 & 0.00 \end{pmatrix}$	$\begin{pmatrix} 0.00 & 0.00 \\ 0.00 & 0.00 \end{pmatrix}$
8	$\begin{pmatrix} 0.00 & 0.00 \\ 0.00 & 0.00 \end{pmatrix}$	$\begin{pmatrix} 0.00 & 0.00 \\ 0.00 & 0.00 \end{pmatrix}$	$\begin{pmatrix} -0.00 & -0.00 \\ -0.00 & -0.00 \end{pmatrix}$	$\begin{pmatrix} 0.00 & -0.00 \\ 0.00 & -0.01 \end{pmatrix}$	$\begin{pmatrix} 0.00 & 0.00 \\ 0.00 & 0.00 \end{pmatrix}$	$\begin{pmatrix} -0.00 & -0.00 \\ -0.00 & -0.00 \end{pmatrix}$	$\begin{pmatrix} 0.00 & 0.00 \\ -0.00 & 0.00 \end{pmatrix}$	$\begin{pmatrix} 0.00 & 0.00 \\ 0.00 & 0.00 \end{pmatrix}$	$\begin{pmatrix} -0.00 & -0.00 \\ -0.00 & -0.00 \end{pmatrix}$	$\begin{pmatrix} 0.00 & 0.00 \\ -0.00 & 0.00 \end{pmatrix}$	$\begin{pmatrix} 0.00 & 0.00 \\ 0.00 & 0.00 \end{pmatrix}$
9	$\begin{pmatrix} 0.00 & 0.00 \\ -0.00 & -0.01 \end{pmatrix}$	$\begin{pmatrix} 0.00 & -0.00 \\ -0.00 & -0.00 \end{pmatrix}$	$\begin{pmatrix} 0.00 & 0.00 \\ 0.00 & 0.00 \end{pmatrix}$	$\begin{pmatrix} -0.00 & -0.00 \\ -0.00 & 0.25 \end{pmatrix}$	$\begin{pmatrix} 0.00 & -0.00 \\ -0.00 & -0.00 \end{pmatrix}$	$\begin{pmatrix} 0.00 & 0.00 \\ 0.00 & 0.00 \end{pmatrix}$	$\begin{pmatrix} 0.00 & -0.00 \\ 0.00 & -0.00 \end{pmatrix}$	$\begin{pmatrix} -0.00 & -0.00 \\ -0.00 & -0.00 \end{pmatrix}$	$\begin{pmatrix} 0.00 & 0.00 \\ 0.00 & 0.00 \end{pmatrix}$	$\begin{pmatrix} 0.00 & -0.00 \\ 0.00 & -0.00 \end{pmatrix}$	$\begin{pmatrix} -0.00 & -0.00 \\ -0.00 & -0.00 \end{pmatrix}$
10	$\begin{pmatrix} -0.00 & -0.00 \\ 0.00 & 0.04 \end{pmatrix}$	$\begin{pmatrix} -0.00 & -0.00 \\ 0.00 & 0.25 \end{pmatrix}$	$\begin{pmatrix} 0.00 & 0.00 \\ -0.00 & -0.01 \end{pmatrix}$	$\begin{pmatrix} 0.00 & -0.00 \\ -0.00 & -0.01 \end{pmatrix}$	$\begin{pmatrix} 0.00 & -0.00 \\ -0.00 & 0.01 \end{pmatrix}$	$\begin{pmatrix} 0.00 & 0.00 \\ -0.00 & -0.00 \end{pmatrix}$	$\begin{pmatrix} 0.00 & -0.00 \\ -0.00 & 0.00 \end{pmatrix}$	$\begin{pmatrix} -0.00 & -0.00 \\ 0.00 & 0.00 \end{pmatrix}$	$\begin{pmatrix} 0.00 & 0.00 \\ -0.00 & -0.00 \end{pmatrix}$	$\begin{pmatrix} 0.00 & -0.00 \\ -0.00 & 0.00 \end{pmatrix}$	$\begin{pmatrix} -0.00 & -0.00 \\ 0.00 & 0.00 \end{pmatrix}$
11	$\begin{pmatrix} 0.00 & 0.00 \\ -0.00 & -0.00 \end{pmatrix}$	$\begin{pmatrix} 0.00 & -0.00 \\ -0.00 & 0.01 \end{pmatrix}$	$\begin{pmatrix} 0.00 & 0.00 \\ -0.00 & -0.00 \end{pmatrix}$	$\begin{pmatrix} -0.00 & 0.00 \\ 0.02 & 2.69 \end{pmatrix}$	$\begin{pmatrix} -0.00 & -0.00 \\ 0.00 & 0.25 \end{pmatrix}$	$\begin{pmatrix} 0.00 & 0.00 \\ -0.00 & -0.01 \end{pmatrix}$	$\begin{pmatrix} 0.00 & -0.00 \\ -0.00 & -0.00 \end{pmatrix}$	$\begin{pmatrix} -0.00 & -0.00 \\ 0.00 & 0.00 \end{pmatrix}$	$\begin{pmatrix} 0.00 & 0.00 \\ -0.00 & -0.01 \end{pmatrix}$	$\begin{pmatrix} 0.00 & -0.00 \\ -0.00 & -0.00 \end{pmatrix}$	$\begin{pmatrix} -0.00 & -0.00 \\ 0.00 & 0.00 \end{pmatrix}$
12	$\begin{pmatrix} -0.00 & -0.00 \\ 0.06 & 1.07 \end{pmatrix}$	$\begin{pmatrix} -0.00 & 0.01 \\ 0.02 & 2.14 \end{pmatrix}$	$\begin{pmatrix} -0.00 & -0.00 \\ 0.00 & 0.00 \end{pmatrix}$	$\begin{pmatrix} 0.00 & -0.00 \\ -0.00 & -0.00 \end{pmatrix}$	$\begin{pmatrix} -0.00 & -0.00 \\ -0.00 & 0.22 \end{pmatrix}$	$\begin{pmatrix} 0.00 & 0.00 \\ -0.00 & -0.01 \end{pmatrix}$	$\begin{pmatrix} -0.00 & -0.00 \\ -0.00 & -0.00 \end{pmatrix}$	$\begin{pmatrix} -0.00 & -0.00 \\ 0.00 & 0.00 \end{pmatrix}$	$\begin{pmatrix} 0.00 & 0.00 \\ -0.00 & -0.01 \end{pmatrix}$	$\begin{pmatrix} -0.00 & -0.00 \\ -0.00 & -0.00 \end{pmatrix}$	$\begin{pmatrix} -0.00 & -0.00 \\ 0.00 & 0.00 \end{pmatrix}$
13	$\begin{pmatrix} 3.50 & 0.13 \\ 0.45 & 9.34 \end{pmatrix}$	$\begin{pmatrix} -0.00 & 0.00 \\ -0.00 & 0.00 \end{pmatrix}$	$\begin{pmatrix} -0.00 & -0.00 \\ -0.00 & -0.00 \end{pmatrix}$	$\begin{pmatrix} 0.00 & 0.00 \\ 0.00 & 0.00 \end{pmatrix}$	$\begin{pmatrix} 0.00 & -0.00 \\ 0.00 & -0.01 \end{pmatrix}$	$\begin{pmatrix} 0.00 & 0.00 \\ 0.00 & 0.00 \end{pmatrix}$	$\begin{pmatrix} -0.00 & -0.00 \\ 0.00 & -0.00 \end{pmatrix}$	$\begin{pmatrix} -0.00 & -0.00 \\ -0.00 & -0.00 \end{pmatrix}$	$\begin{pmatrix} 0.00 & 0.00 \\ 0.00 & 0.00 \end{pmatrix}$	$\begin{pmatrix} -0.00 & -0.00 \\ 0.00 & -0.00 \end{pmatrix}$	$\begin{pmatrix} -0.00 & -0.00 \\ -0.00 & -0.00 \end{pmatrix}$
14	$\begin{pmatrix} -0.00 & -0.00 \\ 0.00 & 0.00 \end{pmatrix}$	$\begin{pmatrix} 3.84 & 0.06 \\ 0.18 & 10.38 \end{pmatrix}$	$\begin{pmatrix} -0.00 & -0.00 \\ 0.06 & 1.07 \end{pmatrix}$	$\begin{pmatrix} -0.00 & -0.00 \\ -0.00 & 0.23 \end{pmatrix}$	$\begin{pmatrix} -0.00 & 0.01 \\ 0.02 & 2.15 \end{pmatrix}$	$\begin{pmatrix} -0.00 & -0.00 \\ 0.00 & 0.00 \end{pmatrix}$	$\begin{pmatrix} -0.00 & 0.00 \\ 0.00 & 0.00 \end{pmatrix}$	$\begin{pmatrix} 0.00 & 0.00 \\ -0.00 & -0.00 \end{pmatrix}$	$\begin{pmatrix} -0.00 & -0.00 \\ 0.00 & 0.00 \end{pmatrix}$	$\begin{pmatrix} -0.00 & 0.00 \\ 0.00 & 0.00 \end{pmatrix}$	$\begin{pmatrix} 0.00 & 0.00 \\ -0.00 & -0.00 \end{pmatrix}$
15	$\begin{pmatrix} -0.00 & -0.00 \\ -0.00 & -0.00 \end{pmatrix}$	$\begin{pmatrix} -0.00 & 0.02 \\ -0.00 & 0.92 \end{pmatrix}$	$\begin{pmatrix} 3.50 & 0.13 \\ 0.45 & 9.34 \end{pmatrix}$	$\begin{pmatrix} 0.00 & -0.00 \\ 0.00 & -0.01 \end{pmatrix}$	$\begin{pmatrix} -0.00 & 0.00 \\ -0.00 & 0.00 \end{pmatrix}$	$\begin{pmatrix} -0.00 & -0.00 \\ -0.00 & -0.00 \end{pmatrix}$	$\begin{pmatrix} 0.00 & 0.00 \\ -0.00 & 0.00 \end{pmatrix}$	$\begin{pmatrix} 0.00 & 0.00 \\ 0.00 & 0.00 \end{pmatrix}$	$\begin{pmatrix} -0.00 & -0.00 \\ -0.00 & -0.00 \end{pmatrix}$	$\begin{pmatrix} 0.00 & 0.00 \\ -0.00 & 0.00 \end{pmatrix}$	$\begin{pmatrix} 0.00 & 0.00 \\ 0.00 & 0.00 \end{pmatrix}$
16	$\begin{pmatrix} 0.00 & 0.00 \\ 0.00 & 0.00 \end{pmatrix}$	$\begin{pmatrix} -0.00 & -0.00 \\ -0.00 & 0.23 \end{pmatrix}$	$\begin{pmatrix} 0.00 & 0.00 \\ -0.00 & -0.01 \end{pmatrix}$	$\begin{pmatrix} 3.84 & 0.06 \\ 0.19 & 10.22 \end{pmatrix}$	$\begin{pmatrix} -0.00 & 0.01 \\ 0.02 & 2.14 \end{pmatrix}$	$\begin{pmatrix} -0.00 & -0.00 \\ 0.00 & 0.00 \end{pmatrix}$	$\begin{pmatrix} -0.00 & 0.00 \\ 0.00 & 0.00 \end{pmatrix}$	$\begin{pmatrix} 0.00 & 0.00 \\ -0.00 & -0.00 \end{pmatrix}$	$\begin{pmatrix} -0.00 & -0.00 \\ 0.00 & 0.00 \end{pmatrix}$	$\begin{pmatrix} -0.00 & 0.00 \\ 0.00 & 0.00 \end{pmatrix}$	$\begin{pmatrix} 0.00 & 0.00 \\ -0.00 & -0.00 \end{pmatrix}$
17	$\begin{pmatrix} 0.00 & 0.00 \\ -0.00 & -0.01 \end{pmatrix}$	$\begin{pmatrix} -0.00 & 0.01 \\ 0.02 & 2.15 \end{pmatrix}$	$\begin{pmatrix} -0.00 & -0.00 \\ 0.00 & 0.00 \end{pmatrix}$	$\begin{pmatrix} -0.00 & 0.01 \\ 0.02 & 2.14 \end{pmatrix}$	$\begin{pmatrix} 3.84 & 0.06 \\ 0.18 & 10.55 \end{pmatrix}$	$\begin{pmatrix} -0.00 & -0.00 \\ 0.06 & 1.17 \end{pmatrix}$	$\begin{pmatrix} -0.00 & 0.00 \\ 0.01 & 0.18 \end{pmatrix}$	$\begin{pmatrix} 0.00 & 0.00 \\ 0.00 & 0.01 \end{pmatrix}$	$\begin{pmatrix} -0.00 & -0.00 \\ 0.06 & 1.17 \end{pmatrix}$	$\begin{pmatrix} -0.00 & 0.00 \\ 0.01 & 0.18 \end{pmatrix}$	$\begin{pmatrix} 0.00 & 0.00 \\ 0.00 & 0.01 \end{pmatrix}$
18	$\begin{pmatrix} 0.00 & 0.00 \\ 0.00 & 0.00 \end{pmatrix}$	$\begin{pmatrix} -0.00 & 0.00 \\ -0.00 & 0.00 \end{pmatrix}$	$\begin{pmatrix} -0.00 & -0.00 \\ -0.00 & -0.00 \end{pmatrix}$	$\begin{pmatrix} -0.00 & 0.00 \\ -0.00 & 0.00 \end{pmatrix}$	$\begin{pmatrix} -0.00 & 0.01 \\ -0.00 & 1.00 \end{pmatrix}$	$\begin{pmatrix} 3.50 & 0.12 \\ 0.43 & 10.71 \end{pmatrix}$	$\begin{pmatrix} -0.00 & -0.01 \\ 0.15 & 3.02 \end{pmatrix}$	$\begin{pmatrix} 0.00 & 0.00 \\ 0.03 & 0.44 \end{pmatrix}$	$\begin{pmatrix} -0.00 & -0.00 \\ -0.00 & -0.06 \end{pmatrix}$	$\begin{pmatrix} -0.00 & 0.00 \\ -0.00 & -0.00 \end{pmatrix}$	$\begin{pmatrix} 0.00 & 0.00 \\ 0.00 & 0.00 \end{pmatrix}$
19	$\begin{pmatrix} -0.00 & 0.00 \\ -0.00 & -0.00 \end{pmatrix}$	$\begin{pmatrix} -0.00 & 0.00 \\ 0.00 & 0.00 \end{pmatrix}$	$\begin{pmatrix} 0.00 & -0.00 \\ 0.00 & 0.00 \end{pmatrix}$	$\begin{pmatrix} -0.00 & 0.00 \\ 0.00 & 0.00 \end{pmatrix}$	$\begin{pmatrix} -0.00 & 0.00 \\ 0.00 & 0.15 \end{pmatrix}$	$\begin{pmatrix} -0.00 & 0.04 \\ -0.03 & 3.02 \end{pmatrix}$	$\begin{pmatrix} 3.50 & 0.12 \\ 0.41 & 10.28 \end{pmatrix}$	$\begin{pmatrix} -0.00 & -0.01 \\ 0.16 & 2.74 \end{pmatrix}$	$\begin{pmatrix} -0.00 & -0.00 \\ 0.00 & -0.00 \end{pmatrix}$	$\begin{pmatrix} -0.00 & -0.00 \\ -0.00 & -0.00 \end{pmatrix}$	$\begin{pmatrix} -0.00 & 0.00 \\ -0.00 & -0.00 \end{pmatrix}$
20	$\begin{pmatrix} -0.00 & -0.00 \\ -0.00 & -0.00 \end{pmatrix}$	$\begin{pmatrix} 0.00 & -0.00 \\ 0.00 & -0.00 \end{pmatrix}$	$\begin{pmatrix} 0.00 & 0.00 \\ 0.00 & 0.00 \end{pmatrix}$	$\begin{pmatrix} 0.00 & -0.00 \\ 0.00 & -0.00 \end{pmatrix}$	$\begin{pmatrix} 0.00 & 0.00 \\ 0.00 & 0.00 \end{pmatrix}$	$\begin{pmatrix} 0.00 & 0.01 \\ 0.00 & 0.44 \end{pmatrix}$	$\begin{pmatrix} -0.00 & 0.04 \\ -0.03 & 2.74 \end{pmatrix}$	$\begin{pmatrix} 3.50 & 0.12 \\ 0.43 & 9.01 \end{pmatrix}$	$\begin{pmatrix} 0.00 & 0.00 \\ 0.00 & 0.00 \end{pmatrix}$	$\begin{pmatrix} -0.00 & -0.00 \\ 0.00 & -0.00 \end{pmatrix}$	$\begin{pmatrix} -0.00 & -0.00 \\ -0.00 & -0.00 \end{pmatrix}$
21	$\begin{pmatrix} 0.00 & 0.00 \\ 0.00 & 0.00 \end{pmatrix}$	$\begin{pmatrix} -0.00 & 0.00 \\ -0.00 & 0.00 \end{pmatrix}$	$\begin{pmatrix} -0.00 & -0.00 \\ -0.00 & -0.00 \end{pmatrix}$	$\begin{pmatrix} -0.00 & 0.00 \\ -0.00 & 0.00 \end{pmatrix}$	$\begin{pmatrix} -0.00 & 0.01 \\ -0.00 & 1.00 \end{pmatrix}$	$\begin{pmatrix} -0.00 & -0.00 \\ -0.00 & -0.06 \end{pmatrix}$	$\begin{pmatrix} -0.00 & 0.00 \\ -0.00 & -0.00 \end{pmatrix}$	$\begin{pmatrix} 0.00 & 0.00 \\ 0.00 & 0.00 \end{pmatrix}$	$\begin{pmatrix} 3.50 & 0.12 \\ 0.43 & 10.71 \end{pmatrix}$	$\begin{pmatrix} -0.00 & -0.01 \\ 0.15 & 3.02 \end{pmatrix}$	$\begin{pmatrix} 0.00 & 0.00 \\ 0.03 & 0.44 \end{pmatrix}$
22	$\begin{pmatrix} -0.00 & 0.00 \\ -0.00 & -0.00 \end{pmatrix}$	$\begin{pmatrix} -0.00 & 0.00 \\ 0.00 & 0.00 \end{pmatrix}$	$\begin{pmatrix} 0.00 & -0.00 \\ 0.00 & 0.00 \end{pmatrix}$	$\begin{pmatrix} -0.00 & 0.00 \\ 0.00 & 0.00 \end{pmatrix}$	$\begin{pmatrix} -0.00 & 0.00 \\ 0.00 & 0.15 \end{pmatrix}$	$\begin{pmatrix} -0.00 & -0.00 \\ 0.00 & -0.00 \end{pmatrix}$	$\begin{pmatrix} -0.00 & -0.00 \\ -0.00 & -0.00 \end{pmatrix}$	$\begin{pmatrix} -0.00 & 0.00 \\ -0.00 & -0.00 \end{pmatrix}$	$\begin{pmatrix} -0.00 & 0.04 \\ -0.03 & 3.02 \end{pmatrix}$	$\begin{pmatrix} 3.50 & 0.12 \\ 0.41 & 10.28 \end{pmatrix}$	$\begin{pmatrix} -0.00 & -0.01 \\ 0.16 & 2.74 \end{pmatrix}$
23	$\begin{pmatrix} -0.00 & -0.00 \\ -0.00 & -0.00 \end{pmatrix}$	$\begin{pmatrix} 0.00 & -0.00 \\ 0.00 & -0.00 \end{pmatrix}$	$\begin{pmatrix} 0.00 & 0.00 \\ 0.00 & 0.00 \end{pmatrix}$	$\begin{pmatrix} 0.00 & -0.00 \\ 0.00 & -0.00 \end{pmatrix}$	$\begin{pmatrix} 0.00 & 0.00 \\ 0.00 & 0.00 \end{pmatrix}$	$\begin{pmatrix} 0.00 & 0.00 \\ 0.00 & 0.00 \end{pmatrix}$	$\begin{pmatrix} -0.00 & -0.00 \\ 0.00 & -0.00 \end{pmatrix}$	$\begin{pmatrix} -0.00 & -0.00 \\ -0.00 & -0.00 \end{pmatrix}$	$\begin{pmatrix} 0.00 & 0.01 \\ 0.00 & 0.44 \end{pmatrix}$	$\begin{pmatrix} -0.00 & 0.04 \\ -0.03 & 2.74 \end{pmatrix}$	$\begin{pmatrix} 3.50 & 0.12 \\ 0.43 & 9.01 \end{pmatrix}$

Table B.7: Analysis of the asymptotic stability for the overall coupled tanks (Part 1).

j	i	$\epsilon_{ji}^{\mathrm{T}}$	$(M_{\mathrm{uu}ji}\epsilon)^{\mathrm{T}}$	$\epsilon_{ji} > M_{\mathrm{uu}ji}\epsilon$
2	1	(0.001475 0.037891)	(0.000089 0.027618)	True
3	1	(0.000057 0.002501)	(0.000000 0.001747)	True
4	1	(0.000053 0.000918)	(0.000000 0.000000)	True
5	1	(0.000028 0.000710)	(0.000000 0.000000)	True
6	1	(0.000011 0.000500)	(0.000000 0.000001)	True
7	1	(0.000015 0.000704)	(0.000000 0.000001)	True
8	1	(0.000015 0.000704)	(0.000000 0.000004)	True
9	1	(0.000003 0.000355)	(0.000000 0.000007)	True
10	1	(0.000006 0.000860)	(0.000000 0.000000)	True
11	1	(0.000006 0.000858)	(0.000000 0.000000)	True
12	1	(0.000014 0.000652)	(0.000000 0.000000)	True
13	1	(0.000046 0.001003)	(0.000000 0.000000)	True
14	1	(0.000011 0.000504)	(0.000000 0.000000)	True
15	1	(0.000046 0.001001)	(0.000000 0.000000)	True
16	1	(0.000016 0.000702)	(0.000000 0.000000)	True
17	1	(0.000010 0.000434)	(0.000000 0.000000)	True
18	1	(0.000024 0.000633)	(0.000000 0.000000)	True
19	1	(0.000027 0.000556)	(0.000000 0.000000)	True
20	1	(0.000050 0.000941)	(0.000000 0.000000)	True
21	1	(0.000024 0.000611)	(0.000000 0.000000)	True
22	1	(0.000025 0.000517)	(0.000000 0.000000)	True
23	1	(0.000047 0.000886)	(0.000000 0.000000)	True
1	2	(0.002426 0.041736)	(0.000450 0.027529)	True
3	2	(0.000569 0.025012)	(0.000018 0.011867)	True
4	2	(0.000053 0.000918)	(0.000000 0.000000)	True
5	2	(0.000028 0.000710)	(0.000000 0.000003)	True
6	2	(0.000057 0.002501)	(0.000005 0.000031)	True
7	2	(0.000015 0.000704)	(0.000005 0.000044)	True
8	2	(0.000015 0.000704)	(0.000001 0.000125)	True
9	2	(0.000003 0.000355)	(0.000000 0.000105)	True
10	2	(0.000006 0.000860)	(0.000000 0.000000)	True
11	2	(0.000006 0.000858)	(0.000000 0.000000)	True
12	2	(0.000014 0.000652)	(0.000000 0.000003)	True
13	2	(0.000046 0.001003)	(0.000000 0.000000)	True
14	2	(0.000011 0.000504)	(0.000000 0.000000)	True
15	2	(0.000046 0.001001)	(0.000000 0.000000)	True
16	2	(0.000016 0.000702)	(0.000000 0.000003)	True
17	2	(0.000010 0.000434)	(0.000000 0.000000)	True
18	2	(0.000024 0.000633)	(0.000000 0.000000)	True
19	2	(0.000027 0.000556)	(0.000000 0.000000)	True
20	2	(0.000050 0.000941)	(0.000000 0.000000)	True
21	2	(0.000024 0.000611)	(0.000000 0.000000)	True
22	2	(0.000025 0.000517)	(0.000000 0.000000)	True
23	2	(0.000047 0.000886)	(0.000000 0.000000)	True
1	3	(0.000485 0.008347)	(0.000030 0.001483)	True
2	3	(0.001475 0.037891)	(0.000144 0.010100)	True
4	3	(0.000053 0.000918)	(0.000000 0.000001)	True
5	3	(0.000138 0.003552)	(0.000000 0.000027)	True
6	3	(0.000569 0.025012)	(0.000059 0.021478)	True
7	3	(0.000758 0.035198)	(0.000059 0.021352)	True
8	3	(0.000076 0.003520)	(0.000004 0.002169)	True
9	3	(0.000021 0.002839)	(0.000002 0.002655)	True
10	3	(0.000006 0.000860)	(0.000000 0.000000)	True
11	3	(0.000006 0.000858)	(0.000000 0.000000)	True
12	3	(0.000014 0.000652)	(0.000000 0.000030)	True
13	3	(0.000046 0.001003)	(0.000000 0.000002)	True
14	3	(0.000011 0.000504)	(0.000000 0.000000)	True
15	3	(0.000046 0.001001)	(0.000000 0.000000)	True
16	3	(0.000016 0.000702)	(0.000000 0.000031)	True
17	3	(0.000010 0.000434)	(0.000000 0.000000)	True
18	3	(0.000024 0.000633)	(0.000000 0.000000)	True
19	3	(0.000027 0.000556)	(0.000000 0.000000)	True
20	3	(0.000050 0.000941)	(0.000000 0.000000)	True
21	3	(0.000024 0.000611)	(0.000000 0.000000)	True
22	3	(0.000025 0.000517)	(0.000000 0.000000)	True
23	3	(0.000047 0.000886)	(0.000000 0.000000)	True

Table B.8: Analysis of the asymptotic stability for the overall coupled tanks (Part 2).

j	i	$\epsilon_{ji}^{\mathrm{T}}$	$(M_{\mathrm{uu}ji}\epsilon)^{\mathrm{T}}$	$\epsilon_{ji} > M_{\mathrm{uu}ji}\epsilon$
1	4	(0.000049 0.000835)	(0.000000 0.000000)	True
2	4	(0.000029 0.000758)	(0.000000 0.000000)	True
3	4	(0.000011 0.000500)	(0.000000 0.000001)	True
5	4	(0.001382 0.035523)	(0.000089 0.027618)	True
6	4	(0.000057 0.002501)	(0.000000 0.001747)	True
7	4	(0.000015 0.000704)	(0.000000 0.000004)	True
8	4	(0.000015 0.000704)	(0.000000 0.000001)	True
9	4	(0.000003 0.000355)	(0.000000 0.000007)	True
10	4	(0.000006 0.000860)	(0.000000 0.000000)	True
11	4	(0.000006 0.000858)	(0.000000 0.000000)	True
12	4	(0.000014 0.000652)	(0.000000 0.000000)	True
13	4	(0.000046 0.001003)	(0.000000 0.000000)	True
14	4	(0.000011 0.000504)	(0.000000 0.000000)	True
15	4	(0.000046 0.001001)	(0.000000 0.000000)	True
16	4	(0.000016 0.000702)	(0.000000 0.000000)	True
17	4	(0.000010 0.000434)	(0.000000 0.000000)	True
18	4	(0.000024 0.000633)	(0.000000 0.000000)	True
19	4	(0.000027 0.000556)	(0.000000 0.000000)	True
20	4	(0.000050 0.000941)	(0.000000 0.000000)	True
21	4	(0.000024 0.000611)	(0.000000 0.000000)	True
22	4	(0.000025 0.000517)	(0.000000 0.000000)	True
23	4	(0.000047 0.000886)	(0.000000 0.000000)	True
1	5	(0.000049 0.000835)	(0.000000 0.000000)	True
2	5	(0.000029 0.000758)	(0.000000 0.000003)	True
3	5	(0.000057 0.002501)	(0.000005 0.000031)	True
4	5	(0.002669 0.045910)	(0.000450 0.027529)	True
6	5	(0.000569 0.025012)	(0.000018 0.011867)	True
7	5	(0.000015 0.000704)	(0.000001 0.000125)	True
8	5	(0.000015 0.000704)	(0.000005 0.000044)	True
9	5	(0.000003 0.000355)	(0.000000 0.000105)	True
10	5	(0.000006 0.000860)	(0.000000 0.000000)	True
11	5	(0.000006 0.000858)	(0.000000 0.000000)	True
12	5	(0.000014 0.000652)	(0.000000 0.000003)	True
13	5	(0.000046 0.001003)	(0.000000 0.000000)	True
14	5	(0.000011 0.000504)	(0.000000 0.000000)	True
15	5	(0.000046 0.001001)	(0.000000 0.000000)	True
16	5	(0.000016 0.000702)	(0.000000 0.000003)	True
17	5	(0.000010 0.000434)	(0.000000 0.000000)	True
18	5	(0.000024 0.000633)	(0.000000 0.000000)	True
19	5	(0.000027 0.000556)	(0.000000 0.000000)	True
20	5	(0.000050 0.000941)	(0.000000 0.000000)	True
21	5	(0.000024 0.000611)	(0.000000 0.000000)	True
22	5	(0.000025 0.000517)	(0.000000 0.000000)	True
23	5	(0.000047 0.000886)	(0.000000 0.000000)	True
1	6	(0.000243 0.004174)	(0.000000 0.000001)	True
2	6	(0.000029 0.000758)	(0.000000 0.000027)	True
3	6	(0.000569 0.025012)	(0.000059 0.021478)	True
4	6	(0.000534 0.009182)	(0.000030 0.001483)	True
5	6	(0.001382 0.035523)	(0.000144 0.010100)	True
7	6	(0.000076 0.003520)	(0.000004 0.002169)	True
8	6	(0.000758 0.035198)	(0.000059 0.021352)	True
9	6	(0.000021 0.002839)	(0.000002 0.002655)	True
10	6	(0.000006 0.000860)	(0.000000 0.000000)	True
11	6	(0.000006 0.000858)	(0.000000 0.000000)	True
12	6	(0.000014 0.000652)	(0.000000 0.000030)	True
13	6	(0.000046 0.001003)	(0.000000 0.000002)	True
14	6	(0.000011 0.000504)	(0.000000 0.000000)	True
15	6	(0.000046 0.001001)	(0.000000 0.000000)	True
16	6	(0.000016 0.000702)	(0.000000 0.000031)	True
17	6	(0.000010 0.000434)	(0.000000 0.000000)	True
18	6	(0.000024 0.000633)	(0.000000 0.000000)	True
19	6	(0.000027 0.000556)	(0.000000 0.000000)	True
20	6	(0.000050 0.000941)	(0.000000 0.000000)	True
21	6	(0.000024 0.000611)	(0.000000 0.000000)	True
22	6	(0.000025 0.000517)	(0.000000 0.000000)	True
23	6	(0.000047 0.000886)	(0.000000 0.000000)	True

Table B.9: Analysis of the asymptotic stability for the overall coupled tanks (Part 3).

j	i	$\epsilon_{ji}^{\mathrm{T}}$	$(M_{\mathrm{uu}ji}\epsilon)^{\mathrm{T}}$	$\epsilon_{ji} > M_{\mathrm{uu}ji}\epsilon$
1	7	(0.000243 0.004174)	(0.000000 0.000001)	True
2	7	(0.000029 0.000758)	(0.000001 0.000038)	True
3	7	(0.000569 0.025012)	(0.000061 0.021352)	True
4	7	(0.000053 0.000918)	(0.000000 0.000004)	True
5	7	(0.000028 0.000710)	(0.000002 0.000106)	True
6	7	(0.000057 0.002501)	(0.000003 0.002170)	True
8	7	(0.000076 0.003520)	(0.000006 0.000640)	True
9	7	(0.000088 0.011709)	(0.000010 0.011529)	True
10	7	(0.000006 0.000860)	(0.000002 0.000000)	True
11	7	(0.000006 0.000858)	(0.000002 0.000000)	True
12	7	(0.000014 0.000652)	(0.000000 0.000139)	True
13	7	(0.000046 0.001003)	(0.000000 0.000011)	True
14	7	(0.000011 0.000504)	(0.000000 0.000002)	True
15	7	(0.000046 0.001001)	(0.000000 0.000000)	True
16	7	(0.000016 0.000702)	(0.000000 0.000141)	True
17	7	(0.000010 0.000434)	(0.000000 0.000002)	True
18	7	(0.000024 0.000633)	(0.000000 0.000000)	True
19	7	(0.000027 0.000556)	(0.000000 0.000000)	True
20	7	(0.000050 0.000941)	(0.000000 0.000000)	True
21	7	(0.000024 0.000611)	(0.000000 0.000000)	True
22	7	(0.000025 0.000517)	(0.000000 0.000000)	True
23	7	(0.000047 0.000886)	(0.000000 0.000000)	True
1	8	(0.000049 0.000835)	(0.000000 0.000004)	True
2	8	(0.000029 0.000758)	(0.000002 0.000106)	True
3	8	(0.000057 0.002501)	(0.000003 0.002170)	True
4	8	(0.000053 0.000918)	(0.000000 0.000001)	True
5	8	(0.000138 0.003552)	(0.000001 0.000038)	True
6	8	(0.000569 0.025012)	(0.000061 0.021352)	True
7	8	(0.000076 0.003520)	(0.000006 0.000640)	True
9	8	(0.000088 0.011709)	(0.000010 0.011529)	True
10	8	(0.000006 0.000860)	(0.000002 0.000000)	True
11	8	(0.000006 0.000858)	(0.000002 0.000000)	True
12	8	(0.000014 0.000652)	(0.000000 0.000139)	True
13	8	(0.000046 0.001003)	(0.000000 0.000011)	True
14	8	(0.000011 0.000504)	(0.000000 0.000002)	True
15	8	(0.000046 0.001001)	(0.000000 0.000000)	True
16	8	(0.000016 0.000702)	(0.000000 0.000141)	True
17	8	(0.000010 0.000434)	(0.000000 0.000002)	True
18	8	(0.000024 0.000633)	(0.000000 0.000000)	True
19	8	(0.000027 0.000556)	(0.000000 0.000000)	True
20	8	(0.000050 0.000941)	(0.000000 0.000000)	True
21	8	(0.000024 0.000611)	(0.000000 0.000000)	True
22	8	(0.000025 0.000517)	(0.000000 0.000000)	True
23	8	(0.000047 0.000886)	(0.000000 0.000000)	True
1	9	(0.000049 0.000835)	(0.000000 0.000003)	True
2	9	(0.000029 0.000758)	(0.000001 0.000040)	True
3	9	(0.000057 0.002501)	(0.000001 0.001210)	True
4	9	(0.000053 0.000918)	(0.000000 0.000003)	True
5	9	(0.000028 0.000710)	(0.000001 0.000040)	True
6	9	(0.000057 0.002501)	(0.000001 0.001210)	True
7	9	(0.000758 0.035198)	(0.000035 0.005249)	True
8	9	(0.000758 0.035198)	(0.000035 0.005249)	True
10	9	(0.000203 0.028367)	(0.000012 0.017536)	True
11	9	(0.000205 0.028327)	(0.000012 0.017536)	True
12	9	(0.000043 0.001957)	(0.000001 0.001144)	True
13	9	(0.000046 0.001003)	(0.000001 0.000032)	True
14	9	(0.000011 0.000504)	(0.000000 0.000005)	True
15	9	(0.000046 0.001001)	(0.000000 0.000000)	True
16	9	(0.000078 0.003511)	(0.000001 0.001150)	True
17	9	(0.000010 0.000434)	(0.000000 0.000005)	True
18	9	(0.000024 0.000633)	(0.000000 0.000000)	True
19	9	(0.000027 0.000556)	(0.000000 0.000000)	True
20	9	(0.000050 0.000941)	(0.000000 0.000000)	True
21	9	(0.000024 0.000611)	(0.000000 0.000000)	True
22	9	(0.000025 0.000517)	(0.000000 0.000000)	True
23	9	(0.000047 0.000886)	(0.000000 0.000000)	True

Table B.10: Analysis of the asymptotic stability for the overall coupled tanks (Part 4).

j	i	$\epsilon_{ji}^{\mathrm{T}}$	$(M_{\mathrm{uu}ji}\epsilon)^{\mathrm{T}}$	$\epsilon_{ji} > M_{\mathrm{uu}ji}\epsilon$
1	10	(0.000049 0.000835)	(0.000000 0.000000)	True
2	10	(0.000029 0.000758)	(0.000000 0.000000)	True
3	10	(0.000011 0.000500)	(0.000000 0.000000)	True
4	10	(0.000053 0.000918)	(0.000000 0.000000)	True
5	10	(0.000028 0.000710)	(0.000000 0.000000)	True
6	10	(0.000011 0.000500)	(0.000000 0.000000)	True
7	10	(0.000015 0.000704)	(0.000000 0.000000)	True
8	10	(0.000015 0.000704)	(0.000000 0.000000)	True
9	10	(0.000133 0.017741)	(0.000012 0.017536)	True
11	10	(0.000037 0.005150)	(0.000001 0.001503)	True
12	10	(0.000580 0.026090)	(0.000030 0.012287)	True
13	10	(0.000455 0.010028)	(0.000003 0.000157)	True
14	10	(0.000055 0.002518)	(0.000000 0.001156)	True
15	10	(0.000046 0.001001)	(0.000001 0.000034)	True
16	10	(0.000016 0.000702)	(0.000000 0.000030)	True
17	10	(0.000010 0.000434)	(0.000000 0.000026)	True
18	10	(0.000024 0.000633)	(0.000000 0.000002)	True
19	10	(0.000027 0.000556)	(0.000000 0.000000)	True
20	10	(0.000050 0.000941)	(0.000000 0.000000)	True
21	10	(0.000024 0.000611)	(0.000000 0.000002)	True
22	10	(0.000025 0.000517)	(0.000000 0.000000)	True
23	10	(0.000047 0.000886)	(0.000000 0.000000)	True
1	11	(0.000049 0.000835)	(0.000000 0.000000)	True
2	11	(0.000029 0.000758)	(0.000000 0.000000)	True
3	11	(0.000011 0.000500)	(0.000000 0.000000)	True
4	11	(0.000053 0.000918)	(0.000000 0.000000)	True
5	11	(0.000028 0.000710)	(0.000000 0.000000)	True
6	11	(0.000011 0.000500)	(0.000000 0.000000)	True
7	11	(0.000015 0.000704)	(0.000000 0.000000)	True
8	11	(0.000015 0.000704)	(0.000000 0.000000)	True
9	11	(0.000133 0.017741)	(0.000012 0.017536)	True
10	11	(0.000031 0.004298)	(0.000001 0.001503)	True
12	11	(0.000014 0.000652)	(0.000000 0.000030)	True
13	11	(0.000046 0.001003)	(0.000000 0.000000)	True
14	11	(0.000011 0.000504)	(0.000000 0.000027)	True
15	11	(0.000046 0.001001)	(0.000000 0.000002)	True
16	11	(0.000775 0.035109)	(0.000030 0.012256)	True
17	11	(0.000030 0.001303)	(0.000000 0.001149)	True
18	11	(0.000024 0.000633)	(0.000001 0.000035)	True
19	11	(0.000027 0.000556)	(0.000000 0.000003)	True
20	11	(0.000050 0.000941)	(0.000000 0.000000)	True
21	11	(0.000024 0.000611)	(0.000001 0.000035)	True
22	11	(0.000025 0.000517)	(0.000000 0.000003)	True
23	11	(0.000047 0.000886)	(0.000000 0.000000)	True
1	12	(0.000049 0.000835)	(0.000000 0.000000)	True
2	12	(0.000029 0.000758)	(0.000000 0.000003)	True
3	12	(0.000011 0.000500)	(0.000000 0.000030)	True
4	12	(0.000053 0.000918)	(0.000000 0.000000)	True
5	12	(0.000028 0.000710)	(0.000000 0.000003)	True
6	12	(0.000011 0.000500)	(0.000000 0.000030)	True
7	12	(0.000015 0.000704)	(0.000001 0.000139)	True
8	12	(0.000015 0.000704)	(0.000001 0.000139)	True
9	12	(0.000021 0.002839)	(0.000003 0.002510)	True
10	12	(0.000276 0.038682)	(0.000021 0.026970)	True
11	12	(0.000006 0.000858)	(0.000000 0.000066)	True
13	12	(0.002277 0.050138)	(0.000150 0.009175)	True
14	12	(0.000548 0.025184)	(0.000059 0.021444)	True
15	12	(0.000461 0.010015)	(0.000001 0.000029)	True
16	12	(0.000016 0.000702)	(0.000001 0.000015)	True
17	12	(0.000059 0.002605)	(0.000003 0.002241)	True
18	12	(0.000024 0.000633)	(0.000002 0.000111)	True
19	12	(0.000027 0.000556)	(0.000000 0.000003)	True
20	12	(0.000050 0.000941)	(0.000000 0.000002)	True
21	12	(0.000024 0.000611)	(0.000002 0.000111)	True
22	12	(0.000025 0.000517)	(0.000000 0.000003)	True
23	12	(0.000047 0.000886)	(0.000000 0.000002)	True

Table B.11: Analysis of the asymptotic stability for the overall coupled tanks (Part 5).

j	i	$\epsilon_{ji}^{\mathrm{T}}$	$(M_{\mathrm{uu}ji}\epsilon)^{\mathrm{T}}$	$\epsilon_{ji} > M_{\mathrm{uu}ji}\epsilon$
1	13	(0.000049 0.000835)	(0.000000 0.000000)	True
2	13	(0.000029 0.000758)	(0.000000 0.000000)	True
3	13	(0.000011 0.000500)	(0.000000 0.000003)	True
4	13	(0.000053 0.000918)	(0.000000 0.000000)	True
5	13	(0.000028 0.000710)	(0.000000 0.000000)	True
6	13	(0.000011 0.000500)	(0.000000 0.000003)	True
7	13	(0.000015 0.000704)	(0.000000 0.000013)	True
8	13	(0.000015 0.000704)	(0.000000 0.000013)	True
9	13	(0.000003 0.000355)	(0.000000 0.000084)	True
10	13	(0.000012 0.001719)	(0.000002 0.000405)	True
11	13	(0.000006 0.000858)	(0.000000 0.000001)	True
12	13	(0.000435 0.019567)	(0.000018 0.010797)	True
14	13	(0.000055 0.002518)	(0.000005 0.000025)	True
15	13	(0.000046 0.001001)	(0.000000 0.000003)	True
16	13	(0.000016 0.000702)	(0.000000 0.000001)	True
17	13	(0.000010 0.000434)	(0.000000 0.000124)	True
18	13	(0.000024 0.000633)	(0.000000 0.000013)	True
19	13	(0.000027 0.000556)	(0.000000 0.000001)	True
20	13	(0.000050 0.000941)	(0.000000 0.000000)	True
21	13	(0.000024 0.000611)	(0.000000 0.000013)	True
22	13	(0.000025 0.000517)	(0.000000 0.000001)	True
23	13	(0.000047 0.000886)	(0.000000 0.000000)	True
1	14	(0.000049 0.000835)	(0.000000 0.000000)	True
2	14	(0.000029 0.000758)	(0.000000 0.000000)	True
3	14	(0.000011 0.000500)	(0.000000 0.000000)	True
4	14	(0.000053 0.000918)	(0.000000 0.000000)	True
5	14	(0.000028 0.000710)	(0.000000 0.000000)	True
6	14	(0.000011 0.000500)	(0.000000 0.000000)	True
7	14	(0.000015 0.000704)	(0.000000 0.000002)	True
8	14	(0.000015 0.000704)	(0.000000 0.000002)	True
9	14	(0.000003 0.000355)	(0.000000 0.000012)	True
10	14	(0.000037 0.005158)	(0.000002 0.002537)	True
11	14	(0.000006 0.000858)	(0.000000 0.000059)	True
12	14	(0.000580 0.026090)	(0.000059 0.021444)	True
13	14	(0.000455 0.010028)	(0.000000 0.000022)	True
15	14	(0.002303 0.050075)	(0.000150 0.009178)	True
16	14	(0.000078 0.003511)	(0.000003 0.002287)	True
17	14	(0.000494 0.021709)	(0.000059 0.021513)	True
18	14	(0.000122 0.003165)	(0.000000 0.000024)	True
19	14	(0.000027 0.000556)	(0.000000 0.000001)	True
20	14	(0.000050 0.000941)	(0.000000 0.000000)	True
21	14	(0.000118 0.003057)	(0.000000 0.000024)	True
22	14	(0.000025 0.000517)	(0.000000 0.000001)	True
23	14	(0.000047 0.000886)	(0.000000 0.000000)	True
1	15	(0.000049 0.000835)	(0.000000 0.000000)	True
2	15	(0.000029 0.000758)	(0.000000 0.000000)	True
3	15	(0.000011 0.000500)	(0.000000 0.000000)	True
4	15	(0.000053 0.000918)	(0.000000 0.000000)	True
5	15	(0.000028 0.000710)	(0.000000 0.000000)	True
6	15	(0.000011 0.000500)	(0.000000 0.000000)	True
7	15	(0.000015 0.000704)	(0.000000 0.000000)	True
8	15	(0.000015 0.000704)	(0.000000 0.000000)	True
9	15	(0.000003 0.000355)	(0.000000 0.000001)	True
10	15	(0.000006 0.000860)	(0.000000 0.000087)	True
11	15	(0.000006 0.000858)	(0.000000 0.000006)	True
12	15	(0.000014 0.000652)	(0.000005 0.000033)	True
13	15	(0.000046 0.001003)	(0.000000 0.000003)	True
14	15	(0.000548 0.025184)	(0.000018 0.010802)	True
16	15	(0.000016 0.000702)	(0.000000 0.000130)	True
17	15	(0.000030 0.001303)	(0.000005 0.000026)	True
18	15	(0.000024 0.000633)	(0.000000 0.000003)	True
19	15	(0.000027 0.000556)	(0.000000 0.000000)	True
20	15	(0.000050 0.000941)	(0.000000 0.000000)	True
21	15	(0.000024 0.000611)	(0.000000 0.000003)	True
22	15	(0.000025 0.000517)	(0.000000 0.000000)	True
23	15	(0.000047 0.000886)	(0.000000 0.000000)	True

Table B.12: Analysis of the asymptotic stability for the overall coupled tanks (Part 6).

j	i	$\epsilon_{ji}^{\mathrm{T}}$	$(M_{\mathrm{uu}ji}\epsilon)^{\mathrm{T}}$	$\epsilon_{ji} > M_{\mathrm{uu}ji}\epsilon$
1	16	(0.000049 0.000835)	(0.000000 0.000000)	True
2	16	(0.000029 0.000758)	(0.000000 0.000003)	True
3	16	(0.000011 0.000500)	(0.000000 0.000031)	True
4	16	(0.000053 0.000918)	(0.000000 0.000000)	True
5	16	(0.000028 0.000710)	(0.000000 0.000003)	True
6	16	(0.000011 0.000500)	(0.000000 0.000031)	True
7	16	(0.000015 0.000704)	(0.000001 0.000141)	True
8	16	(0.000015 0.000704)	(0.000001 0.000141)	True
9	16	(0.000021 0.002839)	(0.000003 0.002524)	True
10	16	(0.000006 0.000860)	(0.000000 0.000066)	True
11	16	(0.000279 0.038627)	(0.000021 0.026903)	True
12	16	(0.000014 0.000652)	(0.000001 0.000015)	True
13	16	(0.000046 0.001003)	(0.000000 0.000001)	True
14	16	(0.000055 0.002518)	(0.000003 0.002287)	True
15	16	(0.000046 0.001001)	(0.000002 0.000111)	True
17	16	(0.000494 0.021709)	(0.000060 0.021446)	True
18	16	(0.000122 0.003165)	(0.000001 0.000030)	True
19	16	(0.000027 0.000556)	(0.000000 0.000001)	True
20	16	(0.000050 0.000941)	(0.000000 0.000001)	True
21	16	(0.000118 0.003057)	(0.000001 0.000030)	True
22	16	(0.000025 0.000517)	(0.000000 0.000001)	True
23	16	(0.000047 0.000886)	(0.000000 0.000001)	True
1	17	(0.000049 0.000835)	(0.000000 0.000000)	True
2	17	(0.000029 0.000758)	(0.000000 0.000000)	True
3	17	(0.000011 0.000500)	(0.000000 0.000000)	True
4	17	(0.000053 0.000918)	(0.000000 0.000000)	True
5	17	(0.000028 0.000710)	(0.000000 0.000000)	True
6	17	(0.000011 0.000500)	(0.000000 0.000000)	True
7	17	(0.000015 0.000704)	(0.000000 0.000002)	True
8	17	(0.000015 0.000704)	(0.000000 0.000002)	True
9	17	(0.000003 0.000355)	(0.000000 0.000011)	True
10	17	(0.000006 0.000860)	(0.000000 0.000057)	True
11	17	(0.000037 0.005150)	(0.000002 0.002521)	True
12	17	(0.000087 0.003913)	(0.000003 0.002241)	True
13	17	(0.000046 0.001003)	(0.000002 0.000105)	True
14	17	(0.000548 0.025184)	(0.000058 0.021514)	True
15	17	(0.000461 0.010015)	(0.000000 0.000023)	True
16	17	(0.000775 0.035109)	(0.000058 0.021446)	True
18	17	(0.001219 0.031652)	(0.000141 0.010020)	True
19	17	(0.000273 0.005559)	(0.000028 0.001545)	True
20	17	(0.000050 0.000941)	(0.000002 0.000048)	True
21	17	(0.001177 0.030568)	(0.000141 0.010020)	True
22	17	(0.000254 0.005173)	(0.000028 0.001545)	True
23	17	(0.000047 0.000886)	(0.000002 0.000048)	True
1	18	(0.000049 0.000835)	(0.000000 0.000000)	True
2	18	(0.000029 0.000758)	(0.000000 0.000000)	True
3	18	(0.000011 0.000500)	(0.000000 0.000000)	True
4	18	(0.000053 0.000918)	(0.000000 0.000000)	True
5	18	(0.000028 0.000710)	(0.000000 0.000000)	True
6	18	(0.000011 0.000500)	(0.000000 0.000000)	True
7	18	(0.000015 0.000704)	(0.000000 0.000000)	True
8	18	(0.000015 0.000704)	(0.000000 0.000000)	True
9	18	(0.000003 0.000355)	(0.000000 0.000001)	True
10	18	(0.000006 0.000860)	(0.000000 0.000006)	True
11	18	(0.000006 0.000858)	(0.000000 0.000092)	True
12	18	(0.000014 0.000652)	(0.000000 0.000130)	True
13	18	(0.000046 0.001003)	(0.000000 0.000013)	True
14	18	(0.000055 0.002518)	(0.000005 0.000028)	True
15	18	(0.000046 0.001001)	(0.000000 0.000003)	True
16	18	(0.000078 0.003511)	(0.000005 0.000034)	True
17	18	(0.000296 0.013025)	(0.000017 0.011775)	True
19	18	(0.001502 0.030573)	(0.000430 0.030231)	True
20	18	(0.000496 0.009413)	(0.000089 0.004417)	True
21	18	(0.000118 0.003057)	(0.000011 0.000615)	True
22	18	(0.000025 0.000517)	(0.000001 0.000012)	True
23	18	(0.000047 0.000886)	(0.000000 0.000011)	True

Table B.13: Analysis of the asymptotic stability for the overall coupled tanks (Part 7).

j	i	$\epsilon_{ji}^{\mathrm{T}}$	$(M_{\mathrm{uu}ji}\epsilon)^{\mathrm{T}}$	$\epsilon_{ji} > M_{\mathrm{uu}ji}\epsilon$
1	19	(0.000049 0.000835)	(0.000000 0.000000)	True
2	19	(0.000029 0.000758)	(0.000000 0.000000)	True
3	19	(0.000011 0.000500)	(0.000000 0.000000)	True
4	19	(0.000053 0.000918)	(0.000000 0.000000)	True
5	19	(0.000028 0.000710)	(0.000000 0.000000)	True
6	19	(0.000011 0.000500)	(0.000000 0.000000)	True
7	19	(0.000015 0.000704)	(0.000000 0.000000)	True
8	19	(0.000015 0.000704)	(0.000000 0.000000)	True
9	19	(0.000003 0.000355)	(0.000000 0.000000)	True
10	19	(0.000006 0.000860)	(0.000000 0.000000)	True
11	19	(0.000006 0.000858)	(0.000000 0.000007)	True
12	19	(0.000014 0.000652)	(0.000000 0.000003)	True
13	19	(0.000046 0.001003)	(0.000000 0.000001)	True
14	19	(0.000011 0.000504)	(0.000000 0.000001)	True
15	19	(0.000046 0.001001)	(0.000000 0.000000)	True
16	19	(0.000016 0.000702)	(0.000000 0.000001)	True
17	19	(0.000049 0.002171)	(0.000001 0.001818)	True
18	19	(0.001219 0.031652)	(0.000087 0.030295)	True
20	19	(0.002481 0.047067)	(0.000446 0.027445)	True
21	19	(0.000024 0.000611)	(0.000000 0.000013)	True
22	19	(0.000025 0.000517)	(0.000000 0.000012)	True
23	19	(0.000047 0.000886)	(0.000000 0.000000)	True
1	20	(0.000049 0.000835)	(0.000000 0.000000)	True
2	20	(0.000029 0.000758)	(0.000000 0.000000)	True
3	20	(0.000011 0.000500)	(0.000000 0.000000)	True
4	20	(0.000053 0.000918)	(0.000000 0.000000)	True
5	20	(0.000028 0.000710)	(0.000000 0.000000)	True
6	20	(0.000011 0.000500)	(0.000000 0.000000)	True
7	20	(0.000015 0.000704)	(0.000000 0.000000)	True
8	20	(0.000015 0.000704)	(0.000000 0.000000)	True
9	20	(0.000003 0.000355)	(0.000000 0.000000)	True
10	20	(0.000006 0.000860)	(0.000000 0.000000)	True
11	20	(0.000006 0.000858)	(0.000000 0.000001)	True
12	20	(0.000014 0.000652)	(0.000000 0.000002)	True
13	20	(0.000046 0.001003)	(0.000000 0.000000)	True
14	20	(0.000011 0.000504)	(0.000000 0.000000)	True
15	20	(0.000046 0.001001)	(0.000000 0.000000)	True
16	20	(0.000016 0.000702)	(0.000000 0.000001)	True
17	20	(0.000010 0.000434)	(0.000000 0.000057)	True
18	20	(0.000244 0.006330)	(0.000004 0.004442)	True
19	20	(0.001366 0.027794)	(0.000087 0.027527)	True
21	20	(0.000024 0.000611)	(0.000000 0.000011)	True
22	20	(0.000025 0.000517)	(0.000000 0.000000)	True
23	20	(0.000047 0.000886)	(0.000000 0.000000)	True
1	21	(0.000049 0.000835)	(0.000000 0.000000)	True
2	21	(0.000029 0.000758)	(0.000000 0.000000)	True
3	21	(0.000011 0.000500)	(0.000000 0.000000)	True
4	21	(0.000053 0.000918)	(0.000000 0.000000)	True
5	21	(0.000028 0.000710)	(0.000000 0.000000)	True
6	21	(0.000011 0.000500)	(0.000000 0.000000)	True
7	21	(0.000015 0.000704)	(0.000000 0.000000)	True
8	21	(0.000015 0.000704)	(0.000000 0.000000)	True
9	21	(0.000003 0.000355)	(0.000000 0.000001)	True
10	21	(0.000006 0.000860)	(0.000000 0.000006)	True
11	21	(0.000006 0.000858)	(0.000000 0.000092)	True
12	21	(0.000014 0.000652)	(0.000000 0.000130)	True
13	21	(0.000046 0.001003)	(0.000000 0.000013)	True
14	21	(0.000011 0.000504)	(0.000005 0.000028)	True
15	21	(0.000046 0.001001)	(0.000000 0.000003)	True
16	21	(0.000078 0.003511)	(0.000005 0.000034)	True
17	21	(0.000296 0.013025)	(0.000017 0.011775)	True
18	21	(0.000122 0.003165)	(0.000011 0.000615)	True
19	21	(0.000027 0.000556)	(0.000001 0.000012)	True
20	21	(0.000050 0.000941)	(0.000000 0.000011)	True
22	21	(0.001525 0.031037)	(0.000430 0.030231)	True
23	21	(0.000467 0.008860)	(0.000089 0.004417)	True

Table B.14: Analysis of the asymptotic stability for the overall coupled tanks (Part 8).

j	i	$\epsilon_{ji}^{\mathrm{T}}$	$(M_{\mathrm{uu}ji}\epsilon)^{\mathrm{T}}$	$\epsilon_{ji} > M_{\mathrm{uu}ji}\epsilon$
1	22	(0.000049 0.000835)	(0.000000 0.000000)	True
2	22	(0.000029 0.000758)	(0.000000 0.000000)	True
3	22	(0.000011 0.000500)	(0.000000 0.000000)	True
4	22	(0.000053 0.000918)	(0.000000 0.000000)	True
5	22	(0.000028 0.000710)	(0.000000 0.000000)	True
6	22	(0.000011 0.000500)	(0.000000 0.000000)	True
7	22	(0.000015 0.000704)	(0.000000 0.000000)	True
8	22	(0.000015 0.000704)	(0.000000 0.000000)	True
9	22	(0.000003 0.000355)	(0.000000 0.000000)	True
10	22	(0.000006 0.000860)	(0.000000 0.000000)	True
11	22	(0.000006 0.000858)	(0.000000 0.000007)	True
12	22	(0.000014 0.000652)	(0.000000 0.000003)	True
13	22	(0.000046 0.001003)	(0.000000 0.000001)	True
14	22	(0.000011 0.000504)	(0.000000 0.000001)	True
15	22	(0.000046 0.001001)	(0.000000 0.000000)	True
16	22	(0.000016 0.000702)	(0.000000 0.000001)	True
17	22	(0.000049 0.002171)	(0.000001 0.001818)	True
18	22	(0.000024 0.000633)	(0.000000 0.000013)	True
19	22	(0.000027 0.000556)	(0.000000 0.000012)	True
20	22	(0.000050 0.000941)	(0.000000 0.000000)	True
21	22	(0.001295 0.033625)	(0.000087 0.030295)	True
23	22	(0.002568 0.048728)	(0.000446 0.027445)	True
1	23	(0.000049 0.000835)	(0.000000 0.000000)	True
2	23	(0.000029 0.000758)	(0.000000 0.000000)	True
3	23	(0.000011 0.000500)	(0.000000 0.000000)	True
4	23	(0.000053 0.000918)	(0.000000 0.000000)	True
5	23	(0.000028 0.000710)	(0.000000 0.000000)	True
6	23	(0.000011 0.000500)	(0.000000 0.000000)	True
7	23	(0.000015 0.000704)	(0.000000 0.000000)	True
8	23	(0.000015 0.000704)	(0.000000 0.000000)	True
9	23	(0.000003 0.000355)	(0.000000 0.000000)	True
10	23	(0.000006 0.000860)	(0.000000 0.000000)	True
11	23	(0.000006 0.000858)	(0.000000 0.000001)	True
12	23	(0.000014 0.000652)	(0.000000 0.000002)	True
13	23	(0.000046 0.001003)	(0.000000 0.000000)	True
14	23	(0.000011 0.000504)	(0.000000 0.000000)	True
15	23	(0.000046 0.001001)	(0.000000 0.000000)	True
16	23	(0.000016 0.000702)	(0.000000 0.000001)	True
17	23	(0.000010 0.000434)	(0.000000 0.000057)	True
18	23	(0.000024 0.000633)	(0.000000 0.000011)	True
19	23	(0.000027 0.000556)	(0.000000 0.000000)	True
20	23	(0.000050 0.000941)	(0.000000 0.000000)	True
21	23	(0.000235 0.006114)	(0.000004 0.004442)	True
22	23	(0.001398 0.028450)	(0.000087 0.027527)	True

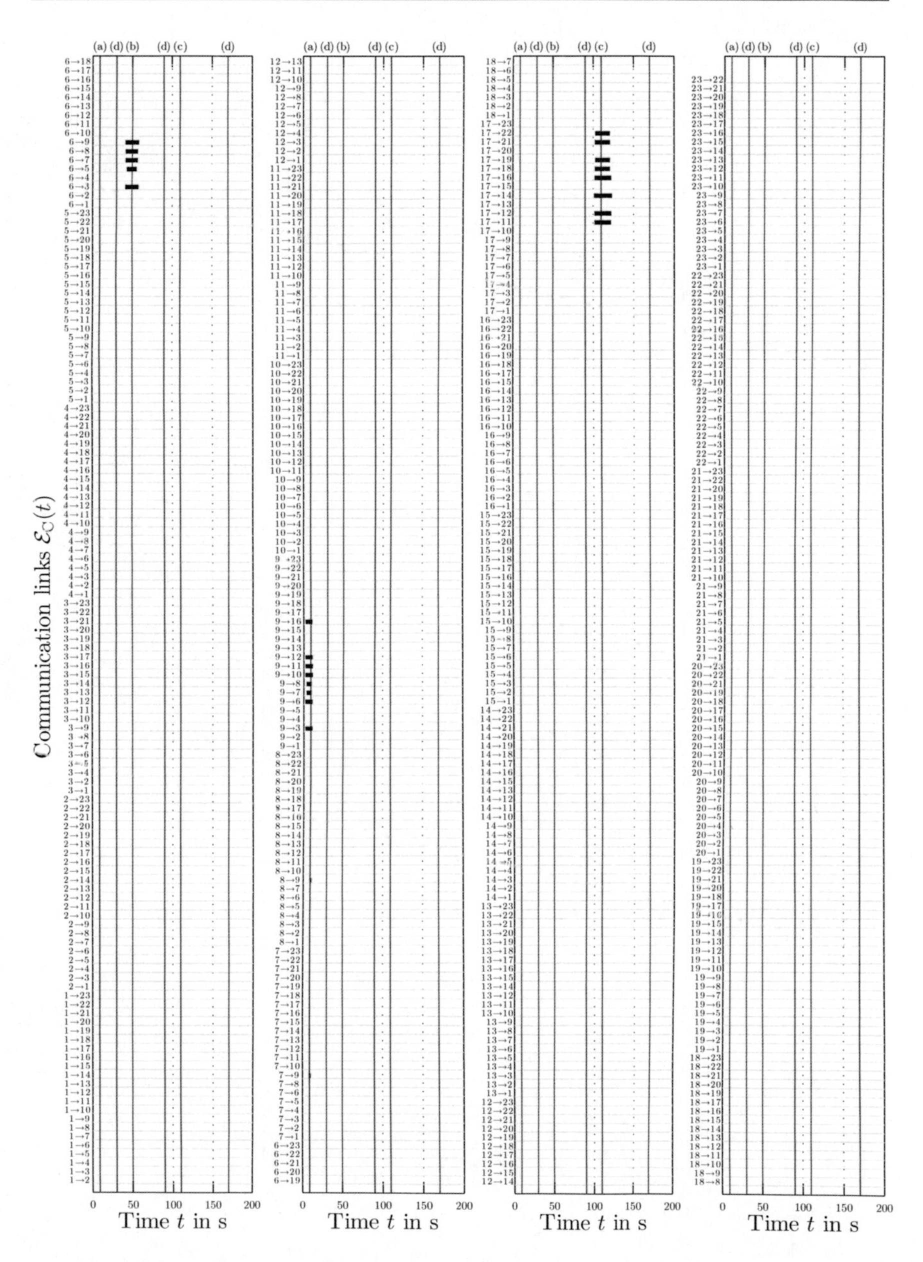

Figure B.3: All communication links $\mathcal{E}_{\mathrm{C}}(t)$ for different disturbance situations in Fig. 5.18.

Table B.15: Dynamics and position of the large robot formation.

i	Dynamics	Position X_i in m	Position Y_i in m
0	average	0	0
1	average	1	0
2	fast	2	0
3	average	3	0
4	average	4	0
5	average	5	0
6	average	6	0
7	average	0	1
8	slow	1	1
9	average	2	1
10	fast	3	1
11	average	4	1
12	fast	5	1
13	slow	6	1
14	slow	0	2
15	average	1	2
16	average	2	2
17	average	3	2
18	fast	4	2
19	average	5	2
20	average	6	2

Table B.16: Communication costs c_{ij} among the local control units of the large robot formation.

$i \backslash j$	0	1	2	3	4	5	6	7	8	9	10	11	12	13	14	15	16	17	18	19	20
1	1																				
2	4	1																			
3	9	4	1																		
4	16	9	4	1																	
5	25	16	9	4	1																
6	36	25	16	9	4	1															
7	1	2	5	10	17	26	37														
8	2	1	2	5	10	17	26	1													
9	5	2	1	2	5	10	17	4	1												
10	10	5	2	1	2	5	10	9	4	1											
11	17	10	5	2	1	2	5	16	9	4	1										
12	26	17	10	5	2	1	2	25	16	9	4	1									
13	37	26	17	10	5	2	1	36	25	16	9	4	1								
14	4	5	8	13	20	29	40	1	2	5	10	17	26	37							
15	5	4	5	8	13	20	29	2	1	2	5	10	17	26	1						
16	8	5	4	5	8	13	20	5	2	1	2	5	10	17	4	1					
17	13	8	5	4	5	8	13	10	5	2	1	2	5	10	9	4	1				
18	20	13	8	5	4	5	8	17	10	5	2	1	2	5	16	9	4	1			
19	29	20	13	8	5	4	5	26	17	10	5	2	1	2	25	16	9	4	1		
20	40	29	20	13	8	5	4	37	26	17	10	5	2	1	36	25	16	9	4	1	

C List of symbols

General conventions

- **Scalars** are represented by lower-case italic letters (x, u, d).
- **Vectors** are represented by lower-case bold italic letters $(\boldsymbol{x}, \boldsymbol{u}, \boldsymbol{d})$.
- **Matrices** are represented by upper-case bold italic letters $(\boldsymbol{A}, \boldsymbol{B}, \boldsymbol{E})$.

Indices

$(.)^{-1}$	Inverse of a matrix
$(.)^{+}$	Pseudoinverse of a matrix
$(.)^{\mathrm{T}}$	Transpose of a vector or matrix
$(.)^{*}$	Desired value
$\dot{(.)}$	Time derivative of a signal
$(.)_0$	Initial value of a signal at time $t = 0$
$(.)_{ij}$	Block-diagonal entry of the i-th row and j-th column of a matrix
$(.)_{\Delta}$	Difference signal
$(.)_{\mathrm{os}}$	Matrix, vector or signal of an overall system

Systems

P	Plant
C	Self-organizing controller \ Networked controller
Σ	Overall closed-loop system \ Self-organizing control system
Σ_i	i-th subsystem \ i-th (follower) agent
Σ_0	Leader (agent)
$\bar{\Sigma}_i$	Controlled subsystem Σ_i
Σ_{C}	Centralized control system \ Overall closed-loop system with permanent communication

Σ_{D}	Decentralized control system \ Overall closed-loop system without communication
Σ_{Δ}	Difference system
$\tilde{\Sigma}_{\Delta}$	Transformed difference system
C_i	Local control unit of subsystem Σ_i
F_i	Feedback unit within the local control unit C_i
O_i	Observation unit within the local control unit C_i
D_i	Decision unit within the local control unit C_i

Scalars

n	Dimension of the state vector
m	Dimension of the input vector
n_{d}	Dimension of the disturbance vector
ϵ	Switching threshold
$\bar{d}$	Upper bound on the disturbance
t	Time
$\bar{t}$	Deactivation time for the overall communication
$\bar{t}_{ji}$	Deactivation time for the communication from C_i to C_j
δ_{ji}	Distance from agent Σ_i to agent Σ_j
c_{ji}	Communication cost from C_i to C_j
$y_{\Delta i}$	(Local) difference output (PIS) \ Control error of an agent (MAS)
e_i	Local control error (MAS)
$y_{\Delta \mathrm{s}i}$	Local reference error (MAS)
$y_{\mathrm{s}i}$	Local reference signal (MAS)
$\bar{y}_{\Delta i}$	Maximal control error
$\bar{y}_{\Delta \mathrm{B}i}$	Basic maximal control error
$y_{\mathrm{d}i}$	Disturbance impact on agent Σ_i
$\hat{y}_{\mathrm{d}i}$	Reconstructed disturbance impact on agent Σ_i

Vectors

$\boldsymbol{x}$	Overall state
$\boldsymbol{u}$	Overall (control) input
$\boldsymbol{d}$	Overall disturbance
$\boldsymbol{s}$	Overall coupling input
$\boldsymbol{z}$	Overall coupling output
$\boldsymbol{e}$	Overall local control error (MAS)
$\boldsymbol{b}$	Input vector
$\boldsymbol{c}^{\mathrm{T}}$	Output vector
$\boldsymbol{g}$	Disturbance input vector
$\boldsymbol{b}_{\mathrm{s}}$	Input vector of the leader (agent) Σ_0
$\boldsymbol{c}_{\mathrm{s}}^{\mathrm{T}}$	Output vector of the leader (agent) Σ_0
$\boldsymbol{y}_{\Delta}$	Overall (measured) difference output (PIS) \ Overall control error (MAS)
$\boldsymbol{x}_{\Delta}$	Overall difference state ($\boldsymbol{x}_{\Delta} := \boldsymbol{x} - \boldsymbol{x}_{\mathrm{C}}$)
$\boldsymbol{x}_{\Delta i}$	(Local) difference state ($\boldsymbol{x}_{\Delta i} := \boldsymbol{x}_i - \boldsymbol{x}_{\mathrm{C}i}$)
$\mathbf{1}$	Vector composed of ones of appropriate dimension
$\boldsymbol{s}_{ji}$	Coupling input of Σ_i on Σ_j
$\hat{\boldsymbol{s}}_{ji}$	Determined coupling input of Σ_i on Σ_j
$\tilde{\boldsymbol{s}}_{ji}$	Transmitted coupling input of Σ_i on Σ_j
$\tilde{\boldsymbol{s}}_i$	Overall transmitted coupling input at C_i
$\boldsymbol{l}_i$	Observer feedback gain (MAS)
$\tilde{\boldsymbol{u}}_i$	Communicated control input
$\hat{\boldsymbol{u}}_{ji}$	Determined control signal by C_i for C_j
$\tilde{\boldsymbol{u}}_{ji}$	Transmitted control signal from C_i to C_j

Matrices

$\boldsymbol{A}$	System matrix
$\boldsymbol{B}$	Input matrix
$\boldsymbol{G}$	Disturbance input matrix
$\boldsymbol{E}$	Coupling input matrix

$\boldsymbol{C}$	Output matrix
$\boldsymbol{C}_{\mathrm{z}}$	Coupling output matrix
$\boldsymbol{L}$	Physical interconnection matrix
$\boldsymbol{K}$	Controller matrix \ Weighting matrix of the communication (MAS)
$\tilde{\boldsymbol{K}}$	Situation-dependent communication matrix
$\boldsymbol{K}_{\mathrm{B}}$	Basic communication matrix
$\boldsymbol{K}_{\mathrm{A}}$	Adjusted communication matrix
$\boldsymbol{I}_n$	Identity matrix of size n
$\boldsymbol{T}$	Transformation matrix
$\boldsymbol{A}_{\mathrm{s}}$	System matrix of the leader (agent) Σ_0

Sets

$\mathbb{R}^n$	Set of real vectors with dimension n
$\mathbb{R}^{n\times m}$	Set of real matrices with n rows and m columns
$\mathcal{G}$	Graph
$\mathcal{G}_{\mathrm{P}}$	Physical coupling graph
$\mathcal{G}_{\mathrm{C}}$	Communication graph
$\mathcal{G}_{\mathrm{B}}$	Basic communication graph
$\mathcal{G}_{\mathrm{A}}$	Adjusted communication graph
$\bar{\mathcal{G}}_{\mathrm{C}}$	Maximal communication graph
$\underline{\mathcal{G}}_{\mathrm{C}}$	Minimal communication graph
$\mathcal{E}$	Set of edges of a graph $\mathcal{G}$
$\mathcal{E}_{\mathrm{P}}$	Set of physical coupling edges of $\mathcal{G}_{\mathrm{P}}$
$\mathcal{E}_{\mathrm{C}}$	Set of communicational edges \ communication links of $\mathcal{G}_{\mathrm{C}}$
$\bar{\mathcal{E}}_{\mathrm{C}}$	Set of maximal communicational edges \ communication links of $\mathcal{G}_{\mathrm{C}}$
$\underline{\mathcal{E}}_{\mathrm{C}}$	Set of minimal communicational edges \ communication links of $\mathcal{G}_{\mathrm{C}}$
$\mathcal{V}$	Set of vertices of a graph $\mathcal{G}$
$\mathcal{V}$	Set of communicational vertices of a graph $\mathcal{G}_{\mathrm{C}}$
$\mathcal{P}_{\mathrm{P}i}$	Set of physical predecessors
$\mathcal{F}_{\mathrm{P}i}$	Set of physical followers
$\mathcal{P}_{\mathrm{C}i}$	Set of (communicational) predecessors

$\mathcal{P}_{\mathrm{S}i}$	Set of substitutional predecessors
$\mathcal{F}_{\mathrm{C}i}$	Set of (communicational) followers
$\bar{\mathcal{P}}_{\mathrm{C}i}$	Set of maximal (communicational) predecessors
$\underline{\mathcal{P}}_{\mathrm{C}i}$	Set of minimal (communicational) predecessors
$\bar{\mathcal{F}}_{\mathrm{C}i}$	Set of maximal (communicational) followers
$\underline{\mathcal{F}}_{\mathrm{C}i}$	Set of minimal (communicational) followers
$\mathcal{D}$	Set of disturbed subsystems \ agents
$\mathcal{U}$	Set of undisturbed subsystems \ agents
$\mathcal{N}$	Set of subsystems, agents or local control units
$\mathcal{N}_{\mathrm{F}}$	Set of faulty subsystems \ agents
$\mathcal{N}_{\mathrm{average}}$	Set of average robots
$\mathcal{N}_{\mathrm{fast}}$	Set of fast robots
$\mathcal{N}_{\mathrm{slow}}$	Set of slow robots
$\mathcal{N}_{\mathrm{large}}$	Set of large tanks
$\mathcal{N}_{\mathrm{middle}}$	Set of middle tanks
$\mathcal{N}_{\mathrm{small}}$	Set of small tanks
$\mathcal{I}$	Set of independent subsystem (PIS) \ Set of interrupting local control units (MAS)

Abbreviations

PIS	Physically interconnected system
MAS	Multi-agent system

Persönliche Daten:

Name	René Schuh
Geburtsdatum	16.05.1985
Geburtsort	Teterow

Bildungs- und Berufsweg:

seit 1 / 2017	Ingenieur für Automatisierungstechnik bei der Windmöller & Hölscher KG, Lengerich.
3 / 2011 – 6 / 2016	Wissenschaftlicher Mitarbeiter am Lehrstuhl für Automatisierungstechnik und Prozessinformatik (Prof. Dr.-Ing. Jan Lunze), Ruhr-Universität Bochum.
10 / 2005 – 10 / 2010	Maschinenbaustudium an der Universität Rostock mit den Schwerpunkten Systemdynamik und Regelungstechnik sowie Antriebssysteme, Titel der Diplomarbeit: *Modellierung, Regelung und Untersuchung von Störfällen des Dampfsystems einer Ammoniak-Anlage*.
04 / 2010 – 10 / 2010	Praktikum und Diplomarbeit bei der Uhde GmbH, Dortmund.
10 / 2008 – 2 / 2009	Praktikum und Studienarbeit bei der Robert Bosch GmbH, Schwieberdingen.
10 / 2004 – 6 / 2005	Zivildienst in den Güstrower Werkstätten, Teterow.
09 / 2001 – 06 / 2004	Fachgymnasium Technik und Wirtschaft - Johann Heinrich von Thünen, Güstrow, Abitur.
08 / 1995 – 06 / 2001	Warbel-Schule, Gnoien, Realschulabschluss.